Embedded and Real-Time Operating Systems

K.C. Wang

Embedded and Real-Time Operating Systems

K.C. Wang
School of Electrical Engineering
and Computer Science
Washington State University
Pullman, WA
USA

ISBN 978-3-319-84672-9 ISBN 978-3-319-51517-5 (eBook)
DOI 10.1007/978-3-319-51517-5

Printed on acid-free paper

This Springer imprint is published by Springer Nature
The registered company is Springer International Publishing AG
The registered company address is: Gewerbestrasse 11, 6330 Cham, Switzerland

Dedicated to

Cindy

Preface

Since the publication of my first book on the Design and Implementation of the MTX Operating System by Springer in 2015, I have received inquiries from many enthusiastic readers about how to run the MTX OS on their ARM based mobile devices, such as iPods or iPhones, etc. which motivated me to write this book.

The purpose of this book is to provide a suitable platform for teaching and learning the theory and practice of embedded and real-time operating systems. It covers the basic concepts and principles of operating systems, and it shows how to apply them to the design and implementation of complete operating systems for embedded and real-time systems. In order to do these in a concrete and meaningful way, it uses the ARM toolchain for program development, and it uses ARM virtual machines to demonstrate the design principles and implementation techniques.

Due to its technical contents, this book is not intended for entry-level courses that teach only the concepts and principles of operating systems without any programming practice. It is intended for technically oriented Computer Science/Engineering courses on embedded and real-time systems that emphasize both theory and practice. The book's evolutional style, coupled with detailed source code and complete working sample systems, make it especially suitable for self-study.

Undertaking this book project has proved to be yet another very demanding and time-consuming endeavor, but I enjoyed the challenges. While preparing the book manuscripts for publication, I have been blessed with the encouragements and helps from numerous people, including many of my former TaiDa EE60 classmates. I would like to take this opportunity to thank all of them. I am also grateful to Springer International Publishing AG for allowing me to disclose the source code of this book to the public for free, which are available at http://www.eecs.wsu.edu/~cs460/ARMhome for download.

Special thanks go to Cindy for her continuing support and inspirations, which have made this book possible. Last but not least, I would like to thank my family again for bearing with me with endless excuses of being busy all the time.

Pullman, WA
October, 2016

K.C. Wang

Contents

About the Author

K.C. Wang received the BSEE degree from National Taiwan University, in 1960 and the Ph.D. degree in Electrical Engineering from Northwestern University, Evanston, Ill in 1965. He is currently a Professor in the School of Electrical Engineering and Computer Science at Washington State University. His academic interests are in Operating Systems, Distributed Systems and Parallel Computing.

1 Introduction

1.1 About This Book

This book is about the design and implementation of embedded and real-time operating systems (Gajski et al. 1994). It covers the basic concepts and principles of operating systems (OS) (Silberschatz et al. 2009; Stallings 2011; Tanenbaum and Woodhull 2006; Wang 2015), embedded systems architecture (ARM Architecture 2016), embedded systems programming (ARM Programming 2016), real-time system concepts and real-time system requirements (Dietrich and Walker 2015). It shows how to apply the theory and practice of OS to the design and implementation of operating systems for embedded and real-time systems.

1.2 Motivations of This Book

In the early days, most embedded systems were designed for special applications. An embedded system usually consists of a microcontroller and a few I/O devices, which is designed to monitor some input sensors and generate signals to control external devices, such as to turn on LEDs or activate some switches, etc. For this reason, control programs of early embedded systems are also very simple. They are usually written in the form of a super-loop or a simple event-driven program structure. However, as the computing power of embedded systems increases in recent years, embedded systems have undergone a tremendous leap in both complexity and areas of applications. As a result, the traditional approaches to software design for embedded systems are no longer adequate. In order to cope with the ever increasing system complexity and demands for extra functionality, embedded systems need more powerful software. As of now, many embedded systems are in fact high-power computing machines with multicore processors, gigabytes memory and multi-gigabyte storage devices. Such systems are intended to run a wide range of application programs. In order to fully realize their potential, modern embedded systems need the support of multi-functional operating systems. A good example is the evolution of earlier cell phones to current smart phones. Whereas the former were designed to perform the simple task of placing or receiving phone calls only, the latter may use multicore processors and run adapted versions of Linux, such as Android, to perform multitasks. The current trend of embedded system software design is clearly moving in the direction of developing multi-functional operating systems suitable for future mobile environment. The purpose of this book is to show how to apply the theory and practice of operating systems to develop OS for embedded and real-time systems.

1.3 Objective and Intended Audience

The objective of this book is to provide a suitable platform for teaching and learning the theory and practice of embedded and real-time operating systems. It covers embedded system architecture, embedded system programming, basic concepts and principles of operating systems (OS) and real-time systems. It shows how to apply these principles and programming techniques to the design and implementation of real OS for both embedded and real-time systems. This book is intended for computer science students and computer professionals, who wish to study the internal details of embedded and real-time operating systems. It is suitable as a textbook for courses on embedded and real-time systems in technically oriented Computer Science/Engineering curriculums that strive for a balance between theory and practice. The book's evolutional

K.C. Wang, *Embedded and Real-Time Operating Systems*,
DOI 10.1007/978-3-319-51517-5_1

style, coupled with detailed example code and complete working sample systems, make it especially suitable for self-study by computer enthusiasts.

The book covers the entire spectrum of software design for embedded and real-time systems, ranging from simple super-loop and event-driven control programs for uniprocessor (UP) systems to complete Symmetric Multiprocessing (SMP) operating systems on multicore systems. It is also suitable for advanced study of embedded and real-time operating systems.

1.4 Unique Features of This Book

This book has many unique features, which distinguish it from other books.

1. This book is self-contained. It includes all the foundation and background information for studying embedded systems, real-time systems and operating systems in general. These include the ARM architecture, ARM instructions and programming (ARM architecture 2016; ARM926EJ-S 2008), toolchain for developing programs (ARM toolchain 2016), virtual machines (QEMU Emulators 2010) for software implementation and testing, program execution image, function call conventions, run-time stack usage and link C programs with assembly code.
2. Interrupts and interrupts processing are essential to embedded systems. This book covers interrupt hardware and interrupts processing in great detail. These include non-vectored interrupts, vectored interrupts (ARM PL190 2004), non-nested interrupts, nested interrupts (Nesting Interrupts 2011) and programming the Generic Interrupt Controller (GIC) (ARM GIC 2013) in ARM MPcore (ARM Cortex-A9 MPCore 2012) based systems. It shows how to apply the principles of interrupts processing to develop interrupt-driven device drivers and event-driven embedded systems.
3. The book presents a general framework for developing interrupt-driven device drivers, with an emphasis on the interaction and synchronization between interrupt handlers and processes. For each device, it explains the principles of operation and programming techniques before showing the actual driver implementation, and it demonstrates the device drivers by complete working sample programs.
4. The book shows the design and implementation of complete OS for embedded systems in incremental steps. First, it develops a simple multitasking kernel to support process management and process synchronization. Then it incorporates the Memory Management Unit (MMU) (ARM MMU 2008) hardware into the system to provide virtual address mappings and extends the simple kernel to support user mode processes and system calls. Then it adds process scheduling, signal processing, file system and user interface to the system, making it a complete operating system. The book's evolutional style helps the reader better understand the material.
5. Chapter 9 covers Symmetric Multiprocessing (SMP) (Intel 1997) embedded systems in detail. First, it explains the requirements of SMP systems and compares the ARM MPcore architecture (ARM11 2008; ARM Cortex-A9 MPCore 2012) with the SMP system architecture of Intel. Then it describes the SMP features of ARM MPcore processors, which include the SCU and GIC for interrupts routing and interprocessor communication and synchronization by Software Generated Interrupts (SGIs). It uses a series of programming examples to show how to start up the ARM MPcore processors and points out the need for synchronization in a SMP environment. Then it explains the ARM LDREX/STREX instructions and memory barriers and uses them to implement spinlocks, mutexes and semaphores for process synchronization in SMP systems. It presents a general methodology for SMP kernel design and shows how to apply the principles to adapt a uniprocessor (UP) kernel for SMP. In addition, it also shows how to use parallel algorithms for process and resource management to improve the concurrency and efficiency in SMP systems.
6. Chapter 10 covers real-time operating systems (RTOS). It introduces the concepts and requirements of real-time systems. It covers the various kinds of task scheduling algorithms in RTOS, and it shows how to handle priority inversion and task preemption in real-time systems. It includes case studies of several popular RTOS and formulates a set of general guidelines for RTOS design. It shows the design and implementation of a UP_RTOS for uniprocessor (UP) systems. Then it extends the UP_RTOS to SMP_RTOS for SMP, which supports nested interrupts, preemptive task scheduling, priority inheritance and inter-processor synchronization by SGI.
7. Throughout the book, it uses completely working sample systems to demonstrate the design principles and implementation techniques. It uses the ARM toolchain under Ubuntu (15.10) Linux to develop software for embedded systems, and it uses emulated ARM virtual machines under QEMU as the platform for implementation and testing.

1.5 Book Contents

This book is organized as follows.

Chapter 2 covers the ARM architecture, ARM instructions, ARM programming and development of programs for execution on ARM virtual machines. These include ARM processor modes, register banks in different modes, instructions and basic programming in ARM assembly. It introduces the ARM toolchain under Ubuntu (15.10) Linux and emulated ARM virtual machines under QEMU. It shows how to use the ARM toolchain to develop programs for execution on the ARM Versatilepb virtual machine by a series of programming examples. It explains the function call convention in C and shows how to interface assembly code with C programs. Then it develops a simple UART driver for I/O on serial ports, and a LCD driver for displaying both graphic images and text. It also shows the development of a generic printf() function for formatted printing to output devices that support the basic print char operation.

Chapter 3 covers interrupts and exceptions processing. It describes the operating modes of ARM processors, exception types and exception vectors. It explains the functions of interrupt controllers and the principles of interrupts processing in detail. Then it applies the principles of interrupts processing to the design and implementation of interrupt-driven device drivers. These include drivers for timers, keyboard, UARTs and SD cards (SDC 2016), and it demonstrates the device drivers by example programs. It explains the advantages of vectored interrupts over non-vectored interrupts. It shows how to configure the Vector Interrupt Controllers (VICs) for vectored interrupts, and demonstrates vectored interrupts processing by example programs. It also explains the principles of nested interrupts and demonstrates nested interrupts processing by example programs.

Chapter 4 covers models of embedded systems. First, it explains and demonstrates the simple super-loop system model and points out its shortcomings. Then it discusses the event-driven model and demonstrates both periodic and asynchronous event-driven systems by example programs. In order to go beyond the simple super-loop and event-driven system models, it justifies the needs for processes or tasks in embedded systems. Then it introduces the various kinds of process models, which are used as the models of developing embedded systems in the book. Lastly, it presents the formal methodologies for embedded systems design. It illustrates the Finite State Machine (FSM) (Katz and Borriello 2005) model by a complete design and implementation example in detail.

Chapter 5 covers process management. It introduces the process concept and the basic principle of multitasking. It demonstrates the technique of multitasking by context switching. It shows how to create processes dynamically and discusses the goals, policy and algorithms of process scheduling. It covers process synchronization and explains the various kinds of process synchronization mechanisms, which include sleep/wakeup, mutexes and semaphores. It shows how to use process synchronization to implement event-driven embedded systems. It discusses the various schemes for inter-process communication, which include shared memory, pipes and message passing. It shows how to integrate these concepts and techniques to implement a uniprocessor (UP) kernel for process management, and it demonstrates the system requirements and programming techniques for both non-preemptive and preemptive process scheduling. The UP kernel serves as the foundation for developing complete operating systems in later chapters.

Chapter 6 covers the ARM memory management unit (MMU) and virtual address space mappings. It explains the ARM MMU in detail and shows how to configure the MMU for virtual address mapping using both one-level and two-level paging. In addition, it also explains the distinction between low VA space and high VA space mappings and their implications on system implementations. Rather than only discussing the principles of memory management, it demonstrates the various kinds of virtual address mappings schemes by complete working example programs.

Chapter 7 covers user mode processes and system calls. First it extends the basic UP kernel of Chapter 5 to support additional process management functions, which include dynamic process creation, process termination, process synchronization and wait for child process termination. Then it extends the basic kernel to support user mode processes. It shows how to use memory management to provide each process with a private user mode virtual address space that is isolated from other processes and protected by the MMU hardware. It covers and demonstrates the various kinds of memory management schemes, which include one-level sections and two-level static and dynamic paging. It covers the advanced concepts and techniques of fork, exec, vfork and threads. In addition, it shows how to use SD cards for storing both kernel and user mode image files in a SDC file system. It also shows how to boot up the system kernel from SDC partitions. This chapter serves as a foundation for the design and implementation of general purpose OS for embedded systems.

Chapter 8 presents a fully functional general purpose OS (GPOS), denoted by EOS, for uniprocessor (UP) ARM based embedded systems. The following is a brief summary of the organization and capabilities of the EOS system.

1. System Images: Bootable kernel image and User mode executables are generated from a source tree by the ARM toolchain under Ubuntu (15.10) Linux and reside in an (EXT2 2001) file system on a SDC partition. The SDC contains stage-1 and stage-2 booters for booting up the kernel image from the SDC partition. After booting up, the EOS kernel mounts the SDC partition as the root file system.
2. Processes: The system supports NPROC = 64 processes and NTHRED = 128 threads per process, both can be increased if needed. Each process (except the idle process P0) runs in either Kernel mode or User mode. Memory management of process images is by 2-level dynamic paging. Process scheduling is by dynamic priority and timeslice. It supports inter-process communication by pipes and message passing. The EOS kernel supports process management functions of fork, exec, vfork, threads, exit and wait for child termination.
3. Device drivers: It contains device drivers for the most commonly used I/O devices, which include LCD display, timer, keyboard, UART and SDC.
4. File system: EOS supports an EXT2 file system that is totally Linux compatible. It shows the principles of file operations, the control path and data flow from user space to kernel space down to the device driver level. It shows the internal organization of file systems, and it describes the implementation of a complete file system in detail.
5. Timer service, exceptions and signal processing: It provides timer service functions, and it unifies exceptions handling with signal processing, which allows users to install signal catchers to handle exceptions in User mode.
6. User Interface: It supports multi-user logins to the console and UART terminals. The command interpreter sh supports executions of simple commands with I/O redirections, as well as multiple commands connected by pipes.
7. Porting: The EOS system runs on a variety of ARM virtual machines under QEMU, mainly for convenience. It should also run on real ARM based system boards that support suitable I/O devices. Porting EOS to some popular ARM based systems, e.g. Raspberry PI-2 is currently underway. The plan is to make it available for readers to download as soon as it is ready.

Chapter 9 covers multiprocessing in embedded systems. It explains the requirements of Symmetric Multiprocessing (SMP) systems and compares the approach to SMP of Intel with that of ARM. It lists ARM MPcore processors and describes the components and functions of ARM MPcore processors in support of SMP. All ARM MPcore based systems depends on the Generic Interrupt Controller (GIC) for interrupts routing and inter-processor communication. It shows how to configure the GIC to route interrupts and demonstrates GIC programming by examples. It shows how to start up ARM MPcores and points out the need for synchronization in a SMP environment. It shows how to use the classic test-and-set or equivalent instructions to implement atomic updates and critical regions and points out their shortcomings. Then it explains the new features of ARM MPcores in support of SMP. These include the ARM LDRES/STRES instructions and memory barriers. It shows how to use the new features of ARM MPcore processors to implement spinlocks, mutexes and semaphores for process synchronization in SMP. It defines conditional spinlocks, mutexes and semaphores and shows how to use them for deadlock prevention in SMP kernels. It also covers the additional features of the ARM MMU for SMP. It presents a general methodology for adapting uniprocessor OS kernel to SMP. Then it applies the principles to develop a complete SMP_EOS for embedded SMP systems, and it demonstrates the capabilities of the SMP_EOS system by example programs.

Chapter 10 covers real-time operating systems (RTOS). It introduces the concepts and requirements of real-time systems. It covers the various kinds of task scheduling algorithms in RTOS, which include RMS, EDF and DMS. It explains the problem of priority inversion due to preemptive task scheduling and shows how to handle priority inversion and task preemption. It includes case studies of several popular real-time OS and presents a set of general guidelines for RTOS design. It shows the design and implementation of a UP_RTOS for uniprocessor (UP) systems. Then it extends the UP_RTOS to a SMP_RTOS, which supports nested interrupts, preemptive task scheduling, priority inheritance and inter-processor synchronization by SGI.

1.6 Use This Book as a Textbook for Embedded Systems

This book is suitable as a textbook for technically oriented courses on embedded and real-time systems in Computer Science/Engineering curricula that strive for a balance between theory and practice. A one-semester course based on this book may include the following topics.

1. Introduction to embedded systems, ARM architecture, basic ARM programming, ARM toolchain and software development for embedded systems, interface assembly code with C programs, execution image and run-time stack usage, program execution on ARM virtual machines, simple device drivers for I/O (Chap. 2).
2. Exceptions and interrupts, interrupts processing, design and implementation of interrupt-driven device drivers (Chap. 3).
3. Embedded system models; super-loop, event-driven embedded systems, process and formal models of embedded systems (Chap. 4).
4. Process management and process synchronization (Chap. 5).
5. Memory management, virtual address mappings and memory protection (Chap. 6).
6. Kernel mode and user mode processes, system calls (Chap. 7).
7. Introduction to real-time embedded systems (parts of Chap. 10).
8. Introduction to SMP embedded systems (parts of Chaps. 9 and 10).

1.7 Use This Book as a Textbook for Operating Systems

This book is also suitable as a textbook for technically oriented courses on general purpose operating systems. A one-semester course based on this book may include the following topics.

1. Computer architecture and systems programming: the ARM architecture, ARM instructions and programming, ARM toolchain, virtual machines, interface of assembly code with C, run-time images and stack usage, simple device drivers (Chap. 2).
2. Exceptions and interrupts, interrupts processing and interrupt-driven device drivers (Chap. 3).
3. Process management and process synchronization (Chap. 5).
4. Memory management, virtual address mappings and memory protection (Chap. 6).
5. Kernel mode and user mode processes, system calls (Chap. 7).
6. General purpose operating system kernel, process management, memory management, device drivers, file system, signal processing and user interface (Chap. 8).
7. Introduction to SMP, inter-processor communication and synchronization of multicore processors, process synchronization in SMP, SMP kernel design and SMP operating systems (Chap. 9).
8. Introduction to real-time systems (parts of Chap. 10).

The problems section of each chapter contains questions designed to review the concepts and principles presented in the chapter. While some of the questions involve only simple modifications of the example programs to let the students experiment with alternative design and implementation, many other questions are suitable for advanced programming projects.

1.8 Use This Book for Self-study

Judging from the large number of OS developing projects, and many popular sites on embedded and real-time systems posted on the Internet and their enthusiastic followers, there is a tremendous number of computer enthusiasts who wish to learn the practical side of embedded and real-time operating systems. The evolutional style of this book, coupled with ample code and demonstration system programs, make it especially suitable for self-study. It is hoped that this book will be useful and beneficial to such readers.

References

ARM Architectures 2016: http://www.arm.products/processors/instruction-set-architectures, ARM Information Center, 2016
ARM Cortex-A9 MPCore: Technical Reference Manual Revision: r4p1, ARM information Center, 2012
ARM GIC: ARM Generic Interrupt Controller (PL390) Technical Reference Manual, ARM Information Center, 2013
ARM MMU: ARM926EJ-S, ARM946E-S Technical Reference Manuals, ARM Information Center, 2008

ARM Programming 2016: "ARM Assembly Language Programming", http://www.peter-cockerell.net/aalp/html/frames.html, 2016
ARM PL190 2004: PrimeCell Vectored Interrupt Controller (PL190), http://infocenter.arm.com/help/topic/com.arm.doc.ddi0181e/DDI0181.pdf, 2004
ARM toolchain 2016: http://gnutoolchains.com/arm-eabi, 2016
ARM11 2008: ARM11 MPcore Processor Technical Reference Manual, r2p0, ARM Information Center, 2008
ARM926EJ-S 2008: "ARM926EJ-S Technical Reference Manual", ARM Information Center, 2008
Dietrich, S., Walker, D., 2015 "The evolution of Real-Time Linux", http://www.cse.nd.edu/courses/cse60463/www/amatta2.pdf, 2015
EXT2 2001: www.nongnu.org/ext2-doc/ext2.html, 2001
Gajski, DD, Vahid, F, Narayan, S, Gong, J, 1994 "Specification and design of embedded systems", PTR Prentice Hall, 1994
Intel: MultiProcessor Specification, v1.4, Intel, 1997
Katz, R. H. and G. Borriello 2005, "Contemporary Logic Design", 2nd Edition, Pearson, 2005.
Nesting Interrupts 2011: Nesting Interrupts, ARM Information Center, 2011
QEMU Emulators 2010: "QEMU Emulator User Documentation", http://wiki.qemu.org/download/qemu-doc.htm, 2010.
SDC: Secure Digital cards: SD Standard Overview-SD Association https://www.sdcard.org/developers/overview, 2016
Silberschatz, A., P.A. Galvin, P.A., Gagne, G, "Operating system concepts, 8th Edition", John Wiley & Sons, Inc. 2009
Stallings, W. "Operating Systems: Internals and Design Principles (7th Edition)", Prentice Hall, 2011
Tanenbaum, A.S., Woodhull, A.S., 2006 "Operating Systems, Design and Implementation, third Edition", Prentice Hall, 2006
Wang, K.C., 2015 "Design and Implementation of the MTX Operating Systems, Springer International Publishing AG, 2015

2 ARM Architecture and Programming

ARM (ARM Architecture 2016) is a family of Reduced Instruction Set Computing (RISC) microprocessors developed specifically for mobile and embedded computing environments. Due to their small sizes and low power requirements, ARM processors have become the most widely used processors in mobile devices, e.g. smart phones, and embedded systems. Currently, most embedded systems are based on ARM processors. In many cases, embedded system programming has become almost synonymous with ARM processor programming. For this reason, we shall also use ARM processors for the design and implementation of embedded systems in this book. Depending on their release time, ARM processors can be classified into classic cores and the more recent (since 2005) Cortex cores. Depending on their capabilities and intended applications, ARM Cortex cores can be classified into three categories (ARM Cortex 2016).

.The Cortex-M series: these are microcontroller-oriented processors intended for Micro Controller Unit (MCU) and System on Chip (SoC) applications.
.The Cortex-R series: these are embedded processors intended for real-time signal processing and control applications.
.The Cortex-A series: these are applications processors intended for general purpose applications, such as embedded systems with full featured operating systems.

The ARM Cortex-A series processors are the most powerful ARM processors, which include the Cortex-A8 (ARM Cortex-A8 2010) single core and the Cortex A9-MPcore (ARM Cortex A9 MPcore 2016) with up to 4 CPUs. Because of their advanced capabilities, most recent embedded systems are based on the ARM Cortex-A series processors. On the other hand, there are also a large number of embedded systems intended for dedicated applications that are based on the classic ARM cores, which proved to be very cost-effective. In this book, we shall cover both the classic and ARM-A series processors. Specifically, we shall use the classic ARM926EJ-S core (ARM926EJ-ST 2008, ARM926EJ-ST 2010) for single CPU systems and the Cortex A9-MPcore for multiprocessor systems. A primary reason for choosing these ARM cores is because they are available as emulated virtual machines (VMs). A major goal of this book is to show the design and implementation of embedded systems in an integrated approach. In addition to covering the theory and principles, it also shows how to apply them to the design and implementation of embedded systems by programming examples. Since most readers may not have access to real ARM based systems, we shall use emulated ARM virtual machines under QEMU for implementation and testing. In this chapter, we shall cover the following topics.

. ARM architecture
. ARM instructions and basic programming in ARM assembly
. Interface assembly code with C programs
. ARM toolchains for (cross) compile-link programs
. ARM system emulators and run programs on ARM virtual machines
. Develop simple I/O device drivers. These include UART drivers for serial ports and LCD driver for displaying both images and text.

K.C. Wang, *Embedded and Real-Time Operating Systems*,
DOI 10.1007/978-3-319-51517-5_2

2.1 ARM Processor Modes

The ARM processor has seven (7) operating modes, which are specified by the 5 mode bits [4:0] in the Current Processor Status Register (CPSR). The seven ARM processor modes are:

USR mode: unprivileged User mode
SYS mode: System mode using the same set of registers as User mode
FIQ mode: Fast interrupt request processing mode
IRQ mode: normal interrupt request processing mode
SVC mode: Supervisor mode on reset or SWI (Software Interrupt)
ABT mode: data exception abort mode
UND mode: undefined instruction exception mode

2.2 ARM CPU Registers

The ARM processor has 37 registers in total, all of which are 32-bits wide. These include

1 dedicated program counter (PC)
1 dedicated current program status register (CPSR)
5 dedicated saved program status registers (SPSR)
30 general purpose registers

The registers are arranged into several banks. The accessibility of the banked registers is governed by the processor mode. In each mode the ARM CPU can access

A particular set of R0-R12 registers
A particular R13 (stack pointer), R14 (link register) and SPSR (saved program status)
The same R15 (program counter) and CPSR (current program status register)

2.2.1 General Registers

Figure 2.1 shows the organization of general registers in the ARM processor.

```
|User| SYS | SVC | ABT | UND | IRQ | FIQ |
-------------------------------------------
|              R0 - R7                    |
-------------------------------------------
|              R8 - R12           |R8-R12|
-------------------------------------------
|    R13    | R13 | R13 | R13 | R13 | R13 |
-------------------------------------------
|    R14    | R14 | R14 | R14 | R14 | R14 |
-------------------------------------------
|                 R15                     |
-------------------------------------------
|                 CPSR                    |
-------------------------------------------
|           |SPSR |SPSR |SPSR |SPSR |SPSR |
-------------------------------------------
```

Fig. 2.1 Register banks in ARM processor

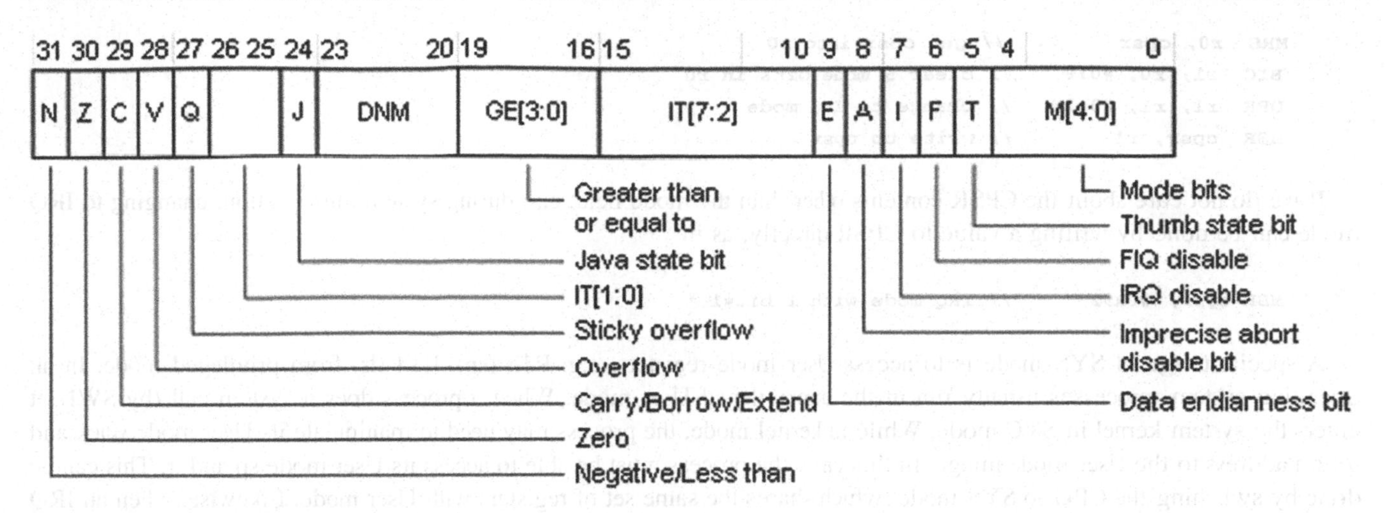

Fig. 2.2 Status register of ARM processor

User and System modes share the same set of registers. Registers R0-R12 are the same in all modes, except for the FIQ mode, which has its own separate registers R8-R12. Each mode has its own stack pointer (R13) and link register (R14). The Program Counter (PC or R15) and Current Status Register (CPSR) are the same in all modes. Each privileged mode (SVC to FIQ) has its own Saved Processor Status Register (SPSR).

2.2.2 Status Registers

In all modes, the ARM processor has the same Current Program Status Register (CPSR). Figure 2.2 shows the contents of the CPSR register.

In the CPSR register, NZCV are the condition bits, I and F are IRQ and FIQ interrupt mask bits, T = Thumb state, and M[4:0] are the processor mode bits, which define the processor mode as

```
USR: 10000 (0x10)
FIQ: 10001 (0x11)
IRQ: 10010 (0x12)
SVC: 10011 (0x13)
ABT: 10111 (0x17)
UND: 11011 (0x1B)
SYS: 11111 (0x1F)
```

2.2.3 Change ARM Processor Mode

All the ARM modes are privileged except the User mode, which is unprivileged. Like most other CPUs, the ARM processor changes mode in response to exceptions or interrupts. Specifically, it changes to FIQ mode when a FIQ interrupt occurs. It changes to IRQ mode when an IRQ interrupt occurs. It enters the SVC mode when power is turned on, following reset or executing a SWI instruction. It enters the Abort mode when a memory access exception occurs, and it enters the UND mode when it encounters an undefined instruction. An unusual feature of the ARM processor is that, while in a privileged mode, it can change mode freely by simply altering the mode bits in the CPSR, by using the MSR and MRS instructions. For example, when the ARM processor starts or following a reset, it begins execution in SVC mode. While in SVC mode, the system initialization code must set up stack pointers of other modes. To do these, it simply changes the processor to an appropriate mode, initialize the stack pointer (R13_mode) and the saved program status register (SPSR) of that mode. The following code segment shows how to switch the processor to IRQ mode while preserving other bits, e.g. F and I bits, in the CPSR.

```
MRS r0, cpsr          // get cpsr into r0
BIC r1, r0, #01F      // clear 5 mode bits in r0
ORR r1, r1, #0x12     // change to IRQ mode
MSR cpsr, r1          // write to cpsr
```

If we do not care about the CPSR contents other than the mode field, e.g. during system initialization, changing to IRQ mode can be done by writing a value to CPSR directly, as in

```
MSR cpsr, #0x92       // IRQ mode with I bit=1
```

A special usage of SYS mode is to access User mode registers, e.g. R13 (sp), R14 (lr) from privileged mode. In an operating system, processes usually run in the unprivileged User mode. When a process does a system call (by SWI), it enters the system kernel in SVC mode. While in kernel mode, the process may need to manipulate its User mode stack and return address to the User mode image. In this case, the process must be able to access its User mode sp and lr. This can be done by switching the CPU to SYS mode, which shares the same set of registers with User mode. Likewise, when an IRQ interrupt occurs, the ARM processor enters IRQ mode to execute an interrupt service routine (ISR) to handle the interrupt. If the ISR allows nested interrupts, it must switch the processor from IRQ mode to a different privileged mode to handle nested interrupts. We shall demonstrate this later in Chap. 3 when we discuss exceptions and interrupts processing in ARM based systems.

2.3 Instruction Pipeline

The ARM processor uses an internal pipeline to increase the rate of instruction flow to the processor, allowing several operations to be undertaken simultaneously, rather than serially. In most ARM processors, the instruction pipeline consists of 3 stages, FETCH-DECODE-EXECUTE, as shown below.

```
PC     FETCH    Fetch instruction from memory
PC-4   DECODE   Decode registers used in instruction
PC-8   EXECUTE  Execute the instruction
```

The Program Counter (PC) actually points to the instruction being fetched, rather than the instruction being executed. This has implications to function calls and interrupt handlers. When calling a function using the BL instruction, the return address is actually PC-4, which is adjusted by the BL instruction automatically. When returning from an interrupt handler, the return address is also PC-4, which must be adjusted by the interrupt handler itself, unless the ISR is defined with the **__attribute__((interrupt))** attribute, in which case the compiled code will adjust the link register automatically. For some exceptions, such as Abort, the return address is PC-8, which points to the original instruction that caused the exception.

2.4 ARM Instructions

2.4.1 Condition Flags and Conditions

In the CPSR of ARM processors, the highest 4 bits, NZVC, are condition flags or simply condition code, where

```
N = negative,  Z = zero,  V = overflow,  C = carry bit out
```

Condition flags are set by comparison and TST operations. By default, data processing operations do not affect the condition flags. To cause the condition flags to be updated, an instruction can be postfix with the S symbol, which sets the S bit in the instruction encoding. For example, both of the following instructions add two numbers, but only the **ADDS** instruction affects the condition flags.

```
ADD r0, r1, r2    ; r0 = r1 + r2
ADDS r0, r1, r2   ; r0 = r1 + r2 and set condition flags
```

In the ARM 32-bit instruction encoding, the leading 4 bits [31:28] represent the various combinations of the condition flag bits, which form the condition field of the instruction (if applicable). Based on the various combinations of the condition flag bits, conditions are defined mnemonically as EQ, NE, LT, GT, LE, GE, etc. The following shows some of the most commonly used conditions and their meanings.

```
0000 : EQ  Equal                          (Z set)
0001 : NE  Not equal                      (Z clear)
0010 : CS  Carry set                      (C set)
0101 : VS  Overflow set                   (V set)
1000 : HI  Unsigned higher                (C set and Z clear)
1001 : LS  Unsigned lower or same         (C clear or Z set)
1010 : GE  Signed greater than or equal   (C  = V)
1011 : LT  Signed less than               (C != V)
1100 : GT  Signed greater than            (Z=0 and N=V)
1101 : LE  Signed less than or equal      (Z=1 or N!=V)
1110 : AL  Always
```

A rather unique feature of the ARM architecture is that almost all instructions can be executed conditionally. An instruction may contain an optional condition suffix, e.g. EQ, NE, LT, GT, GE, LE, GT, LT, etc. which determines whether the CPU will execute the instruction based on the specified condition. If the condition is not met, the instruction will not be executed at all without any side effects. This eliminates the need for many branches in a program, which tend to disrupt the instruction pipeline. To execute an instruction conditionally, simply postfix it with the appropriate condition. For example, a non-conditional ADD instruction has the form:

```
ADD r0, r1, r2    ; r0 = r1 + r2
```

To execute the instruction only if the zero flag is set, append the instruction with EQ.

```
ADDEQ r0, r1, r2  ; If zero flag is set then r0 = r1 + r2
```

Similarly for other conditions.

2.4.2 Branch Instructions

Branching instructions have the form

```
B{<cond>} label              ; branch to label
BL{<cond>} subroutine        ; branch to subroutine with link
```

The Branch (B) instruction causes a direct branch to an offset relative to the current PC. The Branch with link (BL) instruction is for subroutine calls. It writes PC-4 into LR of the current register bank and replaces PC with the entry address of the subroutine, causing the CPU to enter the subroutine. When the subroutine finishes, it returns by the saved return address in the link register R14. Most other processors implement subroutine calls by saving the return address on stack. Rather than saving the return address on stack, the ARM processor simply copies PC-4 into R14 and branches to the called subroutine. If the called subroutine does not call other subroutines, it can use the LR to return to the calling place quickly. To return from a subroutine, the program simply copies LR (R14) into PC (R15), as in

```
MOV PC, LR  or  BX LR
```

However, this works only for one-level subroutine calls. If a subroutine intends to make another call, it must save and restore the LR register explicitly since each subroutine call will change the current LR. Instead of the MOV instruction, the MOVS instruction can also be used, which restores the original flags in the CPSR.

2.4.3 Arithmetic Operations

The syntax of arithmetic operations is:

```
<Operation>{<cond>}{S} Rd, Rn, Operand2
```

The instruction performs an arithmetic operation on two operands and places the results in the destination register Rd. The first operand, Rn, is always a register. The second operand can be either a register or an immediate value. In the latter case, the operand is sent to the ALU via the **barrel shifter** to generate an appropriate value.
Examples:

```
ADD r0, r1, r2    ; r0 = r1 + r2
SUB r3, r3, #1    ; r3 = r3 - 1
```

2.4.4 Comparison Operations

CMP: operand1—operand2, but result not written
TST: operand1 AND operand2, but result not written
TEQ: operand1 EOR operand2, but result not written

Comparison operations update the condition flag bits in the status register, which can be used as conditions in subsequent instructions. Examples:

```
CMP   r0, r1    ; set condition bits in CPSR by r0-r1
TSTEQ r2, #5    ; test r2 and 5 for equal and set Z bit in CPSR
```

2.4.5 Logical Operations

```
AND: operand1 AND operand2          ; bit-wise AND
EOR: operand1 EOR operand2          ; bit-wise exclusive OR
ORR: operand1 OR operand2           ; bit-wise OR
BIC: operand1 AND (NOT operand2)    ; clear bits
```

Examples:

```
AND   r0, r1, r2     ; r0 = r1 & r2
ORR   r0, r0, #0x80  ; set bit 7 of r0 to 1
BIC   r0, r0, #0xF   ; clear low 4 bits of r0
EORS  r1, r3, r0     ; r1 = r3 ExOR r0 and set condition bits
```

2.4.6 Data Movement Operations

```
MOV    operand1, operand2
MVN    operand1, NOT operand2
```

Examples:

```
MOV     r0, r1        ; r0 = r1 : Always execute
MOVS    r2, #10       ; r2 = 10 and set condition bits Z=0 N=0
MOVNEQ r1, #0         ; r1 = 0 only if Z bit != 0
```

2.4.7 Immediate Value and Barrel Shifter

The Barrel shifter is another unique feature of ARM processors. It is used to generate shift operations and immediate operands inside the ARM processor. The ARM processor does not have actual shift instructions. Instead, it has a barrel shifter, which performs shifts as part of other instructions. Shift operations include the conventional shift left, right and rotate, as in

```
MOV r0, r0, LSL #1  ; shift r0 left by 1 bit (multiply r0 by 2)
MOV r1, r1, LSR #2  ; shift r1 right by 2 bits (divide r1 by 4)
MOV r2, r2, ROR #4  ; swap high and low 4 bits of r2
```

Most other processors allow loading CPU registers with immediate values, which form parts of the instruction stream, making the instruction length variable. In contrast, all ARM instructions are 32 bits long, and they do not use the instruction stream as data. This presents a challenge when using immediate values in instructions. The data processing instruction format has 12 bits available for operand2. If used directly, this would only give a range of 0–4095. Instead, it is used to store a 4-bit rotate value and an 8-bit constant in the range of 0–255. The 8 bits can be rotated right an even number of positions (i.e. RORs by 0, 2, 4,…,30). This gives a much larger range of values that can be directly loaded. For example, to load r0 with the immediate value 4096, use

```
MOV r0, #0x40, 26   ; generate 4096 (0x1000) by 0x40 ROR 26
```

To make this feature easier to use, the assembler will convert to this form if given the required constant in an instruction, e.g.

```
MOV r0, #4096
```

The assembler will generate an error if the given value can not be converted this way. Instead of MOV, the LDR instruction allows loading an arbitrary 32-bit value into a register, e.g.

```
LDR rd, =numeric_constant
```

If the constant value can be constructed by using either a MOV or MVN, then this will be the instruction actually generated. Otherwise, the assembler will generate an LDR with a PC-relative address to read the constant from a literal pool.

2.4.8 Multiply Instructions

```
MUL{<cond>}{S} Rd, Rm, Rs          ; Rd = Rm * Rs
MLA{<cond>}{S} Rd, Rm, Rs,Rn       ; Rd = (Rm * Rs) + Rn
```

2.4.9 LOAD and Store Instructions

The ARM processor is a Load/ Store Architecture. Data must be loaded into registers before using. It does not support memory to memory data processing operations. The ARM processor has three sets of instructions which interact with memory. These are:

- Single register data transfer (LDR/STR).
- Block data transfer (LDM/STM).
- Single Data Swap (SWP).

The basic load and store instructions are: Load and Store Word or Byte:

```
LDR / STR / LDRB / STRB
```

2.4.10 Base Register

Load/store instructions may use a base register as an index to specify the memory location to be accessed. The index may include an offset in either pre-index or post-index addressing mode. Examples of using index registers are

```
STR r0, [r1]       ; store r0 to location pointed by r1.
LDR r2, [r1]       ; load r2 from memory pointed to by r1.
STR r0, [r1, #12]  ; pre-index  addressing :STR r0 to [r1+12]
STR r0, [r1], #12  ; Post-index addressing :STR r0 to [r1],r1+12
```

2.4.11 Block Data Transfer

The base register is used to determine where memory access should occur. Four different addressing modes allow increment and decrement inclusive or exclusive of the base register location. Base register can be optionally updated following the data transfer by appending it with a '!' symbol. These instructions are very efficient for saving and restoring execution context, e.g. to use a memory area as stack, or move large blocks of data in memory. It is worth noting that, when using these instructions to save/restore multiple CPU registers to/from memory, the register order in the instruction does not matter. Lower numbered registers are always transferred to/from lower addresses in memory.

2.4.12 Stack Operations

A stack is a memory area which grows as new data is "pushed" onto the "top" of the stack, and shrinks as data is "popped" off the top of the stack. Two pointers are used to define the current limits of the stack.

- A base pointer: used to point to the "bottom" of the stack (the first location).
- A stack pointer: used to point the current "top" of the stack.

A stack is called **descending** if it grows downward in memory, i.e. the last pushed value is at the lowest address. A stack is called **ascending** if it grows upward in memory. The ARM processor supports both descending and ascending stacks. In addition, it also allows the stack pointer to either point to the last occupied address (**Full stack**), or to the next occupied address (**Empty stack**). In ARM, stack operations are implemented by the STM/LDM instructions. The stack type is determined by the postfix in the STM/LDM instructions:

- STMFD/LDMFD: Full Descending stack
- STMFA/LDMFA: Full Ascending stack
- STMED/LDMED: Empty Descending stack
- STMEA/LDMEA: Empty Ascending stack

The C compiler always uses **Full Descending** stack. Other forms of stacks are rare and almost never used in practice. For this reason, we shall only use **Full Descending stacks** throughout this book.

2.4.13 Stack and Subroutines

A common usage of stack is to create temporary workspace for subroutines. When a subroutine begins, any registers that are to be preserved can be pushed onto the stack. When the subroutine ends, it restores the saved registers by popping them off the stack before returning to the caller. The following code segment shows the general pattern of a subroutine.

```
STMFD sp!, {r0-r12, lr} ; save all registers and return address
 { Code of subroutine } ; subroutine code
LDMFD sp!, {r0-r12, pc} ; restore saved registers and return by lr
```

If the pop instruction has the 'S' bit set (by the '^' symbol), then transferring the PC register while in a privileged mode also copies the saved SPSR to the previous mode CPSR, causing return to the previous mode prior to the exception (by SWI or IRQ).

2.4.14 Software Interrupt (SWI)

In ARM, the SWI instruction is used to generate a software interrupt. After executing the SWI instruction, the ARM processor changes to SVC mode and executes from the SVC vector address 0x08, causing it to execute the SWI handler, which is usually the entry point of system calls to the OS kernel. We shall demonstrate system calls in Chap. 5.

2.4.15 PSR Transfer Instructions

The MRS and MSR instructions allow contents of CPSR/SPSR to be transferred from appropriate status register to a general purpose register. Either the entire status register or only the flag bits can be transferred. These instructions are used mainly to change the processor mode while in a privileged mode.

```
MRS{<cond>}  Rd, <psr>   ; Rd = <psr>
MSR{<cond>} <psr>, Rm    ; <psr> = Rm
```

2.4.16 Coprocessor Instructions

The ARM architecture treats many hardware components, e.g. the Memory Management Unit (MMU) as coprocessors, which are accessed by special coprocessor instructions. We shall cover coprocessors in Chap. 6 and later chapters.

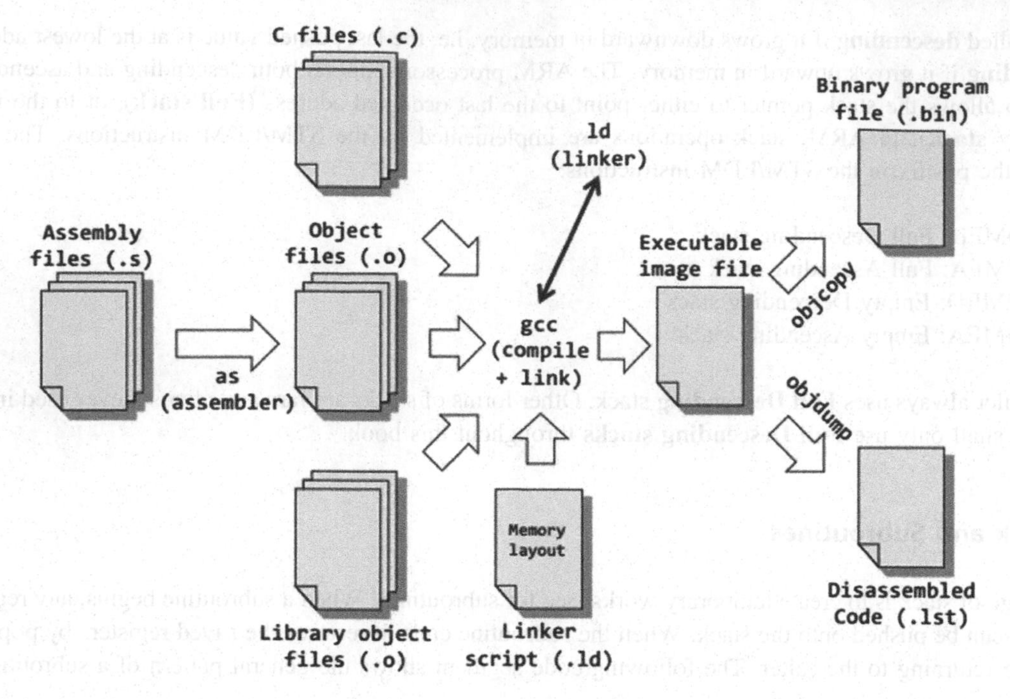

Fig. 2.3 Toolchain components

2.5 ARM Toolchain

A **toolchain** is a collection of programming tools for program development, from source code to binary executable files. A toolchain usually consists of an assembler, a compiler, a linker, some utility programs, e.g. objcopy, for file conversions and a debugger. Figure 2.3 depicts the components and data flows of a typical toolchan.

A toolchain runs on a host machine and generates code for a target machine. If the host and target architectures are different, the toolchain is called a cross toolchain or simply a **cross compiler**. Quite often, the toolchain used for embedded system development is a cross toolchain. In fact, this is the standard way of developing software for embedded systems. If we develop code on a Linux machine based on the Intel x86 architecture but the code is intended for an ARM target machine, then we need a Linux-based ARM-targeting cross compiler. There are many different versions of Linux based toolchains for the ARM architecture (ARM toolchains 2016). In this book, we shall use the arm-none-eabi toolchain under Ubuntu Linux Versions 14.04/15.10. The reader can get and install the toolchain, as well as the qemu-system-arm for ARM virtual machines, on Ubuntu Linux as follows.

```
sudo apt-get install gcc-arm-none-eabi
sudo apt-get install qemu-system-arm
```

In the following sections, we shall demonstrate how to use the ARM toolchain and ARM virtual machines under QEMU by programming examples.

2.6 ARM System Emulators

QEMU supports many emulated ARM machines (QEMU Emulator 2010). These includes ARM Integrator/CP board, ARM Versatile baseboard, ARM RealView baseboard and several others. The supported ARM CPUs include ARM926E, ARM1026E, ARM946E, ARM1136 or Cortex-A8. All these are uniprocessor (UP) or single CPU systems.

To begin with, we shall consider only uniprocessor (UP) systems. Multiprocessor (MP) systems will be covered later in Chap. 9. Among the emulated ARM virtual machines, we shall choose the ARM Versatilepb baseboard (ARM926EJ-S 2016) as the platform for implementation and testing, for the following reasons.

(1). It supports many peripheral devices. According to the QEMU user manual, the ARM Versatilepb baseboard is emulated with the following devices:

- ARM926E, ARM1136 or Cortex-A8 CPU.
- PL190 Vectored Interrupt Controller.
- Four PL011 UARTs.
- PL110 LCD controller.
- PL050 KMI with PS/2 keyboard and mouse.
- PL181 MultiMedia Card Interface with SD card.
- SMC 91c111 Ethernet adapter.
- PCI host bridge (with limitations).
- PCI OHCI USB controller.
- LSI53C895A PCI SCSI Host Bus Adapter with hard disk and CD-ROM devices.

(2). The ARM Versatile board architecture is well documented by on-line articles in the ARM Information Center.
(3). QEMU can boot up the emulated ARM Versatilepb virtual machine directly. For example, we may generate a binary executable file, t.bin, and run it on the emulated ARM VersatilepbVM by

qemu-system-arm –M versatilepb –m 128M –kernel t.bin –serial mon:stdio

QEMU will load the t.bin file to 0x10000 in RAM and executes it directly. This is very convenient since it eliminates the need for storing the system image in a flash memory and relying on a dedicated booter to boot up the system image.

2.7 ARM Programming

2.7.1 ARM Assembly Programming Example 1

We begin ARM programming by a series of example programs. For ease of reference, we shall label the example programs by C2.x, where C2 denotes the chapter number and x denotes the program number. The first example program, C2.1, consists of a ts.s file in ARM assembly. The following shows the steps of developing and running the example program.

(1). **ts.s file of C2.1**

```
/************ ts.s file of C2.1 ***********/
        .text
        .global start
start:
        mov r0, #1          @  r0 = 1
        MOV R1, #2          @  r1 = 2
        ADD R1, R1, R0      // r1 = r1 + r0
        ldr r2, =result     // r2 = &result
        str r1, [r2]        /* result = r1 */
stop:   b   stop
        .data
result:    .word   0        /* a word location */
```

The program code loads the CPU registers r0 with the value 1, r1 with the value 2. Then it adds r0 to r1 and stores the result into the memory location labeled result.

Before continuing, it is worth noting the following. First, in an assembly code program, instructions are case insensitive. An instruction may use uppercase, lowercase or even mixed cases. For better readability, the coding style should be consistent, either all lowercase or all uppercase. However, other symbols, e.g. memory locations, are case sensitive. Second, as shown in the program, we may use the symbol @ or // to start a comment line, or include comments in matched pairs of /*

and */. Which kind of comment lines to use is a matter of personal preference. In this book, we shall use // for single comment lines, and use matched pairs of /* and */ for comment blocks that may span multiple lines, which are applicable to both assembly code and C programs.

(2). The mk script file:

A sh script, mk, is used to (cross) compile-link ts.s into an ELF file. Then it uses objcopy to convert the ELF file into a binary executable image named t.bin.

```
arm-none-eabi-as -o ts.o ts.s                     # assemble ts.s to ts.o
arm-none-eabi-ld -T t.ld -o t.elf ts.o            # link ts.o to t.elf file
arm-none-eabi-nm t.elf                            # show symbols in t.elf
arm-none-eabi-objcopy -O binary t.elf t.bin # objcopy t.elf to t.bin
```

(3). linker script file:

In the linking step, a linker script file, t.ld, is used to specify the entry point and the layout of the program sections.

```
ENTRY(start)          /* define start as the entry address */
SECTIONS              /* program sections */
{
   . = 0x10000;       /* loading address, required by QEMU */
   .text : { *(.text) }    /* all text in .text section */
   .data : { *(.data) }    /* all data in .data section */
   .bss  : { *(.bss) }     /* all bss  in .bss section */
    =ALIGN(8);
    . =. + 0x1000;          /* 4 KB stack space */
   stack_top =.;   /* stack_top is a symbol exported by linker */
}
```

The linker generates an ELF file. If desired, the reader may view the ELF file contents by

arm-none-eabi-readelf -a t.elf # display all information of t.elf
arm-none-eabi-objdump -d t.elf # disassemble t.elf file

The ELF file is not yet executable. In order to execute, it must be converted to a binary executable file by objcopy, as in

```
arm-none-eabi-objcopy -O binary t.elf t.bin # convert t.elf to t.bin
```

(3) **Run the binary executable:** To run t.bin on the ARM Versatilepb virtual machine, enter the command

```
qemu-system-arm -M versatilepb -kernel t.bin -nographic -serial /dev/null
```

The reader may include all the above commands in a mk script, which will compile-link and run the binary executable by a single script command.

```
root@D632: ~/KCW1/ch2/C2.1
info registers
R00=00000001 R01=00000003 R02=0001001c R03=00000000
R04=00000000 R05=00000000 R06=00000000 R07=00000000
R08=00000000 R09=00000000 R10=00000000 R11=00000000
R12=00000000 R13=00000000 R14=00000000 R15=00010014
```

Fig. 2.4 Register contents of program C2.1

(4). **Check Results:** To check the results of running the program, enter the QEMU monitor commands:

```
info registers         :  display CPU registers
xp  /wd [address]      :  display memory contents in 32-bit words
```

Figure 2.4 shows the register contents of running the C2.1 program. As the figure shows, the register R2 contains 0x0001001C, which is the address of result. Alternatively, the command line **arm-none-eabi-nm t.elf** in the mk script also shows the locations of symbols in the program. The reader may enter the QEMU monitor command

```
xp  /wd  0x1001C
```

to display the contents of result, which should be 3. To exit QEMU, enter Control-a x, or Control-C to terminate the QEMU process.

2.7.2 ARM Assembly Programming Example 2

The next example program, denoted by C2.2, uses ARM assembly code to compute the sum of an integer array. It shows how to use a stack to call subroutine. It also demonstrates indirect and post-index addressing modes of ARM instructions. For the sake of brevity, we only show the ts.s file. All other files are the same as in the C2.1 program.

C2.2: ts.s file:

```
       .text
       .global start
start: ldr  sp,  =stack_top // set stack pointer
       bl   sum            // call sum
stop:  b    stop           // looping
sum: // int sum(): compute the sum of an int array in Result
       stmfd sp!, {r0-r4, lr} // save r0-r4, lr on stack
       mov  r0, #0          // r0 = 0
       ldr  r1, =Array      // r1 = &Array
       ldr  r2, =N          // r2 = &N
       ldr  r2, [r2]        // r2 = N
loop:  ldr  r3, [r1], #4    // r3 = *(r1++)
       add  r0, r0, r3      // r0 += r3
       sub  r2, r2, #1      // r2—
       cmp  r2, #0          // if (r2 != 0 )
       bne  loop            //    goto loop;
       ldr  r4, =Result     // r4 = &Result
       str  r0, [r4]        // Result = r0
```

```
root@D632: ~/KCW1/ch2/C2.2
info registers
R00=00000037 R01=00010078 R02=00000000 R03=0000000a
R04=00010078 R05=00000000 R06=00000000 R07=00000000
R08=00000000 R09=00000000 R10=00000000 R11=00000000
R12=00000000 R13=00011090 R14=00010008 R15=00010008
```

Fig. 2.5 Register contents of program C2.2

```
        ldmfd sp!, {r0-r4, pc}   // pop stack, return to caller

                .data
N:              .word 10              // number of array elements
Array:          .word 1,2,3,4,5,6,7,8,9,10
Result:         .word 0
```

The program computes the sum of an integer array. The number of array elements (10) is defined in the memory location labeled N, and the array elements are defined in the memory area labeled Array. The sum is computed in R0, which is saved into the memory location labeled Result. As before, run the mk script to generate a binary executable t.bin. Then run t.bin under QEMU. When the program stops, use the monitor commands info and xp to check the results. Figure 2.5 shows the results of running the C2.2 program. As the figure shows, the register R0 contains the computed result of 0x37 (55 in decimal). The reader may use the command

arm-none-eabi-nm t.elf

to display symbols in an object code file. It lists the memory locations of the global symbols in the t.elf file, such as

```
0001004C   N
00010050   Array
00010078   Result
```

Then, use xp/wd 0x10078 to see the contents of Result, which should be 55 in decimal.

2.7.3 Combine Assembly with C Programming

Assembly programming is indispensable, e.g. when accessing and manipulating CPU registers, but it is also very tedious. In systems programming, assembly code should be used as a tool to access and control low-level hardware, rather than as a means of general programming. In this book, we shall use assembly code only if absolutely necessary. Whenever possible, we shall implement program code in the high-level language C. In order to integrate assembly and C code in the same program, it is essential to understand program execution images and the calling convention of C.

2.7.3.1 Execution Image

An executable image (file) generated by a complier-linker consists of three logical parts.

Text section: also called Code section containing executable code
Data section: initialized global and static variables, static constants
BSS section: un-initialized global and static variables. (BSS is not in the image file)

During execution, the executable image is loaded into memory to create a run-time image, which looks like the following.

```
              ------------------------------------------------
(Low Address)    | Code | Data | BSS | Heap | Stack |    (High address)
              ------------------------------------------------
```

A run-time image consists of 5 (logically) contiguous sections. The Code and Data sections are loaded directly from the executable file. The BSS section is created by the BSS section size in the executable file header. Its contents are usually cleared to zero. In the execution image, the Code, Data and BSS sections are fixed and do not change. The Heap area is for dynamic memory allocation within the execution image. The stack is for function calls during execution. It is logically at the high (address) end of the execution image, and it grows downward, i.e. from high address toward low address.

2.7.3.2 Function Call Convention in C

The function call convention of C consists of the following steps between the calling function (the caller) and the called function (the callee).

----------------------------------- Caller -----------------------------------

(1). load first 4 parameters in r0-r3; push any extra parameters on stack
(2). transfers control to callee by BL callee

----------------------------------- Callee -----------------------------------

(3). save LR, FP(r12) on stack, establish stack frame (FP point at saved LR)
(4). shift SP downward to allocate local variables and temp spaces on stack
(4). use parameters, locals, (and globals) to perform the function task
(5). compute and load return value in R0, pop stack to return control to caller

----------------------------------- Caller -----------------------------------

(6). get return value from R0
(7). Clean up stack by popping off extra parameters, if any.

The function call convention can be best illustrated by an example. The following t.c file contains a C function func(), which calls another function g().

```
/****************** t.c file ********************/
extern int g(int x, in y);  // an external function

int func(int a, int b, int c, int d, int e, int f)
{
  int x, y, z;              // local variables
  x = 1; y =2; z = 3;       // access locals
  g(x, y);                  // call g(x,y)
  return a + e;             // return value
}
```

Use the ARM cross compiler to generate an assembly code file named t.s by

arm-none-eabi-gcc –S –mcpu=arm926ej-s **t.c**

The following shows the assembly code generated by the ARM GCC compiler, in which the symbol sp is the stack pointer (R13) and fp is the stack frame pointer (R12).

```
        .global func              // export func as global symbol
func:
(1). Establish stack frame
        stmfd sp!, {fp, lr}  // save lr, fp in stack
        add   fp, sp, #4       // FP point at saved LR
(2). Shift SP downward 8 (4-byte) slots for locals and temps
        sub   sp, sp, #32
(3). Save r0-r3 (parameters a,b,c,d) in stack at -offsets(fp)
        str   r0, [fp, #-24]     // save r0 a
        str   r1, [fp, #-28]     // save r1 b
        str   r2, [fp, #-32]     // save r2 c
        str   r3, [fp, #-36]     // save r3 d
(4). Execute x=1; y=2; z=3; show their locations on stack
        mov   r3, #1
        str   r3, [fp, #-8]      // x=1 at  -8(fp);
        mov   r3, #2
        str   r3, [fp, #-12]     // y=2 at -12(fp)
        mov   r3, #3
        str   r3, [fp, #-16]     // z=3 at -16(fp)
(5). Prepare to call g(x,y)
        ldr   r0, [fp, #-8]      // r0 = x
        ldr   r1, [fp, #-12]     // r1 = y
        bl    g                  // call g(x,y)
(6). Compute a+e as return value in r0
        ldr   r2, [fp, #-24]     // r2 = a (saved at -24(fp)
        ldr   r3, [fp, #4]       // r3 = e        at  +4(fp)
        add   r3, r2, r3         // r3 = a+e
        mov   r0, r3             // r0 = return value in r0
(7). Return to caller
        sub   sp, fp, #4         // sp=fp-4 (point at saved FP)
        ldmfd sp!, {fp, pc}      // return to caller
```

When calling the function func(), the caller must pass (6) parameters (a, b, c, d, e, f) to the called function. The first 4 parameters (a, b, c, d) are passed in registers r0–r3. Any extra parameters are passed via the stack. When control enters the called function, the stack top contains the extra parameters (in reverse order). For this example, the stack top contains the 2 extra parameters e and f. The initial stack looks like the following.

```
  High       SP                       low
----|---|---|--------------------
    | f | e |
----|exParam|--------------------
```

(1). Upon entry, the called function first establishes the stack frame by pushing LR, FP into stack and letting FP (r12) point at the saved LR.
(2). Then it shifts SP downward (toward low address) to allocate space for local variables and temporary working area, etc. The stack becomes

```
                        SP                          SP
High          |-push->|---- subtract SP by #32 ------>|----- Low
---|+8 |+4 | 0 |-4 |-8 |-12|-16|-20|-24|-28|-32|-36|
   | f  | e |LR |FP |   |   |   |   |   |   |   |   |
---|exParam|-|-|---| local variables, working space|-----
              FP
```

In the diagram, the (byte) offsets are relative to the location pointed by the FP register. While execution is inside a function, the extra parameters (if any) are at [fp, +offset], local variables and saved parameters are at [fp, –offset], all are referenced by using FP as the base register. From the assembly code lines (3) and (4), which save the first 4 parameters passed in r0-r3 and assign values to local variables x, y, z, we can see that the stack contents become as shown in the next diagram.

```
High                                                  SP  Low
----|+8 |+4 | 0 |-4 |-8 |-12|-16|-20|-24|-28|-32|-36|---
    | f | e |LR |FP | x | y | z | ? | a | b | c | d |
----|exParam|-|-|---|- locals --|---| saved r0-r3 --|---
               FP
    |< ----------- stack frame of func ----------- >|
```

Although the stack is a piece of contiguous memory, logically each function can only access a limited area of the stack. The stack area visible to a function is called the **stack frame** of the function, and FP (r12) is called the **stack frame pointer**.

At the assembly code lines (5), the function calls g(x, y) with only two parameters. It loads x into r0, y into r1 and then BL to g.

At the assembly code lines (6), it computes a + e as the return value in r0.

At the assembly code lines (7), it deallocates the space in stack, pops the saved FP and LR into PC, causing execution return to the caller.

The ARM C compiler generated code only uses r0–r3. If a function defines any register variables, they are assigned the registers r4–r11, which are saved in stack first and restored later when the function returns. If a function does not call out, there is no need to save/restore the link register LR. In that case, the ARM C compiler generated code does not save and restore the link register LR, allowing faster entry/exit of function calls.

2.7.3.3 Long Jump

In a sequence of function calls, such as

```
main() -> A() -> B()->C();
```

when a called function finishes, it normally returns to the calling function, e.g. C() returns to B(), which returns to A(), etc. It is also possible to return directly to an earlier function in the calling sequence by a long jump. The following program demonstrates long jump in Unix/Linux.

```
/***** longjump.c demonstrating long jump in Linux *****/
  #include <stdio.h>
  #include <setjmp.h>
jmp_buf env;        // for saving longjmp environment
main()
 {
   int r, a=100;
   printf("call setjmp to save environment\n");
   if ((r = setjmp(env)) == 0){
      A();
      printf("normal return\n");
   }
   else{
```

```
        printf("back to main() via long jump, r=%d a=%d\n", r, a);
    }
}
int A()
{   printf("enter A()\n");
    B();
    printf("exit A()\n");
}
int B()
{
    printf("enter B()\n");
    printf("long jump? (y|n) ");
    if (getchar()=='y')
        longjmp(env, 1234);
    printf("exit B()\n");
}
```

In the above longjump.c program, the main() function first calls setjmp(), which saves the current execution environment in a jmp_buf structure and returns 0. Then it proceeds to call A(), which calls B(). While in the function B(), if the user chooses not to return by long jump, the functions will show the normal return sequence. If the user chooses to return by longjmp(env, 1234), execution will return to the last saved environment with a nonzero value. In this case, it causes B() to return to main() directly, bypassing A().

The principle of long jump is very simple. When a function finishes, it returns by the (callerLR, callerFP) in the current stack frame, as shown in the following diagram.

```
-----------------------------------------------
|params|callerLR|callerFP|.............|
-------------|------------------------|---
          CPU.FP                   CPU.SP
```

If we replace (callerLR, callerFP) with (savedLR, savedFP) of an earlier function in the calling sequence, execution would return to that function directly. For example, we may implement setjmp(int env[2]) and longjmp(int env[2], int value) in assembly as follows.

```
          .global setjmp, longjmp
setjmp:   // int setjmp(int env[2]); save LR, FP in env[2]; return 0
          stmfd sp!, {fp, lr}
          add   fp, sp, #4
          ldr   r1, [fp]          // caller's return LR
          str   r1, [r0]          // save LR in env[0]
          ldr   r1, [fp, #-4]     // caller's FP
          str   r1, [r0, #4]      // save FP in env[1]
          mov   r0, #0            // return 0 to caller
          sub   sp, fp, #4
          ldmfd sp!, {fp, pc}

longjmp: // int longjmp(int env[2], int value)
          stmfd sp!, {fp, lr}
          add   fp, sp, #4
          ldr   r2, [r0]           // return function's LR
          str   r2, [fp]           // replace saved LR in stack frame
          ldr   r2, [r0, #4]       // return function's FP
          str   r2, [fp, #-4]      // replace saved FP in stack frame
          mov   r0, r1             // return value
          sub   sp, fp, #4
```

```
        ldmfd sp!, {fp, pc}       // return via REPLACED LR and FP
```

Long jump can be used to abort a function in a calling sequence, causing execution to resume to a known environment saved earlier. In addition to the (savedLR, savedFP), setjmp() may also save other CPU registers and the caller's SP, allowing longjmp() to restore the complete execution environment of the original function. Although rarely used in user mode programs, long jump is a common technique in systems programming. For example, it may be used in a signal catcher to bypass a user mode function that caused an exception or trap error. We shall demonstrate this technique later in Chap. 8 on signals and signal processing.

2.7.3.4 Call Assembly Function from C

The next example program C2.3 shows how to call assembly function from C. The main() function in C calls the assembly function sum() with 6 parameters, which returns the sum of all the parameters. In accordance with the calling convention of C, the main() function passes the first 4 parameters a, b, c, d in r0–r3 and the remaining parameters e, f, on stack. Upon entry to the called function, the stack top contains the parameters e, f, in the order of increasing addresses. The called function first establish the stack frame by saving LR, FP on stack and letting FP (r12) point at the save LR. The parameters e and f are now at FP + 4 and FP + 8, respectively. The sum function simply adds all the parameters in r0 and returns to the caller.

```
(1)./************* t.c file of Program C2.3 ***********/
int g;                              // un-initialized global
int main()
{
    int a, b, c, d, e, f;           // local variables
    a = b = c =d =e = f = 1;       // values do not matter
    g = sum(a,b,c,d,e,f);          // call sum(), passing a,b,c,d,e,f
}

(2)./*********** ts.s file of Program C2.3 **********/
        .global start, sum
start:  ldr sp, =stack_top
        bl  main                // call main() in C
stop:   b stop

sum:  // int sum(int a,b,c,d,e,f){ return a+b+e+d+e+f;}
// upon entry, stack top contains e, f, passed by main()in C
// Establish stack frame
      stmfd sp!, {fp, lr}  // push fp, lr
      add   fp, sp, #4     // fp -> saved lr on stack
// Compute sum of all (6) parameters
      add   r0, r0, r1     // first 4 parameters are in r0-r3
      add   r0, r0, r2
      add   r0, r0, r3
      ldr   r3, [fp, #4]   // load e into r3
      add   r0, r0, r3     // add to sum in r0
      ldr   r3, [fp, #8]   // load f into r3
      add   r0, r0, r3     // add to sum in r0
// Return to caller
      sub   sp, fp, #4     // sp=fp-4 (point at saved FP)
      ldmfd sp!, {fp, pc}  // return to caller
```

It is noted that in the C2.3 program, the sum() function does not save r0–r3 but use them directly. Therefore, the code should be more efficient than that generated by the ARM GCC compiler. Does this mean we should write all programs in assembly? The answer is, of course, a resounding NO. It should be easy for the reader to figure out the reasons.

(3). Compile-link t.c and ts.s into an executable file

```
arm-none-eabi-as -o ts.o ts.s                       # assemble ts.s
arm-none-eabi-gcc -c t.c                            # compile t.c into t.o
arm-none-eabi-ld -T t.ld -o t.elf t.o ts.o # link to t.elf file
arm-none-eabi-objcopy -O binary t.elf t.bin # convert t.elf to t.bin
```

(4). Run t.bin and check results on an ARM virtual machine as before.

2.7.3.5 Call C Function from Assembly

Calling C functions with parameters from assembly is also easy if we follow the calling convention of C. The following program C2.4 shows how to call a C function from assembly.

```
/*********** t.c file of Program 2.4 ****************/
int sum(int x, int y){ return x + y; }    // t.c file

/*********** ts.s file of Program C2.4 *************/
        .text
        .global start, sum
start:
        ldr sp,    = stack_top // need a stack to make calls
        ldr r2, =a
        ldr r0, [r2]          // r0 = a
        ldr r2, =b
        ldr r1, [r2]          // r1 = b
        bl sum                // c = sum(a,b)
        ldr r2, =c
        str r0, [r2]          // store return value in c
stop: b    stop

        .data
a:      .word 1
b:      .word 2
c:      .word 0
```

2.7.3.6 Inline Assembly

In the above examples, we have written assembly code in a separate file. Most ARM tool chains are based on GCC. The GCC compiler supports inline assembly, which is often used in C code for convenience. The basic format of inline assembly is

```
__asm__("assembly code");   or simply  asm("assembly code");
```

If the assembly code has more than one line, the statements are separated by \n\t; as in

```
asm("mov %r0, %r1\n\t;  add %r0,#10,r0\n\t");
```

Inline assembly code can also specify operands. The template of such inline assembly code is

```
asm ( assembler template
  : output operands
  : input operands
  : list of clobbered registers
  );
```

The assembly statements may specify output and input operands, which are referenced as %0, %1. For example, in the following code segment,

```
int a, b=10;
asm("mov %1,%%r0; mov %%r0,%0;"   // use %%REG for registers
      :"=r"(a)                    // output MUST have =
      :"r"(b)                     // input
      :"%r0"                      // clobbered registers
   );
```

In the above code segment, %0 refers to a, %1 refers to b, %%r0 refers to the r0 register. The constraint operator "r" means to use a register for the operand. It also tells the GCC compiler that the r0 register will be clobbered by the inline code. Although we may insert fairly complex inline assembly code in a C program, overdoing it may compromise the readability of the program. In practice, inline assembly should be used only if the code is very short, e.g. a single assembly instruction or the intended operation involves a CPU control register. In such cases, inline assembly code is not only clear but also more efficient than calling an assembly function.

2.8 Device Drivers

The emulated ARM Versatilepb board is a virtual machine. It behaves just like a real hardware system, but there are no drivers for the emulated peripheral devices. In order to do any meaningful programming, whether on a real or virtual system, we must implement device drivers to support basic I/O operations. In this book, we shall develop drivers for the most commonly used peripheral devices by a series of programming examples. These include drivers for UART serial ports, timers, LCD display, keyboard and the Multimedia SD card, which will be used later as a storage device for file systems. A practical device driver should use interrupts. We shall show interrupt-driven device drivers in Chap. 3 when we discuss interrupts and interrupts processing. In the following, we shall show a simple UART driver by polling and a LCD driver, which does not use interrupts. In order to do these, it is necessary to know the ARM Versatile system architecture.

2.8.1 System Memory Map

The ARM system architecture uses memory-mapped-I/O. Each I/O device is assigned a block of contiguous memory in the system memory map. Internal registers of each I/O device are accessed as offsets from the device base address. Table 2.1 shows the (condensed) memory map of the ARM Versatile/926EJ-S board (ARM 926EJ-S 2016). In the memory map, I/O devices occupy a 2 MB area beginning from 256 MB.

2.8.2 GPIO Programming

Most ARM based system boards provide General Purpose Input-Output (GPIO) pins as I/O interface to the system. Some of the GPIO pins can be configured for inputs. Other pins can be configured for outputs. In many beginning level embedded system courses, the programming assignments and course project are usually to program the GPIO pins of a small embedded system board to interface with some real devices, such as switches, sensors, LEDs and relays, etc. Compared with other I/O devices, GPIO programming is relatively simple. A GPIO interface, e.g. the LPC2129 GPIO MCU used in many early embedded system boards, consists of four 32-bit registers.

GPIODIR: set pin direction; 0 for input, 1 for output
GPIOSET: set pin voltage level to high (3.3 V)
GPIOCLR: set pin voltage level to low (0 V)
GPIOPIN: read this register returns the states of all pins

Table 2.1 Memory map of ARM versatile/ARM926EJ-S

MPMC Chip Select 0, 128 MB SRAM	0x00000000	128 MB
MPMC Chip Select 1, 128 MB expansion SRAM	0x08000000	128 MB
System registers	0x10000000	4 KB
Secondary Interrupt Controller (SIC)	0x10003000	4 KB
Multimedia Card Interface 0 (MMCID)	0x10005000	4 KB
Keyboard/Mouse Interface 0 (keyboard)	0x10006000	4 KB
Reserved (UART3 Interface)	0x10009000	4 KB
Ethernet Interface	0x10010000	64 KB
USB Interface	0x10020000	64 KB
Color LCD Controller	0x10120000	64 KB
DMA Controller	0x10130000	64 KB
Vectored Interrupt Controller (PIC)	0x10140000	64 KB
System Controller	0x101E0000	4 KB
Watchdog Interface	0x101E1000	4 KB
Timer modules 0 and 1 interface	0x101E2000	4 KB
(Timer 1 at 0x101E2020)	0x101E2FFF	
Timer modules 2 and 3 interface	0X101E3000	4 KB
(Timer 3 at 0x101E3020)	0X101E3FFF	
GPIO Interface (port 0)	0X101E4000	4 KB
GPIO Interface (port 1)	0x101E5000	4 KB
GPIO Interface (port 2)	0X101E6000	4 KB
UART 0 Interface	0x101E1000	4 KB
UART 1 Interface	0X101E2000	4 KB
UART 2 Interface	0x101F3000	4 KB
SSMC static expansion memory	0x20000000	256 MB

The GPIO registers can be accessed as word offsets from a (memory mapped) base address. In the GPIO registers, each bit corresponds to a GPIO pin. Depending on the direction setting in the IODIR, each pin can be connected to an appropriate I/O device.

As a specific example, assume that we want to use the GPIO pin0 for input, which is connected to a (de-bounced switch), and pin1 for output, which is connected to the (ground side) of an LED with its own +3.3 V voltage source and a current-limiting resistor. We can program the GPIO registers as follows.

GPIODIR: bit0=0 (input), bit1=1 (output);
GPIOSET: all bits=0 (no pin is set to high);
GPIOCLR: bit1=1 (set to LOW or ground);
GPIOPIN: read pin state, check pin0 for any input.

Similarly, we may program other pins for desired I/O functions. Programming the GPIO registers can be done in either assembly code or C. Given the GPIO base address and the register offsets, it should be fairly easy to write a GPIO control program, which

- turn on the LED if the input switch is pressed or closed, and
- turn off the LED if the input switch is released or open.

We leave this and other GPIO programming cases as exercises in the Problem section. In some systems, the GPIO interface may be more sophisticated but the programming principle remains the same. For example, on the ARM Versatile-PB board, GPIO interfaces are arranged in separate groups called ports (Port0 to Port2), which are at the base addresses 0x101E4000-0x101E6000. Each port provides 8 GPIO pins, which are controlled by a (8-bit) GPIODIR register and a (8-bit) GPIODATA register. Instead of checking the input pin states, GPIO inputs may use interrupts. Although interesting and inspiring to students, GPIO programming can only be performed on real hardware systems. Since the emulated ARM VMs do not have GPIO pins, we can only describe the general principles of GPIO programming. However, all the ARM VMs support a variety of other I/O devices. In the following sections, we shall show how to develop drivers for such devices.

2.8.3 UART Driver for Serial I/O

Relying on the QEMU monitor commands to display register and memory contents is very tedious. It would be much better if we can develop device drivers to do I/O directly. In the next example program, we shall write a simple UART driver for I/O on emulated serial terminals. The ARM Versatile board supports four PL011 UART devices for serial I/O (ARM PL011 2016). Each UART device has a base address in the system memory map. The base addresses of the 4 UARTs are

```
UART0: 0x101F1000
UART1: 0x101F2000
UART2: 0x101F3000
UART3: 0x10090000
```

Each UART has a number of registers, which are byte offsets from the base address. The following lists the most important UART registers.

```
0x00 UARTDR   Data register: for read/write chars
0x18 UARTFR   Flag register: TxEmpty, RxFull, etc.
0x24 UARIBRD  Baud rate register: set baud rate
0x2C UARTLCR  Line control register: bits per char, parity, etc.
0x38 UARTIMIS Interrupt mask register for TX and RX interrupts
```

In general, a UART must be initialized by the following steps.

(1). Write a divisor value to the baud rate register for a desired baud rate. The ARM PL011 technical reference manual lists the following integer divisor values (based on 7.38 MHz UART clock) for the commonly used baud rates:

```
0x4 = 1152000, 0xC = 38400, 0x18 = 192000, 0x20 = 14400, 0x30 = 9600
```

(2). Write to Line Control register to specify the number of bits per char and parity, e.g. 8 bits per char with no parity.
(3). Write to Interrupt Mask register to enable/disable RX and TX interrupts

When using the emulated ARM Versatilepb board, it seems that QEMU automatically uses default values for both baud rate and line control parameters, making steps (1) and (2) either optional or unnecessary. In fact, it is observed that writing any value to the integer divisor register (0x24) would work but this is not the norm for UARTs in real systems. For the emulated Versatilepb board, all we need to do is to program the Interrupt Mask register (if using interrupts) and check the Flag register during serial I/O. To begin with, we shall implement the UART I/O by polling, which only checks the Flag status register. Interrupt-driven device drivers will be covered later in Chap. 3 when we discuss interrupts and interrupts processing. When developing device drivers, we may need assembly code in order to access CPU registers and the interface hardware. However, we shall use assembly code only if absolutely necessary. Whenever possible, we shall implement the driver code in C, thus keeping the amount of assembly code to a minimum. The UART driver and test program, C2.5, consists of the following components.

(1). ts.s file: when the ARM CPU starts, it is in the Supervisor or SVC mode. The ts.s file sets the SVC mode stack pointer and calls main() in C.

```
    .global start, stack_top  // stack_top defined in t.ld
start:
    ldr sp, =stack_top // set SVC mode stack pointer
    bl  main           // call main() in C
    b .                // if main() returns, just loop
```

(2). t.c file: this file contains the main() function, which initializes the UARTs and uses UART0 for I/O on the serial port.

```
/************* t.c file of C2.5 **************/
int v[] = {1,2,3,4,5,6,7,8,9,10}; // data array
int sum;

#include "string.c"   // contains strlen(), strcmp(), etc.
#include "uart.c"     // UART driver code file

int main()
{
  int i;
  char string[64];
  UART *up;
  uart_init();          // initialize UARTs
  up = &uart[0];        // test UART0
  uprints(up, "Enter lines from serial terminal 0\n\r");
  while(1){
     ugets(up, string);
     uprints(up, "    ");
     uprints(up, string);
     uprints(up, "\n\r");
     if (strcmp(string, "end")==0)
        break;
  }
  uprints(up, "Compute sum of array:\n\r");
  sum = 0;
  for (i=0; i<10; i++)
    sum += v[i];
  uprints(up, "sum = ");
  uputc(up, (sum/10)+'0'); uputc(up, (sum%10)+'0');
  uprints(up, "\n\rEND OF RUN\n\r");
}
```

(3). uart.c file: this file implements a simple UART driver. The driver uses the UART data register for input/output chars, and it checks the flag register for device readiness. The following lists the meaning of the UART register contents.

Data register (offset 0x00): data in (READ)/data out (WRITE)
Flag register (offset 0x18): status of UART port

```
       7    6    5    4    3      2     1     0
    | TXFE RXFF TXFF RXFE BUSY    -     -     -  |
```

where TXFE=Tx buffer empty, RXFF=Rx buffer full, TXFF=Tx buffer full,
RXFE=Rx buffer empty, BUSY=device busy.

A standard way to access the individual bits of a register is to define them as symbolic constants, as in

```
#define TXFE 0x80
#define RXFF 0x40
#define TXFF 0x20
#define RXFE 0x10
#define BUSY 0x08
```

Then use them as bit masks to test the various bits of the flag register. The following shows the UART driver code.

```
/******** uart.c file of C2.5 : UART Driver Code ********/
/*** bytes offsets of UART registers from char *base ***/
#define UDR   0x00
#define UFR   0x18
typedef volatile struct uart{
  char *base;             // base address; as char *
  int  n;                 // uart number 0-3
}UART;
UART uart[4];             // 4 UART structures
int uart_init()           // UART initialization function
{
  int i; UART *up;
  for (i=0; i<4; i++){    // uart0 to uart2 are adjacent
    up = &uart[i];
    up->base = (char *)(0x101F1000 + i*0x1000);
    up->n = i;
  }
  uart[3].base = (char *)(0x10009000); // uart3 at 0x10009000
}

int ugetc(UART *up)       // input a char from UART pointed by up
{
  while (*(up->base+UFR) & RXFE);   // loop if UFR is REFE
  return *(up->base+UDR);           // return a char in UDR
}

int uputc(UART *up, charc)          // output a char to UART pointed by up
{
  while (*(up->base+UFR) & TXFF);   // loop if UFR is TXFF
  *(up->base+UDR) = c;              // write char to data register
}

int upgets(UART *up, char *s)       // input a string of chars
{
  while ((*s = ugetc(up)) != '\r') {
    uputc(up, *s);
    s++;
  }
  *s = 0;
}

int uprints(UART *up, char *s)      // output a string of chars
{
  while (*s)
    uputc(up, *s++);
}
```

(4). link-script file: The linker script file, t.ld, is the same as in the program C2.2.
(5). mk and run script file: The mk script file is also the same as in C2.2. For one serial port, the run script is

qemu-system-arm -M versatilepb -m 128M -kernel t.bin **-serial mon:stdio**

For more serial ports, add –serial /dev/pts/1 –serial /dev/pts/2, etc. to the command line. Under Linux, open xterm(s) as pseudo terminals. Enter the Linux ps command to see the pts/n numbers of the pseudo terminals, which must match the pts/n numbers in the –serial /dev/pts/n option of QEMU. On each pseudo terminal, there is a Linux sh process running, which will grab all the inputs to the terminal. To use a pseudo terminal as serial port, the Linux sh process must be made inactive. This can be done by entering the Linux sh command

sleep 1000000

which lets the Linux sh process sleep for a large number of seconds. Then the pseudo terminal can be used as a serial port of QEMU.

2.8.3.1 Demonstration of UART Driver

In the uart.c file, each UART device is represented by a UART data structure. As of now, the UART structure only contains a base address and a unit ID number. During UART initialization, the base address of each UART structure is set to the physical address of the UART device. The UART registers are accessed as *(up->base+OFFSET) in C. The driver consists of 2 basic I/O functions, ugetc() and uputc().

(1). int ugetc(UART *up): this function returns a char from the UART port. It loops until the UART flag register is no longer RXFE, indicating there is a char in the data register. Then it reads the data register, which clears the RXFF bit and sets the RXFE bit in FR, and returns the char.
(2). int uputc(UART *up, c): this function outputs a char to the UART port. It loops until the UART's flag register is no longer TXFF, indicating the UART is ready to transmit another char. Then it writes the char to the data register for transmission out.

The functions ugets() and uprints() are for I/O of strings or lines. They are based on ugetc() and uputc(). This is the typical way of how I/O functions are developed. For example, with gets(), we can implement an int itoa(char *s) function which converts a sequence of numerical digits into an integer. Similarly, with putc(), we can implement a printf() function for formatted printing, etc. We shall develop and demonstrate the printf() function in the next section on LCD driver. Figure 2.6 shows the outputs of running the C2.5 program, which demonstrates UART drivers.

2.8.3.2 Use TCP/IP Telnet Session as UART Port

In addition to pseudo terminals, QEMU also supports TCP/IP telnet sessions as serial ports. First run the program as

```
qemu-system-arm -M versatilepb -m 128M -kernel t.bin \
-serial telnet:localhost:1234,server
```

When QEMU starts, it will wait until a telnet connection is made. From another (X-window) terminal, enter **telnet localhost 1234** to connect. Then, enter lines from the telnet terminal.

```
Enter lines from serial termianl 0
test UART driver    test UART driver
a new test line    a new test line
end    end
Compute sum of 10 integers:
sum = 55
END OF RUN
```

Fig. 2.6 Demonstration of UART driver program

2.8.4 Color LCD Display Driver

The ARM Versatile board supports a color LCD display, which uses the ARM PL110 Color LCD controller (ARM PrimeCell Color LCD Controller PL110, ARM Versatile Application Baseboard for ARM926EF-S). On the Versatile board, the LCD controller is at the base address 0x10120000. It has several timing and control registers, which can be programmed to provide different display modes and resolutions. To use the LCD display, the controller's timing and control registers must be set up properly. ARM's Versatile Application Baseboard manual provides the following timing register settings for VGA and SVGA modes.

```
Mode  Resolution  OSC1     timeReg0     timeReg1     timeReg2
------------------------------------------------------------
VGA   640x480    0x02C77  0x3F1F3F9C  0x090B61DF  0x067F1800
SVGA  800x600    0x02CAC  0x1313A4C4  0x0505F6F7  0x071F1800
------------------------------------------------------------
```

The LCD's frame buffer address register must point to a frame buffer in memory. With 24 bits per pixel, each pixel is represented by a 32-bit integer, in which the low 3 bytes are the BGR values of the pixel. For VGA mode, the needed frame buffer size is 1220 KB bytes. For SVGA mode, the needed frame buffer size is 1895 KB. In order to support both VGA and SVGA modes, we shall allocate a frame buffer size of 2 MB. Assuming that the system control program runs in the lowest 1 MB of physical memory, we shall allocate the memory area from 2 to 4 MB for the frame buffer. In the LCD Control register (0x1010001C), bit0 is LCD enable and bit11 is power-on, both must be set to 1. Other bits are for byte order, number of bits per pixel, mono or color mode, etc. In the LCD driver, bits3–1 are set to 101 for 24 bits per pixel, all other bits are 0s for little-endian byte order by default. The reader may consult the LCD technical manual for the meanings of the various bits. It should be noted that, although the ARM manual lists the LCD Control register at 0x1C, it is in fact at 0x18 on the emulated Versatilepb board of QEMU. The reason for this discrepancy is unknown.

2.8.4.1 Display Image Files

As a memory mapped display device, the LCD can display both images and text. It is actually much easier to display images than text. The principle of displaying images is rather simple. An image consists of H (height) by W (width) pixels, where H<=480 and W<=640 (for VGA mode). Each pixel is specified by a 3-byte RGB color values. To display an image, simply extract the RGB values of each pixel and write them to the corresponding pixel location in the display frame buffer. There are many different image file formats, such as BMP, JPG, PNG, etc. Applications for Microsoft Windows typically use BMP images. JPG images are popular with Internet Web pages due to their smaller size. Each image file has a header which contains information of the image. In principle, it should be fairly easy to read the file header and then extract the pixels of the image file. However, many image files are often in compressed format, e.g. JPG files, which must be uncompressed first. Since our purpose here is to show the LCD display driver, rather than manipulation of image files, we shall only use 24-bit color BMP files due to their simple image format. Table 2.2 shows the format of BMP files.

A 24-bit color BMP image file is uncompressed. It begins with a 14-byte file header, in which the first two bytes are BMP file signature 'M' and 'B', indicating that it is a BMP file. Following the file header is a 40-byte image header, which contains the width (W) and height (H) of the image in number of pixels at the byte offsets 18 and 22, respectively. The image header also contains other information, which can be ignored for simple BMP files. Immediately following the image header are 3-byte BGR values of the image pixels arranged in H rows. In a BMP file, the image is stored upside down. The first row in the image file is actually the bottom row of the image. Each row contains (W*3) raised to a multiple of 4 bytes. The example program reads BMP images and displays them to the LCD screen. Since the LCD can only display 640x480 pixels in VGA mode, larger images can be displayed in reduced size, e.g. 1/2 or 1/4 of their original size.

2.8.4.2 Include Binary Data Sections

Raw image files can be included as binary data sections in an executable image. Assume that IMAGE is a raw image file. The following steps show how to include it as a binary data section in an executable image.

(1). Convert raw data into object code by objcopy

```
arm-none-eabi-objcopy -I binary -O elf32-littlearm -B arm IMAGE image.o
```

Table 2.2 BMP file format

Offset	Size	Description
--------------- 14-byte file header ------------------		
0	2	Signature ('MB')
2	4	Size of BMP file in bytes
6	2	Reserved (0)
8	2	Reserved (0)
10	4	Offset to start of image data in bytes
---------------- 40-byte image header ------------------		
14	4	Size of image header (40)
18	4	Image width in pixels
22	4	Image height in pixels
26	2	Number of image planes (1)
28	2	Number of bits per pixel (24)
------------ Other fields -------------------------		
50	4	Number of important colors (0)
----------------- Rows of image image ------------------		
54 to end of file: rows of images		

(2). Include object code as binary data sections in linker script

```
#---- linker script file t.ld -------
ENTRY(resset_start)
SECTIONS
{  . = 0x10000;
   .text : { ts.o  *( .text) }
   .data : { *(.data) }
   .bss  : { *(.bss)  }
   .data : { *(image.o) } /* include image.o as a data section */
   /* stack areas */
}
```

2.8.4.3 Programming Example C2.6: LCD Driver

The example program C2.6 implements an LCD driver which displays raw image files. The program consists of the following components.

(1). The ts.s file: Since the driver program does not use interrupts, nor tries to handle any exceptions, there is no need to install the exception vectors. Upon entry, it sets up the SVC mode stack and calls main() in C.

```
/*********** ts.s file of C2.6 *********/
        .global reset_start
reset_start:
  LDR sp, =stack_top    // set SVC stack pointer
  BL main
  B .
```

(2). The vid.c file: This is the LCD driver. It initializes the LCD registers to VGA mode of 640x480 resolutions and sets the frame buffer at 2 MB. It also includes code for SVGA mode with 800x600 resolutions but they are commented out.

```
/************** vid.c file of C2.6 ******************/
int volatile *fb;
int WIDTH = 640; // default to VGA mode for 640x480
int fbuf_init(int mode)
{
   fb = (int *)(0x200000); // at 2 MB to 4 MB
   //***************** for 640x480 VGA ****************/
    *(volatile unsigned int *)(0x1000001c) = 0x2C77;
    *(volatile unsigned int *)(0x10120000) = 0x3F1F3F9C;
    *(volatile unsigned int *)(0x10120004) = 0x090B61DF;
    *(volatile unsigned int *)(0x10120008) = 0x067F1800;
   }
    /***************** for 800X600 SVGA ****************
    *(volatile unsigned int *)(0x1000001c) = 0x2CAC;
    *(volatile unsigned int *)(0x10120000) = 0x1313A4C4;
    *(volatile unsigned int *)(0x10120004) = 0x0505F6F7;
    *(volatile unsigned int *)(0x10120008) = 0x071F1800;
   }
    ***************************************************/
   *(volatile unsigned int *)(0x10120010) = 0x200000; // fbuf
   *(volatile unsigned int *)(0x10120018) = 0x82B;
}
```

(3). uart.c file: This is the same UART driver in Example C2.5, except that it uses the basic uputc() function to implement a uprintf() function for formatted printing.
(4). The t.c file: This file contains the main() function, which calls the show_bmp() function to display images. In the linker script, two image files, image1 and image2, are included as binary data sections in the executable image. The start position of an image file can be accessed by the symbols _binary_imageI_start generated by the linker.

```
/************** t.c file of program C2.6 ******************/
#include "defines.h"   // device base addresses, etc.
#include "vid.c"       // LCD driver
#include "uart.c"      // UART driver
extern char _binary_image1_start, _binary_image2_start;
#define WIDTH 640
int show_bmp(char *p, int start_row, int start_col)
{
   int h, w, pixel, rsize, i, j;
   unsigned char r, g, b;
   char *pp;
   int *q = (int *)(p+14);     // skip over 14-byte file header
   w = *(q+1);                 // image width in pixels
   h = *(q+2);                 // image height in pixels
   p += 54;                    // p-> pixels in image
   //BMP images are upside down, each row is a multiple of 4 bytes
   rsize = 4*((3*w + 3)/4);    // multiple of 4
   p += (h-1)*rsize;           // last row of pixels
   for (i=start_row; i<start_row + h; i++){
      pp = p;
      for (j=start_col; j<start_col + w; j++){
```

```
        b = *pp; g = *(pp+1); r = *(pp+2); // BRG values
        pixel = (b<<16) | (g<<8) | r;      // pixel value
        fb[i*WIDTH + j] = pixel; // write to frame buffer
        pp += 3;                 // advance pp to next pixel
      }
      p -= rsize;                // to preceding row
    }
    uprintf(\nBMP image height=%d width=%d\n", h, w);
}
int main()
{
    char c,* p;
    uart_init();              // initialize UARTs
    up = upp[0];              // use UART0
    fbuf_init();              // default to VGA mode
    while(1){
      p = &_binary_image1_start;
      show_bmp(p, 0, 80); // display image1
      uprintf("enter a key from this UART : ");
      ugetc(up);
      p = &_binary_image2_start;
      show_bmp(p,120, 0); // display image2
    }
    while(1);                 // loop here
}
```

(5). The mk script file: The mk script generates object code for the image files, which are included as binary data sections in the executable image.
 # mk and run script file of C2.6: The only thing new is to convert images to object files.

```
arm-none-eabi-objcopy -I binary -O elf32-littlearm -B arm image1 image1.o
arm-none-eabi-objcopy -I binary -O elf32-littlearm -B arm image2 image2.o

arm-none-eabi-as -mcpu=arm926ej-s ts.s -o ts.o
arm-none-eabi-gcc -c -mcpu=arm926ej-s t.c -o t.o
arm-none-eabi-ld -T t.ld ts.o t.o -o t.elf
arm-none-eabi-objcopy -O binary t.elf t.bin
echo ready to go?
read dummy

qemu-system-arm -M versatilepb -m 128 M -kernel t.bin -serial mon:stdio
```

2.8.4.4 Demonstration of Display Images on LCD

Figure 2.7 shows the sample outputs of running the C2.6 program in VGA mode.

The top part of Fig. 2.7 shows the UART port I/O. The bottom part of the figure shows the LCD display. When the program starts, it first displays image1 to (row = 0, col = 80) on the LCD, and it also prints the image size to UART0. Entering an input key from UART0 will let it display image2 to (row = 120, col = 0), etc. Displaying image files can be very interesting, which is the basis of computer animation. Variations to the image displaying program are listed as exercises in the Problems section for interested readers.

2.8.4.5 Display Text

In order to display text, we need a font file, which specifies the fonts or bit patterns of the ASCII chars. The font file, font.bin, is a raw bitmap of 128 ASCII chars, in which each char is represented by a 8x16 bitmap, i.e. each char is represented by 16 bytes, each byte specifies the pixels of a scan line of the char. To display a char, use the char's ASCII code value (multiplied by 16) as an offset to access its bytes in the bitmap. Then scan the bits in each byte. For each 0 bit, write BGR = 0x000000 (black) to the corresponding pixel. For each 1 bit, write BRG = 0x111111 (white) to the pixel. Instead of black and white, each char can also be displayed in color by writing different RGB values to the pixels.

Like image files, raw font files can be included as binary data sections in the executable image. An alternative way is to convert bitmaps to char maps first. As an example, the following program, bitmap2charmap.c, converts a font bitmap to a char map.

```
/**** bitmap2charmap.c: run as a.out font.bin  >  font ****/
  #include <stdio.h>
  main(int argc, char *argv[ ])
  {
    int i, n; u8 buf[16];
    FILE *fp = fopen(argv[1], "r");     // fopen file for READ
    while((n = fread(buf, 1, 16, fp)){ // read 16 bytes
      for (i=0; i<n; i++)              // write each byte as 2 hex
         printf("0x%2x ", buf[i]);
    }
    printf("\n");
  }
```

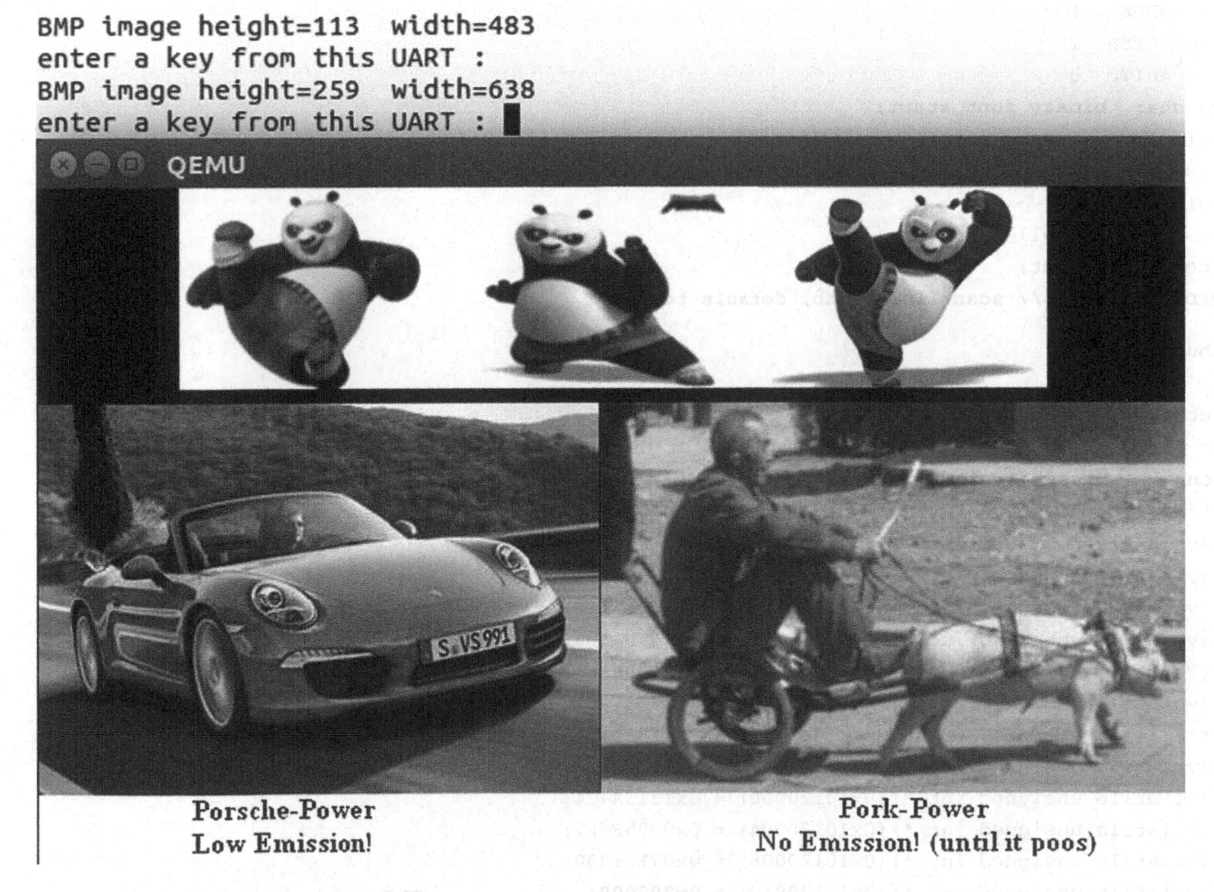

Fig. 2.7 Demonstration of display images on LCD

Unlike raw bitmap files, which must be converted to object files first, char map files are larger in size but they can be included directly in the C code.

2.8.4.6 Color LCD Display Driver Program

The example program C2.7 demonstrates an LCD driver for displaying text. The program consists of the following components.

(1). ts.s file: The ts.s file is the same as in the example program C2.6.
(2). vid.c file: The vid.c file implements a driver for the ARM PL110 LCD display [ARM PL110, 2016]. On the Versatilepb board the base address of the Color LCD is at 0x10120000. Other registers are (u32) offsets from the base address.

```
/******** vid.c file : LCD driver ********/
00  timing0
04  timing1
08  timing2
0C  timing3
10  upperPanelframeBaseAddressRegister // use the upper panel
14  lowerPanelFrameBaseAddressRegister // NOT use lower panel
18  controlRegister // NOTE: QEMU emulated PL110 CR is at 0x18
*******************************************************/
#define RED     0
#define BLUE    1
#define GREEN   2
#define WHITE   3
extern char _binary_font_start;
int color;
u8 cursor;
int volatile *fb;
int row, col, scroll_row;
unsigned char *font;
int WIDTH = 640;   // scan line width, default to 640

int fbuf_init()
{
    int i;
    fb = (int *)0x200000;          // frame buffer at 2MB-4MB
    font = &_binary_font_start;  // font bitmap
    /********* for 640x480 VGA mode *******************/
    *(volatile unsigned int *)(0x1000001c) = 0x2C77;
    *(volatile unsigned int *)(0x10120000) = 0x3F1F3F9C;
    *(volatile unsigned int *)(0x10120004) = 0x090B61DF;
    *(volatile unsigned int *)(0x10120008) = 0x067F1800;
    *(volatile unsigned int *)(0x10120010) = 0x200000; // at 2MB
    *(volatile unsigned int *)(0x10120018) = 0x82B;
    /********** for 800X600 SVGA mode ******************
    *(volatile unsigned int *)(0x1000001c) = 0x2CAC; // 800x600
    *(volatile unsigned int *)(0x10120000) = 0x1313A4C4;
    *(volatile unsigned int *)(0x10120004) = 0x0505F6F7;
    *(volatile unsigned int *)(0x10120008) = 0x071F1800;
    *(volatile unsigned int *)(0x10120010) = 0x200000;
    *(volatile unsigned int *)(0x10120018) = 0x82B;
    **********/
    cursor = 127; // cursor = row 127 in font bitmap
}
```

```
int clrpix(int x, int y)    // clear pixel at (x,y)
{
  int pix = y*640 + x;
  fb[pix] = 0x00000000;
}
int setpix(int x, int y)    // set pixel at (x,y)
{
  int pix = y*640 + x;
  if (color==RED)
     fb[pix] = 0x000000FF;
  if (color==BLUE)
     fb[pix] = 0x00FF0000;
  if (color==GREEN)
     fb[pix] = 0x0000FF00;
}
int dchar(unsigned char c, int x, int y) // display char at (x,y)
{
  int r, bit;
  unsigned char *caddress, byte;
  caddress = font + c*16;
  for (r=0; r<16; r++){
    byte = *(caddress + r);
    for (bit=0; bit<8; bit++){
      if (byte & (1<<bit))
        setpix(x+bit, y+r);
    }
  }
}
int undchar(unsigned char c, int x, int y) // erase char at (x,y)
{
  int row, bit;
  unsigned char *caddress, byte;
  caddress = font + c*16;
  for (row=0; row<16; row++){
    byte = *(caddress + row);
    for (bit=0; bit<8; bit++){
      if (byte & (1<<bit))
        clrpix(x+bit, y+row);
    }
  }
}
int scroll() // scrow UP one line (the hard way)
{
  int i;
  for (i=64*640; i<640*480; i++){
    fb[i] = fb[i + 640*16];
  }
}
int kpchar(char c, int ro, int co) // print char at (row, col)
{
   int x, y;
   x = co*8;
   y = ro*16;
   dchar(c, x, y);
}
```

```
int unkpchar(char c, int ro, int co) // erase char at (row, col)
{
    int x, y;
    x = co*8;
    y = ro*16;
    undchar(c, x, y);
}
int erasechar()  // erase char at (row,col)
{
  int r, bit, x, y;
  unsigned char *caddress, byte;
  x = col*8;
  y = row*16;
  for (r=0; r<16; r++){
     for (bit=0; bit<8; bit++){
        clrpix(x+bit, y+r);
     }
  }
}
int clrcursor() // clear cursor at (row, col)
{
  unkpchar(127, row, col);
}
int putcursor(unsigned char c) // set cursor at (row, col)
{
  kpchar(c, row, col);
}

int kputc(char c)  // print char at cursor position
{
  clrcursor();
  if (c=='\r'){     // return key
    col=0;
    putcursor(cursor);
    return;
  }
  if (c=='\n'){     // new line key
    row++;
    if (row>=25){
      row = 24;
      scroll();
    }
    putcursor(cursor);
    return;
  }
  if (c=='\b'){     // backspace key
    if (col>0){
      col--;
      erasechar();
      putcursor(cursor);
    }
    return;
  }
  // c is ordinary char case
  kpchar(c, row, col);
```

```
  col++;
  if (col>=80){
    col = 0;
    row++;
    if (row >= 25){
      row = 24;
      scroll();
    }
  }
  putcursor(cursor);
}

// The following implements kprintf() for formatted printing
int kprints(char *s)
{
  while(*s){
    kputc(*s);
    s++;
  }
}
int krpx(int x)
{
  char c;
  if (x){
     c = tab[x % 16];
     krpx(x / 16);
  }
  kputc(c);
}
int kprintx(int x)
{
  kputc('0'); kputc('x');
  if (x==0)
    kputc('0');
  else
    krpx(x);
  kputc(' ');
}
int krpu(int x)
{
  char c;
  if (x){
     c = tab[x % 10];
     krpu(x / 10);
  }
  kputc(c);
}
int kprintu(int x)
{
  if (x==0)
    kputc('0');
  else
    krpu(x);
  kputc(' ');
}
```

```
int kprinti(int x)
{
  if (x<0){
    kputc('-');
    x = -x;
  }
  kprintu(x);
}
int kprintf(char *fmt,...)
{
  int *ip;
  char *cp;
  cp = fmt;
  ip = (int *)&fmt + 1;

  while(*cp){
    if (*cp != '%'){
      kputc(*cp);
      if (*cp=='\n')
      kputc('\r');
      cp++;
      continue;
    }
    cp++;
    switch(*cp){
    case 'c': kputc((char)*ip);        break;
    case 's': kprints((char *)*ip);    break;
    case 'd': kprinti(*ip);            break;
    case 'u': kprintu(*ip);            break;
    case 'x': kprintx(*ip);            break;
    }
    cp++; ip++;
  }
}
```

2.8.4.7 Explanations of the LCD Driver Code

The LCD screen may be regarded as a rectangular box consisting of 480x640 pixels. Each pixel has a (x,y) coordinate on the screen. Correspondingly, the frame buffer, u32 fbuf[ ], is a memory area containing 480*640 u32 integers, in which the low 24 bits of each integer represents the BGR values of the pixel. The linear address or index of a pixel in fbuf[ ] at the coordinate (x,y) = (column, row) is given by the Mailman's algorithm (Chap. 2, Wang 2015).

$$\text{pixel\_index} = x + y*640;$$

The basic display functions of the LCD driver are

(1). setpix(x,y): set the pixel at (x, y) to BGR values (by a global color variable).
(2). clrpix(x,y): clear the pixel at (x, y) by setting the BGR to background color (black).
(3). dchar(char, x, y): display char at coordinate (x, y). Each char is represented by a 8x16 bitmap. For a given char value (0 to 127), dchar() fetches the 16 bytes of the char from the bitmap. For each bit in a byte, it calls clrpix(x+bitNum, y +byteNum) to clear the pixel first. This erases the old char, if any, at (x, y). Otherwise, it will display the composite bit patterns of the chars, making it unreadable. Then it calls setpix(x+bitNum, y+byteNum) to set the pixel if the bit is 1.
(4). erasechar(): erase the char at (x, y). For a memory mapped display device, once a char in written to the frame buffer, it will remain there (and hence be rendered on the screen) until it is erased. For ordinary chars, dchar() automatically erases the

original char. For special chars like the cursor, it is necessary to erase it first before moving it to a different position. This is done by the erasechar() operation.
(5). kputc(char c): display a char at the current (row, col) and move the cursor, which may cause scroll-up the screen.
(6). scroll(): scroll screen up or down by one line.
(7). The cursor: When displaying text, the cursor allows the user to see where the next char will be displayed. The ARM LCD controller does not have a cursor generator. In the LCD driver, the cursor is simulated by a special char (ASCII code 127) in which all the pixels are 1's, which defines a solid rectangular box as the cursor. The cursor may be made to blink if it is turned on/off periodically, e.g. every 0.5 s, which requires a timer. We shall show the blinking cursor later in Chap. 3 when we implement timers with timer interrupts. The putcursor() function draws the cursor at the current (row, col) position on the screen, and the erasecursor() function erases the cursor from its current position.
(8). **The printf() Function**

For any output device that supports the basic printing char operation, we can implement a printf() function for formatted printing. The following shows how to develop such a generic printf() function, which can be used for both UART and the LCD display. First, we implement a printu() function, which prints unsigned integers.

```
char *ctable = "0123456789ABCDEF";
int BASE = 10; // for decimal numbers
int rpu(u32 x)
{
    char c;    // local variable
    if (x){
        c = ctable[x % BASE];
        rpu(x / BASE);
        putc(c);
    }
}
int printu(u32 x)
{
    (x==0)? putc('0') : rpu(x);
    putc(' ');
}
```

The function rpu(x) generates the digits of x % 10 in ASCII recursively and prints them on the return path. For example, if x=123, the digits are generated in the order of '3', '2', '1', which are printed as '1', '2', '3' as they should. With printu(), writing a printd() function to print signed integers becomes trivial. By setting BASE to 16, we can print in hex. Assume that we have prints(), printd(), printu() and printx() already implemented. Then we can write a

```
int printf(char *fmt, ...)          // NOTE the 3 dots in function heading
```

function for formatted printing, where fmt is a format string containing conversion symbols %c, %s, %u, %d, %x.

```
int printf(char *fmt, ...) // most C compilers require the 3 dots
{
    char *cp = fmt;              // cp points to the fmt string
    int *ip = (int *)&fmt +1;    // ip points to first item in stack
    while (*cp){                 // scan the format string
      if (*cp != '%'){           // spit out ordinary chars
        putc(*cp);
        if (*cp=='\n')           // for each '\n'
          putc('\r');            // print a '\r'
        cp++;
        continue;
      }
      cp++;          // cp points at a conversion symbol
```

```
      switch(*cp){ // print item by %FORMAT symbol
         case 'c' :   putc((char   )*ip);  break;
         case 's' : prints((char *)*ip);  break;
         case 'u' : printu((u32    )*ip);  break;
         case 'd' : printd((int    )*ip);  break;
         case 'x' : printx((u32    )*ip);  break;
      }
     cp++; ip++;                  // advance pointers
   }
}
```

(5). The t.c file: The t.c file contains the main() function and the show_bmp() function. It first initializes both the UART and LCD drivers. The UART driver is used for I/O from the serial port. For demonstration purpose, it displays outputs to both the serial port and the LCD. When the program starts, it displays a small logo image at the top of the screen. The scroll upper limit is set to a line below the logo image, so that the logo will remain on the screen when the screen is scrolled upward.

```
/************** t.c file of C2.7 *************/
#include "defines.h"
#include "uart.c"
#include "vid.c"
extern char _binary_panda1_start;
int show_bmp(char *p, int startRow, int startCol){// SAME as before}
int main()
{
   char line[64];
   fbuf_init();
   char *p = &_binary_panda1_start;
   show_bmp(p, 0, 0); // display a logo
   uart_init();
   UART *up = upp[0];
   while(1){
     color = GREEN;
     kprintf("enter a line from UART port : ");
     uprintf("enter line from UART : ");
     ugets(up, line);
     uprintf(" line=%s\n", line);
     color = RED;
     kprintf("line=%s\n",line);
   }
}
```

(6). t.ld file: The linker script includes the object files of a font and an image as binary data sections, similar to that of the example program C2.6.
(7). mk and run script file: This is similar to that of C2.6.

2.8.4.8 Demonstration of LCD Driver Program

Figure 2.8 shows the sample outputs of running the example program C2.7. It uses the LCD driver to display both images and text on the LCD screen. In addition, it also uses the UART driver to do I/O from the serial port.

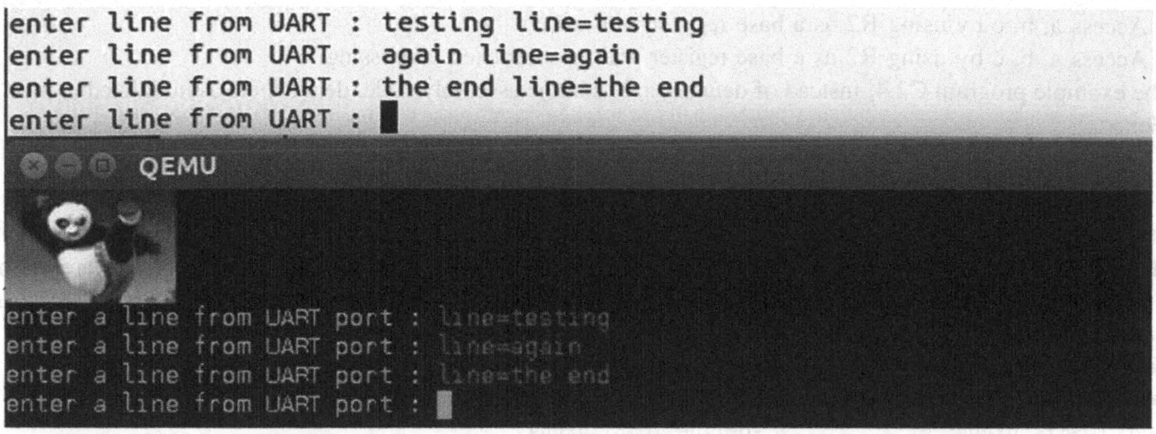

```
enter line from UART : testing line=testing
enter line from UART : again line=again
enter line from UART : the end line=the end
enter line from UART :
```

QEMU

```
enter a line from UART port : line=testing
enter a line from UART port : line=again
enter a line from UART port : line=the end
enter a line from UART port :
```

Fig. 2.8 Demonstration of display text on LCD

2.9 Summary

This chapter covers the ARM architecture, ARM instructions, programming in ARM assembly and development of programs for execution on ARM virtual machines. These include ARM processor modes, banked registers in different modes, instructions and basic programming in ARM assembly. Since most software for embedded systems are developed by cross-compiling, it introduces the ARM toolchain, which allows us to develop programs for execution on emulated ARM virtual machines. We choose the Ubuntu (14.04/15.0) Linux as the program development platform because it supports the most complete ARM toolchains. Among the ARM virtual machines, we choose the emulated ARM Versatilepb board under QEMU because it supports many commonly used peripheral devices found in real ARM based systems. Then it shows how to use the ARM toolchain to develop programs for execution on the ARM Versatilepb virtual machine by a series of programming examples. It explains the function call convention in C and shows how to interface assembly code with C programs. Then it develops a simple UART driver for I/O on serial ports, and a LCD driver for displaying both graphic images and text. It also shows the development of a generic printf() function for formatted printing to output devices that support the basic print char operation.

List of Sample Programs

C2.1: ARM assembly programming
C2.2: Sum of integer array in assembly
C2.3: Call assembly function from C
C2.4: Call C function from assembly
C2.5: UART driver
C2.6: LCD driver for displaying images
C2.7: LCD driver for displaying text

Problems

1. The example program C2.2 contains 3 highlighted instructions

```
sub  r2, r2, #1      // r2--
cmp  r2, #0          // if (r2 != 0)
bne  loop            //     goto loop;
```

   (1). If you delete the cmp instruction, the program would not work. Explain why?
   (2). However, replacing the 3 lines with

```
subs  r2, r2, #1
bne loop
```

   would work. Explain why?
2. In the assembly code of the program C2.4, a, b, c are adjacent. Modify the assembly code by letting R2 point at the first word a. Then

(1). Access a, b, c by using R2 as a base register with offsets.
(b). Access a, b, c by using R2 as a base register with post-indexed addressing.

3. In the example program C2.4, instead of defining a, b, c in the assembly code, define them as initialized globals in the t. c file:

$$int\ a = 1, \quad b = 2, \quad c = 0;$$

Declare them as global symbols in the ts.s file. Compile and run the program again.

4. GPIO programming: Assume that BASE is the base address of a GPIO, and the GPIO registers are at the offset addresses IODIR, IOSET, IOCLR, IOPIN. The following shows how to define them as constants in ARM assembly code and C.

```
.set BASE, 0x101E4000          // #define BASE 0x101E4000
.set IODIR, 0x000              // #define IODIR 0x000
.set IOSET, 0x004              // #define IOSET 0x004
.set IOCLR, 0x008              // #define IOCLR 0x008
.set IOPIN, 0x00C              // #define IOPIN 0x00C
```

Write a GPIO control program in both assembly and C to perform the following tasks.
(1). Program the GPIO pins as specified in Sect. 2.8.1.
(2). Determine the state of the GPIO pins.
(3). Modify the control program to make the LED blink while the input switch is closed.

5. The example program C2.6 assumes that every image size is h<=640 and width<=480 pixels. Modify the program to handle BMP images of larger sizes by
(1). Cropping: display at most 480x640 pixels.
(2). Shrinking: reduce the image size by a factor, e.g. 2 but keep the same 4:3 aspect ratio.
6. Modify the example program C2.6 to display a sequence of slightly different images to do animation.
7. Many image files, e.g. JPG images, are compressed, which must be uncompressed first. Modify the example program C2.6 to display (compressed) images of different format, e.g. JPG image files.
8. In the LCD display driver program C2.7, define tab_size = 8. Each tab key (\t) expands to 8 spaces. Modify the LCD driver to support tab keys. Test the modified LCD driver by including \t in printf() calls.
9. In the LCD driver of program C2.7, scroll-up one line is implemented by simply copying the entire frame buffer. Devise a more efficient way to implement the scroll operation. HINT: The display memory may be regarded as a circular buffer. To scroll up one line, simply increment the frame buffer pointer by line size.
10. Modify the example program C2.7 to display text with different fonts.
11. Modify the example program C2.8 to implement long jump by using the code of Sect. 2.7.3.3. Use UART0 to get user inputs but display outputs to the LCD. Verify that long jump works.
12. In the LCD driver, the generic printf() function is defined as

```
int printf(char *fmt, ...);              // note the 3 dots
```

In the implementation of printf(), it assumes that all the parameters are adjacent on stack, so that they can be accessed linearly. This seems to be inconsistent with the calling convention of ARM C, which passes the first 4 parameter in r0–r3 and extra parameters, if any, on stack. Compile the printf() function code to generate an assembly code file. Examine the assembly code to verify that the parameters are indeed adjacent on stack.

References

ARM Architectures: http://www.arm.products/processors/instruction-set-architectures, ARM Information Center, 2016
ARM Cortex-A8: "ARM Cortex-A8 Technical Reference Manual", ARM Information Center, 2010
ARM Cortex A9 MPcore: "Cortex A9 MPcore Technical Reference Manual", ARM Information Center, 2016
ARM926EJ-ST: "ARM926EJ-S Technical Reference Manual", ARM Information Center, 2008
ARM926EJ-ST: "Versatile Application Baseboard for ARM926EJ-S User guide", ARM Information Center, 2010
ARM PL011: "PrimeCell UART (PL011) Technical Reference Manual", ARM Information Center, 2016
ARM PrimeCell Color LCD Controller PL110: "ARM Versatile Application Baseboard for ARM926EF-S", ARM Information Center, 2016
ARM Programming: "ARM Assembly Language Programming", http://www.peter-cockerell.net/aalp/html/frames.html
ARM toolchain: http://gnutoolchains.com/arm-eabi, 2016
QEMU Emulators: "QEMU Emulator User Documentation", http://wiki.qemu.org/download/qemu-doc.htm, 2010

3 Interrupts and Exceptions Processing

In every computer system, the CPU is designed to continually execute instructions. An exception is an event recognized by the CPU, which diverts the CPU from its normal executions to do something else, called exception processing. An interrupt is an external event, which diverts the CPU from its normal executions to do interrupt processing. In a broader sense, interrupts are special kinds of exceptions. The only difference between exceptions and interrupts is that the former may originate from the CPU itself but the latter always originate from external sources. Interrupts are essential to every computer system. Without interrupts, a computer system would be unable to respond to external events, such as user inputs, timer events and requests for service from I/O devices, etc. Most embedded systems are designed to respond to external events and handle such events when they occur. For this reason, interrupts and interrupts processing are especially important to embedded systems. In this chapter, we shall discuss exceptions, interrupts and interrupts processing in ARM based systems.

In Chap. 2, we developed simple drivers for the LCD display and UARTs. The LCD is a memory mapped device, which does not use interrupts. UARTs support interrupts but the simple UART driver uses polling, not interrupts, for I/O. The main disadvantage of I/O by polling is that it does not use the CPU efficiently. While the CPU is doing I/O by polling, it is constantly busy and can't do anything else. In a computer system, I/O should be done by interrupts whenever possible. In this Chapter, we shall show how to apply the principle of interrupts processing to design and implement interrupt-driven device drivers.

3.1 ARM Exceptions

3.1.1 ARM Processor Modes

The ARM processor has seven different operating modes, which are determined by the 5 mode bits [4:0] in the current processor status register (CSPR) (ARM Architecture 2016; ARM Processor Architecture 2016). Table 3.1 shows the seven modes of the ARM processor.

Among the seven modes, only the User mode is non-privileged. All other modes are privileged. An unusual feature of the ARM architecture is that, while the CPU is in a privileged mode, it can change to any other mode, by simply altering the mode bits in the CPSR. When the CPU is in the un-privileged User mode, the only way to change to a privileged mode is through exceptions, interrupts, or the SWI instruction. Each privileged mode has its own banked registers, except System mode, which shares the same set of registers with the User mode, e.g. they have the same stack pointer (R13) and the same link register (R14).

3.1.2 ARM Exceptions

An exception is an event recognized by the processor, which diverts the processor from its normal executions to handle the exception. In a general sense, interrupts are also exceptions. In ARM, there are seven exception types (excluding the Reserved type) (ARM Processor Architecture 2016), which are shown in Table 3.2.

When an exception occurs, the ARM processor does the following.

K.C. Wang, *Embedded and Real-Time Operating Systems*,
DOI 10.1007/978-3-319-51517-5_3

Table 3.1 ARM Processor Modes

Mode	ModeBits	Mode Usage
USR	0x10	for running tasks in user mode
FIQ	0x11	for fast interrupt processing
IRQ	0x12	for ordinary interrupts processing
SVC	0x13	supervisor mode for OS kernel
ABT	0x17	for handling prefetch or data aborts
UND	0x1B	for handling undefined instructions
SYS	0x1F	privileged system mode

Table 3.2 ARM Exceptions

Name	Vector	Description	Mode	Priority
Reset	0x00	reset	SVC	1
UND	0x04	undefined instruction	UND	6
SWI	0x08	software interrupt	SVC	6
PAB	0x0C	prefetch abort	ABT	5
DAB	0x10	data exception abort	ABT	2
Reserved	0x14	reserved	-	-
IRQ	0x18	interrupt request	IRQ	4
FIQ	0x1C	fast interrupt request	FIRQ	3

(1). Copy CPSR into SPSR for the mode in which the exception is to be handled.
(2). Change CPSR mode bits to the appropriate mode, map in the banked registers and disable interrupts. IRQ is always disabled, FIQ is disabled only when an FIQ occurs and on reset.
(3). Set the LR_mode register to the return address.
(4). Set the Program Counter (PC) to the vector address of the exception. This forces a branch to the appropriate exception handler.

If multiple exceptions occur simultaneously, they are handled in accordance with the priority order shown in Table 3.2. The following is a list of the exception events and how they are handled by the ARM processor.

- A Reset event occurs when the processor is powering up. This is the highest priority event and shall be taken whenever it is signaled. Upon entry to the reset handler, the CPSR is in SVC mode and both IRQ and FIQ interrupts are masked out. The task of the reset handler is to initialize the system. This includes setting up stacks of various modes, configuring the memory and initializing device drivers, etc.
- A Data Abort (DAB) events occur when the memory controller or MMU indicates that an invalid memory address has been accessed. For example, if there is no physical memory for an address, or the processor does not have access permission to a region of memory, the data abort exception is raised. Data aborts have the second highest priority. This means that the processor will handle data abort exceptions first, before handling any interrupts.
- A FIQ interrupt occurs when an external peripheral sets the FIQ pin to nFIQ. A FIQ interrupt is the highest priority interrupt. Upon entry to the FIQ handler, both IRQ and FIQ interrupts are disabled. This means that while handling an FIQ interrupt, no other interrupts can occur unless they are explicitly enabled by the software. In ARM based systems, FIQ is usually used to handle interrupts from a single interrupt source of extreme urgency. Allowing multiple FIQ sources would defeat the purpose of FIQ.

- An IRQ interrupt occurs when an external peripheral device sets the IRQ pin. An IRQ interrupt is the second highest priority interrupt. The processor will handle an IRQ interrupt if there is no FIQ interrupt or data abort exception. Upon entry to the IRQ handler, IRQ interrupts are masked out. The CSPR's I-bit should remain set until the current interrupt source has been cleared.
- A Pre-fetch Abort (PFA) event occurs when an attempt to load an instruction results in a memory fault. The exception occurs if the instruction reaches the execution stage of the pipeline and none of the higher exceptions/interrupts have been raised. Upon entry to the PFA handler, IRQ is disabled but FIQ remains enabled, so that any FIQ interrupt will be taken immediately while processing a PFA exception.
- A SWI interrupt occurs when the SWI instruction has been fetched and decoded successfully and none of the other higher priority exceptions/interrupts have been raised. Upon entry to the SWI handler, the CPSR is set to SVC mode. SWI interrupts are usually used to implement system calls from User mode to an OS kernel in SVC mode.
- An Undefined Instruction event occurs when an instruction not in the ARM/Thumb instruction set has been fetched and decoded successfully, and none of the other exceptions/interrupts have been flagged. In an ARM based system with coprocessors, the ARM processor will poll the coprocessors to see if they can handle the instruction. If no coprocessor claims the instruction, then an undefined instruction exception is raised. SWI and Undefined Instruction have the same priority since they cannot occur at the same time. In other words, the instruction being executed can not be both a SWI and an undefined instruction. In practice, undefined instructions can be used to provide software breakpoints when debugging ARM programs.

3.1.3 Exceptions Vector Table

The ARM processor uses a vector table to handle exceptions and interrupts. Table 3.3 shows the ARM vector table contents.

The vector table defines the entry points of exceptions and interrupts handlers. The vector table is located at the physical address 0. Many ARM based systems begins execution from a flash memory or ROM, which may be remapped to 0xFFFF0000 during booting. If the initial vector table is not at 0 in SRAM, it must be copied to SRAM before remap it to 0x00000000. This is normally done during system initialization.

3.1.4 Exception Handlers

Each vector table entry contains an ARM instruction (B, BL or LDR), which causes the processor to load the PC with the entry address of an exception handler routine. The following code segment shows the typical vector table contents, in which each LDR instruction loads the PC with the entry address of an exception handler function. For the reserved vector entry (0x14), a branch to itself loop is sufficient since the exception can never occur.

Table 3.3 ARM Vector Table

Address	Exception	Mode
0x00	Reset	SVC
0x04	Undefined instruction	UND
0x08	Software interrupt (SWI)	SVC
0x0C	Prefetch abort (PFA)	ABT
0x10	Data abort	ABT
0x14	Reserved	N/A
0x18	IRQ	IRQ
0x1C	FIQ	FIQ

```
0x00 LDR PC, reset_handler_addr
0x04 LDR PC, undef_handler_addr
0x08 LDR PC, swi_handler_addr
0x0C LDR PC, prefetch_abort_handler_addr
0x10 LDR PC, data_abort_handler_addr
0x14 B .
0x18 LDR PC, irq_handler_addr
0x1C LDR PC, fiq_handler_addr
reset_handler_addr:          .word reset_handler
undef_handler_addr:          .word undef_handler
swi_handler_addr:            .word swi_handler
prefetch_abort_handler_addr: .word prefetch_abort_handler
data_abort_handler_addr:     .word data_abort_handler
irq_handler_addr:            .word irq_handler
fiq_handler_addr:            .word fiq_handler
```

When a vector uses LDR to load the handler entry address, the handler will be called indirectly. LDR must load a constant located within 4 kB from the vector table but it can branch to a full 32-bit address. A B (branch) instruction will go directly to the handler, but it can only branch to a 24-bit address. Since the FIQ vector is the last entry in the vector table, the FIQ handler code can be placed at the FIQ vector location directly, allowing the FIQ handler to be executed quickly.

3.1.5 Return from Exception Handlers

Assume that exception/interrupt handlers are entered by LDR PC instructions, as in

```
LDR PC, handler_entry_address
```

Upon entry to an exception/interrupt handler routine, the processor automatically stores the return address in the link register r14 of the current mode. Due to the instruction pipeline in the ARM processor, the return address stored in the link register includes an offset, which must be subtracted from the link register in order to return to the correct location priori to the exception or interrupt. Table 3.4 shows the program counter offsets of different exceptions.

For Data Abort, the return address is PC-8, which points to the original instruction that caused the exception. For interrupts and Prefetch Abort, the return address is PC-4 since the instruction to be executed before the exception is at current PC-4. For SWI and Undefined instruction, the return address is the current LR because the ARM processor has already executed the SWI or undefined instruction. To most beginners, the non-uniform treatment of the IRQ and SWI return addresses is often a source of confusion. The 'S' suffix at the end of the MOV instruction specifies that, if the destination registers involve loading the PC, the CPSR shall be restored from the saved SPSR also.

A typical method of return from an interrupt handler is to execute the following instruction at the end of the interrupt handler

Table 3.4 Program Counter Offsets

Exception	Offset	Return PC
Reset	n/a	n/a
Data Abort	-8	SUBS pc,lr,#8
FIQ	-4	SUBS pc,lr,#4
IRQ	-4	SUBS pc,lr,#4
Prefetch Abort	-4	SUBS pc,lr,#4
SWI	0	MOVS pc,lr
Undefined	0	MOVS pc,lr

```
SUBS pc, r14_irq, #4
```

which loads the PC with r14_irq – 4, assuming that r14_irq has not been altered in the interrupt handler. Alternatively, the link register can be adjusted at the beginning of the interrupt handler, as in.

```
SUB  lr, lr, #4
   <handler code>
MOVS pc, lr
```

A more widely used method is as follows.

```
SUB  lr, lr, #4              // subtract 4 form LR
STMFD sp_irq!,{r0-r12,lr}   // push r0-r12,LR
    <handler code>
LDMFD sp_irq!,{r0-r12,pc}^ // pop  r0-r12,PC,SPSR
```

The interrupt handler first subtracts 4 from the link register, saves it in the stack, then it executes the handler code. When the handler finishes, it returns to the interrupted point by the LDMFD instruction with the ^ symbol, which loads PC with the saved LR and restores SPSR, causing return to the previous mode priori to the interrupt. Instead of subtracting 4 from the link register manually, an interrupt handler may be written in C with the interrupt attribute, as in

```
void __attribute__((interrupt))handler()
{
   // actual handler code
}
```

In that case, the compiler generated code will adjust the link register automatically.

3.2 Interrupts and Interrupts Processing

3.2.1 Interrupt Types

The ARM processor accepts only two external interrupt requests, FIQ and IRQ. Both are level-sensitive active low signals to the processor. For an interrupt to be accepted by the CPU, the appropriate interrupt mask (I or F) bit in the CPSR must be cleared to 0. FIQ has higher priority than IRQs, so that FIQ will be handled first when multiple interrupts occur. Handling an FIQ causes IRQs and subsequent FIQ to be disabled, preventing them from being taken until after the FIQ handler exits or explicitly enables them. This is usually done by restoring the CPSR from the SPSR at the end of the FIQ handler.

The FIQ vector is the last entry in the vector table. The FIQ handler code can be placed directly at the vector location and run sequentially from that address. This avoids a branch instruction and its associated delay. If the system has a cache memory, the vector table and FIQ handler might all be locked down in one block within the cache. This is important because FIQ is designed to handle the interrupt as quickly as possible. Each privileged mode has its own banked registers r13, r14 and SPSR. The FIQ mode has 5 extra banked register (r8–r12), which can be used to contain information between calls in the FIQ handler, further increasing the execution speed of the FIQ handler.

3.2.2 Interrupt Controllers

An ARM based system usually supports only one FIQ interrupt source but it may support many IRQ requests from different sources. In order to support multiple IRQ interrupts, an interrupt controller is necessary, which sorts out different IRQ sources and presents only one IRQ request to the CPU. Most ARM boards include a Vectored Interrupt Controller (VIC), which is either the ARM PL190 or PL192 (ARM PL190, PL192 2016). A VIC provides the following functions.

. Prioritize the interrupt sources
. Support vectored interrupts

3.2.2.1 ARM PL190/192 Interrupt Controller

Figure 3.1 shows the block diagram of the PL190 VIC. It supports 16 vectored interrupts. The PL192 VIC is similar but it supports 32 vectored interrupts.

3.2.2.2 Vectored and Non-vectored IRQs

The VIC takes all the interrupt requests from different sources and arranges them into three categories, FIQ, vectored IRQ and non-vectored IRQ. FIQ has the highest priority. In the case of the PL190 VIC, it takes 16 interrupt requests and prioritizes them within sixteen vectored IRQ slots, denoted by IRQ0-IRQ15. Each vectored interrupt has a vector address, denoted by VectAddr0-VectAddr15.

3.2.2.3 Interrupt Priorities

Among the vectored interrupts, IRQ0 has the highest priority and IRQ15 has the lowest priority. Non-vectored IRQ's have the lowest priority. The VIC OR's the requests from both the vectored and non-vectored IRQ's to generate the IRQ signal to the ARM core, which is shown as nVICIRQ line in Fig. 3.1.

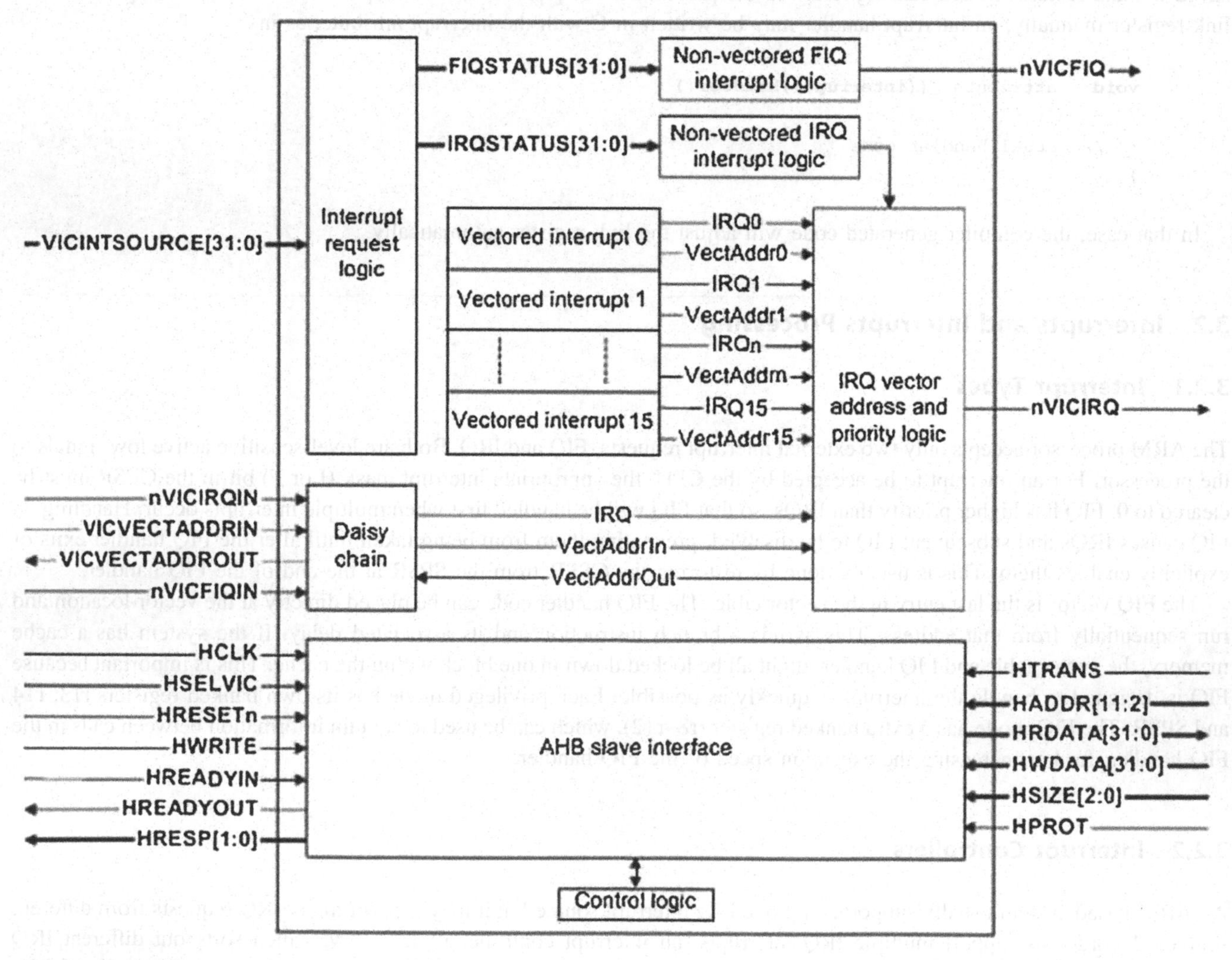

Fig. 3.1 ARM PL190 VIC

3.2.3 Primary and Secondary Interrupt Controllers

Some ARM boards may contain more than one VIC. For example, the ARM926EJ-S board has two VICs, a primary VIC (PIC) and a secondary VIC (SIC), which are shown in Fig. 3.2. Most inputs to the PIC are dedicated to high priority interrupts, such as timers, GPIO and UARTs. Low priority interrupt sources, such as USB, Ethernet, keyboard and mouse, are fed to the SIC. Some of these interrupts may be routed to the PIC at IRQs 21 to 26. Lower priority interrupts, such as touch-screen, keyboard and mouse are collectively routed to IRQ 31 of the PIC.

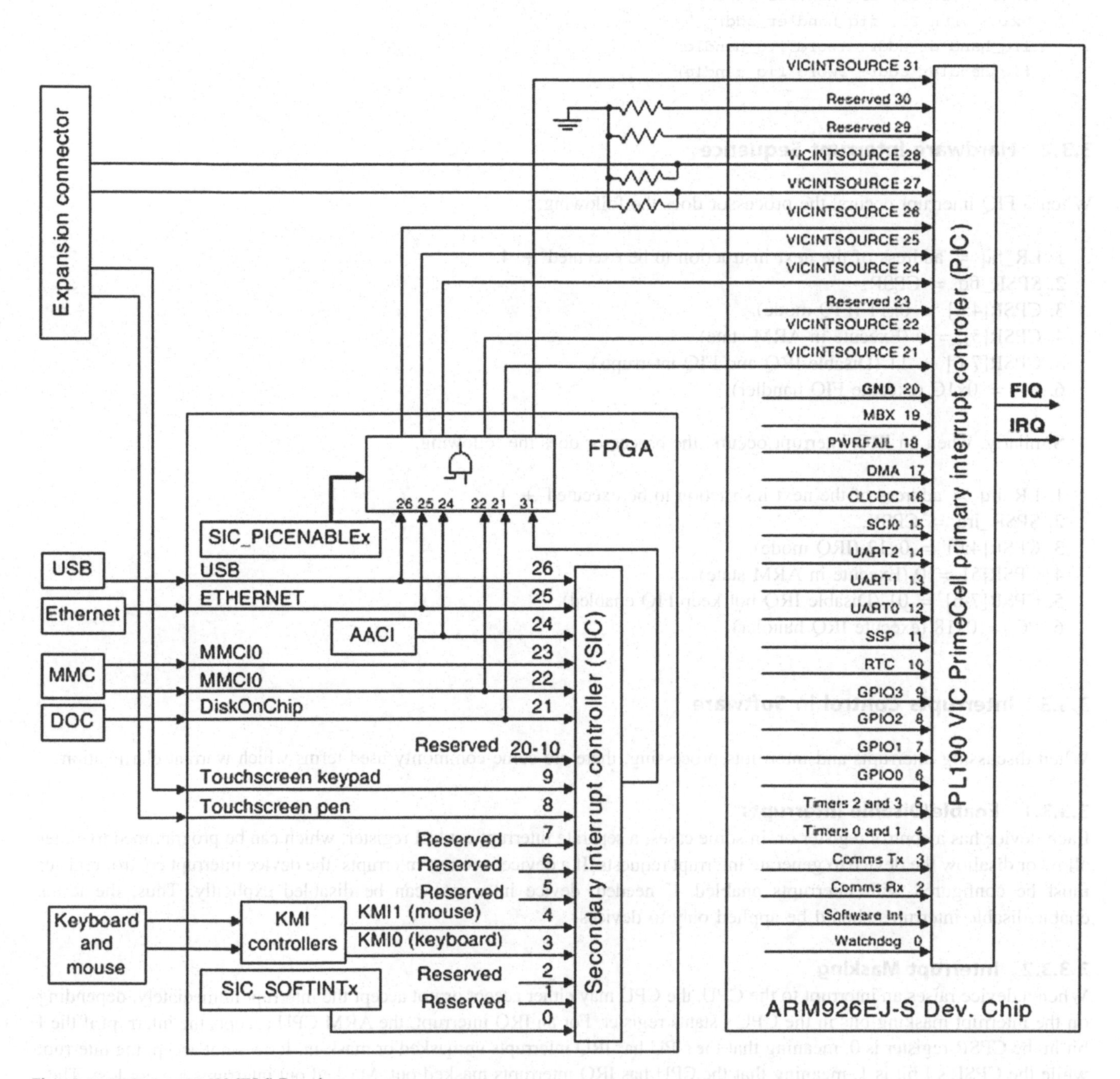

Fig. 3.2 VICs in ARM926EJ-S Board

3.3 Interrupt Processing

3.3.1 Vector Table Contents

Interrupt vectors are in the exception vector table. Each interrupt vector location contains an instruction which loads the PC with the entry address of an interrupt handler. For FIQ and IRQ interrupts, the vector contents are

```
0x18:   LDR PC, irq_handler_addr
0x1C:   LDR PC, fiq_handler_addr
irq_handler_addr: .word irq_handler
fiq_handler_addr: .word fiq_handler
```

3.3.2 Hardware Interrupt Sequence

When a FIQ interrupt occurs, the processor does the following.

1. LR_fiq = address of the next instruction to be executed + 4.
2. SPSR_fiq = CPSR.
3. CPSR[4:0] = 0x11 (FIQ mode).
4. CPSR[5] = 0 (Execute in ARM state).
5. CPSR[7-6] = 11 (Disable IRQ and FIQ interrupts).
6. PC = 0x1C (execute FIQ handler).

Similarly, when an IRQ interrupt occurs, the processor does the following.

1. LR_irq = address of the next instruction to be executed + 4.
2. SPSR_irq = CPSR.
3. CPSR[4:0] = 0x12 (IRQ mode).
4. CPSR[5] = 0 (Execute in ARM state).
5. CPSR[7-6] = 01 (Disable IRQ but keep FIQ enabled).
6. PC = 0x18 (execute IRQ handler).

3.3.3 Interrupts Control in Software

When discussing interrupts and interrupts processing, there are some commonly used terms which warrant clarification.

3.3.3.1 Enable/Disable Interrupts

Each device has a control register or, in some cases, a separate interrupt control register, which can be programmed to either allow or disallow the device to generate interrupt requests. If a device is to use interrupts, the device interrupt control register must be configured with interrupts enabled. If needed, device interrupts can be disabled explicitly. Thus, the terms enable/disable interrupts should be applied only to devices.

3.3.3.2 Interrupt Masking

When a device raises an interrupt to the CPU, the CPU may either accept or not accept the interrupt immediately, depending on the interrupt masking bits in the CPU's status register. For an IRQ interrupt, the ARM CPU accepts the interrupt if the I bit in the CPSR register is 0, meaning that the CPU has IRQ interrupts unmasked or mask-in. It does not accept the interrupt while the CPSR's I bit is 1, meaning that the CPU has IRQ interrupts masked out. Masked out interrupts are not lost. They are kept pending until the CPSR's I bit is changed to 0, at which time the CPU will accept the interrupt. Thus, when applied to the CPU, enable/disable interrupts really mean mask-in/mask-out interrupts. In most literatures these terms are used interchangeably but the reader should be aware of their differences.

3.3.3.3 Clear Device Interrupt Request

When the CPU accepts an IRQ interrupt, it starts to execute the interrupt handler for that device. At the end of the interrupt handler it must clear the interrupt request, which causes the device to drop its interrupt request, allowing it to generate the next interrupt. This is usually done by accessing some of the device interface registers. For example, reading the data register of an input device clears the device interrupt request. For some output devices, it may be necessary to disable the device interrupt explicitly when there is no more data to output.

3.3.3.4 Send EOI to Vectored Interrupt Controller

In a system with multiple interrupt sources, a Vectored Interrupt Controller (VIC) is usually used to prioritize the device interrupts, each with a dedicated vector address. At the end of handling the current interrupt, the interrupt handler must inform the VIC that it has finished processing the current interrupt (of the highest priority), allowing the VIC to re-prioritize pending interrupts requests. This is referred to as sending an End-of-Interrupt (EOI) to the interrupt controller. For the ARM PL190, this is done by writing an arbitrary value to the VIC's vector address register at base +0x30.

The ARM processor has a simple way to enable/disable (unmask/mask) interrupts while in privileged mode. The following code segments show how to enable/disable IRQ interrupts of the ARM processor.

```
lock:
        MRS  r1, CPSR           ; read CPSR into r1
        ORR  r1, r1, #0x80      ; set bit-7(I bit) to 1
        MSR  CPSR, r1           ; write r1 back to CPSR

unlock:
        MRS  r1, CPSR           ; read CPSR into r1
        BIC  r1, r1, #0x80      ; clear bit-7 (I bit) in r1
        MSR  CPSR r1            ; write r1 back to CPSR
```

To enable IRQ interrupts, first copy CPSR into a working register, clear the I_bit (7) in the working register to 0. Then copy the updated register back to CPSR, which enables IRQ interrupts. Similarly, setting CPSR's I_bit disables IRQ interrupts. Similar code segments can be used to enable/disable FIQ interrupts (by clear/set bit-6 in CPSR).

3.3.4 Interrupt Handlers

Interrupt handlers are also known as Interrupt Service Routines (ISRs). Interrupt handlers can be classified into three different types.

3.3.4.1 Interrupt Handler Types

. Non-nested interrupt handler: handles one interrupt at a time. Interrupts are not enabled until execution of the current ISR has finished.
. Nested interrupt handler: While inside an ISR, enable IRQ to allow IRQ interrupts of higher priorities to occur. This implies that while executing the current ISR, it may be interrupted to execute another ISR of higher priority.
. Re-entrant interrupt handler: Enable IRQ as soon as possible, allowing the same ISR to be executed again.

3.3.5 Non-nested Interrupt Handler

When the ARM CPU accepts an IRQ interrupt, it enters the IRQ mode with IRQ interrupts masked out, preventing it from accepting other IRQ interrupts. The CPU sets PC to point to the IRQ entry in the vector table and executes that instruction. The instruction loads PC with the entry address of the interrupt handler, causing execution to enter the interrupt handler. The interrupt handler first saves the execution context before the interrupt. Then it determines the interrupt source and calls the appropriate ISR. After servicing the interrupt, it restores the saved context and sets the PC to point back to the next instruction prior to the interruption. Then it returns to the original place of the interruption. The simplest interrupt handler serves only one interrupt at a time. While executing the interrupt handler, IRQ interrupts are masked out until control is returned back to the interrupted point. The algorithm and control flow of a non-nested interrupt handler is as follows.

```
non_nested_interrupt_handler()
{
   // IRQ interrupt masked out in CPSR
   1. Save Context
   2. Determine interrupt source
   3. Call ISR to handle the interrupt
   4. Restore Context
   5. Restore CPSR by SPSR and return to interrupt point
}
```

The following code segment shows the organization of a simple IRQ interrupt handler. It assumes that an IRQ mode stack has been set up properly, which is typically done in the reset handler during system initialization.

```
SUB   lr, lr, #4
STMFD sp_irq!, {r0-r12, lr}
 {specific interrupt service routine code}
LDMFD sp_irq!, {r0-r12, pc}^
```

The first instruction adjusts the link register (r14) for return to the interrupted point. The STMFD instruction saves the context at the point of interruption by pushing CPU registers that must be preserved onto the stack. The time taken to execute a STMFD or LDMFD instruction is proportionally to the number of registers being transferred. To reduce interrupt processing latency, a minimum number of registers should be saved. When writing ISR in a high-level programming language, such as C, it is important to know the calling convention of the compiler generated code as this will affect the decision on which registers should be saved on the stack. For instance, the ARM compiler generated code preserves r4–r11 during function calls, so there is no need to save these registers unless they are going to be used by the interrupt handler. Once the registers have been saved, it is now safe to call C functions to process the interrupt. At the end of the interrupt handler the LDMFD instruction restores the saved context and return from the interrupt handler. The '^' symbol at the end of the LDMFD instruction means that the CPSR will be restored from the saved SPSR. As noted in the Chap. 2 on ARM instructions, the '^' restores the saved SPSR only if the PC is loaded at the same time. Otherwise, it only restores the bank registers of the previous mode, excluding the saved SPSR. This special feature can be used to access User mode registers while in a privileged mode.

The organization of the simple interrupt handler is suitable for handling both FIQ and IRQ interrupts one at a time without interrupt nesting. After saving the execution context, the interrupt handler must determine the interrupt source. In a simple ARM system that does not use vectored interrupts, the interrupt source is in the interrupt status register, denoted by IRQstatus, located at a known (memory-mapped) address. To determine the interrupt source, simply read the IRQstatus register and scan the contents for any bit or bits that are set. Each non-zero bit stands for an active interrupt request. The interrupt handler may scan the bits in a specific order, which determines the interrupt processing priority in software.

With the above background information on interrupts and interrupts processing, we are ready to write some real programs that use interrupts. In the following, we shall show how to write interrupt handlers for I/O devices. Device drivers using interrupts are called **interrupt-driven device drivers**. Specifically, we shall show how to implement interrupt-driven drivers for timer, keyboard, UART and Secure Digital Cards (SDC).

3.4 Timer Driver

3.4.1 ARM Versatile 926EJS Timers

The ARM Versatile 926EJS board contains two ARM SB804 dual timer modules [ARM Timers 2004]. Each timer module contains two timers, which are driven by the same clock. The base addresses of the timers are at

```
Timer0: 0x101E2000, Timer1: 0x101E2020
Timer2: 0x101E3000, Timer3: 0x101E3020
```

Timer0 and Timer1 interrupt at IRQ4. Timer2 and Timer3 interrupt at IRQ5, both on the primary Vectored Controller (VIC). To begin with, we shall not use vectored interrupts. Vectored interrupts will be discussed later. From a programming point of view, the most important timer registers are the control and counter registers. The following lists the meanings of the timer control register bits.

```
Bit   --------- Function -------        ----- Setting=0x66 --------
 7    timer disable/enable              0 set to 1 in timer_start()
 6    free running/periodic mode        1 for periodic mode
 5    interrupt disable/enable          1 for interrupt enabled
 4    NOT used                          0 (use 0 for default)
 3:2  divider:00=1,01=8,10=256          01 for divide by 8
 1    16/32-bit counter value           1 for 32-bit counter
 0    wraparound/oneshot mode           0 for wraparound mode
-------------------------------------------------------------
```

3.4.2 Timer Driver Program

The sample program C3.1 demonstrates a timer driver for the ARM Versatile board. The program consists of the following components.

(1). t.ld file: The linker script file specifies reset_handler as the entry address. It defines two 4 KB areas as SVC and IRQ stacks.

```
ENTRY(reset_handler)
SECTIONS
{
  . = 0x10000;            /* loading address */
  .text : { ts.o *(.text) }
  .data : { *(.data) }
  .bss : { *(.bss) }
  . = ALIGN(8);
  . = . + 0x1000;         /* 4kB of SVC stack space */
  svc_stack_top = .;
  . = . + 0x1000;         /* 4kB of IRQ stack space */
  irq_stack_top = .;
}
```

(2). ts.s file: The assembly code file, ts.s, defines the entry point and the reset code.

```
/****************** ts.s file of C3.1 *******************/
.text
.code 32
.global reset_handler, vectors_start, vectors_end
```

```
reset_handler:
  LDR sp, =svc_stack_top    // set SVC mode stack
  BL copy_vectors           // copy vector table to address 0
  MSR cpsr, #0x92           // to IRQ mode
  LDR sp, =irq_stack_top    // set IRQ mode stack
  MSR cpsr, #0x13           // go back to SVC mode with IRQ on
  BL main                   // call main() in C
  B .                       // loop if main ever return

irq_handler:
  sub lr, lr, #4
  stmfd sp!, {r0-r12, lr}     // stack r0-r12 and lr
  bl IRQ_handler             // call IRQ_hanler() in C
  ldmfd sp!, {r0-r12, pc}^ // return

vectors_start:
  LDR PC, reset_handler_addr
  LDR PC, undef_handler_addr
  LDR PC, swi_handler_addr
  LDR PC, prefetch_abort_handler_addr
  LDR PC, data_abort_handler_addr
  B .
  LDR PC, irq_handler_addr
  LDR PC, fiq_handler_addr
  reset_handler_addr:            .word reset_handler
  undef_handler_addr:            .word undef_handler
  swi_handler_addr:              .word swi_handler
  prefetch_abort_handler_addr: .word prefetch_abort_handler
  data_abort_handler_addr:       .word data_abort_handler
  irq_handler_addr:              .word irq_handler
  fiq_handler_addr:              .word fiq_handler
vectors_end:
```

When the program starts, QEMU loads the executable image, t.bin, to 0x10000. Upon entry to ts.s, execution begins from the label reset_handler. First, it sets the SVC mode stack pointer and calls copy_vector() in C to copy the vectors to address 0. It switches to IRQ mode to set up the IRQ stack pointer. Then it switches back to SVC mode with IRQ interrupts enabled and calls main() in C. The main program normally runs in SVC mode. It enters IRQ mode only to handle interrupts. Since we are not using FIQ, nor trying to deal with any exceptions at this moment, all other exception handlers (in exceptions.c file) are while(1) loops.

(3). t.c file: this file contains main(), copy_vector() and IRQ_handler() in C

```
#include "defines.h"     // LCD, TIMER and UART addresses
#include "string.c"      // strcmp, strlen, etc
#include "timer.c"       // timer handler file
#include "vid.c"         // LCD driver file
#include "exceptions.c" // other exception handlers

void copy_vectors(void){// copy vector table in ts.s to 0x0
    extern u32 vectors_start, vectors_end;
    u32 *vectors_src = &vectors_start;
    u32 *vectors_dst = (u32 *)0;
    while (vectors_src < &vectors_end)
      *vectors_dst++ = *vectors_src++;
}
```

```
void timer_handler();
TIMER *tp[4];                // 4 TIMER structure pointers

void IRQ_handler()           // IRQ interrupt handler in C
{
    // read VIC status registers to determine interrupt source
    int vicstatus = VIC_STATUS;
    // VIC status BITs: timer0,1=4, uart0=13, uart1=14
    if (vicstatus & (1<<4)){          // bit4=1:timer0,1
      if (*(tp[0]->base+TVALUE)==0)  // timer 0
         timer_handler(0);
      if (*(tp[1]->base+TVALUE)==0)  // timer 1
         timer_handler(1);
    }
    if (vicstatus & (1<<5)){          // bit5=1:timer2,3
      if (*(tp[2]->base+TVALUE)==0)  // timer 2
         timer_handler(2);
      if (*(tp[3]->base+TVALUE)==0)  // timer 3
         timer_handler(3);
    }
}

int main()
{
   int i;
   color = RED;                 // int color in vid.c file
   fbuf_init();                 // initialize LCD driver
   printf("main starts\n");
   /* enable VIC for timer interrupts */
   VIC_INTENABLE  = 0;
   VIC_INTENABLE |= (1<<4);  // timer0,1 at VIC.bit4
   VIC_INTENABLE |= (1<<5);  // timer2,3 at VIC.bit5
   timer_init();
   for (i=0; i<4; i++){         // start all 4 timers
      tp[i] = &timer[i];
      timer_start(i);
   }
   printf("Enter while(1) loop, handle timer interrupts\n");
}
```

(4). **timer.c file:** timer.c implements the timer handlers

```
// timer register u32 offsets from base address
#define TLOAD   0x0
#define TVALUE  0x1
#define TCNTL   0x2
#define TINTCLR 0x3
#define TRIS    0x4
#define TMIS    0x5
#define TBGLOAD 0x6
```

```
typedef volatile struct timer{
  u32 *base;             // timer's base address; as u32 pointer
  int tick, hh, mm, ss; // per timer data area
  char clock[16];
}TIMER;

volatile TIMER timer[4]; //4 timers; 2 per unit; at 0x00 and 0x20

void timer_init()
{
  int i; TIMER *tp;
  printf("timer_init()\n");
  for (i=0; i<4; i++){
    tp = &timer[i];
    if (i==0) tp->base = (u32 *)0x101E2000;
    if (i==1) tp->base = (u32 *)0x101E2020;
    if (i==2) tp->base = (u32 *)0x101E3000;
    if (i==3) tp->base = (u32 *)0x101E3020;
    *(tp->base+TLOAD) = 0x0;   // reset
    *(tp->base+TVALUE)= 0xFFFFFFFF;
    *(tp->base+TRIS)  = 0x0;
    *(tp->base+TMIS)  = 0x0;
    *(tp->base+TLOAD) = 0x100;
    // CntlReg=011-0010=|En|Pe|IntE|-|scal=01|32bit|0=wrap|=0x66
    *(tp->base+TCNTL) = 0x66;
    *(tp->base+TBGLOAD) = 0x1C00; // timer counter value
    tp->tick = tp->hh = tp->mm = tp->ss = 0; // initialize wall clock
    strcpy((char *)tp->clock, "00:00:00");
  }
}

void timer_handler(int n) {
    int i;
    TIMER *t = &timer[n];
    t->tick++;          // Assume 120 ticks per second
    if (t->tick==120){
      t->tick = 0; t->ss++;
      if (t->ss == 60){
         t->ss = 0; t->mm++;
         if (t->mm == 60){
            t->mm = 0; t->hh++; // no 24 hour roll around
         }
      }
      t->clock[7]='0'+(t->ss%10); t->clock[6]='0'+(t->ss/10);
      t->clock[4]='0'+(t->mm%10); t->clock[3]='0'+(t->mm/10);
      t->clock[1]='0'+(t->hh%10); t->clock[0]='0'+(t->hh/10);
    }
    color = n;                  // display in different color
    for (i=0; i<8; i++){
        kpchar(t->clock[i], n, 70+i); // to line n of LCD
    }
    timer_clearInterrupt(n); // clear timer interrupt
}

void timer_start(int n) // timer_start(0), 1, etc.
{
```

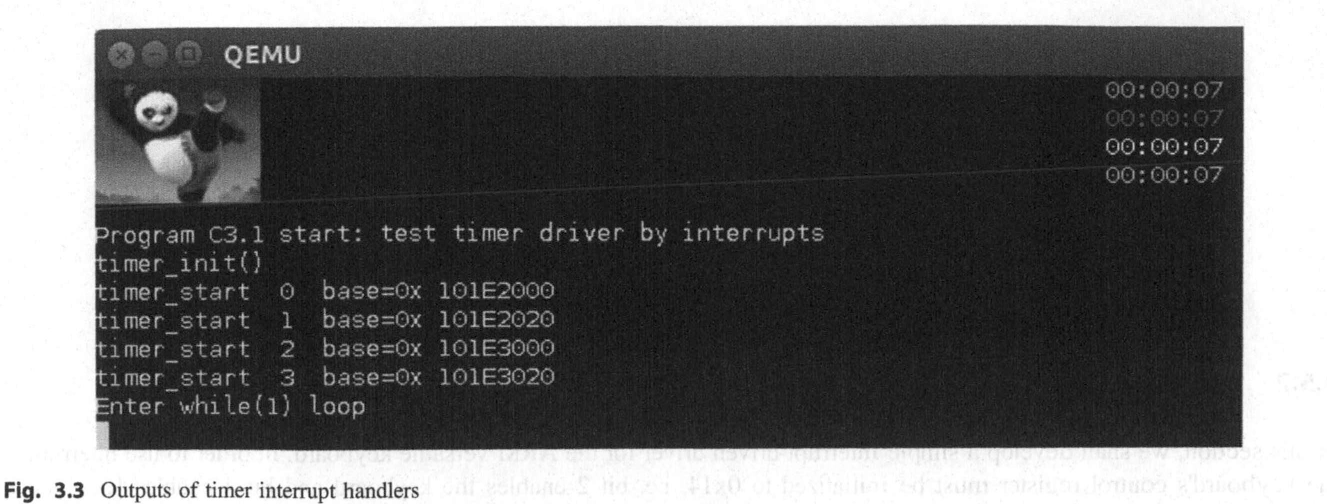

Fig. 3.3 Outputs of timer interrupt handlers

```
  TIMER *tp = &timer[n];
  kprintf("timer_start %d base=%x\n", n, tp->base);
  *(tp->base+TCNTL) |= 0x80;     // set enable bit 7
}
int timer_clearInterrupt(int n) // timer_start(0), 1, etc.
{
  TIMER *tp = &timer[n];
  *(tp->base+TINTCLR) = 0xFFFFFFFF;
}
void timer_stop(int n)          // stop a timer
{
  TIMER *tp = &timer[n];
  *(tp->base+TCNTL) &= 0x7F; // clear enable bit 7
}
```

Figure 3.3 shows the LCD screen of running the program C3.1. Each timer displays a wall clock at the right-top corner of the screen. In the LCD driver, the scroll-up limit is set to a line below the logo and the wall clocks, so that they will not be affected during scroll-up. The wall clocks are updated on each second. As exercises, the reader may change the starting values of the wall clocks to display local time in different time zones, or change the timer counter values to generate timer interrupts with different frequencies.

3.5 Keyboard Driver

3.5.1 ARM PL050 Mouse-Keyboard Interface

The ARM Versatile board includes an ARM PL050 Mouse-Keyboard Interface (MKI) which provides support for a mouse and a PS/2 compatible keyboard [ARM PL050 MKI 1999]. The keyboard's base address is at 0x1000600. It has several 32-bit registers, which are at offsets from the base address.

```
Offset   Register        Bits Assignment
------   ----------   ----------------------------
0x00     Control      bit 5=0(AT) 4=IntEn  2=Enable
0x04     Status       bit 4=RXF    3=RXBUSY
0x08     Data         input scan code
0x0C     ClkDiv       (a value between 0-15)
0x10     IntStatus    bit 0=RX interrupt
---------------------------------------------------
```

3.5.2 Keyboard Driver

In this section, we shall develop a simple interrupt-driven driver for the ARM versatile keyboard. In order to use interrupts, the keyboard's control register must be initialized to 0x14, i.e. bit 2 enables the keyboard and bit 4 enables Rx (input) interrupts. The keyboard interrupts at IRQ3 on the Secondary VIC, which is routed to IRQ31 on the Primary VIC. Instead of ACSII code, the keyboard generates scan codes. A complete listing of scan codes is included in the keyboard driver. Translation of scan code to ASCII is done by mapping tables in software. This allows the same keyboard to be used for different languages. For each key typed, the keyboard generates two interrupts; one when the key is pressed and another one when the key is released. The scan code of key release is 0x80 + the scan code of key press, i.e. bit7 is 0 for key press and 1 for key release. When the keyboard interrupts, the scan code is in the data register (0x08). The interrupt handler must read the data register to get the scan code, which also clears the keyboard interrupt. Some special keys generate escape key sequences, e.g. the UP arrow key generates 0xE048, where 0xE0 is the escape key itself. The following shows the mapping tables for translating scan codes into ASCII. The keyboard has 105 keys. Scan codes above 0x39 (57) are special keys, which cannot be mapped directly, so they are not shown in the key maps. Such special keys are recognized by the driver and handled accordingly. Figure 3.4 shows the key mapping tables.

3.5.3 Interrupt-Driven Driver Design

Every interrupt-driven device driver consists of three parts; a lower-half part, which is the interrupt handler, an upper-half part, which is called by the application program, and a common data area containing a buffer for data and control variables for synchronization, which are shared by the lower and upper parts. Figure 3.5 shows the organization of the keyboard driver. The top part of the figure shows kbd_init(), which initialize the KBD driver when the system starts. The middle part shows the control and data flow path from the KBD device to a program. The bottom part shows the lower-half, input buffer, and the upper-half organization of the KBD driver.

When the main program starts, it must initialize the keyboard driver control variables. When a key is pressed, the KBD generates an interrupt, causing the interrupt handler to be executed. The interrupt handler fetches the scan code from KBD data port. For normal key presses, it translates the scan code into ASCII, enters the ASCII char into an input buffer, buf[N],

```
#define NSCAN 58
/* Scan codes to ASCII for unshifted keys */
char unshift[NSCAN] = { // NSCAN=58
  0, 033,'1','2','3','4','5','6','7','8','9','0','-','=','\b','\t',
 'q','w','e','r','t','y','u','i','o','p','[',']', '\r', 0,'a','s',
 'd','f','g','h','j','k','l',';', 0,  0,  0,  0, 'z', 'x','c','v',
 'b','n','m',',','.','/', 0, '*', 0, ' ' };
/* Scan codes to ASCII for shifted keys */
char shift[NSCAN] = {
  0, 033,'!','@','#','$','%','^','&','*','(',')','_','+','\b','\t',
 'Q','W','E','R','T','Y','U','I','O','P','{','}', '\r', 0,'A','S',
 'D','F','G','H','J','K','L',':', 0, '~', 0,  '|','Z','X','C','V',
 'B','N','M','<','>','?', 0, '*', 0, ' ' };
```

Fig. 3.4 Key mapping tables

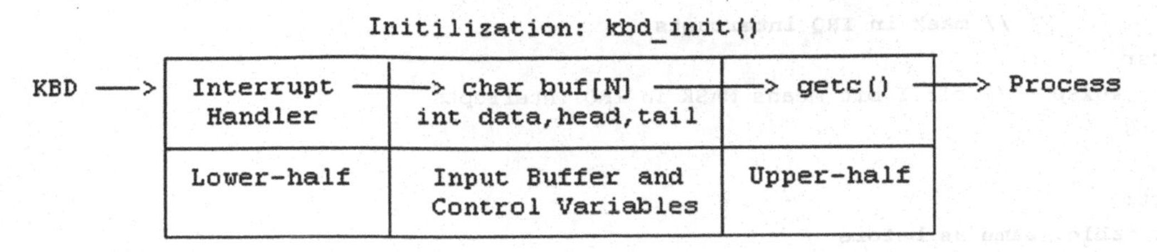

Fig. 3.5 KBD driver organization

and notifies the upper-half of the input char. When the program side needs an input char, it calls getc() of the upper-half driver, trying to get a char from buf[N]. The program waits if there is no char in buf[N]. The control variable, data, is used to synchronize the interrupt handler and the main program. The choice of the control variable depends on the synchronization tool used. The following shows the C code of a simple KBD driver. The driver handles only lower case keys. Extending the driver to handle upper case and special keys is left as an exercise in the Problem section.

3.5.4 Keyboard Driver Program

The sample program C3.2 demonstrates a simple interrupt-driven keyboard driver. It consists of the following components.

(1). **t.ld file:** The linker script file is the same as in C3.1.
(2). **ts.s file:** The ts.s file is almost the same as in C3.1, except that it adds the lock and unlock functions for enable/disable IRQ interrupts.

```
/********** ts.s file ************/
.text
.code 32
.global reset, vectors_start, vectors_end
.global lock, unlock
reset_handler:    // entry point of program
// set SVC stack
  LDR sp, =svc_stack_top
// copy vector table to address 0
  BL copy_vectors
// go in IRQ mode to set IRQ stack
  MSR cpsr, #0x92
  LDR sp, =irq_stack_top
// go back in SVC mode with IRQ interrupts enabled
  MSR cpsr, #0x13
// call main() in SVC mode
  BL main
  B .
irq_handler:
  sub lr, lr, #4
  stmfd sp!, {r0-r12, lr}  // stack ALL registers
  bl  IRQ_handler           // call IRQ_hanler() in C
  ldmfd sp!, {r0-r3, r12, pc}^ // return
lock:                   // mask out IRQ interrupts
  MRS r0, cpsr
  ORR r0, r0, #0x80    // set I bit means MASK out IRQ interrupts
  MSR cpsr, r0
  mov pc, lr
```

```
unlock:                    // mask in IRQ interrupts
  MRS r0, cpsr
  BIC r0, r0, #0x80   // clr I bit means MASK in IRQ interrupts
  MSR cpsr, r0
mov pc, lr
vectors_start:
   // vector table: same as before
vectors_end:
```

(3). **t.c file:** This file implements the main(), copy_vector() and IRQ_handler() functions. For simplicity, we shall only use timer0 to display a single wall clock.

```
/*************** t.c file *************/
#include "defines.h"
#include "string.c"
void timer_handler();
#include "kbd.c"
#include "timer.c"
#include "vid.c"
#include "exceptions.c"
void copy_vectors(){ // same as before }
void IRQ_handler()        // IRQ interrupt handler in C
{
  // read VIC status registers to find out which interrupt
  int vicstatus = VIC_STATUS;
  // VIC status BITs: timer0,1=4, uart0=13, uart1=14
  if (vicstatus & (1<<4)){     // bit4=1:timer0,1
     timer_handler(0);         // timer0 only
  }
  if (vicstatus & (1<<31)){    // PIC.bit31= SIC interrupts
     if (sicstatus & (1<<3)){ // SIC.bit3 = KBD interrupt
        kbd_handler();
     }
  }
}

int main()
{
  int i;
  char line[128];
  color = RED;    // int color in vid.c file
  fbuf_init();    // initialize LCD display
  /* enable VIC interrupts: timer0 at IRQ3, SIC at IRQ31 */
  VIC_INTENABLE  = 0;
  VIC_INTENABLE |= (1<<4);     // timer0,1 at PIC.bit4
  VIC_INTENABLE |= (1<<5);     // timer2,3 at PIC.bit5
  VIC_INTENABLE |= (1<<31);    // SIC to PIC.bit31
  /* enable KBD IRQ on SIC */
  SIC_INTENABLE = 0;
  SIC_INTENABLE |= (1<<3);     // KBD int=SIC.bit3
  timer_init();   // initialize timer
  timer_start(0); // start timer0
```

```
  kbd_init();       // initialize keyboard driver
  printf("C3.2 start: test KBD and TIMER drivers\n");
  while(1){
     color = CYAN;
     printf("Enter a line from KBD\n");
     kgets(line);
     printf("line = %s\n", line);
  }
}
```

In order to use interrupts, the devices must be configured to generate interrupts. This is done in the device initialization code, in which each device is initialized with interrupts enabled. In addition, both the Primary and Secondary Interrupt Controllers (PIC and SIC) must be configured to enable the device interrupts. The keyboard interrupts at IRQ3 on the SIC, which is routed to IRQ31 on the PIC. To enable keyboard interrupts, both bit 3 of the SIC_INTENABLE and bit 31 of the VIC_INTENABLE registers must be set to 1.

These are done in main(). After initializing the devices for interrupts, the main program executes a while(1) loop, in which it prompts for an input line from the KBD and prints the line to the LCD display.

(4). **kbd.c file:** The kbd.c file implements a simple keyboard driver.

```
/******** kbd.c file *******************/
#include "keymap"
/******** KBD register byte offsets; for char *base *****/
#define KCNTL 0x00 // 7-6- 5(0=AT)4=RxIntEn 3=TxIntEn
#define KSTAT 0x04 // 7-6=TxE 5=TxBusy 4=RXFull 3=RxBusy
#define KDATA 0x08 // data register;
#define KCLK  0x0C // clock divisor register;   (not used)
#define KISTA 0x10 // interrupt status register;(not used)

typedef volatile struct kbd{ // base = 0x10006000
  char *base;       // base address of KBD, as char *
  char buf[128];    // input buffer
  int head, tail, data, room; // control variables
}KBD;
volatile KBD kbd;  // KBD data structure

int kbd_init()
{
  KBD *kp = &kbd;
  kp->base = (char *)0x10006000;
  *(kp->base+KCNTL) = 0x14; // 00010100=INTenable, Enable
  *(kp->base+KCLK)  = 8;    // PL051 manual says a value 0 to 15
  kp->data = 0; kp->room = 128; // counters
  kp->head = kp->tail = 0;  // index to buffer
}
void kbd_handler()               // KBD interrupt handler in C
{
  u8 scode, c;
  int i;
  KBD *kp = &kbd;
  color = RED;                   // int color in vid.c file
  scode = *(kp->base+KDATA); // read scan code in data register
  if (scode & 0x80)              // ignore key releases
```

```
    return;
  c = unsh[scode];               // map scan code to ASCII
  if (c != '\r')
    printf("kbd interrupt: c=%x %c\n", c, c);
  kp->buf[kp->head++] = c;  // enter key into CIRCULAR buf[ ]
  kp->head %= 128;
  kp->data++; kp->room--;   // update counters
}

int kgetc() // main program calls kgetc() to return a char
{
  char c;
  KBD *kp = &kbd;
  unlock();                 // enable IRQ interrupts
  while(kp->data <= 0);     // wait for data; READONLY
  lock();                   // disable IRQ interrupts
    c = kp->buf[kp->tail++];// get a c and update tail index
    kp->tail %= 128;
    kp->data--; kp->room++; // update with interrupts OFF
  unlock();                 // enable IRQ interrupts
  return c;
}

int kgets(char s[ ])        // get a string from KBD
{
  char c;
  while((c=kgetc()) != '\r'){
    *s++ = c;
  }
  *s = 0;
  return strlen(s);
}
```

(5). Interrupt Processing Sequence

(5).1. When an interrupt occurs, the CPU follows the IRQ vector at 0x18 to enter

```
irq_handler:
    sub lr, lr, #4              // adjust lr
    stmfd sp!, {r0-r12, lr}     // save regs in IRQ stack
    bl IRQ_handler              // call IRQ_hanler() in C
    ldmfd sp!, {r0-r12, pc}^    // return
```

irq_handler first subtract 4 form the link register lr_irq. The adjusted lr is the correct return address to the point of interruption. It pushes registers r0–r12 and lr into the IRQ stack. Then it call IRQ_handler() in C. Upon return from IRQ_handler(), it pops the stack, which loads PC with the saved lr and also restores the SPSR, causing control return to the original point of interruption. To speed up interrupt processing, irq_handler may only save registers that must be preserved. Since our interrupt handler is written in C, not in assembly, it suffices to save only r0–r3, r12 (stack frame pointer) and lr (link register).

(5).2. IRQ_handler(): IRQ_handler() first reads the status registers of both PIC and SIC. The interrupt handler must scan the bits of the status register to determine the interrupt source. Each bit = 1 in the status register represents an active interrupt. The scanning order should follow the interrupt priority order, i.e. from bit 0 to bit 31.

```
    vicstatus = VIC_STATUS;
  sicstatus = SIC_STATUS;
  // VIC status BITs: timer0=4, uart0=13, uart1=14, SIC=31: KBD at 3
  if (vicstatus & (1<<4))      // bit 4 => timer interrupt
     timer_handler(0);
  if (vicstatus & (1<<31)){    // SIC interrupts=bit_31=>KBD at bit 3
     if (sicstatus & (1<<3)){ // SIC bit 3 => KBD interrupt
        kbd_handler();
     }
  }
```

(5).3. kbd_handler(): The kbd_handler() reads the scan code from the KBD data register, which clears the KBD interrupt. It ignores any key release, so the driver can only handle lower case keys and no special keys at all. For each key pressed, it prints a "kbd interrupt key" message to the LCD display. As noted before, we have adapted the generic printf() function for formatted printing to the LCD screen. Then, it maps the scan code to a (lower case) ASCII char and enters the char into the input buffer. The control variable, data, represents the number of chars in the input buffer.
(5).4. kgetc() and kgets() functions: kgetc() is for getting an input char from the keyboard. kgets() is for getting an input line ended with the \r key. The simple KBD driver is intended mainly to illustrate the design principle of an interrupt-driven input device driver. The driver's buffer and control variables form a critical region since they are accessed by both the main program and KBD interrupt handler. When the interrupt handler executes, the main program is logically not executing. So the main program can not interfere with the interrupt handler. However, while the main program executes, interrupts may occur, which diverts the program to execute the interrupt handler, which may interfere with the main program. For this reason, when a program calls kgetc(), which may modify the shared variables in the driver, it must mask out interrupts to prevent keyboard interrupts from occurring. In kgetc(), the main program first enables interrupts, which is optional if the program is already running with interrupts enabled. Then it loops until the variable kp->data is nonzero, meaning that there are chars in the input buffer. Then it disables interrupts, gets a char from the input buffer and updates the shared variables. The code segment which ensures shared variables can only be updated by one execution entity at a time is commonly known as a **Critical Region or Critical Section** (Silberschatz et al. 2009; Stallings 2011; Wang 2016). Finally, it enables interrupts and returns a char. On the ARM CPU it is not possible to mask out only the keyboard interrupts. The lock() operation masks out all IRQ interrupts, which is a little overkill but it gets the job done. Alternatively, we may write to the keyboard's control register to explicitly disable/enable keyboard interrupts. The disadvantage is that it must access memory mapped locations, which is much slower than masking out interrupts via the CPU's CPSR register.

Figure 3.6 shows the KBD driver using interrupts. As the figure shows, the main program only prints complete lines, but each input key generates an interrupt and prints a "kbd interrupt" message, along with the input char in ASCII.

3.6 UART Driver

3.6.1 ARM PL011 UART Interface

The ARM Versatile board supports four PL011 UART devices for serial I/O (ARM PL011 2005). Each UART device has a base address in the system memory map.

```
UART0: 0x101F1000
UART1: 0x101F2000
UART2: 0x101F3000
UART3: 0x10009000
```

```
QEMU
00:00:14
C3.2 start: test timer KBD drivers by interrupts
timer_init()
timer_start  0  base=0x 101E2000
enter a line from KBD
kbd interrupt: c= t
kbd interrupt: c= e
kbd interrupt: c= s
kbd interrupt: c= t
line = test
enter a line from KBD
```

Fig. 3.6 KBD driver using interrupts

The first 3 UARTs, UART0 to UART2 are adjacent in the system memory map. They interrupt at IRQ12 to IRQ14 on the primary PIC. UART4 in located at 0x10000900 and it interrupts at IRQ6 on the SIC. In general, a UART must be initialized by the following steps.

(1). Write a divisor value to the baud rate register for a desired baud rate. The ARM PL011 technical reference manual lists the following integer divisor values (based on 7.38 MHz UART clock) for the commonly used baud rates:

```
0x4=1152000, 0xC=38400, 0x18=192000, 0x20=14400, 0x30=9600
```

(2). Write to Line Control register to specify the number of bits per char and parity, e.g. 8 bits per char with no parity, etc.
(3). Write to Interrupt Mask register to enable/disable RX and TX interrupts

When using the emulated ARM Versatilepb board under QEMU, it seems that QEMU automatically uses default values for both baud rate and line control parameters, making steps (1) and (2) either optional or unnecessary. In fact, it is observed that writing any value to the integer divisor register (0x24) would work, but the reader should be aware this is not the norm for UARTs in real systems. In this case, we only need to program the Interrupt Mask register (if using interrupts) and check the Flag register during serial I/O.

3.6.2 UART Registers

Each UART interface contains several 32-bit registers. The most important UART registers are

```
Offset    Name        Function
------    --------    ------------------------------
 0x00     UARTDR      data register: for read/write chars
 0x04     UARTSR      receive status/clear error
 0x18     UARTFR      TxEmpty, RxFull, etc.
 0x24     UARIBRD     set baud rate
 0x2c     UARTLCR     line control register
 0x30     UARTCR      control register
 0x38     UARTIMSC    interrupt mask for TX and RX
 0x40     UARTMIS     interrupt status
```

Some of the UART registers are already explained in the program C2.3 of Chap. 2. Here, we shall focus on the registers that are related to interrupts. The ARM UART interface supports many kinds of interrupts. For simplicity, we shall only consider Rx (Receiving) and Tx (Transmitting) interrupts for data transfer and ignore such interrupts as modem status and error conditions. To allow UART Rx and Tx interrupts, bits 4 and 5 of the interrupt mask register (UARTIMSC) must be set to 1. When a UART interrupts, the masked interrupt status register (UARTMIS) contains the interrupt identification, e.g. bit 4 = 1 if it it's an Rx interrupt and bit 5 = 1 if it's a Tx interrupt. Depending on the interrupt type, the interrupt handler can branch to a corresponding ISR to handle the interrupt.

3.6.3 Interrupt-Driven UART Driver Program

This section shows the design and implementation of an interrupt-driven UART driver for serial I/O. In order to keep the diver simple, we shall only use UART0 and UART1 but the same code is also applicable to other UARTs. The UART driver program is denoted by C3.3, which is organized as follows.

3.6.3.1 The uart.c File

This file implements the UART driver using interrupts. Each UART is represented by a UART structure. It contains the UART base address, a unit number, an input buffer, an output buffer and control variables. Both buffers are circular, with head pointers for entering chars and tail pointers for removing chars. Among the control variables, data is the number of chars in the buffer and room is the number of empty spaces in the buffer. For outputs, txon is a flag indicating whether the UART is already in the transmission state. To use interrupts, the UART Interrupt Mask Set/Clear register (IMSC) must be set up properly. The UART supports many kinds of interrupts. For simplicity, we shall only consider TX (bit 5) and RX (bit 4) interrupts. In uart_init(), the C statement

```
*(up->base+IMSC) |= 0x30; // bits 4,5 = 1
```

sets bits 4 and 5 of IMSC to 1, which enable TX and RX interrupts of the UART. For simplicity, we shall only use UART0 and UART1. UART0 interrupts at IRQ12 and UART1 interrupts at IRQ13, both on the primary VIC. In order to allow UART interrupts, bits 12 and 13 of the PIC must be set to 1. These are done in main() by the statements

```
VIC_INTENABLE |= (1<<12); // UART0 at bit12
VIC_INTENABLE |= (1<<13); // UART1 at bit13
```

A special feature of the ARM PL011 UART is that it supports FIFO buffers in hardware for both send and receive operations. It can be programmed to raise interrupts when the FIFO buffers are at different levels between FULL and EMPTY. In order to keep the UART driver simple, we shall not use the hardware FIFO buffers. They are disabled by the statement

```
*(up->base+CNTL) &= ~0x10; // disable UART FIFO
```

This makes the UART operate in single char mode. Also, TX interrupt is triggered only after writing a char to the data register. When the system has no more outputs to a UART, it must disable the TX interrupt. The following shows the uart.c driver code.

```
/************* uart.c file ****************/
#define UDR      0x00
#define UDS      0x04
#define UFR      0x18
#define CNTL     0x2C
#define IMSC     0x38
#define MIS      0x40
#define SBUFSIZE 128
```

```
typedef volatile struct uart{
   char *base; // base address; as char *
   int n;      // uart number 0-3
   char inbuf[SBUFSIZE];
   int indata, inroom, inhead, intail;
   char outbuf[SBUFSIZE];
   int outdata, outroom, outhead, outtail;
   volatile int txon; // 1=TX interrupt is on
}UART;
UART uart[4]; // 4 UART structures
int uart_init()
{
   int i; UART *up;
   for (i=0; i<4; i++){ // uart0 to uart2 are adjacent
   up = &uart[i];
   up->base = (char *)(0x101F1000 + i*0x1000);
   *(up->base+CNTL) &= ~0x10; // disable UART FIFO
   *(up->base+IMSC) |= 0x30;
   up->n = i; UART ID number
   up->indata = up->inhead = up->intail = 0;
   up->inroom = SBUFSIZE;
   up->outdata = up->outhead = up->outtail = 0;
   up->outroom = SBUFSIZE;
   up->txon = 0;
 }
 uart[3].base = (char *)(0x10009000); // uart3 at 0x10009000
}
void uart_handler(UART *up)
{
 u8 mis = *(up->base + MIS); // read MIS register
 if (mis & (1<<4)) // MIS.bit4=RX interrupt
 do_rx(up);
 if (mis & (1<<5)) // MIS.bit5=TX interrupt
 do_tx(up);
}
int do_rx(UART *up) // RX interrupt handler
{
 char c;
 c = *(up->base+UDR);
 printf("rx interrupt: %c\n", c);
 if (c==0xD)
 printf("\n");
 up->inbuf[up->inhead++] = c;
 up->inhead %= SBUFSIZE;
 up->indata++; up->inroom--;
}
int do_tx(UART *up) // TX interrupt handler
{
 char c;
 printf("TX interrupt\n");
 if (up->outdata <= 0){ // if outbuf[ ] is empty
 *(up->base+MASK) = 0x10; // disable TX interrupt
 up->txon = 0; // turn off txon flag
 return;
 }
```

```
c = up->outbuf[up->outtail++];
up->outtail %= SBUFSIZE;
*(up->base+UDR) = (int)c; // write c to DR
up->outdata--; up->outroom++;
}

int ugetc(UART *up)            // return a char from UART
{
  char c;
  while(up->indata <= 0); // loop until up->data > 0 READONLY
  c = up->inbuf[up->intail++];
  up->intail %= SBUFSIZE;
  // updating variables: must disable interrupts
  lock();
    up->indata--; up->inroom++;
  unlock();
  return c;
}
int uputc(UART *up, char c)    // output a char to UART
{
  kprintf("uputc %c ", c);
  if (up->txon){ //if TX is on, enter c into outbuf[]
     up->outbuf[up->outhead++] = c;
     up->outhead %= 128;
     lock();
      up->outdata++; up->outroom--;
     unlock();
     return;
  }
  // txon==0 means TX is off => output c & enable TX interrupt
  // PL011 TX is riggered only if write char, else no TX interrupt
  int i = *(up->base+UFR);           // read FR
  while( *(up->base+UFR) & 0x20 ); // loop while FR=TXF
  *(up->base+UDR) = (int)c;          // write c to DR
  UART0_IMSC |= 0x30; // 0000 0000: bit5=TX mask bit4=RX mask
  up->txon = 1;
}
int ugets(UART *up, char *s)         // get a line from UART
{
  kprintf("%s", "in ugets: ");
  while ((*s = (char)ugetc(up)) != '\r'){
    uputc(up, *s++);
  }
 *s = 0;
}
int uprints(UART *up, char *s)       // print a line to UART
{
  while(*s)
    uputc(up, *s++);
}
```

3.6.3.2 Explanations of the UART Driver Code

(1). uart_handler(UART *up): The uart interrupt handler reads the MIS register to determine the interrupt type. It's a RX interrupt if bit 4 of MIS is 1. It's a TX interrupt if bit 5 of MIS is 1. Depending on the interrupt type, it calls do_rx() or do_tx () to handle the interrupt.
(2). do_rx(): This is an input interrupt. It reads the ASCII char from the data register, which clears the RX interrupt of the UART. Then it enters the char into the circular input buffer and increment the data variable by 1. The data variable represents the number of chars in the input buffer.
(3). do_tx(): This is an output interrupt, which is triggered by writing the last char to the output data register and transmission of that char has finished. The handler checks whether there are any chars in the output buffer. If the output buffer is empty, it disables the UART TX interrupt and returns. If the TX interrupt is not disabled, it would cause an infinite sequence of TX interrupts. If the output buffer is not empty, it takes a char from the buffer and writes it to the data register for transmission out.
(4). ugetc(): ugetc is for the main program to get a char from a UART port. Its logic and synchronization with the RX interrupt handler are the same as kgetc() of the keyboard driver. So we shall not repeat them here again.
(5). uputc(): uputc() is for the main program to output a char to a UART port. If the UART port is not transmitting (the txon flag is off), it writes the char to the data register, enables TX interrupt and sets the txon flag. Otherwise, it enters the char into the output buffer, updates the data variable and returns. The TX interrupt handler will output the chars from the output buffer on each successive interrupt.
(6). Formatted printing: uprintf(UART *up, char *fmt,...) is for formatted printing to a UART port. It is based on uputc().

3.6.3.3 Demonstration of KBD and UART Drivers

The t.c file contains the IRQ_handler() and the main() function. The main() function first initializes the UART devices and the VIC for interrupts. Then it tests the UART driver by issuing serial I/O on the UART terminals. For each IRQ interrupt, the IRQ_handler() determines the interrupt source and calls an appropriate handler to handle the interrupt. For clarity, UART related code are shown in bold faced lines.

```
/*************** C3.3. t.c file *******************/
#include "defines.h"
#include "string.c"
#include "uart.c"   // UART driver
#include "kbd.c"    // keyboard driver
#include "timer.c"  // timer
#include "vid.c"    // LCD display driver
#include "exceptions.c"

void copy_vectors(void){ // same as before }

void IRQ_handler()         // IRQ interrupt handler in C
{
  // read VIC status registers to find out which interrupt
  int vicstatus = VIC_STATUS;
  // VIC status BITs: timer0,1=4,uart0=13,uart1=14
  if (vicstatus & (1<<5)){            // bit5:timer2,3
     if (*(tp[2]->base+TVALUE)==0) // timer 2
        timer_handler(2);
     if (*(tp[3]->base+TVALUE)==0) // timer 3
        timer_handler(3);
  }
  if (vicstatus & (1<<4)){            // bit4=1:timer0,1
     timer_handler(0);
  }
  if (vicstatus & (1<<12))            // bit12=1: uart0
     uart_handler(&uart[0]);
```

```
   if (vicstatus & (1<<13))              // bit13=1: uart1
      uart_handler(&uart[1]);
   if (vicstatus & (1<<31)){    // PIC.bit31= SIC interrupts
      if (sicstatus & (1<<3)){ // SIC.bit3 = KBD interrupt
         kbd_handler();
      }
   }
}

int main()
{
   char line[128];
   UART *up;
   KBD  *kp;
   fbuf_init();     // initialize LCD display
   printf("C3.4 start: test KBD TIMER UART drivers\n");

   /* enable timer0,1, uart0,1 SIC interrupts */
   VIC_INTENABLE = 0;
   VIC_INTENABLE |= (1<<4);   // timer0,1 at bit4
   VIC_INTENABLE |= (1<<5);   // timer2,3 at bit5
   VIC_INTENABLE |= (1<<12); // UART0 at bit12
   VIC_INTENABLE |= (1<<13); // UART1 at bit13
   VIC_INTENABLE |= (1<<31); // SIC to VIC's IRQ31
   SIC_INTENABLE = 0;
   SIC_INTENABLE |= (1<<3);   // KBD int=bit3 on SIC
   kbd_init();        // initialize keyboard
   uart_init();       // initialize UARTs
   up = &uart[0];     // test UART0 I/O
   kp = &kbd;
   timer_init();
   timer_start(0);    // timer0 only
   while(1){
     kprintf("Enter a line from KBD\n");
     kgets(line);
     kprintf("Enter a line from UART0\n");
     uprints("Enter a line from UARTS\n\r");
     ugets(up, line);
     uprintf("%s\n", line);
   }
}
```

Figure 3.7 shows the outputs of the interrupt-driven KBD and UART drivers. The figure shows that UART inputs are by rx interrupts and outputs are by tx interrupts.

3.7 Secure Digital (SD) Cards

For most embedded systems, the primary mass storage devices are Secure Digital (SD) cards (SDC 2016) due to their compact size, low power consumption and compatibility with other kinds of mobile devices. Many embedded systems may not have any mass storage device to provide file system support, but they usually start up from either a flash memory card or a SD card. A good example is the Raspberry Pi (Raspberry_Pi 2016). It requires a SD card to boot up an operating system, which is usually a version of Linux, called Raspbian, adapted for the ARM architecture. Most ARM based systems include the ARM PrimeCell PL180/ PL181 multimedia card interface (ARM PL180 1998; ARM PL181 2001) to provide support for

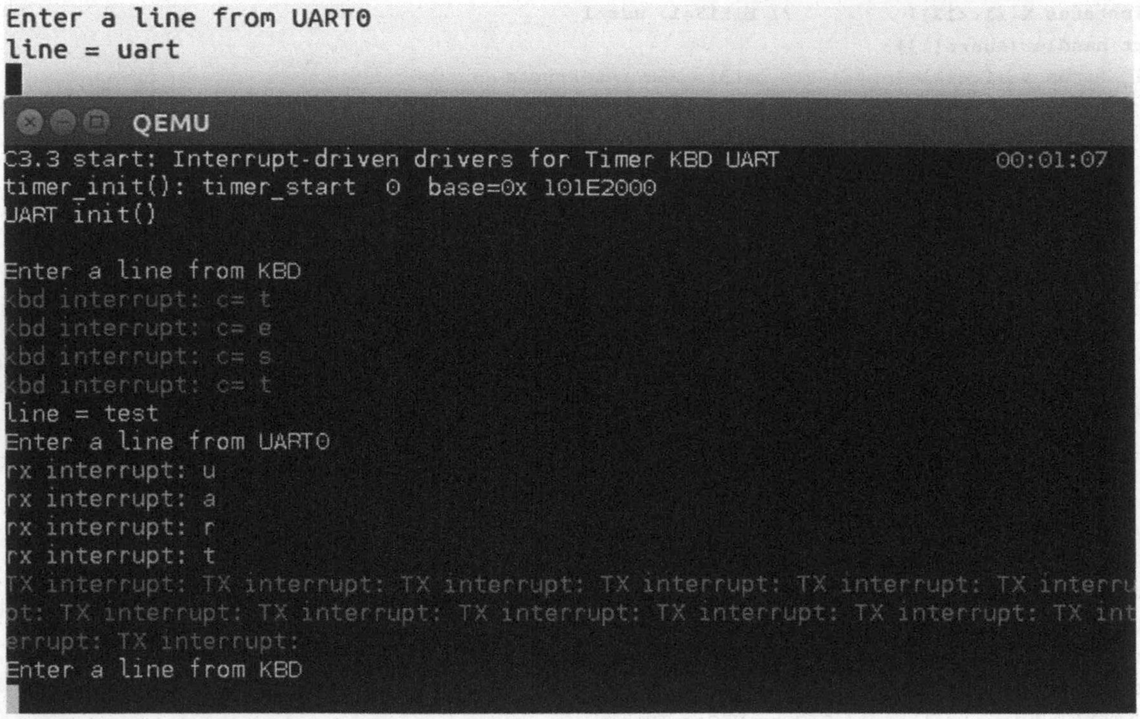

```
Enter a line from UART0
line = uart

QEMU
C3.3 start: Interrupt-driven drivers for Timer KBD UART          00:01:07
timer_init(): timer_start  0  base=0x 101E2000
UART init()

Enter a line from KBD
kbd interrupt: c= t
kbd interrupt: c= e
kbd interrupt: c= s
kbd interrupt: c= t
line = test
Enter a line from UART0
rx interrupt: u
rx interrupt: a
rx interrupt: r
rx interrupt: t
TX interrupt: TX interrupt: TX interrupt: TX interrupt: TX interrupt: TX interru
pt: TX interrupt: TX interrupt: TX interrupt: TX interrupt: TX interrupt: TX int
errupt: TX interrupt:
Enter a line from KBD
```

Fig. 3.7 Demonstration of KBD and UART drivers

both multimedia and SD cards. The emulated ARM Versatilepb virtual machine under QEMU includes the PL180 multimedia interface also but it only supports SD cards.

3.7.1 SD Card Protocols

The simplest SD card protocol is the Serial Peripheral Interface (SPI) (SPI 2016). The SPI protocol requires the host machine to have a SPI port, which is available in many ARM based systems. For host machines without SPI ports, SD cards must use the native SD protocol [SD specification 2016], which is more capable and therefore more complex than the SPI protocol. QEMU's multimedia interface supports SD cards in native mode but not in SPI mode. For this reason, we shall develop a SD card driver that operates in the native SD mode.

3.7.2 SD Card Driver

The sample program C3.4 implements an interrupt-driven SD card driver. It demonstrates the SD driver by writing to the sectors of a SD card and then reading back the sectors to verify the results. The program consists of the following components.

(1). **sdc.h:** This header file defines the PL180 Multi-Media Card (MMC) registers and bit masks. For the sake of brevity, we only show the PL180 registers.

```
/** PL180 registers from BASE **/
#define   POWER          0x00
#define   CLOCK          0x04
#define   ARGUMENT       0x08
#define   COMMAND        0x0C
#define   RESPCOMMAND    0x10
#define   RESPONSE0      0x14
#define   RESPONSE1      0x18
```

```
#define    RESPONSE2      0x1C
#define    RESPONSE3      0x20
#define    DATATIMER      0x24
#define    DATALENGTH     0x28
#define    DATACTRL       0x2C
#define    DATACOUNT      0x30
#define    STATUS         0x34
#define    STATUS_CLEAR   0x38
#define    MASK0          0x3C
#define    MASK1          0x40
#define    CARD_SELECT    0x44
#define    FIFO_COUNT     0x48
#define    FIFO           0x80
/** more ID registers NOT USED **/
```

(2). **sdc.c file:** This file implements the SDC driver.

```
/*********** sdc.c ARM PL180 SDC driver **********/
#include "sdc.h"
u32 base;
#define printf kprintf

int delay(){ int i; for (i=0; i<100; i++); }

int do_command(int cmd, int arg, int resp)
{
  *(u32 *)(base + ARGUMENT) = (u32)arg;
  *(u32 *)(base + COMMAND)  = 0x400 | (resp<<6) | cmd;
  delay();
}
int sdc_init()
{
  u32 RCA = (u32)0x45670000; // QEMU's hard-coded RCA
  base    = (u32)0x10005000; // PL180 base address
  printf("sdc_init : ");
  *(u32 *)(base + POWER) = (u32)0xBF; // power on
  *(u32 *)(base + CLOCK) = (u32)0xC6; // default CLK

  // send init command sequence
  do_command(0,  0,   MMC_RSP_NONE);// idle state
  do_command(55, 0,   MMC_RSP_R1);  // ready state
  do_command(41, 1,   MMC_RSP_R3);  // argument must not be zero
  do_command(2,  0,   MMC_RSP_R2);  // ask card CID
  do_command(3,  RCA, MMC_RSP_R1);  // assign RCA
  do_command(7,  RCA, MMC_RSP_R1);  // transfer state: must use RCA
  do_command(16, 512, MMC_RSP_R1);  // set data block length

  // set interrupt MASK0 registers bits = RxAvail|TxEmpty
  *(u32 *)(base + MASK0) = (1<<21)|(1<<18); //0x00240000;
  printf("done\n");
}

// shared variables between SDC driver and interrupt handler
volatile char *rxbuf, *txbuf;
```

```
volatile int  rxcount, txcount, rxdone, txdone;
int sdc_handler()
{
  u32 status, err;
  int i;
  u32 *up;

  // read status register to find out RxDataAvail or TxBufEmpty
  status = *(u32 *)(base + STATUS);

  if (status & (1<<21)){ // RxDataAvail: read data
    printf("SDC RX interrupt: );
    up = (u32 *)rxbuf;
    err = status & (DCRCFAIL | DTIMEOUT | RXOVERR);
    while (!err && rxcount) {
       printf("R%d ", rxcount);
       *(up) = *(u32 *)(base + FIFO);
       up++;
       rxcount -= sizeof(u32);
       status = *(u32 *)(base + STATUS);
       err = status & (DCRCFAIL | DTIMEOUT | RXOVERR);
    }
    rxdone = 1;
  }
  else if (status & (1<<18)){ // TxBufEmpty: send data
    printf("SDC TX interrupt: );
    up = (u32 *)txbuf;
    status_err = status & (DCRCFAIL | DTIMEOUT);

    while (!status_err && txcount) {
       printf("W%d ", txcount);
       *(u32 *)(base + FIFO) = *up;
       up++;
       txcount -= sizeof(u32);
       status = *(u32 *)(base + STATUS);
       status_err = status & (DCRCFAIL | DTIMEOUT);
    }
    txdone = 1;
  }
  //printf("write to clear register\n");
  *(u32 *)(base + STATUS_CLEAR) = 0xFFFFFFFF;
  printf("SDC interrupt handler return\n");
}

int get_sector(int sector, char *buf)
{
  u32 cmd, arg;

  //printf("get_sector %d %x\n", sector, buf);
  rxbuf = buf; rxcount = 512; rxdone = 0;
  *(u32 *)(base + DATATIMER) = 0xFFFF0000;
  // write data_len to datalength reg
  *(u32 *)(base + DATALENGTH) = 512;

  //printf("dataControl=%x\n", 0x93);
  // 0x93=|9|0011|=|9|DMA=0,0=BLOCK,1=Host<-Card,1=Enable
```

```
   *(u32 *)(base + DATACTRL) = 0x93;
   cmd = 17;        // CMD17 = READ single block
   arg = (sector*512);
   do_command(cmd, arg, MMC_RSP_R1);
   while(rxdone == 0);
   printf("get_sector return\n");
}

int put_sector(int sector, char *buf)
{
   u32 cmd, arg;

   //printf("put_sector %d %x\n", sector, buf);
   txbuf = buf; txcount = 512; txdone = 0;
   *(u32 *)(base + DATATIMER) = 0xFFFF0000;
   *(u32 *)(base + DATALENGTH) = 512;

   cmd = 24;       // CMD24 = Write single block
   arg = (u32)(sector*512);
   do_command(cmd, arg, MMC_RSP_R1);

   //printf("dataControl=%x\n", 0x91);
   // 0x91=|9|0001|=|9|DMA=0,BLOCK=0,0=Host->Card,1=Enable
   *(u32 *)(base + DATACTRL) = 0x91; // Host->card
   while(txdone == 0);
   printf("put_sector return\n");
}
```

The SDC driver consists of 4 major parts, which are explained below.

(1). sdc_init(): This is called from the main program to initialize the SDC. When writing SDC drivers, the most crucial step is the SD card initialization, which comprises the following steps.

. send CMD0 to get the SDC into idle state
. send CMD8 to determine SDC type and voltage range
. send CMD55 followed by CMD41 to get the SDC into ready state
. send CMD2 to get Card Identification (CID)
. send CMD3 to get/set Card Relative Address (RCA)
. send CMD7 to select the SDC by RCA to transfer data state.
. set Interrupt Mask register MASK0 bits to enable Rx and Tx interrupts

In general, every command except CMD0 expects a response of different type. After sending a command, the responses are in the response registers, but they are ignored in the SD driver. It is observed that in the emulated PL180 MMC of QEMU, the Relative Card Address (RCA) assigned to the SDC is hard-coded as 0x4567. In fact, each CMD3 command increases the RCA by 0x4567. The reason for this rather peculiar behavior is probably due to a typo in the PL180 emulator of QEMU. It should just set the RCA once by RCA = 0x4567, rather than RCA += 0x4567, which increments it by 0x4567 on each CMD3 command. After initialization, the driver may issue a CMD17 to read a block or a CMD24 to write a block. For SDC, the default block (sector) size is 512 bytes. The SDC also supports read/write multiple sectors by CMD18 and CMD25, respectively. Data transfer may use one of three different schemes: polling, interrupts or DMA. DMA is suitable for transferring large amounts of data. In order to keep the driver code simple, the SDC driver only uses interrupts, not DMA, to read/write one sector of 512 bytes at a time. It will be extended to read/write multi-sectors later. SDC interrupts are enabled by setting bits in the Interrupt Mask registers MASK0. In sdc_init(), the interrupt mask bits are set to RxDataAvail (bit 21) and TxBufEmpty (bit 18). The MMC will generate a SDC interrupt when either there are data in the input buffer or there are rooms in the output buffer. It disables and ignores other kinds of SDC interrupts, e.g. interrupts due to error conditions.

(2). get_sector(int sector, char *buf): get_sector() is for reading a sector of 512 bytes from the SDC. The algorithm of get_sector() is as follows.
. Set global rxbuf = buf and rxdone = 0 for the Rx interrupt handler to use;
. Set DataTimer to default, set DataLength to 512;
. Set DataCntl to 0x93 (block size = 2**9, respR1, SDC to Host, and Enable);
. Send CMD17 with argument = sector*512 (byte address on SDC);
. (Busy) wait for the Rx interrupt handler to finish reading data;

Upon receiving a CMD17, the PL180 MultiMedia Controller (MMC) starts to transfer data from SDC to its internal input buffer. The MMC has a 16x32 bits FIFO input data buffer. When data becomes available, it generates an Rx interrupt, causing SDC_handler() to be executed, which actually transfers data from the MMC to rxbuf. After sending CMD17, the main program busily waits for a volatile rxdone flag, which will be set by the interrupt handler when data transfer completes.

(3). put_sector(int sector, char *buf): put_sector() is for writing a block of data to the SDC. The algorithm of put_sector() is as follows.

. Set global txbuf=buf and txdone=0 for the Tx interrupt handler to use;
. Set DataTimer to default and DataLength to 512;
. Send CMD24 with argument=sector*512 (byte address on SDC);
. Set DataCntl to 0x91 (block size=2**9, respR1, Host to SDC, Enable);
. (Busy) wait for the Tx interrupt handler to finish writing data;

Upon receiving a CMD24, the PL180 MMC starts to transfer data. The MMC has a 16x32 bits FIFO output data buffer. If the Tx buffer is empty, it generates an SDC interrupt, causing SDC_handler() to be executed, which actually transfers data from buf to the MMC. After sending CDM24, the main program busily waits for a volatile txdone flag, which will be set by the interrupt handler when data transfer completes.

(4). sdc_handler(): This is the SDC interrupt handler. It first checks the status register to determine the interrupt source. If it is an RxDataAvail interrupt (bit 21 is set), it transfers data from the MMC controller to rxbuf by a loop.

```
while (!err && rxcount) {
  printf("R%d ", rxcount);
  *(up) = *(u32 *)(base + FIFO);
  up++;
  rxcount -= sizeof(u32);
  status = *(u32 *)(base + STATUS);
  err = status & (SDI_STA_DCRCFAIL | SDI_STA_DTIMEOUT |
                  SDI_STA_RXOVERR);
}
```

Barring any errors, each iteration of the loop reads a u32 (4 bytes) from the MMC's FIFO input buffer, decrements rxcount by 4 until rxcount reaches 0. Then it sets the rxdone flag to 1, allowing the main program in get_sector() to continue. If the SDC interrupt is a Tx interrupt (bit 18 is set), it writes data from txbuf to the MMC's FIFO by a loop.

```
while (!err && txcount){
   printf("W%d ", txcount);
   *(u32 *)(base + FIFO) = *up;
   up++;
   txcount -= sizeof(u32);
   status = *(u32 *)(base + STATUS);
   err = status & (SDI_STA_DCRCFAIL | SDI_STA_DTIMEOUT);
}
```

Barring any errors, each iteration of the loop writes a u32 (4 bytes) to the MMC's FIFO output buffer, decrements txcount by 4 until txcount reaches 0. Then it sets the txdone flag to 1, allowing the main program in put_sector() to continue.

(3). **t.c file:** The t.c file is the same as in C3.3, except for the added code for SDC initialization and testing. For clarity, the modified lines of t.c are shown in bold face letters.

```
#include "defines.h"
#include "string.c"
#define printf kprintf
char *tab = "0123456789ABCDEF";
#include "uart.c"
#include "kbd.c"
#include "timer.c"
#include "vid.c"
#include "exceptions.c"
#include "sdc.c"

void copy_vectors(void) { // same as before }

void IRQ_handler()
{
   int vicstatus, sicstatus;
   int ustatus, kstatus;
   // read VIC SIV status registers to find out which interrupt
   vicstatus = VIC_STATUS;
   sicstatus = SIC_STATUS;
   // VIC status BITs: timer0=4, uart0=13, uart1=14, SIC=31: KBD at 3
   if (vicstatus & (1<<4)){   // bit 4: timer0
      timer_handler(0);
   }
   if (vicstatus & (1<<12)){ // Bit 12: UART0
      uart_handler(&uart[0]);
   }
   if (vicstatus & (1<<13)){ // bit 13: UART1
      uart_handler(&uart[1]);
   }
   if (vicstatus & (1<<31)){ // SIC interrupts=bit_31 on VIC
      if (sicstatus & (1<<3)){   // KBD at IRQ3 of SIC
         kbd_handler();
      }
      if (sicstatus & (1<<22)){ // SDC at IRQ22 of SIC
         sdc_handler();
      }
   }
}

char rbuf[512], wbuf[512];
char *line[2] = {"THIS IS A TEST LINE", "this is a test line"};

int main()
{
   int i, sector, N;
   fbuf_init();
   kbd_init();
   uart_init();
```

```
    /* enable timer0,1, uart0,1 SIC interrupts */
    VIC_INTENABLE = (1<<4);   // timer0,1 at bit4
    VIC_INTENABLE |= (1<<12); // UART0 at bit12
    VIC_INTENABLE |= (1<<13); // UART1 at bit13
    VIC_INTENABLE |= (1<<31); // SIC to VIC's IRQ31
    /* enable KBD and SDC IRQ */
    SIC_INTENABLE = (1<<3);   // KBD int=bit3 on SIC
    SIC_INTENABLE |= (1<<22); // SDC int=bit22 on SIC
    SIC_ENSET = (1<<3);       // KBD int=3 on SIC
    SIC_ENSET |= (1<<22);     // SDC int=22 on SIC
    timer_init();
    timer_start(0);
    /* Code for testing UART and KBD drivers are omitted */
    printf("test SDC DRIVER\n");
    sdc_init();
    N = 1;          // Write|Read N sectors of SDC
    for (sector=0; sector < N; sector++){
      printf("WRITE sector %d: ", sector);
      memset(wbuf, ' ', 512);  // blank out wbuf
      for (i=0; i<12; i++)     // write lines to wbuf
          strcpy(wbuf+i*40, line[sector % 2]);
      put_sector(sector, wbuf);
    }
    printf("\n");

    for (sector=0; sector < N; sector++){
      printf("READ  sector %d\n", sector);
      get_sector(sector, rbuf);
      for (i=0; i<512; i++){
        printf("%c", rbuf[i]);
      }
      printf("\n");
    }
    printf("in while(1) loop: enter keys from KBD or UART\n");
    while(1);
}
```

The ARM Versatile user manual specifies that the MMCI0 interrupts at IRQ22 on both the VIC and SIC. However, in the PL180 emulated by QEMU, it actually interrupts at IRQ22 of the SIC, which is routed to IRQ31 of the VIC. The reason for this discrepancy is unknown. Other than this minor discrepancy, the emulated PL180 works as expected. Figure 3.8 shows the outputs of running the SDC driver program C3.4.

3.7.3 Improved SDC Driver

In the SDC driver, the interrupt handler performs all the data transfer on a single interrupt. Since data transfer from the MMC to the SDC may be slow, the interrupt handler must execute the data transfer loop many times while waiting for the MMC to become ready to provide or accept data. The drawback of this scheme is that it is essentially the same as, or even worse than, I/O by polling. In general, an interrupt handler should be completed as soon as possible. Any excessive checking and waiting inside an interrupt handler must be avoided or eliminated. It is therefore desirable to minimize the number of interrupts and maximize the amount of data transfer on each interrupt. This leads us to an improved SDC driver. In the improved SDC driver, we program the MMC to generate interrupts only when the Rx FIFO is full or the Tx FIFO is empty. For each interrupt, we transfer 16 u32 data on each interrupt. The following code segments show the improved SDC driver, in which the modifications are shown in bold faced lines.

```
QEMU
C10 start: test KBD TIMER UART SDC drivers                    00:00:18
timer_init()
timer_start 0  base=0x 101E2000
test SDC DRIVER
sdc_init : done
WRITE sector 0 :
SDC TX interrupt: status=0x 455C0 : SDC interrupt handler done
putSector return

READ  sector 0
SDC RX interrupt: status=0x 22A040 : SDC interrupt handler done
getSector return
THIS IS A TEST LINE                         THIS IS A TEST LINE
THIS IS A TEST LINE                         THIS IS A TEST LINE
THIS IS A TEST LINE                         THIS IS A TEST LINE
THIS IS A TEST LINE                         THIS IS A TEST LINE
THIS IS A TEST LINE                         THIS IS A TEST LINE
THIS IS A TEST LINE                         THIS IS A TEST LINE

in while(1) loop: enter keys from KBD or UART
```

Fig. 3.8 Outputs of SDC driver program

```
// shared variables between SDC driver and interrupt handler
volatile char *rxbuf, *txbuf;
volatile int  rxcount, txcount, rxdone, txdone;

int sdc_handler()
{
  u32 status, status_err;
  int i;
  u32 *up;
  // read status register to find out TXempty or RxFull interrupt
  status = *(u32 *)(base + STATUS);

  if (status & (1<<17)){ // RxFull: read 16 u32 at a time;
     printf("RX interrupt: ");
     up = (u32 *)rxbuf;
     status_err = status & (DCRCFAIL | DTIMEOUT | RXOVERR);
     if (!status_err && rxcount) {
        printf("R%d ", rxcount);
        for (i = 0; i < 16; i++)
           *(up + i) = *(u32 *)(base + FIFO);
        up += 16;
        rxcount -= 64;
        rxbuf += 64;
        status = *(u32 *)(base + STATUS); // clear Rx interrupt
     }
     if (rxcount == 0)
        rxdone = 1;
  }
  else if (status & (1<<18)){ // TXempty: write 16 u32 at a time
    printf("TX interrupt: ");
```

```
    up = (u32 *)txbuf;
    status_err = status & (DCRCFAIL | DTIMEOUT);
    if (!status_err && txcount) {
       printf("W%d ", txcount);
       for (i = 0; i < 16; i++)
           *(u32 *)(base + FIFO) = *(up + i);
       up += 16;
       txcount -= 64;
       txbuf += 64;              // advance txbuf for next write
       status = *(u32 *)(base + STATUS); // clear Tx interrupt
    }
    if (txcount == 0)
       txdone = 1;
  }
  //printf("write to clear register\n");
  *(u32 *)(base + STATUS_CLEAR) = 0xFFFFFFFF;
  // printf("SDC interrupt handler done\n");
}
```

Figure 3.9 shows the outputs of running the improved SDC driver program. As the figure shows, each interrupt transfers 16 4-byte data, so that the byte transfer count decrements by 64 on each interrupt. As a further improvement, the reader may program the MMC to generate interrupts when the Rx FIFO is half-full and Tx FIFO is half-empty. In that case, each interrupt can transfer 8 4-byte data. It improves the data transfer rate at the expense of more interrupts and hence more overhead due to interrupt processing.

3.7.4 Multi-sector Data Transfer

The above SD drivers transfer one sector (512 bytes) of data at a time. An embedded system may support file systems, which usually use 1 or 4 kB file block size. In that case, it would be more efficient to transfer data from/to SD cards in multi-sectors that matches file block size. The following code segments show the modified SD driver which transfers data in multi-sectors.

To read multi-sectors, issue the command CMD18. To write multi-sectors, issue the command CMD25. In both cases, the data length is the file block size. For multi-sector data transfers, data transmission must be terminated by a stop transmission command CMD12, which is issued in the interrupt handler when the byte count (rxcount or txcount) reaches 0. The modified lines of the driver are shown in bold face. In the code segment, FBLK_SIZE is defined as 4096. Each get_block()/put_block() call read/write a (file) block of 4 KB data.

```
// shared variables between SDC driver and interrupt handler
#define FBLK_SIZE 4096
volatile char *rxbuf, *txbuf;
volatile int  rxcount, txcount, rxdone, txdone;

int sdc_handler()
{
  u32 status, err, *up;
  int i;
  // read status register to find out TXempty or RxAvail
  status = *(u32 *)(base + STATUS);
  if (status & (1<<17)){ // RxFull: read 16 u32 at a time;
     up = (u32 *)rxbuf;
     err = status & (DRCFAIL | DTIMEOUT | RXOVERR);
     if (!err && rxcount){
        for (i = 0; i < 16; i++)
```

Fig. 3.9 Outputs of the improved SDC driver

```
            *(up + i) = *(u32 *)(base + FIFO);
        up += 16;
        rxcount -= 64;
        rxbuf += 64;
        status = *(u32 *)(base + STATUS); // read status
     }
     if (rxcount == 0){
        do_command(12, 0, MMC_RSP_R1); // stop transmission
        rxdone = 1;
     }
  }
  else if (status & (1<<18)){ // TXempty: write 16 u32 at a time
     up = (u32 *)txbuf;
     err = status & (DCRCFAIL | DTIMEOUT);
     if (!err && txcount) {
        for (i = 0; i < 16; i++)
            *(u32 *)(base + FIFO) = *(up + i);
        up += 16;
        txcount -= 64;
        txbuf += 64;           // advance txbuf for next write
        status = *(u32 *)(base + STATUS); // read status
     }
     if (txcount == 0){
        do_command(12, 0, MMC_RSP_R1); // stop transmission
        txdone = 1;
     }
  }
  *(u32 *)(base + STATUS_CLEAR) = 0xFFFFFFFF;
```

```
}
  int get_block(int blk, char *buf)
{
  u32 cmd, arg;
  rxbuf = buf; rxcount = FBLK_SIZE; rxdone = 0;
  *(u32 *)(base + DATATIMER) = 0xFFFF0000;
  // write data_len to datalength reg
  *(u32 *)(base + DATALENGTH) = FBLK_SIZE;
  // 0x93=|9|0011|=|9|DMA=0,0=BLOCK,1=Host<-Card,1=Enable
  *(u32 *)(base + DATACTRL) = 0x93; // Card to Host

  cmd = 18;          // CMD18 = READ multi-sectors
  arg = (blk*FBLK_SIZE);
  do_command(cmd, arg, MMC_RSP_R1);

  while(rxdone == 0);
  printf("get_block return\n");
}

int put_block(int blk, char *buf)
{
  u32 cmd, arg;
  txbuf = buf; txcount = FBLK_SIZE; txdone = 0;
  *(u32 *)(base + DATATIMER) = 0xFFFF0000;
  *(u32 *)(base + DATALENGTH) = FBLK_SIZE;

  cmd = 25;        // CMD25 = Write multi-sectors
  arg = (u32)(blk*fBLK_SIZE);
  do_command(cmd, arg, MMC_RSP_R1);

  // 0x91=|9|0001|=|9|DMA=0,0=BLOCK,0=Host->Card,1=Enable
  *(u32 *)(base + DATACTRL) = 0x91; // Host to Card

  while(txdone == 0);
  printf("put_block return\n");
}
```

The reader may replace the SDC driver in the sample program C3.4 with the above code and test run the program to verify multi-sector data transfers. The reader may also change FBLK_SIZE to suit other block sizes, which must be a multiple of sector size (512).

3.8 Vectored Interrupts

So far, all the example programs use non-vectored interrupts. The disadvantage of non-vectored interrupts is that the interrupt handler must scan the interrupt status register for non-zero bits to determine the interrupt sources, which is time-consuming. In many other computer systems, such as the Intel x86 based PCs, interrupts are vectored by hardware. In the vectored interrupt scheme, each interrupt is assigned a vector number determined by the interrupt priority. When an interrupt occurs, the CPU can get the vector number of the interrupt from the interrupt controller hardware and uses it to invoke a corresponding interrupt service routine directly. The ARM PL190 Vectored Interrupt Controller (VIC) also has this capability. In this section, we show how to program the PL190 VIC for vectored interrupts processing.

3.8.1 ARM PL190 Vectored Interrupts Controller (VIC)

The PL190 VIC of the ARM Versatile/926EJ-S board supports vectored interrupts. The VIC technical manuals [ARM PL190 2004] contain the following information on how to program the VIC for vectored interrupts.

. VectorAddr Register (0x30): The VectorAddr register contains the ISR address of current active IRQ. At the end of current ISR, write a value to this register to clear the current interrupt. The PL192 VIC has the additional capability of prioritizing the IRQ sources by writing values 0–15 to the IntPriority registers. At the end of the current ISR, writing to the VectorAddr register allows the VIC to re-prioritize pending IRQs.
. DefaultVecAddr Register (0x34): This register contains the ISR address of a default interrupt, e.g. for any spurious interrupt.
. VectorAddress Registers [0–15] (0x100-0x13C): Each of these registers contains the ISR address of IRQ0 to IRQ15. The PL192 has 32 vectorAddress registers for 32 ISRs.
. VectorControl Registers [0–15] (0x200-0x23C): Each of these registers contains the interrupt source (bits 4–1) and an Enable bit (bit 5).

In order to use vectored interrupts, each device must be enabled for interrupts at both the device level and also on the VIC. We demonstrate vectored interrupts by the sample program C3.5 for the following devices.

```
. Timer0:      VectorInt0
. UART0:       VectorInt1
. UART1:       VectorInt2
. Keyboard:    VectorInt3
```

The code segments used to program the VIC for vectored interrupts are listed below.

3.8.2 Configure VIC for Vectored Interrupts

C3.5.1 : The function vecotrInt_init() configures the VIC for vectored interrupts.

```
int vectorInt_init() / Use vectored interrupts of PL190 VIC
{
  printf("vectorInterrupt_init()\n");

  /*********** set up vectored interrupts *****************
  (1). write to vectoraddr0 (0x100) with ISR of timer0
                vectoraddr1 (0x104) with ISR of UART0
                vectoraddr2 (0x108) with ISR of UART1
                vectoraddr3 (0x10C) with ISR of KBD
  **********************************************************/
  *((int *)(VIC_BASE_ADDR+0x100)) = (int)timer0_handler;
  *((int *)(VIC_BASE_ADDR+0x104)) = (int)uart0_handler;
  *((int *)(VIC_BASE_ADDR+0x108)) = (int)uart1_handler;
  *((int *)(VIC_BASE_ADDR+0x10C)) = (int)kbd_handler;
  // (2). write to intControlRegs  = E=1|IRQ#=1xxxxx
  *((int *)(VIC_BASE_ADDR+0x200)) = 0x24;  //100100 at IRQ 4
  *((int *)(VIC_BASE_ADDR+0x204)) = 0x2C;  //101100 at IRQ 12
  *((int *)(VIC_BASE_ADDR+0x208)) = 0x2D;  //101101 at IRQ 13
  *((int *)(VIC_BASE_ADDR+0x20C)) = 0x3F;  //111111 at IRQ 31
  // (3). write 0's to IntSelectReg to generate IRQ interrupts
  //(any bit=1 generates FIQ interrupt)
  *((int *)(VIC_BASE_ADDR+0x0C)) = 0;
}
```

3.8.3 Vectored Interrupts Handlers

C3.5.2. Rewrite IRQ_handler() for Vectored Interrupts. When using vectored interrupts, any IRQ interrupt still comes to the IRQ_handler() as usual. However, we must rewrite IRQ_handler() to use vectored interrupts. Upon entry to IRQ_handler(), we must read the VectorAddr register to acknowledge the interrupt first. Unlike the non-vectored interrupt case, there is no need to read the status registers to determine the interrupt source. Instead, we can get the address of the current IRQ handler from the vectorAddr register directly. Then simply invoke the handler by its entry address. In addition, the interrupt source can also be determined from the VectorStatus register. Upon return from the handler, send EOI to the VIC controller by writing a (any) value to the vectorAddr register, allowing it to re-prioritize pending interrupt requests. The following shows the modified IRQ_handler() function for vectored interrupts.

```
void IRQ_handler( )
{
  void *(*f)( );        // f as a function pointer
  int status = *(int *)(VIC_BASE_ADDR+0x30);
  f =(void *)*((int *)(VIC_BASE_ADDR+0x30));
  f();                  // call the ISR function
  *((int *)(VIC_BASE_ADDR+0x30)) = 1; // write to vectorAddr as EOI
}
```

3.8.4 Demonstration of Vectored Interrupts

C3.5.3. t.c file: For the sake of brevity, we only show the relevant code of main() that are used to test the various device interrupts.

```
int main()
{
   fbuf_init();  // LCD display
   kbd_init();   // KBD
   uart_init();  // UARTs
   timer_init(); // Timer
   up = &uart[0];
   kp = &kbd;
   /* enable timer0,1, uart0,1 SIC interrupts */
   VIC_INTENABLE = 0;
   VIC_INTENABLE |= (1<<4);   // timer0,1 at bit4
   VIC_INTENABLE |= (1<<12); // UART0 at bit12
   VIC_INTENABLE |= (1<<13); // UART1 at bit13
   VIC_INTENABLE |= (1<<31); // SIC to VIC's IRQ31
   /* enable KBD IRQ from SIC */
   SIC_INTENABLE = 0;
   SIC_INTENABLE |= (1<<3);   // KBD int=bit3 on SIC
   printf("Program start: test Vectored Interrupts\n");
   vectorInt_init(); // must do this AFTER driver_init()
   timer_start(0);    // start Timer0
   printf("test UART0 I/O: enter text from UART 0\n");
   while(1){
     ugets(up, line);
     uprintf("  line=%s\n", line);
     if (strcmp(line, "end")==0)
         break;
   }
```

```
QEMU
Program start: test Vectored Interrupts                00:00:20
vectorInterrupt_init()
timer_init()
timer_start 0  base=0x 101E2000
test uart0 I/O: enter text from UART 0 (end to exit)
rx interrupt: e uputc to UART0  e TX interrupt
rx interrupt: n uputc to UART0  n TX interrupt
rx interrupt: d uputc to UART0  d TX interrupt
   interrupt:
uputc to UART0     TX interrupt
uputc to UART0   e TX interrupt
uputc to UART0   n TX interrupt
uputc to UART0   d TX interrupt
uputc to UART0
                  TX interrupt
 TX interrupt
test KBD inputs (end to exit)
kbd interrupt: c=0x 65  e
kbd interrupt: c=0x 6E  n
kbd interrupt: c=0x 64  d
kbd interrupt: c=0x 0D <cr>
line=end
END OF run  1234
```

Fig. 3.10 Demonstration of vectored interrupts

```
   printf("test UART1 I/O: enter text from UART 1\n");
   up = &uart[1];
   while(1){
     ugets(up, line);
     ufprintf(up, "  line=%s\n", line);
     if (strcmp(line, "end")==0)
        break;
   }
   printf("test KBD inputs\n"); // print to LCD
   while(1){
     kgets(line);
     printf("line=%s\n", line);
     if (strcmp(line, "end")==0)
        break;
   }
   printf("END OF run %d\n", 1234);
}
```

Figure 3.10 shows the sample outputs of running the program C3.5, which demonstrates timer, UART and KBD drivers using vectored interrupts.

3.9 Nested Interrupts

3.9.1 Why Nested Interrupts

In the above example programs, C3.1–C3.5, which demonstrate interrupts processing, every IRQ interrupt handler begins execution with IRQ interrupts disabled (masked out). IRQ interrupts are not enabled (masked-in) until the interrupt handler has finished. This implies that interrupts can only be handled one at a time. The disadvantage of this scheme is that it may lead to **interrupt priority inversion**, in which processing a low priority interrupt may block or delay processing of higher priority interrupts. Interrupt priority inversion may increase the system response time to interrupts, which is undesirable in embedded systems with critical timing requirements. To remedy this, embedded systems should allow nested interrupts. In the nested interrupts scheme, a higher priority interrupt may preempt the processing of lower priority interrupts, i.e. before the current interrupt handler finishes, it can accept and handle interrupts of higher priorities, thereby reducing interrupts processing latency and improving the system response to interrupts.

3.9.2 Nested Interrupts in ARM

The ARM processor is not designed to support nested interrupts efficiently, due to the following properties of the ARM processor architecture.

When the ARM CPU accepts an IRQ interrupt, it switches to IRQ mode, which has its own banked registers lr_irq, sp_irq, spsr and cpsr. The CPU saves the return address (with an +4 offset) into lr_irq, saves the previous mode CPSR into spsr and enters the interrupt handler to handle the current interrupt. In order to support nested interrupts, the interrupt handler must unmask interrupts at some point in time to allow interrupts of higher priorities to occur. However, this creates two problems.

(1). Accepting another interrupt may corrupt the link register: Assume that, after enabling IRQ interrupts, the interrupt handler calls an ISR to handle the current interrupt, as in

```
irq_handler:
     // find out the IRQ source to determine its ISR
     // enable IRQ interrupts
     bl  ISR
HERE:
```

When calling the ISR to handle the current interrupt, the link register lr_irq contains the return address to the label HERE. While executing the ISR, if another interrupt occurs, the CPU would re-enter irq_handler, which changes lr_irq to the return address of the new interrupt point. This corrupts the original link register lr_irq, causing the ISR to return to the wrong address when it finishes.

(2). Over-writing saved CPSR: When the ARM processor accepts an interrupt, it saves the CPSR of the interrupted point in the (banked) SPSR_irq, The saved SPSR may be in USER or SVC mode if the interrupted code was executing in USER or SVC mode. While executing an ISR in IRQ mode, if another interrupt occurs the CPU would over-write the SPSR_irq with the CPSR in IRQ mode, which would cause the first ISR to return to the wrong mode when it finishes.

It is fairly easy to deal with the problem in (2). Upon entry to the interrupt handler but before enabling IRQ for further interrupts, we can save SPSR into the IRQ mode stack. When the ISR finishes, we restore the saved SPSR from the stack. However, if we allow nested interrupts in IRQ mode, there is no way to overcome the problem in (1). It will cause an infinite loop since every ISR would return to the beginning of the interrupt handler again. The only way to alleviate this problem is to steer the CPU away from IRQ mode. For this reason, ARM introduced the SYS mode, which is a privileged mode but has a different link register than the IRQ mode. Assume that, before enabling further IRQ interrupts, we switch the CPU to SYS mode and call an ISR in SYS mode. If another interrupt occurs, it would alter the IRQ mode link register lr_irq but not the link register lr_sys in SYS mode. This allows the ISR to return to the correct address when it finishes. So the scheme of handling nested interrupt is as follows.

3.9.3 Handle Nested Interrupts in SYS Mode

irq_handler: // entry point of IRQ interrupts
(1). Save context of interrupted point and lr into IRQ mode stack

(2). Save SPSR into IRQ mode stack
(3). Switch to SYS mode with IRQ disabled
(4). Determine interrupt source and clear interrupt.
(5). Enable IRQ interrupts
(6). Call ISR in SYS mode for the current IRQ interrupt
(7). Switch back to IRQ mode with IRQ disabled
(8). Restore saved SPSR form IRQ mode stack
(9). Return by the saved context and lr in IRQ mode stack

Many ARM processors require 8-byte aligned stacks. When changing the CPU to SYS mode, it may be necessary to check and adjust the SYS mode stack for proper alignment first. Since the SYS mode stack begins at an 8-byte boundary, we may assume that the checking and adjustment is unnecessary. The following list the irq_handler code which implements the above algorithm.

```
irq_handler:
  stmfd sp!, {r0-r3, r12, lr} // save context in IRQ stack
  mrs   r12, spsr             // copy spsr into r12
  stmfd sp!, {r12}            // save SPSR in IRQ stack
  msr cpsr, #0x9F          // switch to SYS mode with IRQ disabled
  ldr r1, =vectorAddr      // read vectorAddr reg to ack interrupt
  ldr r0, [r1]
  msr   cpsr, #0x1F        // enable IRQ in SYS mode

  bl    IRQ_handler        // handle current IRQ in SYS mode

  msr   cpsr, #0x92        // switch back to IRQ mode with IRQ disabled
  ldmfd sp!, {r12}         // get saved SPSR from IRQ stack
  msr   spsr, r12          // restore spsr
  ldr   r1, =vectorAddr    // write to vectorAddr as EOI
  str   r0, [r1]
  ldmfd sp!, {r0-r3, r12, lr} // restore saved context from IRQ stack
  subs pc, r14, #4            // return to interrupted point
```

ARM recommends handling nested interrupts in SYS mode (Nesting Interrupts 2011) but it can also be done in SVC mode, which is used in the demonstration program.

3.9.4 Demonstration of Nested Interrupts

C3.6.1. ts.s file: First, we show the modified irq_handler for nested interrupts. Instead of SYS mode, it handles nested interrupts in SVC mode. The get_cpsr() function returns the processor mode in CPSR. It is for displaying the current mode of the CPU.

```
.set vectorAddr, 0x10140030  // VIC vectorAddress register
irq_handler:
  stmfd sp!, {r12, lr}  // save r12, lr in IRQ stack
  mrs   r12, spsr       // copy spsr into r12
  stmfd sp!, {r12}      // save spsr in IRQ stack
  msr cpsr, #0x93       // switch to SVC mode with IRQ disabled
  stmfd sp!, {r0-r3, r12, lr} // save context in SVC mode stack
  ldr r1, =vectorAddr
  ldr r0, [r1]             // read VIC vectAddr to ACK interrupt
  // bl enterINT           // see comments
  msr cpsr, #0x13          // enable IRQ in SVC mode
```

```
  bl IRQ_handler          // handle IRQ in SVC mode
  ldmfd sp!, {r0-r3, r12, lr} // restore context from SVC stack
  msr cpsr, #0x92         // to IRQ mode with IRQ disabled
  ldmfd sp!, {r12}        // pop saved spsr
  msr spsr, r12           // restore spsr
  // bl exitINT           // see comment
  ldr  r1, =vectorAddr    // issue EOI to VIC
  str  r0, [r1]
  ldmfd sp!, {r12, r14}   // return
  subs pc, r14, #4
get_cpsr:                 // return CPSR
  mrs r0, cpsr
  mov pc, lr
```

C3.6.2. t.c file: The t.c file is the same as in C3.5, except for the added functions enterINT() and exitINT(). In the C3.5 program, which uses vectored interrupts, the interrupts priorities are (from high to low)

Timer0, UART0, UART1, KBD

Without nested interrupts, each interrupt is processed from start to finish without preemption. With nested interrupts, processing a low priority interrupt may be preempted by a higher priority interrupt. In order to demonstrate this, we add the following code to the C3.6 program.

(1). In the irq_handler code, before enabling IRQ interrupts and calling ISR for the current interrupt, we let the interrupt handler call enterINT(), which reads the VICstatus register to determine the interrupt source. If it's a KBD interrupt, it sets a volatile global inKBD flag to 1, clears a volatile global tcount to 0 and prints an enterKBD message. If it's a timer interrupt in the middle of handling a KBD interrupt, it increments tcount by 1.
(2). When an ISR returns to irq_handler, it calls exitINT(). If the current interrupt is from KBD, it prints the tcount value and resets tcount to 0.

The tcount value represents the number of high priority timer interrupts serviced while executing the low priority KBD handler. The reader may un-comment the statements labeled (1) and (2) in the irq_handler code to verify the effect of nested (timer) interrupts. The following lists the code of enterINT() and exitINT().

```
volatile int status, inKBD, tcount;

int enterINT()
{
  status = *((int *)(VIC_BASE_ADDR)); // read VICstatus register
  if (status & (1<<31)){ // KBD
     tcount = 0;
     inKBD = 1;
     printf("enterKBD ");
  }
  if ((status & (1<<4)) && (inKBD){ // timer0 inside KBD handler
     tcount++;              // tcount=number of timer interrupts
  }
}
int exitINT()
{
  if (status & 0x80000000){ // KBD
     printf("exitKBD=%d\n", tcount); // show tcount in KBD handler
     tcount=0;              // reset tcount to 0
  }
}
```

```
void IRQ_handler( )
{
  void *(*f)();                                  // f as a function pointer
  f =(void *)*((int *)(VIC_BASE_ADDR+0x30)); // read ISR address
  (*f)( );                                       // call the ISR function
  //*((int *)(VIC_BASE_ADDR+0x30)) = 1;       // write EOI to vectorAddr
}

int main()
{
   char line[128];
   UART *up; kp = &kbd;
   fbuf_init();
   kbd_init();
   uart_init();
   // enable timer0,1 uart0,1 SIC and KBD interrupts as in C8
   vectorInt_init(); // same as in C8
   timer_init();
   timer_start(0);

   printf("Program C3.6 start: test NESTED Interrupts\n");
   up = &uart[0];
   printf("test uart0 I/O: enter text from UART 0\n");
   while(1){
     ugets(up, line);
     uprintf("  line=%s\n", line);
     if (strcmp(line, "end")==0)
      break;
   }
   printf("test UART1 I/O: enter text from UART 1\n");
   up = &uart[1];
   while(1){
     ugets(up, line);
     ufprintf(up, "  line=%s\n", line);
     if (strcmp(line, "end")==0)
      break;
   }
   printf("test KBD inputs\n"); // print to LCD
   while(1){
      kgets(line);
      printf("line=%s\n", line);
      if (strcmp(line, "end")==0)
         break;
   }
   printf("END OF run %d\n", 1234);
}
```

Figure 3.11 shows the outputs of running the example program C3.6, which demonstrates nested interrupts. As the figure shows, many timer interrupts may occur while handling a single KBD interrupt.

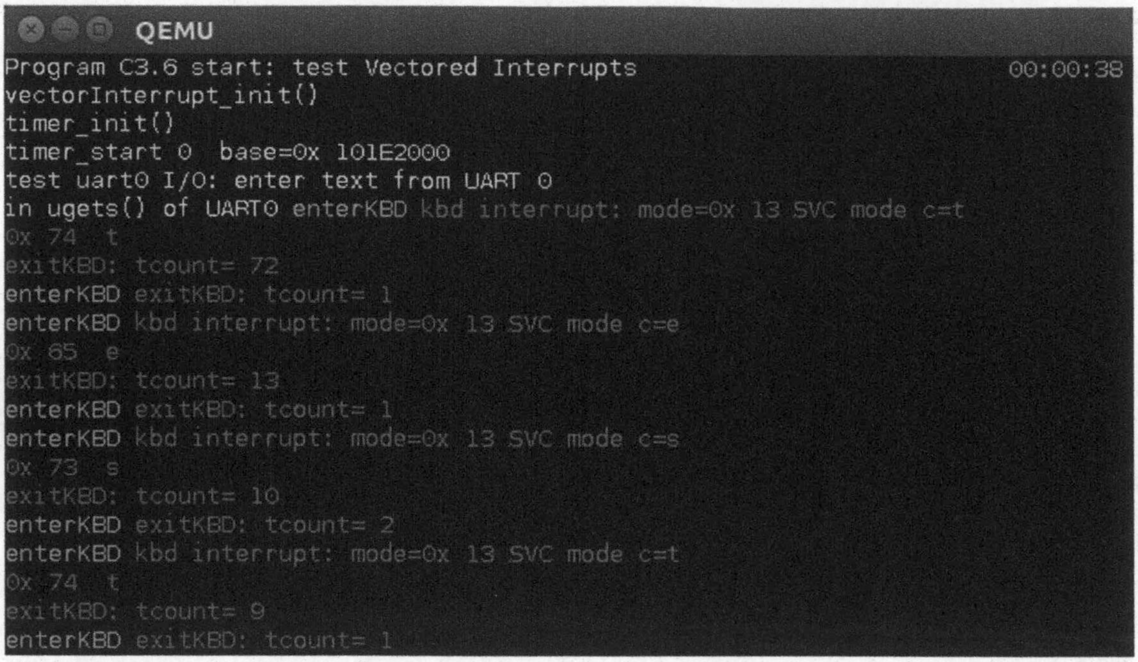

```
QEMU
Program C3.6 start: test Vectored Interrupts                    00:00:38
vectorInterrupt_init()
timer_init()
timer_start 0  base=0x 101E2000
test uart0 I/O: enter text from UART 0
in ugets() of UART0 enterKBD kbd interrupt: mode=0x 13 SVC mode c=t
0x 74  t
exitKBD: tcount= 72
enterKBD exitKBD: tcount= 1
enterKBD kbd interrupt: mode=0x 13 SVC mode c=e
0x 65  e
exitKBD: tcount= 13
enterKBD exitKBD: tcount= 1
enterKBD kbd interrupt: mode=0x 13 SVC mode c=s
0x 73  s
exitKBD: tcount= 10
enterKBD exitKBD: tcount= 2
enterKBD kbd interrupt: mode=0x 13 SVC mode c=t
0x 74  t
exitKBD: tcount= 9
enterKBD exitKBD: tcount= 1
```

Fig. 3.11 Demonstration of nested interrupts

3.10 Nested Interrupts and Process Switch

The ARM scheme of handling nested interrupts in SYS or SVC mode works only if every interrupt handler returns to the original point of interruption, so that no process switch occurs after an interrupt. It would not work if an interrupt may cause context switch to a different process. This is because part of the context belonging to the switched out process still remains in the IRQ stack, which may be over-written when the new process handles another interrupt. If this happens, the switched out process would never be able to resume running again due to corrupted or lost execution context. In order to prevent this, the IRQ stack contents must be transferred to the SVC stack of the switched out process (and reset the IRQ stack pointer to prevent it from growing out of bounds). A possible way to avoid transferring stack contents is to allocate a separate IRQ stack for each process. But this would require a lot of memory spaces dedicated to processes as IRQ stacks, which basically nullifies the advantage of using a single IRQ stack for interrupt processing. Also, in the ARM architecture it is not possible to use the same memory area as both the SVC and IRQ stacks of a process due to the separate stack pointers in SVC and IRQ modes. These necessitate transferring IRQ stack contents during context switch, which seems to be an inherent weakness of the ARM processor architecture in terms of multitasking.

3.11 Summary

Interrupts and interrupts processing are essential to embedded systems. This chapter covers exceptions and interrupts processing. It describes the operating modes of ARM processors, exception types and exception vectors. It explains the functions of interrupt controllers and the principles of interrupts processing in detail. Then it applies the principles of interrupts processing to the design and implementation of interrupt-driven device drivers, which include drivers for timers, keyboard, UART and SD cards, and demonstrates the device drivers by example programs. It explained the advantages of vectored interrupts, showed how to configure the VIC for vectored interrupts, and demonstrated vectored interrupts processing. It also explained the principles of nested interrupts and demonstrated nested interrupts processing by example programs.

List of Sample Programs

C3.1: Timer driver
C3.2: KBD driver
C3.3: UART driver
C3.4: SDC driver: R/W sector
    Improved SDC driver: R/W 64 long per interrupt
    Multi-sector R/W for file BLKSIZE
C3.5: Vectored interrupts for Timer, UART, KBD
C3.6: Nested interrupts for KBD and Timer

Problems

1. In the example program C3.1, the vector table is at the end of the ts.s file.
(1). In the ts.s file, comment out the line

```
BL copy_vector
```

Re-compile and run the program again. The program should NOT work. Explain why?

(2). Move the vector table to the beginning of the ts.s file, as in

```
vectors_start:
    // vector table
vectors_end:
reset_handler:
```

Change the entry point in t.ld to vectors_start. Recompile and run the program again. It should also work. Explain why?

2. In the example 3.2, the irq_handler save all the registers in the IRQ stack. Modify it to save a MINIMUM number of registers. Figure out which registers must be saved to make the program still work.
3. Modify the KBD driver in program C3.2 to support upper case letters, and special control keys, Control-C and Control-D.
4. Modify the UART driver program C3.3 to support UART2 and UART3.
5. Modify the UART driver program C3.3 to support internal FIFO buffers of the UART.
6. The ARM VIC interrupt controller assigns fixed IRQ priorities in the order of IRQ0 (high) to IRQ31 (low). Vectored interrupts allows reordering of the IRQ priorities. In the example program C3.5, the priorities of vectored interrupts are assigned in their original order. Modify vectorInt_init() to assign different vectored interrupt priorities, e.g. in the order of KBD, UART1, UART0, timer0, from high to low. Test whether vectored interrupts still work. Discuss the implications of such an assignment of interrupt priorities.
7. The Example program C3.6 handles nested interrupts in SVC mode. Modify it to handle nested interrupts in SYS mode, as suggested by ARM.
8. In the example program C3.6, which supports nested interrupts, there are two lines in the irq_handler:

```
LDR  r1, =vectorAddr
LDR  r0, [r1]      // read VIC vectAddr to ACK interrupt
```

Comment out these lines to see what would happen and explain why?

9. In embedded systems, the SD card is often used as a booting device for booting up an operating system. During booting, the booter code may read sectors from the SD card by polling since it must wait for data. Instead of using interrupts, rewrite the SD driver using polling.
10. Rewrite the SD driver by using DMA to transfer large amounts of data.

References

ARM Architecture: https://en.wikipedia.org/wiki/ARM_architecture, 2016.
ARM Processor Architecture: http://www.arm.com/products/processors/instruction-set-architectures, 2016.
ARM PL011: ARM PrimeCellUART (PL011) Technical Reference Manual, http://infocenter.arm.com/help/topic/com.arm.doc.ddi0183, 2005.
ARMPL180: ARM PrimeCell Multimedia Card Interface (PL180) Technical Reference Manual, ARM Information Center, 1998.
ARMPL181: ARM PrimeCell Multimedia Card Interface (PL181) Technical Reference Manual, ARM Information Center, 2001.
ARM PL190: PrimeCell Vectored Interrupt Controller (PL190), http://infocenter.arm.com/help/topic/com.arm.doc.ddi0181e/DDI0181.pdf, 2004.
ARM Timers: ARM Dual-Timer Module (SP804) Technical Reference Manual, Arm Information Center, 2004.
ARM PL050: ARM PrimeCell PS2 Keyboard/Mouse Interface (PL050) Technical Reference Manual, ARM Information Center, 1999.
Nesting Interrupts: Nesting Interrupts, ARM Information Center, 2011.
SD specification: Simplified Version of SD Host Controller Spec, https://www.sdcard.org/developers/overview/host_controller/simple_spec.
SDC: Secure Digital cards: SD Standard Overview - SD Association https://www.sdcard.org/developers/overview, 2016.
SPI: Serial Peripheral Interface, https://en.wikipedia.org/wiki/Serial_Peripheral_Interface_Bus, 2016, 2016.
Raspberry_Pi: https://www.raspberrypi.org/products/raspberry-pi-2-model-b, 2016.
Silberschatz, A., P.A. Galvin, P.A., Gagne, G, "Operating system concepts, 8th Edition", John Wiley & Sons, Inc. 2009.
Stallings, W. "Operating Systems: Internals and Design Principles (7th Edition)", Prentice Hall, 2011.
Wang, K.C., "Design and Implementation of the MTX Operating Systems, Springer International Publishing AG, 2015.

4 Models of Embedded Systems

4.1 Program Structures of Embedded Systems

In the early days, most embedded systems were designed for specific applications. An embedded system usually consists of a microcontroller, which is used to monitor a few sensors and generate signals to control a few external devices, such as to turn on LEDs or activates relays and servo motors to control a robot, etc. For this reason, control programs of early embedded systems are also very simple. They are written in the form of either a super-loop or event-driven program structure. However, as computing power and demand for multi-functional systems increase in recent years, embedded systems have undergone a tremendous leap in both applications and complexity. To cope with the ever increasing demands for extra functionality and the resulting system complexity, the super-loop and event driven program structures are no longer adequate. Modern embedded systems need more powerful software. As of now, many embedded systems are in fact high-powered computing machines capable of running full-fledged operating systems. A good example is smart phones, which use the ARM core with gig bytes internal memory and multi-gig bytes micro SD card for storage and run adapted versions of Linux, such as Android (Android 2016). The current trend in embedded systems design is clearly moving in the direction of developing multi-functional operating systems suitable for the mobile environment. In this chapter, we shall discuss the various program structures and programming models that are suitable for current and future embedded systems.

4.2 Super-Loop Model

A super-loop is a program structure composed of an infinite loop, with all the tasks of the system contained in the loop. The general form of a super-loop program is

```
main()
 {
    system_initialization();
    while(1){
       Check_Device_Status();
       Process_Device_Data();
       Output_Response();
    }
 }
```

After system initialization, the program executes an infinite loop, in which it checks the status of a system component, such as that of an input device. When the device indicates there are input data, it collects the input data, processes the data and generates outputs as response. Then it repeats the loop.

K.C. Wang, *Embedded and Real-Time Operating Systems*,
DOI 10.1007/978-3-319-51517-5_4

4.2.1 Super-Loop Program Examples

We illustrate the super-loop program structure by examples. In the first example program, denoted by C4.1, we assume that an embedded system controls an UART for I/O. Our goal here is to develop a control program which continually checks whether there is any input from the UART port. Whenever a key is pressed, it gets the input key, processes the input and generates an output, such as to turn on an LED, flip a switch, etc. When running the program on an emulated ARM virtual machine, which does not have any LED or switch, we shall simply echo the input key to simulate processing the input and generating an output response.

In the program, it checks a UART for inputs. For each alphabetical key in lowercase, it converts the key to uppercase and displays the key to the UART port. In addition, it also handles the return key by outputting a new line char to produce the right visual effects. The program's assembly code is the same as those in Chap. 3. We only show the C code of the program.

Example Program C4.1: Super-loop Program

```
/************* t.c file of a super-loop program C4.1 *************/
#define UDR  0x00
#define UFR  0x18
char *ubase;
int main()
{
   char c;
   ubase = (char *)0x101F1000;        // 1. initialize UART0 base address
   while(1){
      if (*(ubase + UFR) & 0x40){    // 2. check UART0 for RxFull
        c = *(ubase + UDR);          // 3. get UART0 input key
        if (c > = 'a' && c <= 'z')  //   if lowercase key
            c += ('A' - 'a');        //   convert to uppercase
        *(ubase + UDR) = c;          // 4. generate output
        if (c == '\r')
            *(ubase + UDR) = '\n';
      }
   }
}
```

When running the C4.1 program on an ARM virtual machine under QEMU, it echoes each alphabetical key to UART0 in uppercase. Instead of a single device, the program can be generalized to monitor and control several devices, all in the same loop. We demonstrate this technique by the next program, C4.2, which monitors and controls two devices, a UART and a keyboard.

Example Program C4.2: The program monitors and controls 2 devices in a super-loop.

```
/*********** t.c file of Program C4.2 ***********/
#include "keymap"
#define KSTAT 0x04
#define KDATA 0x08
#define UDR   0x00
#define UFR   0x18
char *ubase,*kbase;
int uart_init(){ ubase = (char *)0x101F1000; }
int kbd_init() { kbase = (char *)0x10006000; }
int main()
{
   char c, scode;
   uart_init(); kbd_init();           // 1. initialization
```

```
while(1){                                    // super-loop
  if (*(ubase + UFR) &  0x40){               // 2. if UART RxFull
     c = *(ubase + UDR);                     // 3. handle UART input
     if (c >= 'a' && c <= 'z')
         c += ('A' - 'a');
     *(ubase + UDR) = c;
     if (c == '\r')
        *(ubase + UDR) = '\n';
  }
  if (*(kbase + KSTAT) & 0x10){  // 4. if KBD RxFull
    scode = *(kbase + KDATA);    // 5. handle KBD key (press)
    if (scode & 0x80)
       continue;
    c = unsh[scode];
    *(ubase + UDR) = c;                      // echo KBD key to UART
    if (c == '\r')
       *(ubase + UDR) = '\n';
  }
 }
}
```

When running the program C4.2, it echoes UART0 inputs in uppercase, and keyboard inputs in lowercase, all to UART0.

4.3 Event-Driven Model

4.3.1 Shortcomings of Super-Loop Programs

The drawback of the super-loop program model is that the program must continually check the status of each and every device, even if the device is not ready. This not only wastes CPU time but also causes excessive power consumption. In an embedded system, it is often more desirable to reduce power consumption than to increase CPU utilization. Rather than continually checking the status of every device, an alternative way is to wait for the device to become ready. For example, in the program C4.1, we may replace the checking device status statements with a busy-wait loop, as in

while(device_has_no_data);

But this does not remedy the problem since the CPU is still continually executing the busy-wait loop. Another drawback of this scheme is that the program would be unable to respond to any KBD inputs while it is waiting for UART inputs, and vice versa.

4.3.2 Events

In a programming environment, an event is something that is generated by a source and recognized by a recipient, causing the latter to take action to handle the event. Instead of continually checking for inputs, an embedded system can be designed to be event-driven, i.e. it takes action only in response to events. For this reason, **event-driven systems** (Cheong et al. 2003; Dunkels et al. 2006) are also called **reactive systems**. Events can be synchronous, i.e. they occur in a predictable manner, or asynchronous, i.e. they may occur at any time and in any order. Examples of synchronous events are periodic events from a timer, e.g. when the timer count has reached a certain value. Examples of asynchronous events are user inputs, such as press a key, click a mouse button and flip a switch, etc. Because of their unpredictability, the simple super-loop program structure is unsuited to dealing with asynchronous events. In the event-driven programming model, the main program may executes in a loop or sits in an idle state, waiting for any event to occur. When an event occurs, an event catcher recognizes the event and notifies the main program, causing it to take an appropriate action to handle the event. In an embedded system, events are usually associated with interrupts from hardware devices. In this case, an **event-driven program** becomes a simple

interrupt-driven program. We illustrate the interrupt-driven program structure by two examples. The first example handles periodic events and the second example handles asynchronous events.

4.3.3 Periodic Event-Driven Program

In this example, we assume that an embedded system consists of a timer and a display device, e.g. an LCD. The timer is programmed to generate 60 interrupts per second. The system must react to the following periodic timing events: On each second, it displays a wall clock in hh:mm:ss format to the LCD. Every 5 s, it displays a message string, also to the LCD. There are two possible ways to implement a control program that meets these periodic timing requirements. If the tasks to be performed are short, they can be performed by the timer interrupt handler directly. In this case, after initialization, the main program can execute an idle loop. To reduce power consumption, the idle loop can use Wait-For-Interrupt (WFI) or equivalent instruction, which puts the CPU into a power-saving state, waiting for interrupts. The ARM926EJ-S board does not support the WFI instruction but most ARM Cortex-5 processors (ARM Cortex-5 2010) implement a WFI mode by writing to the coprocessor CP15. The following shows the program code of the first version of Example C4.3. The tasks performed by the timer interrupt handler are shown in bold faced lines.

Version 1 of C4.3:

```
/************* timer.c file *****************/
typedef volatile struct timer{
  u32 *base;              // timer's base address
  int tick, hh, mm, ss; // timer data area
  char clock[16];
}TIMER;
volatile TIMER timer;    // timer structure

void timer_init()
{// initialize timer0 to generate 60 interrupts per second
  // timer.clock=00:00:00
}

void timer_handler() {
  TIMER *t = &timer;
  t->tick++;             // assume 60 ticks per second
  if (t->tick == 60){ // update ss, mm, hh
     t->tick = 0; t->ss++;
     if (t->ss == 60){
        t->ss = 0; t->mm++;
        if (t->mm == 60){
           t->mm = 0; t->hh++; t->hh %= 24;
        }
     }
  }
  if (t->tick == 0){ // on each second, display a wall clock
     for (i=0; i<8; i++)
        unkpchar(t->clock[i], 0, 70+i);
     t->clock[7]='0'+(t->ss%10); t->clock[6]='0'+(t->ss/10);
     t->clock[4]='0'+(t->mm%10); t->clock[3]='0'+(t->mm/10);
     t->clock[1]='0'+(t->hh%10); t->clock[0]='0'+(t->hh/10);
     for (i=0; i<8; i++)
        kpchar(t->clock[i], 0, 70+i); // kputchr(char, row, col)
  }
```

```
  if ((t->ss % 5) == 0) // every 5 seconds, display string
     printf("5 seconds event\n");
  timer_clearInterrupt();
}
/*************** t.c file ********************/
#include "timer.c"
#include "vid.c"
#include "interrupts.c"
void IRQ_handler()
{
  int vicstatus;
  // read VIC SIV status registers to find out which interrupt
  vicstatus = VIC_STATUS;
  if (vicstatus & (1<<4)){    // timer0=bit4
     timer_handler();
  }
}
int main()
{
   fbuf_init();                // LCD driver
   VIC_INTENABLE = (1<<4);  // timer0 enable=bit4
   printf("Timer Event Program start\n");
   timer_init();
   timer_start();
   printf("Enter while(1) loop\n", 1234);
   while(1){
     asm("MOV r0, #0; MCR p15,0,R0,c7,c0,4"); // CPU enter WFI state
     printf("CPU out WFI state\n");
   }
}
```

In general, an interrupt handler should be as short as possible. If the periodic tasks are long, e.g. longer than a timer tick, it is undesirable to perform the timer dependent tasks inside the interrupt handler, unless the system supports nested interrupts. As shown in Chap. 3, the ARM architecture can not handle nested interrupts directly. In this case, it would be better to let the main program perform all the tasks. As before, the main program executes in a loop. When there are no events from the timer, it enters a power-saving state, waiting for the next interrupt. When a timer event occurs, the timer interrupt handler simply sets a global volatile flag variable. After waking up from the power-saving state, the main program can check the flag variables to take appropriate actions.

In the second version of the example program C4.3, the task of displaying the wall clock is re-written as a function, which is executed by the main program on each second. The following shows the second version of the C4.3 program.

Version 2 of C4.3: main program perform periodic tasks.

```
/************ new timer.c file **********/
volatile int one_second=0; five_seconds=0; // flags
void timer_handler() {
   TIMER *t = &timer;
   t->tick++;
   if (t->tick == 60){
      t->tick = 0; t->ss++;
      if (t->ss == 60){
         t->ss = 0; t->mm++;
```

```
            if (t->mm == 60){
                t->mm = 0; t->hh++;
            }
        }
    }
    if (t->tick == 0)           // on each second
        one_second = 1;         // turn on one_second flag
    if ((t->ss % 5)==0)         // every 5 seconds
        five_seconds = 1;       // turn on five_seconds flag
    timer_clearInterrupt();
}

char clock[16] = "00:00:00";
/***************** t.c file ****************/
int wall_clock(TIMER *t)
{
    int i, ss, mm, hh;
    int_off();
     ss = t->ss; mm = t->mm; hh = t->hh; // copy from timer struct
    int_on();
    for (i=0; i<8; i++)
        unkpchar(clock[i], 0, 70+i);
    clock[7]='0'+(ss%10); clock[6]='0'+(ss/10);
    clock[4]='0'+(mm%10); clock[3]='0'+(mm/10);
    clock[1]='0'+(hh%10); clock[0]='0'+(hh/10);
    for (i=0; i<8; i++)
        kpchar(clock[i], 0, 70+i); // kputchr(char, row, col)
}
int main()
{
    fbuf_init();
    VIC_INTENABLE = (1<<4);   // timer0 at bit4
    printf("C4.3: Periodic Events Program\n");
    timer_init();
    timer_start();
    while(1){
      if (one_second){
        wall_clock(&timer);
        one_second = 0;
      }
      if (five_seconds){
        printf("five seconds event\n");
        five_seconds = 0;
      }
      asm("MOV r0, #0; MCR p15,0,R0,c7,c0,4"); // enter WFI mode
      printf("CPU come out WFI state\n");
    }
}
```

Figure 4.1 shows the outputs of running the C4.3 program, which demonstrates periodic events.

```
QEMU
C4.3: Periodic Events Program                                    00:00:06
timer_init()
timer_start: base=0x 101E2000
Enter while(1) loop
CPU come out WFI state
CPU come out WFI state
CPU come out WFI state
CPU come out WFI state
five seconds event
CPU come out WFI state
CPU come out WFI state
```

Fig. 4.1 Periodic Event-Driven Program

4.3.4 Asynchronous Event-Driven Program

Asynchronous events are non-periodic in nature, which may occur at any time and in any order. The next program, denoted by C4.4, demonstrates non-periodic or asynchronous events. In this example, we assume that the program monitors and controls two input devices; a UART and a keyboard. The main program tries to get an input line form either the UART or the KBD and echoes the line. The program is a condensed version of the example program C3.3 of Chap. 3, which implements interrupt-driven UART and KBD drivers. First, it initializes the (volatile) global flag variables, uline and kline to 0. Then it repeatedly checks the flag variables, which will be set by the interrupt handlers. When keys are pressed on the UART terminal or the keyboard, the UART or KBD interrupt handler will get the keys, echo them and enter them into an input buffer. When an ENTER key is pressed on either device, the interrupt handler turns on the corresponding flag variable to signal the occurrence of an event, allowing the main program to continue. When the main program detects an event, it extracts a line from the device driver's input buffer and echoes it to the LCD (for KBD inputs) or the UART. Then, it clears the flag variable and continues the loop. Due to the asynchronous nature of events, it is necessary to disable interrupts when clearing the flag variables in order to prevent race conditions between the main program and the interrupt handler.

```
/** Condensed uart.c, kbd.c and t.c files of Program C4.4. **/
volatile int uline=0, kline=0; // GLOBAL flag variables

void uart_handler(UART *up)
{
  u8 mis = *(up->base + MIS);    // read UART MIS register
  if (mis & 0x10) do_rx(up);
  else            do_tx(up);
}
int do_rx(UART *up)
{
  // get and echo input char c; enter c into inbuf[ ](same as before)
  if (c=='\r'){  // has an input line now
     uprintf("UART interrupt handler: turn on uline flag!\n");
     uline = 1;
  }
}
int do_tx(UART *up){ // output a char from outbuf[ ]; }
int ugetc(UART *up){ // return a char from inbuf[]; }
int uputc(UART *up, char c){ // enter char into outbuf[]; }
int ugets(UART *up, char *s){// get a line from inbuf[]; }
int uprints(UART *up, char *s) { // print a line; }
int uprintf(char *fmt, ...){// formatted print to UART; }

// kbd.c file
int kbd_init(){ // KBD initialization }
```

```
void kbd_handler()
{
  // get and echo KBD key press, enter key into buf[ ](same as before)
  if (c=='\r'){
     kprintf("KBD interrupt handler: turn on kline flag!\n");
     kline = 1;
  }
}
/************ t.c file of main() function *************/
#include "uart.c"           // UART driver
#include "kbd.c"            // KBD driver
#include "vid.c"            // LCD display driver
#include "exceptions.c"

void IRQ_handler()
{ // determine IRQ source; invoke UART or KBD interrupt handler; }

int main()
{
   char line[128];
   fbuf_init();
   uart_init(); // set UART base and enable UART interrupts
   kbd_init();  // set KBD base and enable KBD interrupts
   // code to enable UART and KBD interrupts on VIC and SIC
   while(1){// main program: check event flags; handle events
     if (uline){
        ugets(line);
        uprintf("UART: line=%s\n", line);
        lock(); uline = 0; unlock(); // reset uline to 0
     }
     if (kline){
        gets(line);
        color = GREEN;
        printf("KBD: line=%s\n", line);
        lock(); kline = 0; unlock(); // reset kline to 0
     }
     asm("MOV r0, #0; MCR p15,0,R0,c7,c0,4"); // enter WFI mode
   }
}
```

Figure 4.2 shows the outputs of running the C4.4 program, which demonstrates asynchronous events.

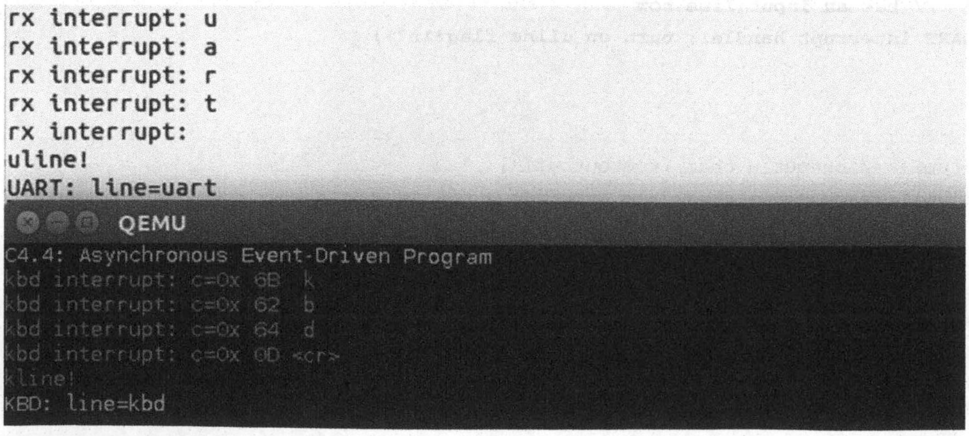

```
rx interrupt: u
rx interrupt: a
rx interrupt: r
rx interrupt: t
rx interrupt:
uline!
UART: line=uart
QEMU
C4.4: Asynchronous Event-Driven Program
kbd interrupt: c=0x 6B  k
kbd interrupt: c=0x 62  b
kbd interrupt: c=0x 64  d
kbd interrupt: c=0x 0D <cr>
kline!
KBD: line=kbd
```

Fig. 4.2 Asynchronous Event-Driven Program

4.4 Event Priorities

In an event-driven system, some events may be more urgent than others. Events can be assigned different priorities according to their urgency and importance. Events should be handled in accordance with their priority order. There are several ways to prioritize the events. First, interrupt related events can be assigned different priorities by an interrupt controller. The event processing order of the main program should be consistent with the interrupts priorities. Second, event handlers can be implemented as independent execution entities called **processes** or **tasks**, which can be scheduled to run by priority. We shall explain the process programming model in the next section. Here, we briefly justify the need for processes first. The example programs of C4.3 and C4.4 have three main shortcomings. First, for fast response to interrupts, interrupt handlers should be as short as possible. This is especially important to timer interrupts in order not to lose timer ticks or require nested interrupts. Even for periodic events, they should be handled by the main program, not in the timer interrupt handler. Second, while in the power-saving state, the main program still needs to come up to execute the loop on each and every interrupt, even though the awaited events, e.g. a complete input line, have not occurred yet. It would be better if the program can come out of the power-saving state only when needed. This way, it does not have to poll every event when it runs. Third, events are not limited to user inputs or device interrupts. They may originate from other execution entities in the system as a means of synchronization and communication. In order to accomplish these, it is necessary to incorporate the notion of processes or tasks into the system. This leads us to the process model for embedded systems.

4.5 Process Models

In the process model, an embedded system comprises many concurrent processes. A process is an execution entity which can be scheduled to run, suspended from running (and yield CPU to other processes), and resumed to run again, etc. Each process is an independent execution unit designed to perform a specific task. Depending on the execution environment of the processes, the process model can be classified into several sub-models.

4.5.1 Uniprocessor Process Model

A uniprocessor (UP) system consists of only one CPU. In a UP system, processes run on the same CPU concurrently through multitasking.

4.5.2 Multiprocessor Process Model

A Multiprocessor (MP) system consists of a multiple number of CPUs, including multi-core processors. In a MP system, processes may run on different CPUs in parallel. In addition, each CPU or processor core may also run processes through multitasking. MP systems will be covered in Chap. 9.

4.5.3 Real Address Space Process Model

In the real address space model, the system either is not equipped with, or does not utilize, the memory management hardware due to timing constraints. Without memory management hardware to provide address mapping, all processes run in the same real address space of the system kernel. The drawback of this model is the lack of memory protection. Its main advantages are simplicity, less hardware resource requirements and high efficiency.

4.5.4 Virtual Address Space Process Model

In the virtual address space model, the system uses memory management hardware to provide each process with a unique virtual address space through address mapping. Processes may run in either kernel mode or user mode. While in kernel mode, all processes share the same address space of the kernel. While in user mode, each process has a distinct virtual address space that is isolated and protected from other processes.

4.5.5 Static Process Model

In the static process model, all processes are created when the system starts, and they remain in the system permanently. Each process may be periodic or event-driven. Process scheduling is usually by static process priority without preemption, i.e. each process runs until it gives up the CPU voluntarily.

4.5.6 Dynamic Process Model

In the dynamic process model, processes can be created dynamically to perform specific tasks on demand. When a process completes its task, it terminates and releases all the resources back to the system for reuse.

4.5.7 Non-preemptive Process Model

In the non-preemptive process model, each process runs until it gives up the CPU voluntarily, e.g. when a process goes to sleep, suspends itself or explicitly yields CPU to another process.

4.5.8 Preemptive Process Model

In the preemptive process model, CPU can be taken away from a process to run another process at any time.

The above classifications of process models are not all mutually exclusive. Depending on the application, an embedded system may be designed as a mixture of appropriate process models. For instance, most existing embedded systems can be classified as the following types.

4.6 Uniprocessor (UP) Kernel Model

In this model, the system has only one CPU with no memory management hardware for address mapping. All processes run in the same address space of a kernel. Processes can be either static or dynamic. Processes are scheduled by static priority without preemption. Most simple embedded systems fit this model. The resulting system is equivalent to the non-preemptive kernel of an operating system. We shall discuss this system model in more detail in Chap. 5.

4.7 Uniprocessor (UP) Operating System Model

This is an extension of the UP Kernel model. In this model, the system uses memory management hardware to support address mapping, thus providing each process with a unique virtual address space. Each process runs in either kernel mode or a separate user mode. While in kernel mode, all processes run in the same address space of the kernel. While in user mode, each process executes in a private address space, which is isolated and protected from other processes. Processes share common data objects only in the protected kernel space. While in kernel mode, a process runs until it gives up the CPU voluntarily without preemption. While in user mode, a process can be preempted to yield CPU to other process of higher priority. Such a system is equivalent to a general purpose UP operating system. We shall discuss general purpose OS in more detail in Chap. 7.

4.8 Multiprocessor (MP) System Model

In this model, the system consists of a multiple number of CPUs or processor cores, which share the same physical memory. In a MP system, processes may run on different CPUs in parallel. Compared with UP systems, MP systems require advanced concurrent programming techniques and tools for process synchronization and protection. We shall discuss MP system in Chap. 9.

4.9 Real-Time (RT) System Model

In general, all embedded systems are designed with some timing requirements, such as quick responses to interrupts and short interrupt processing completion time, etc. An embedded system may only use these timing requirements as guidelines but it does not guarantee that the requirements are always attainable. In contrast, an embedded system intended for real-time applications must meet very stringent timing requirements, such as guaranteed minimal response time to interrupts and completing every requested service within a prescribed time limit. This environment is equivalent to a real-time system. We shall discuss real-time embedded systems in Chap. 10.

4.10 Design Methodology of Embedded System Software

As embedded systems becoming ever more complex, the traditional software design for embedded systems by the ad hoc approach is no longer adequate. As a result, there are many formal design methodologies proposed for embedded system software design, which include

4.10.1 High-Level Language Support for Event-Driven Programming

This design approach focuses on using the events and exceptions support of high-level programming languages, such as JAVA and C++, as a model to develop event-driven programs for embedded systems. A representative work in this area is the task model for event-driven programming (Fischer et al. 2007).

4.10.2 State Machine Model

This design method treats embedded system software as a finite state machine (FSM) (Edwards et al. 1997; Gajski et al. 1994). A **finite state machine (FSM)** is a system

```
FSM = {S, X, O, f}, where S = a finite set of states,
                          X = a finite set of inputs,
                          O = a finite set of outputs,
                          f is a state transition function, which maps S X I into S X O.
```

For each pair of (state, input) = (s, x), f(s, x) = (s', o), where s' is the next state of s, and o is the output generated during the state transition. A FSM is **fully specified** if f(s, x) is defined for every pair of (s, x). A FSM is **deterministic** if, for every pair of (s, x), f(s, x) is unique. A FSM is of the **Mealy model** (Katz and Borriello 2005) if the outputs depend on inputs. A FSM is of the **Moore model** if the outputs depend only on states.

The state machine design method models the specifications of embedded systems by fully specified and deterministic Mealy model FSMs, which allow for formal verification of the resulting systems. It also exploits programming language features to translate state machines into program code. We illustrate the state machine design model by an example.

Example Program C4.5: Assume that comment lines in C programs begin with two adjacent / symbols and end on the same line. Design an embedded system which takes C program source files as inputs and removes comment lines from the C programs. Design and implementation of such a system based on the FSM model consists of three steps.

Step 1: Construct a FSM State Table: The system can be modeled by a FSM with 5 states.

S0 = initial state, which has not seen any input
S1 = has not seen any / symbol yet
S2 = has seen the first / symbol
S3 = has seen 2 adjacent // symbols
S4 = final or termination state

Although each input is a single char, we shall classify the input chars into different cases, which are treated as distinct inputs to the system. Thus, we define inputs as

x1 = '/'
x2 = '\n'
x3 = not in {'/', \n', EOF}
x4 = EOF (end-of-file)

While the system is in a state, each input causes a state transition to a next state and generates an output (string). A FSM can be represented by a **state table**, which specifies the state transitions and outputs due to each input. For this example, the initial state table of the FSM is shown in Table 4.1, in which the output symbol - denotes null string.

In the state table, S0 is the initial state, which represents the condition that the system has not seen any input yet, and S4 is the final or termination state, in which the system has completed its task and halts. The initial state table is constructed in accordance with the problem specification. Starting from the initial state S0, if the input is a '/', it goes to state S2, which represents the condition that the system has seen the first '/', and generates a null output string. This is because this '/' may be the start of a comment line. If so, it should not be emitted as part of the output. If the input is a '\n', it goes to state S1 and generates an output string "\n". If the input is not '/' or '\n' or EOF, it goes to S1 also and generates an output string containing the same input char. If the input is EOF, it goes to the final state S4 with a null output string and terminates. While in the state S2, if the input x is not a '/' or\n or EOF, it goes back to S1 and generates the output string "/x". This is because a '/' followed by an ordinary char is not a comment line, which must be part of the output string. Other entries of the state table are constructed in a similar way.

Step 2: State Table Minimization: When constructing an initial state table from a problem specification, depending on how the states are defined as perceived by the system designer, the number of states may be more than actually needed. Thus, the initial state table may not be minimal. For instance, if we regard the final state S4 as a default condition for the system to terminate, then S4 is redundant, which can be eliminated. An initial state table may also contain many states that are actually equivalent. The second step in the FSM design model is to minimize the state table by eliminating redundant and equivalent states. In order to do this, we first clarify what is meant by equivalent states.

(1) **Equivalence relation**: An equivalent relation R is a binary relation applied to a set of objects, which is

Table 4.1 Initial State Table of FSM

State	x1=/	x2=\n	x3=others	x4=EOF
S0	S2/-	S1/"x2"	S1/"x3"	S4/-
S1	S2/-	S1/"x2"	S1/"x3"	S4/-
S2	S3/-	S1/"/x2"	S1/"/x3"	S4/-
S3	S3/-	S1/"x2"	S3/-	S4/-
S4	S4/-	S4/-	S4/-	S4/-

Reflexive: for any object x, x R x is true.
Symmetric: for any objects x, y, x R y implies y R x.
Transitive: for any objects x, y, z, x R y and y R z imply x R z.

For example, the = relation of real numbers is an equivalence relation. Similarly, for any integer N >0, the modulo-N (% N) relation of nonnegative integers is also an equivalence relation.

(2) **Equivalent class**: An equivalence relation can be used to partition (divide) a set into equivalent classes such that all the objects in the same class are equivalent. As a result, each equivalent class can be represented by a single object of that class.

Example: When applying the % 10 relation to the set of nonnegative integers, it partitions the set into the equivalent classes {0}–{9}. Each class {i} consists of all integers which yield a remainder of i when divided by 10. We may use 0–9 to represent the equivalent classes {0}–{9}.

(3). **Equivalent states**: In a FSM, two states Si and Sj are equivalent if for every input x,

their outputs are identical and their next states are equivalent.

Note that the definition of equivalent states only requires that, for each input their outputs must be identical but not their next states, which only need to be equivalent. This seems to create a chicken-egg problem, but we can handle it easily, as will be shown shortly.

(4) **State table minimization**: This step tries to reduce the number of states in a FSM state table to a minimum. While it may be hard to find all the equivalent states in a state table directly, it is very easy to spot state pairs that can not be equivalent based on their outputs. In elementary logic, we know that

if A then B is **equivalent** to if [not B] then [not A]

When trying to prove something like "if A then B", we may either attack from the front by showing that "if A is true, then B must be true", or from the rear by showing that "if B is not true, then A can not be true" The strategy of attacking from the rear is often known as **proof by contradiction**, which is used in almost all proofs in the computability theory in computer science. So, rather than trying to identify equivalent states in a state table, we shall use a strategy which tries to identify and eliminate all non-equivalent states. The scheme is implemented by an Implication Chart, which is a table containing all pairs of states in a state table. In the implication chart, each cell corresponds to a state-pair (Si, Sj). Since an implication chart is symmetrical and all diagonal cells (Si, Si) are obviously equivalent states, it suffices to show only the lower half of the chart, without the diagonal cells. The algorithm of identifying non-equivalent state-pairs in an implication chart is as follows.

(1) Use the outputs of the states to cross out any cell (Si, Sj) that can not be equivalent.
(2) For each non-crossed out cell (Si, Sj), examine their next state pairs (Si', Sj') under each input. Cross out the cell (Si, Sj) if any of their next state-pair (Si', Sj') has been crossed out.
(3) Repeat (2) until there are no state-pair cells that can be crossed out

When the algorithm ends, each non-crossed out cell (Si, Sj) identifies a pair of equivalent states. Then, use the transitive property of equivalent state-pairs to construct equivalent classes.

For our example, we first construct an implication chart and cross out all the cells of the state-pairs that are non-equivalent. For example, it is obvious that S0 and S2 can not be equivalent since their outputs are not the same for every input. So we cross out the cell of (S0, S2). For the same reason, we can cross out the cell of (S0, S3). Likewise, we can cross out the cells of (S1, S2), (S1, S3) and (S2, S3). Table 4.2 shows the implication cart after applying (1) to cross out the cells of non-equivalent state-pairs.

Table 4.2 Initial Implication Chart of FSM

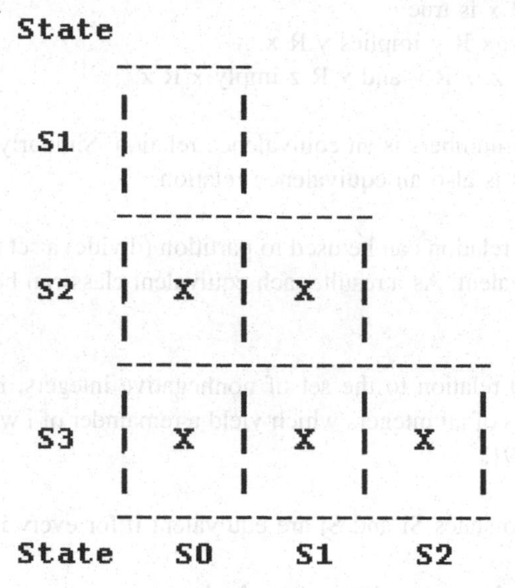

State			
S1			
S2	X	X	
S3	X	X	X
State	S0	S1	S2

Table 4.3 Implication Chart of FSM

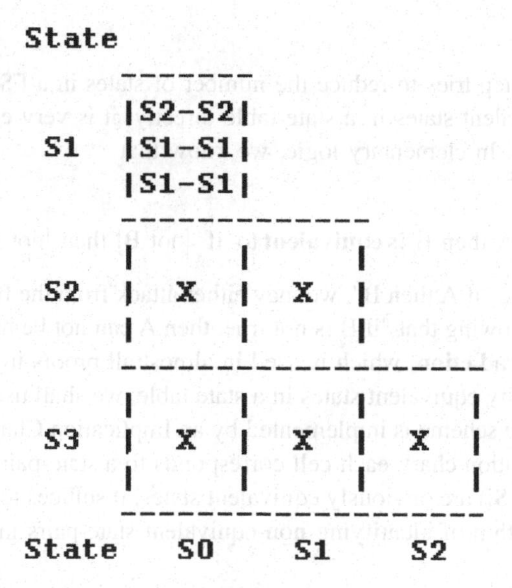

State			
S1	S2-S2 S1-S1 S1-S1		
S2	X	X	
S3	X	X	X
State	S0	S1	S2

From the initial implication chart of Table 4.2, we fill each non-crossed out cell with next state pairs, which is shown in the cell of (S0, S1) of Table 4.3.

Then we apply step (2) of the algorithm, trying to cross out any cells containing state pairs that are already crossed out. In this case, there are none. So the algorithm terminates. The final implication chart of Table 4.3 reveals that (S0, S1) are equivalent states, which can be combined into a single state.

The process of identifying and eliminating equivalent states in state tables is known as the **FSM minimization** problem, which has been thoroughly studied in the design of finite state machines (Katz and Borriello 2005). It suffices to say that we can always reduce a fully specified and deterministic FSM state table to a minimal form, which is unique up to isomorphism (by renaming the states). Furthermore, the algorithm requires only polynomial computing time. For this example, the minimal state table is shown in Table 4.4, which has only 3 non-equivalent states.

Table 4.4 Minimal State Table of FSM

```
State |  x1=/   |   x2=\n      | x3 !={/,\n}
------|---------|--------------|-----------
S1    |  S2/-   |  S1/"x2"     |  S1/"x3"
S2    |  S3/-   |  S1/"/x2"    |  S1/"/x3"
S3    |  S3/-   |  S1/"x2"     |  S3/-
---------------------------------------------
```

The **state diagram** of a FSM is a directed graph, in which each node represents a state and an arc form Si to Sj, denoted by Si->Sj, represents a state transition from state Si to stat Sj. The arc is marked with all the input/output pairs that cause the state transition. State tables and state diagrams are equivalent in the sense that they convey exactly the same information. The reader may draw a state diagram for the state table shown in Table 4.4. This is left as an exercise.

Step 3: Translate State Table/State Diagram into Code: A state table or state diagram can be translated into C code almost directly. Using the switch–case statements of C, each state corresponds to a distinct case in an outer switch statement, and each input corresponds to a distinct case in an inner switch statement. We illustrate the translation by a complete C program, which simulates the intended embedded system.

```
/**************** Example Program C4.5 *****************/
#include <stdio.h>
int main()
{
  int c;                         // input char
  int state = 1;                 // initial current state = S1
  FILE *fp = fopen("cprogram.c", "r"); // input is a C program

  while((c=fgetc(fp))!= EOF){  // if EOF: terminate
    switch(state){               // switch based on current state
      case 1:                    // state S1
        switch(c){               // next states/outputs
          case '/' : state = 2;                      break;
          case '\n': state = 1; printf("%c", c);   break;
          default  : state = 1; printf("%c", c);   break;
        };                                           break;
      case 2:                    // state S2
        switch(c){               // next states/outputs
          case '/' : state = 3;                      break;
          case '\n': state = 1; printf("/%c", c); break;
          default:   state = 1; printf("/%c", c); break;
        };                                           break;
      case 3:                    // state S3
        switch(c){               // next states/outputs
          case '/' : state = 3;                      break;
          case '\n': state = 1; printf("%c", c);   break;
          default  : state = 3;                      break;
      };
    }
  }
}
```

The reader may compile and run the above C4.5 program under Linux on C source files that use // as comment lines. The outputs should show that it removes all comment lines from C source files. The reader may also consult Problem 4.2 to handle a minor design flaw of the program.

It is noted that, when translating a FSM state table or state diagram into code, the resulting C code may not be very pretty nor efficient (in terms of code size), but the translation process is almost mechanical, which can be automated if needed. It makes the coding step an engineering endeavor rather than an art of programming. This is the strongest asset of the FSM model. However, the FSM model does have its limitations in that the number of states can not be too large. Whereas it may be quite easy to handle problems with only a few states, it would be too difficult to manage state tables or state diagrams with hundreds of states. For this reason, it is impractical and nearly impossible to design and implement a complete operating system by the FSM model.

4.10.3 StateChart Model

The StateChart model (Franke B 2016) is based on the finite state machine model. It adds concurrency and communications among the concurrent execution entities. It is intended for modeling complex embedded systems with concurrent tasks. Since this model involves the advanced concepts of concurrent processes, process synchronization and inter-process communication, we shall not discuss it any further.

4.11 Summary

This chapter covers models of embedded systems. It explained the simple super-loop system model and pointed out its shortcomings. It discussed the event-driven model and demonstrated the periodic and asynchronous event-driven system models by example programs. Then it justified the need for processes or tasks in embedded systems, and discussed the various process models. Lastly, it introduced some of the formal design methodologies for embedded systems, and it illustrated the FSM model by a detailed design and implementation example.

PROBLEMS

1. In the example program C4.3, after initializing the system the main() function executes a while(1) loop:

```
int main()
{
   // initiallzation
   while(1){
     asm("MOV r0, #0; MCR p15,0,R0,c7,c0,4");
     printf("CPU out of WFI state\n");
   }
}
```

   (1) Comment out the asm line. Run the program again to see what would happen.
   (2) In terms of CPU power consumption, what difference does the asm statement make?

2. In the Example Program C4.5, it assumes that a comment line starts with 2 adjacent / symbols to the end of line. However, string constants enclosed in matched pairs of double quotes may contain any number of / symbols but they are not comment lines, e.g. printf("this // is not a /// comment line\n"); Modify the state table of program C4.5 to handle this case. Translate the modified state table or state diagram into C code and run the modified program to test whether it works correctly.

3. Assume that comment blocks in C programs begins with /* and ends with */. Nested comment blocks are not allowed, which should result in an error. Write a C program, which detects and removes comment blocks from C program source files.

(1) Model the program as a FSM and construct a state diagram for the FSM.
(2) Write C code to implement the FSM as an event-driven system.

References

Android, https://en.wikipedia.org/wiki/Android_(operating_system), 2016.
ARM Cortex-5, ARM Cortex-A5 Processor, ARM Information Center, 2010.
Cheong, E, Liebman, J, Liu, J, Zhao, F, "TinyGALS: A programming model for event-driven embedded systems", ACM symposium on Applied Computing, 2003.
Dunkels, A, Schmidt, O, Voigt, T, Ali, "MProtothreads: simplifying event-driven programming of memory-constrained embedded systems", SenSys '06 Proc. of the 4th international conference on Embedded networked sensor systems, 2006.
Edwards, S, Lavagno, L, Lee, E.A, Sangiovanni-Vincentelli, A, "Design of Embedded Systems: Formal Models, Validation, and Synthesis", Proc. of the IEEE, Vol. 85, No.3, March, 1997, PP366–390.
Franke, B, "Embedded Systems Lecture 4: Statecharts, University of Edinburgh.
Fischer, J, Majumdar, R, Millstein, T, "Tasks: Language Support for Event-driven.
Programming", ACM SIGPLAN 2007 Workshop on PEPM, 2007.
Gajski, DD, Vahid, F, Narayan, S, Gong, J, "Specification and design of embedded systems", PTR Prentice Hall, 1994.
Katz, R. H. and G. Borriello, "Contemporary Logic Design", 2nd Edition, Pearson, 2005.

Process Management in Embedded Systems

5

5.1 Multitasking

In general, multitasking refers to the ability of performing several independent activities at the same time. For example, we often see people talking on their mobile phones while driving. In a sense, these people are doing multitasking, although a very dangerous kind. In computing, multitasking refers to the execution of several independent tasks at the same time. In a single CPU or uniprocessor (UP) system, only one task can execute at a time. Multitasking is achieved by multiplexing the CPU's execution time among different tasks, i.e. by switching the CPU from one task to another. If the switch is fast enough, it gives the illusion that all the tasks are executing simultaneously. This logical parallelism is called concurrency. In a multiprocessor (MP) system, tasks can execute on different CPUs in parallel in real time. In addition, each processor may also do multitasking by executing different tasks concurrently. Multitasking is the basis of all operating systems, as well as the foundation of concurrent programming in general. For simplicity, we shall consider uniprocessor (UP) systems first. MP systems will be covered later in Chap. 9 on Multiprocessor Systems.

5.2 The Process Concept

A multitasking system supports concurrent executions of many processes. The heart of a multitasking system is a control program, known as the operating system (OS) kernel, which provides functions for process management. In a multitasking system, processes are also called tasks. For all practical purposes, the terms process and task can be used interchangeably. First, we define an execution image as a memory area containing the execution's code, data and stack. Formally, **a process is the execution of an image**. It is a sequence of executions regarded as a single entity by the OS kernel for using system resources. System resources include memory space, I/O devices and, most importantly, CPU time. In an OS kernel, each process is represented by a unique data structure, called the Process Control Block (PCB) or Task Control Block (TCB), etc. In this book, we shall simply call it the PROC structure. Like a personal record, which contains all the information of a person, a PROC structure contains all the information of a process. In a single CPU system, only one process can be executing at a time. The OS kernel usually uses a global PROC pointer, running or current, to point at the PROC that is currently executing. In a real OS, the PROC structure may contain many fields and quite large. To begin with, we shall define a very simple PROC structure to represent processes.

```
typedef struct proc{
        struct proc *next;
        int     *ksp;
        int     kstack[1024];
}PROC;
```

In the PROC structure, the next field is a pointer pointing to the next PROC structure. It is used to maintain PROCs in dynamic data structures, such as link lists and queues. The ksp field is the saved stack pointer of a process when it is not executing, and kstack is the execution stack of a process. As we expand the OS kernel, we shall add more fields to the PROC structure later.

K.C. Wang, *Embedded and Real-Time Operating Systems*,
DOI 10.1007/978-3-319-51517-5_5

5.3 Multitasking and Context Switch

5.3.1 A Simple Multitasking Program

We begin to demonstrate multitasking by a simple program. The program is denoted by C5.1. It consists of a ts.s file in ARM assembly code and a t.c file in C.

(1). ts.s file: the ts.s file defines the program's entry point reset_handler, in which it
(1). set the SVC stack pointer to the high end of proc0.kstack[ ].
(2). add a tswitch() function in assembly code for task switching.

```
// ——————————— ts.s file of C5.1 ———————————
.global main, proc0, procsize  // imported from C code
.global reset_handler, tswitch, scheduler, running

reset_handler:
  LDR r0, =proc0      // r0->proc0
  LDR r1, =procsize  // r1 ->procsize
  LDR r2, [r1, #0]    // r2 = procsize
  ADD r0, r0, r2      // r0 -> high end of proc0
  MOV sp, r0          // sp -> high end of proc0
  BL  main            // call main() in C

// ADD tswitch() function for task switch
tswitch:
SAVE:
  STMFD sp!, {r0-r12, lr}
  LDR r0, =running    // r0=&running
  LDR r1, [r0, #0]    // r1->runningPROC
  STR sp, [r1, #4]    // running->ksp = sp
FIND:
  BL scheduler         // call scheduler() in C
RESUME:
  LDR r0, =running
  LDR r1, [r0, #0]    // r1->running PROC
  LDR sp, [r1, #4]    // restore running->ksp
  LDMFD sp!, {r0-r12, lr} // restore register
  MOV pc, lr          // return
```

(2). t.c file: The t.c file includes the LCD and keyboard drivers for I/O. It defines the PROC structure type, a PROC structure proc0 and a running PROC pointer pointing to the current executing PROC.

```
/************* t.c file of C5.1 *****************/
#include "vid.c"              // LCD display driver
#include "kbd.c"              // KBD driver
#define SSIZE  1024           // stack size per PROC
typedef struct proc{          // process structure
       struct proc *next;     // next PROC pointer
    int *ksp;                 // saved sp when PROC is not running
    int  kstack[SSIZE];       // process kernel mode 4KB stack
}PROC;                        // PROC is a type
int  procSize = sizeof(PROC);
```

```
PROC proc0, *running;          // proc0 structure and running pointer
int scheduler(){ running = &proc0; }
main()                         // called from ts.s
{
  running = &proc0;            // set running PROC pointer
  printf("call tswitch()\n");
  tswitch();                   // call tswitch()
  printf("back to main()\n");
}
```

Use the ARM toolchain (2016) to compile-link ts.s and t.c to generate a binary executable t.bin as usual. Then run t.bin on the Versatilepb VM (Versatilepb 2016) under QEMU, as in

qemu-system-arm –M versatilepb –m 128M **–kernel t.bin**

During booting, QEMU loads t.bin to 0x10000 and jumps to there to execute the loaded image. When execution starts in ts.s, it sets the SVC mode stack pointer to the high end of proc0. This makes proc0's kstack area as the initial stack. Up to this point, the system has no notion of any process because there is none. The assembly code calls main() in C. When control enters main(), we have an image in execution. By the definition of process, which is the execution of an image, we have a process in execution, although the system still does not know which process is executing. In main(), after setting running = &proc0, the system is now executing the process proc0. This is how a typical OS kernel starts to run an initial process when it begins. The initial process is handcrafted or created by brute force. Starting from main(), the run-time behavior of the program can be traced and explained by the execution diagram of Fig. 5.1, in which the key steps are labeled **(1)** to **(6)**.

At (1), it lets running point to proc0, as shown on the right-hand side of Fig. 5.1. Since we assume that running always points at the PROC of the current executing process, the system is now executing the process proc0.
At (2), it calls tswitch(), which loads LR(r14) with the return address and enters tswitch.
At (3), it executes the SAVE part of tswitch(), which saves CPU registers into stack and saves the stack pointer sp into proc0. ksp.
At (4), it calls scheduler(), which sets running to point at proc0 again. For now, this is redundant since running already points at proc0. Then it executes the RESUME part of tswitch().
At (5), it sets sp to proc0.ksp, which is again redundant since they are already the same. Then it pops the stack, which restores the saved CPU registers.
At (6), it executes MOV pc, lr at the end of RESUME, which returns to the calling place of tswitch().

```
Main()                                                         running
{                                                                 |
   (1). running=&proc0;                                         proc0
   (2). tswtich();                                               ksp
      .          (3):SAVE === push ======== >                    |   (4):scheduler()
proc0 -|-|-------------------------------------------------------sp--------
kstack |lr|r12|r11|r10|r9|r8|r7|r6|r5|r4|r3|r2|r1|r0|
       ----------------------------------------------
        1   2   3   4   5   6  7  8  9  10  11 12  13 14
   (6)              <=========== pop ========        (5)     sp (RESUME)
}
```

Fig. 5.1 Execution diagram of proc0

5.3.2 Context Switching

Besides printing a few messages, the program seems useless since it does practically nothing. However, it is the basis of all multitasking programs. To see this, assume that we have another PROC structure, proc1, which called tswitch() and executed the SAVE part of tswitch() before. Then proc1's ksp must point to its stack area, which contains saved CPU registers and a return address from where it called tswitch(), as shown in Fig. 5.2.

In scheduler(), if we let running point to proc1, as shown in the right-hand side of Fig. 5.2, the RESUME part of tswitch () would change sp to proc1's ksp. Then the RESUME code would operate on the stack of proc1. This would restore the saved registers of proc1, causing proc1 to resume execution from where it called tswitch() earlier. This changes the execution environment from proc0 to proc1.

Context Switching

Changing the execution environment of one process to that of another is called context switching, which is the basic mechanism of multitasking.

With context switching, we can create a multitasking environment containing many processes. In the next program, denoted by C5.2, we define NPROC = 5 PROC structures. Each PROC has a unique pid number for identification. The PROCs are initialized as follows.

```
running -> P0 -> P1 -> P2 -> P3 -> P4 ->
           |                          |
           <--------------------------<--
```

P0 is the initial running process. All the PROCs form a circular link list for simple process scheduling. Each of the PROCs, P1 to P4, is initialized in such a way that it is ready to resume running from a body() function. Since the initialization of the PROC stack is crucial, we explain the steps in detail. Although the processes never existed before, we may pretend that they not only existed before but also ran before. The reason why a PROC is not running now is because it called tswitch() to give up CPU earlier. If so, the PROC's ksp must point to its stack area containing saved CPU registers and a return address, as shown in Fig. 5.3, where the index -i means SSIZE-i.

Since the PROC never really ran before, we may assume that its stack was initially empty, so that the return address, rPC=LR, is at the very bottom of the stack. What should be the rPC? It may point to any executable code, e.g. the entry address of a body() function. What about the "saved" registers? Since the PROC never ran before, the register values do not matter, so they can all be set to 0. Accordingly, we initialize each of the PROCs, P1 to P4, as shown in Fig. 5.4.

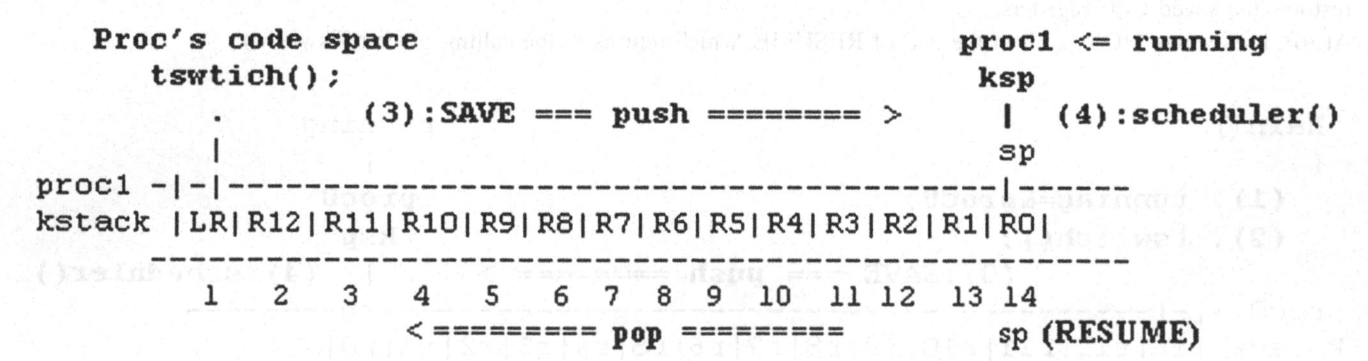

Fig. 5.2 Execution diagram of proc1

```
      rPC                                                         SP
proc  -|-|-------------------------------------------------------|---
kstack |LR|r12|r11|r10|r9 |r8 |r7 |r6 |r5 |r4 |r3 |r2 |r1 |r0 |
       ----------------------------------------------------------------
        -1  -2  -3  -4   -5  -6   -7  -8   -9  -10 -11 -12  -13  -14
```

Fig. 5.3 Process stack contents

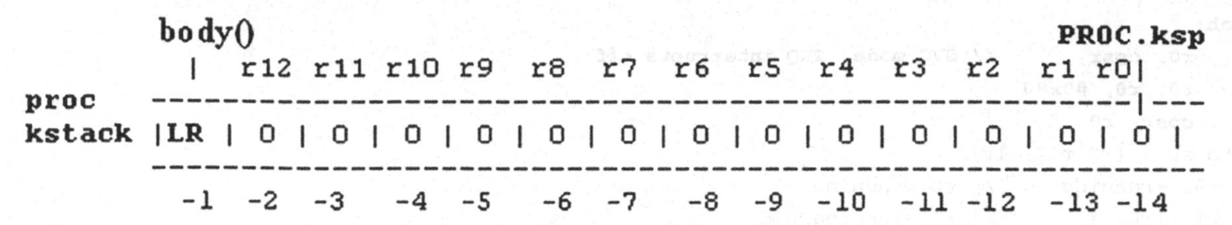

Fig. 5.4 Initial stack contents of process

With this setup, when a PROC becomes running, i.e. when running points to the PROC, it would execute the RESUME part of tswitch(),

```
LDMFD sp!, {r0-r12, lr}
MOV pc, lr
```

which restores the "saved" CPU registers, followed by MOV pc, lr, causing the process to execute the body() function.

After initialization, P0 calls tswitch() to switch process. In tswitch(), P0 saves CPU registers into its own stack, saves the stack pointer in its PROC.ksp and calls scheduler(). We modify the scheduler() function by letting running point to the next PROC, i.e.

```
running = running->next;
```

So P0 switches to P1. P1 begins by executing the RESUME part of tswitch(), causing it to resume to the body() function. While in body(), the running process prints its pid and prompts for an input char. Then it calls tswitch() to switch to the next process, etc. Since the PROCs are in a circular link list, they will take turn to run. The following lists the assembly and C code of C5.2.

(1). ts.s file of C5.2

```
.global vectors_start, vectors_end
.global main, proc, procsize
.global tswitch, scheduler, running
.global lock, unlock
reset_handler:
// set SVC sp to proc[0] high end
  LDR r0, =proc
  LDR r1, =procsize
  LDR r2, [r1, #0]
  ADD r0, r0, r2
  MOV sp, r0
// copy vector table to address 0
  BL copy_vectors
// go in IRQ mode to set IRQ stack
  MSR cpsr, #0x92
  LDR sp, =irq_stack
// go back to SVC mode with IRQ enabled
  MSR cpsr, #0x13
// call main() in SVC mode
  BL main
  B .
```

```
tswitch:
  mrs   r0, cpsr          // SVC mode, IRQ interrupts off
  orr   r0, r0, #0x80
  msr   cpsr, r0
  stmfd sp!, {r0-r12, lr}
  LDR r0, =running        // r0=&running
  LDR r1, [r0, #0]        // r1->runningPROC
  str sp, [r1, #4]        // running->ksp = sp
  bl scheduler
  LDR r0, =running
  LDR r1, [r0, #0]        // r1->runningPROC
  lDR sp, [r1, #4]
  mrs   r0, cpsr          // SVC mode, IRQ interrupts on
  bic   r0, r0, #0x80
  msr   cpsr, r0
  ldmfd sp!, {r0-r12, pc}

irq_handler:
  sub lr, lr, #4
  stmfd sp!, {r0-r12, lr}
  bl irq_chandler
  ldmfd sp!, {r0-r12, pc}^
lock:                     // disable IRQ interrupts
  MRS r0, cpsr
  ORR r0, r0, #0x80       // set IRQ mask bit
  MSR cpsr, r0
  mov pc, lr
unlock:                   // enable IRQ interrupts
  MRS r0, cpsr
  BIC r0, r0, #0x80       // clear IRQ mask bit
  MSR cpsr, r0
  mov pc, lr
vectors_start:            // vector table
  LDR PC, reset_handler_addr
  LDR PC, undef_handler_addr
  LDR PC, swi_handler_addr
  LDR PC, prefetch_abort_handler_addr
  LDR PC, data_abort_handler_addr
  B .
  LDR PC, irq_handler_addr
  LDR PC, fiq_handler_addr
reset_handler_addr:          .word reset_handler
undef_handler_addr:          .word undef_handler
swi_handler_addr:            .word swi_handler
prefetch_abort_handler_addr: .word prefetch_abort_handler
data_abort_handler_addr:     .word data_abort_handler
irq_handler_addr:            .word irq_handler
fiq_handler_addr:            .word fiq_handler
vectors_end:
```

The assembly code is the same as in C5.1, except for the following modifications:

. the initial SVC mode stack pointer is set to proc[0]'s kstack
. copy vector table to address 0
. set IRQ mode stack and install IRQ handler (for keyboard driver)
. call main() in SVC mode with IRQ interrupts enabled

(2). Device Drivers of C5.2: The LCD driver is the same as in C5.1. The keyboard driver is the same as in Sect. 3.5.4 of Chap. 3.
(3). t.c file of C5.2:

```
/************ t.c file of C5.2 **********/
#include "vid.c"        // LCD driver
#include "kbd.c"        // KBD driver
#define  NPROC     5
#define  SSIZE 1024
typedef struct proc{
  struct proc *next;
  int     *ksp;
  int     pid;
  int     kstack[SSIZE];
}PROC;
PROC proc[NPROC], *running;
int procsize = sizeof(PROC);
int body()
{
  char c;
  printf("proc %d resume to body()\n", running->pid);
  while(1){
    printf("proc %d in body() input a char [s] : ", running->pid);
    c = kgetc(); printf("%c\n", c);
    tswitch();
  }
}
int kernel_init()
{
  int i, j;
  PROC *p;
  printf("kernel_init()\n");
  for (i=0; i<NPROC; i++){
    p = &proc[i];
    p->pid = i;
    p->status = READY;
    for (j=1; j<15; j++) // initialize proc.kstack and saved ksp
        p->kstack[SSIZE-j] = 0;        // all saved regs = 0
    p->kstack[SSIZE-1] = (int)body;   // resume point = body
    p->ksp = &(p->kstack[SSIZE-14]); // saved ksp
    p->next = p + 1;                 // point to next PROC
  }
  proc[NPROC-1].next = &proc[0];      // circular PROC list
  running = &proc[0];
}
```

```
int scheduler()
{
  printf("proc %d in scheduler\n", running->pid);
  running = running->next;
  printf("next running = %d\n", running->pid);
}
int main()
{
   char c;
   fbuf_init();              // initialize LCD driver
   kbd_init();               // initialize KBD driver
   printf("Welcome to WANIX in Arm\n");
   kernel_init();
   while(1){
     printf("P0 running input a key : ");
     c = kgetc(); printf("%c\n", c);
     tswitch();
   }
}
```

5.3.3 Demonstration of Multitasking

Figure 5.5 shows the outputs of running the C5.2 multitasking program. It uses the process pid to display in different colors, just for fun.

Before continuing, it is worth noting the following.

(1). In the C5.2 multitasking program, none of processes, P1 to P4, actually calls the body() function. What we have done is to convince each process that it called tswitch() from the entry address of body() to give up CPU earlier, and that is where it shall resume to when it begins to run. Thus, we can fabricate an initial environment for each process to start. The process has no choice but to obey. This is the power (and joy) of systems programming.

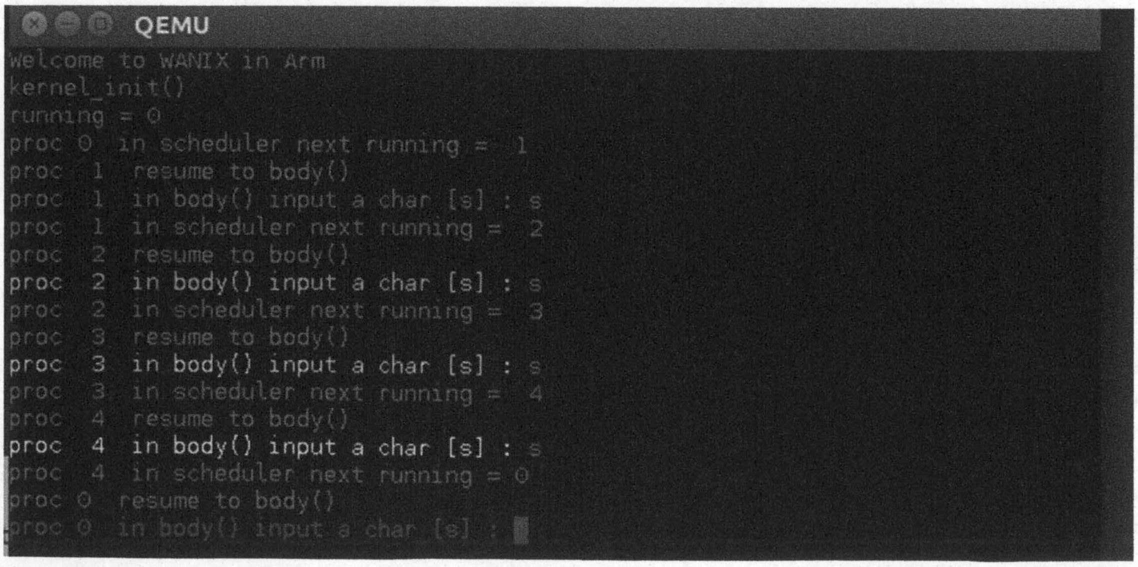

```
QEMU
welcome to WANIX in Arm
kernel_init()
running = 0
proc 0  in scheduler next running =  1
proc  1  resume to body()
proc  1  in body() input a char [s] : s
proc  1  in scheduler next running =  2
proc  2  resume to body()
proc  2  in body() input a char [s] : s
proc  2  in scheduler next running =  3
proc  3  resume to body()
proc  3  in body() input a char [s] : s
proc  3  in scheduler next running =  4
proc  4  resume to body()
proc  4  in body() input a char [s] : s
proc  4  in scheduler next running = 0
proc 0  resume to body()
proc 0  in body() input a char [s] :
```

Fig. 5.5 Demonstration of multitasking

(2). All the processes, P1 to P4, execute the same body() function but each executes in its own environment. For instance, while executing the body() function, each process has its own local variable c in the process stack. This shows the difference between processes and functions. A function is just a piece of passive code, which has no life. Processes are executions of functions, which makes the function code alive.
(3). When a process first enters the body() function, the process stack is logically empty. As soon as execution starts, the process stack will grow (and shrink) by the function calling sequence as described in Sect. 2.7.3.2 of Chap. 2.
(4). The per-process kstack size is defined as 4KB. This implies that the maximal length of function call sequence (and the associated local variable spaces) of every process must never exceed the kstack size. Similar remarks also apply to other privileged mode stacks, e.g. the IRQ mode stack for interrupts processing. All of these are under the planning and control of the kernel designer. So stack overflow should never occur in kernel mode.

5.4 Dynamic Processes

In the program C5.2, P0 is the initial process. All other processes are created statically by P0 in kernel_init(). In the next program, denoted by C5.3, we shall show how to create processes dynamically.

5.4.1 Dynamic Process Creation

(1). First, we add a status and a priority field to the PROC structure, and define the PROC link lists: freeList and readyQueuee, which are explained below.

```
#define NPROC     9
#define FREE      0
#define READY     1
#define SSIZE 1024
typedef struct proc{
   struct proc *next;      // next PROC pointer
   int      *ksp;          // saved sp when NOT running
   int      pid;           // process ID
   int      status;        // FREE|READY, etc.
   int      priority;      // priority value
   int      kstack[SSIZE]; // process kernel mode stack
}PROC;
PROC proc[NPROC], *running, *freeList, *readyQueue;
```

. **freeList** = a (singly) link list containing all FREE PROCs. When the system starts, all PROCs are in the freeList initially. When create a new process, we allocate a free PROC from freeList. When a process terminates, we deallocate its PROC and release it back to the freeList for reuse.
. **readyQueue** = a priority queue of PROCs that are ready to run. PROCs with the same priority are ordered First-in-first-out (FIFO) in the readyQueue.

(2). In the queue.c file, we implement the following functions for list and queue operations.

```
/*************** queue.c file ***************/
PROC *get_proc(PROC **list){ return a PROC pointer from list }
int put_proc(PROC **list, PROC *p){ // enter p into list }
int enqueue(PROC **queue, PROC *p){ // enter p into queue by priority }
PROC *dequeue(PROC **queue){ // remove & return first PROC from queue }
int printList(PROC *p){ // print list elements }
```

(3). In the kernel.c file, kernel_init() initializes the kernel data structures, such as the freeList and readyQueue. It also creates P0 as the initial running process. The function

int pid = kfork(int func, int priority)

creates a new process to execute a function func() with the specified priority. In the example program, every new process begins execution from the same body() function. When a task has completed its work, it may terminate by the function

void kexit()

which releases its PROC structure back to the freeList for reuse. The function scheduler() is for process scheduling. The following lists the C code of kernel.c and t.c files.

```
/************** kernel.c file of Program C5.3 **********/
int kernel_init()
{
  int i, j;
  PROC *p;
  printf("kernel_init()\n");
  for (i=0; i<NPROC; i++){
    p = &proc[i];
    p->pid = i;
    p->status = FREE;
    p->next = p + 1;
  }
  proc[NPROC-1].next = 0;
  freeList = &proc[0];       // all PROCs in freeList
  readyQueue = 0;            // readyQueue empty
  // create P0 as initial running process
  p = get_proc(&freeList);
  p->priority = 0;           // P0 has the lowest priority 0
  running = p;
  printf("running = %d\n", running->pid);
  printList(freeList);
}
int body()  // process code
{
  char c;
  color = running->pid;
  printf("proc %d resume to body()\n", running->pid);
  while(1){
    printf("proc %d in body() input a char [s|f|x] : ", running->pid);
    c = kgetc(); printf("%c\n", c);
    switch(c){
      case 's': tswitch();              break;
      case 'f': kfork((int)body, 1); break;
      case 'x': kexit();                break;
    }
  }
}
// kfork() creates a new process to exceute func with priority
int kfork(int func, int priority)
{
  int i;
  PROC *p = get_proc(&freeList);
```

```
  if (p==0){
     printf("no more PROC, kfork failed\n");
     return -1;                          // return -1 for FAIL
  }
  p->status = READY;
  p->priority = priority;
  // set kstack for proc resume to execute func()
  for (i=1; i<15; i++)
       p->kstack[SSIZE-i] = 0;           // all "saved" regs = 0
  p->kstack[SSIZE-1] = func;             // resume execution address
  p->ksp = &(p->kstack[SSIZE-14]); // saved ksp
  enqueue(&readyQueue, p);               // enter p into readyQueue
  printf("%d kforked a new proc %d\n", running->pid, p->pid);
  printf("freeList = "); printList(readyQueue);
  return p->pid;
}
void kexit()   // called by process to terminate
{
   printf("proc %d kexit\n", running->pid);
   running->status = FREE;
   put_proc(running);
   tswitch();  // give up CPU
}
int scheduler()
{
  if (running->status == READY)
     enqueue(&readyQueue, running);
  running = dequeue(&readyQueue);
}

/************ t.c file of Program C5.3 ***********/
#include "type.h"          // PROC type and constants
#include "string.c"        // string opertion functions
#include "queue.c"         // list and queue operation functions
#include "vid.c"           // LCD driver
#include "kbd.c"           // KBD driver
#include "exceptions.c"
#include "kernel.c"
void copy_vectors(void){// same as before }
void IRQ_handler(){      // handles only KBD interrupts }

int main()
{
   fbuf_init();            // initialize LCD display
   kbd_init();             // initialize KBD driver
   printf("Welcome to Wanix in ARM\n");
   kernel_init();          // initialize kernel, create and run P0
   kfork((int)body, 1); // P0 create P1 in readyQueue
   while(1){
     while(readyQueue==0); // P0 loops if readyQueue empty
     tswitch();
   }
}
```

```
QEMU
Welcome to WANIX in Arm
kernel_init()
running = 0
freeLis    = [ 1 ]->[ 2 ]->[ 3 ]->[ 4 ]->[ 5 ]->[ 6 ]->[ 7 ]->[ 8 ]->NULL
proc0  kforked a new proc 1
readyQueue = [ 1  1 ]->NULL
proc0  in scheduler next running =  1
proc 1  resume to body()
proc 1  in body() input a char [s|f|x] : f
proc 1  kforked a new proc 2
readyQueue = [ 2  1 ]->[0 0 ]->NULL
proc 1  in body() input a char [s|f|x] : s
proc 1  in scheduler next running =  2
proc 2  resume to body()
proc 2  in body() input a char [s|f|x] : x
proc 2  kexit
proc 2  in scheduler next running =  1
proc 1  in body() input a char [s|f|x] :
```

Fig. 5.6 Demonstration of dyamic process

In the t.c file, it first initializes the LCD display and the KBD driver. Then it initializes the kernel to run the initial process P0, which has the lowest priority 0. P0 creates a new process P1 and enters it into readyQueue. Then P0 calls tswitch() to switch process to run P1. Every new process resumes to execute the body() function. While a process runs, the user may enter 's' to switch process, 'f' to create a new process and 'x' to terminate, etc.

5.4.2 Demonstration of Dynamic Processes

Figure 5.6 shows the screen of running the C5.3 program. As the figure shows, a 'f' input causes P1 to kfork a new process P2 in the readyQueue. A 's' input causes P1 to switch process to run P2, which resumes to execute the same body() function. While P2 runs, the reader may enter commands to let P2 switch process or kfork a new process, etc. While a process runs, a 'x' input causes the process to terminate.

5.5 Process Scheduling

5.5.1 Process Scheduling Terminology

In a multitasking operating system, there are usually many processes ready to run. The number of runnable processes is in general greater than the number of available CPUs. Process scheduling is to decide when and on which CPU to run the processes in order to achieve an overall good system performance. Before discussing process scheduling, we first clarify the following terms, which are usually associated with process scheduling.

(1). I/O-bound vs. compute-bound processes:

A process is considered as I/O-bound if it suspends itself frequently to wait for I/O operations. I/O-bound processes are usually from interactive users who expect fast response time. A process is considered as compute-bound if it uses CPU time extensively. Compute-bound processes are usually associated with lengthy computations, such as compiling programs and numerical computations, etc.

(2). Response time vs. throughput:

Response time refers to how fast a system can respond to an event, such as entering a key from the keyboard. Throughput is the number of processes completed per unit time.

(3). Round-robin vs. dynamic priority scheduling:

In round-robin scheduling, processes take turn to run. In dynamic priority scheduling, each process has a priority, which changes dynamically (over time), and the system tries to run the process with the highest priority.

(4). Preemption vs. non-preemption:

Preemption means the CPU can be taken away from a running process at any time. Non-preemption means a process runs until it gives up CPU by itself, e.g. when the process finishes, goes to sleep or becomes blocked.

(5). Real-time vs. time-sharing:

A real-time system must respond to external events, such as interrupts, within a minimum response time, often in the order of a few milliseconds. In addition, the system may also need to complete the processing of such events within a specified time limit. In a time-sharing system, each process runs with a guaranteed time slice so that all processes receive their fair share of CPU time.

5.5.2 Goals, Policy and Algorithms of Process Scheduling

Process scheduling is intended to achieve the following goals.

.high utilization of system resources, especially CPU time,
.fast response to interactive or real-time processes,
.guaranteed completion time of real-time processes,
.fairness to all processes for good throughput, etc.

It is easy to see that some of goals are conflicting to one another. For example, fast response time and high throughput usually cannot be achieved at the same time. A scheduling policy is a set of rules, by which a system tries to achieve all or some of the goals. For a general purpose operating system, the scheduling policy is usually trying to achieve good overall system performance by striving for a balance among the conflicting goals. For embedded and real-time systems, the emphases are usually on fast response to external events and guaranteed process execution time. A scheduling algorithm is a set of methods that implements a scheduling policy. In an OS kernel, the various components, i.e. data structures and code used to implement the scheduling algorithm, are collectively known as the **process scheduler**. It is worth noting that in most OS kernel there is not a single piece of code or module that can be identified as the scheduler. The functions of a scheduler are implemented in many places inside the OS kernel, e.g. when a running process suspends itself or terminate, when a suspended process becomes runnable again and, most notably, in the timer interrupt handler.

5.5.3 Process Scheduling in Embedded Systems

In an embedded system, processes are created to perform specific tasks. Depending on the importance of the task, each process is assigned a priority, which is usually static. Processes run either periodically or in response to external events. The primary goal of process scheduling is to ensure quick response to external events and guarantee process execution time. Resource utilization and throughput are relatively unimportant. For these reasons, the process scheduling policy is usually based on process priority or by round-robin for processes with the same priority. In most simple embedded systems, processes usually execute in the same address space. In this case, the scheduling policy is usually non-preemptive. Each process runs until it gives up the CPU voluntarily, e.g. when the process goes to sleep, becomes suspended or explicitly yields control to another process. Preemptive scheduling is more complex due to the following reasons. With preemption, many processes may run concurrently in the same address space. If a process is in the middle of modifying a shared data object, it must not be preempted unless the shared data object is protected in a critical region. Otherwise, the shared data object may be corrupted by other processes. Protection of critical regions will be discussed in the next section on process synchronization.

5.6 Process Synchronization

When multiple processes execute in the same address space, they may access and modify shared (global) data objects. Process synchronization refers to the rules and mechanisms used to ensure the integrity of shared data objects in a concurrent processes environment. There are many kinds of process synchronization tools. For a detailed list of such tools, their implementation and usage, the reader may consult (Wang 2015). In the following, we shall discuss some simple synchronizing tools that are suitable for embedded systems. In addition to discussing the principles of process synchronization, we shall also show how to apply them to the design and implementation of embedded systems by example programs.

5.6.1 Sleep and Wakeup

The simplest mechanism for process synchronization is the sleep/wakeup operations, which are used in the original Unix kernel. When a process must wait for something, e.g. a resource, that is currently unavailable, it goes to sleep to suspend itself and give up the CPU, allowing the system to run other processes. When the needed resource becomes available, another process or an interrupt handler wakes up the sleeping processes, allowing them to continue. Assume that each PROC structure has an added event field. The algorithms of sleep/wakeup are as follows.

```
sleep(int event)
{
  record event value in running PROC.event;
  change running PROC status to SLEEP;
  switch process;
}
wakeup(int event)
{
  for each PROC *p do{
     if (p->status==SLEEP && p->event==event){
        change p->status to READY;
        enter p into readyQueue;
     }
  }
}
```

In order for the mechanism to work, sleep() and wakeup() must be implemented properly. First, each operation must be atomic (indivisible) from the process point of view. For instance, when a process executes sleep(), it must complete the sleep operation before someone else tries to wake it up. In a non-preemptive UP kernel, only one process runs at a time, so processes can not interfere with one another. However, while a process runs, it may be diverted to handle interrupts, which may interfere with the process. To ensure the atomicity of sleep and wakeup, it suffices to disable interrupts. Thus, we may implement sleep() and wakeup() as follows.

```
int sleep(int event)
{
   int SR = int_off();  // disable IRQ and return CPSR
   running->event = event;
   running->status = SLEEP;
```

```
        tswitch();                // switch process
        int_on(SR);               // restore original CPSR
    }
    int wakeup(int event)
    {
        int SR = int_off();  // disable IRQ and return CPSR
        for each PROC *p do{
            if (p->status==SLEEP && p->event==event){
                p->status = READY;
                enqueue(&readyQueue, p);
            }
        }
        int_on(SR);               // restore original CPSR
    }
```

Note that wakeup() wakes up ALL processes, if any, that are sleeping on an event. If no process is sleeping on the event, wakeup has no effect, i.e. it amounts to a NOP and does nothing. It is also worth noting the interrupt handlers can never sleep or wait (Wang 2015). They can only issue wakeup calls to wake up sleeping processes.

5.6.2 Device Drivers Using Sleep/Wakeup

In Chap. 3, we developed several device drivers using interrupts. The organization of these device drivers exhibits a common pattern. Every interrupt-driven device driver consists of three parts; a lower-half part, which is the interrupt handler, an upper-half part, which is called by a main program, and a data area containing an I/O buffer and control variables, which are shared by the lower and upper parts. Even with interrupts, the main program still must use busy-wait loops to wait for data or room in the I/O buffer, which is essentially the same as polling. In a multitasking system, I/O by polling does not use the CPU effectively. In this section, we shall show how to use processes and sleep/wakeup to implement interrupt-driven device drivers without busy-wait loops.

5.6.2.1 Input Device Drivers

In Chap. 3, the KBD driver uses interrupts but the upper-half uses polling. When a process needs an input key, it executes a busy-wait loop until the interrupt handler puts a key into the input buffer. Our goal here is to replace the busy-wait loop with sleep/wakeup. First, we show the original driver code by polling. Then, we modify it to use sleep/wakeup for synchronization.

(1). KBD structure: The KBD structure is the middle part of the driver. It contains an input buffer and control variables, e.g. data = number of keys in the input buffer.

```
typedef struct kbd{       // base = 0x10006000
  char *base;             // base address of KBD, as char *
  char buf[BUFSIZE];      // input buffer size=128 bytes
  int head, tail, data; // control variables; data=0 initially
}KBD; KBD kbd;
```

(2). kgetc(): This is the (base function) of the upper-half of the KBD driver.

```
int kgetc() // main program calls kgetc() to return a char
{
  char c;
  KBD *kp = &kbd;
  unlock();                    // enable IRQ interrupts
  while(kp->data == 0);        // busy-wait for data;
```

```
  lock();                    // disable IRQ interrupts
    c = kp->buf[kp->tail++];// get a char and update tail index
    kp->tail %= BUFSIZE;
    kp->data--;              // update data with interrupts OFF
  unlock();                  // enable IRQ interrupts
  return c;
}
```

We assume that the main program is now running a process. When a process needs an input key, it calls kgetc(), trying to get a key from the input buffer. Without any means of synchronization, the process must rely on a busy-wait loop

while (kp->data == 0); // busy-wait for data;

which continually checks the data variable for any key in the input buffer.

(3). kbd_handler(): The interrupt handler is the lower-half of the KBD driver.

```
kbd_handler()
{
  struct KBD *kp = &kbd;
  char scode = *(kp->base+KDATA); // read scan code in data register
  if (scode & 0x80)            // ignore key releases
     return;
  if (data == BUFSIZE)         // if input buffer FULL
     return;                   // ignore current key
  c = unsh[scode];             // map scan code to ASCII
  kp->buf[kp->head++] = c;     // enter key into CIRCULAR buf[ ]
  kp->head %= BUFSIZE;
  kp->data++;                  // inc data counter by 1
}
```

For each key press, the interrupt handler maps the scan code to a (lowercase) ASCII char, stores the char in the input buffer and updates the counting variables data. Again, without any means of synchronization, that's all the interrupt handler can do. For instance, it can not notify the process of available keys directly. Consequently, the process must check for input keys by continually polling the driver's data variable. In a multitasking system, the busy-wait loop is undesirable. We can use sleep/wakeup to eliminate the busy-wait loop in the KBD driver as follows.

(1). KBD structure: no need to change.

(2). kgetc(): rewrite kgetc() to let process sleep for data if there are no keys in the input buffer. In order to prevent race conditions between the process and the interrupt handler, the process disables interrupts first. Then it checks the data variable and modifies the input buffer with interrupts disabled, but it must enable interrupts before going to sleep. The modified kgetc () function is

```
int kgetc() // main program calls kgetc() to return a char
{
  char c;
  KBD *kp = &kbd;
  while(1){
    lock();                  // disable IRQ interrupts
    if (kp->data==0){        // check data with IRQ disabled
       unlock();             // enable IRQ interrupts
       sleep(&kp->data);     // sleep for data
```

```
    }
  }
  c = kp->buf[kp->tail++]; // get a c and update tail index
  kp->tail %= BUFSIZE;
  kp->data--;              // update with interrupts OFF
  unlock();                // enable IRQ interrupts
  return c;
}
```

(3). kbd_handler(): rewrite KBD interrupt handler to wake up sleeping processes, if any, that are waiting for data. Since process cannot interfere with interrupt handler, there is no need to protect the data variables inside the interrupt handler.

```
kbd_handler()
{
  struct KBD *kp = &kbd;
  scode = *(kp->base+KDATA); // read scan code in data register
  if (scode & 0x80)          // ignore key releases
    return;
  if (kp->data == BUFSIZE)   // ignore key if input buffer FULL
  c = unsh[scode];           // map scan code to ASCII
  kp->buf[kp->head++] = c;   // enter key into CIRCULAR buf[ ]
  kp->head %= BUFSIZE;
  kp->data++;                // update counter
  wakeup(&kp->data);         // wakeup sleeping process, if any
}
```

5.6.2.2 Output Device Drivers

An output device driver also consists of three parts; a lower-half, which is the interrupt handler, an upper-half, which is called by process to output data, and a middle part containing data buffer and control variables, which are shared by the lower and upper halves. The major difference between an output device driver and an input device driver is that the roles of process and interrupt handler are reversed. In an output device driver, process writes data to the data buffer. If the data buffer is full, it goes to sleep to wait for rooms in the data buffer. The interrupt handler extracts data from the buffer and outputs them to the device. Then it wakes up any process that is sleeping for rooms in the data buffer. A second difference is that for most output devices the interrupt handler must explicitly disable the device interrupts when there are no more data to output. Otherwise, the device will keep generating interrupts, resulting in an infinite loop. The third difference is that it is usually acceptable for several processes to share the same output device, but an input device can only allow one active process at a time. Otherwise, processes may get random inputs from the same input device.

5.7 Event-Driven Embedded Systems Using Sleep/Wakeup

With dynamic process creation and process synchronization, we can implement event-driven multitasking systems without busy-wait loops. We demonstrate such a system by the example program C5.4.

Example Program C5.4: We assume that the system hardware consists of three devices; a timer, a UART and a keyboard. The system software consists of three processes, each controls a device. For convenience, we also include an LCD for displaying outputs from the timer and keyboard processes.

Upon starting up, each process waits for a specific event. A process runs only when the awaited event has occurred. In this case, events are timer counts and I/O activities. At each second, the timer process displays a wall clock on the LCD. Whenever an input line is entered from the UART, the UART process gets the line and echoes it to the serial terminal. Similarly, whenever an input line is entered from the KBD, the KBD process gets the line and echoes it to the LCD. As before, the system runs on an emulated ARM virtual machine under QEMU. The system's startup sequence is identical to that of C5.3. We shall only show how to set up the system to run the required processes and their reactions to events. The system operates as follows.

(1). Initialization:
copy vectors, configure VIC and SIC for interrupts;
run the initial process P0, which has the lowest priority 0;
initialize drivers for LCD, timer, UART and KBD; start the timer;
(2). Create tasks: P0 call kfork(NAME_task, priority) to create the timer, UART and KBD processes and enter them into the readyQueue. Each process executes its own NAME_task() function with a (static) priority, ranging from 3 to 1.
(3). Then P0 executes a while(1) loop, in which it switches process whenever the readyQueue is non-empty.
(4). Each process resumes to execute its own NAME_code() function, which is an infinite loop. Each process calls sleep (event) to sleep on a unique event value (address of the device data structure).
(5). When an event occurs, the device interrupt handler calls wakeup(event) to wake up the corresponding process. Upon waking up, each process resumes running to handle the event. For example, the timer interrupt handler no longer displays the wall clock. It is performed by the timer process on each second.

The following lists the C code of the Example program C5.4.

```
/********** C code of Example Program C5.4 ***********/
#include "type.h"
#include "vid.c"            // LCD driver
#include "kbd.c"            // KBD driver
#include "uart.c"           // UART driver
#include "timer.c"          // timer driver
#include "exceptions.c"     // exception handlers
#include "queue.c"          // queue functions
#include "kernel.c"         // kernel for task management

int copy_vectors() { // copy vectors as before }
int irq_chandler() { // invoke IRQ handler of timer, UART, KBD }

int timer_handler(){ // on each second: kwakeup(&timer); }
int uart_handler() { // on input line:  kwakeup(&uart); }
int kbd_handler()  { // on input line:  kwakeup(&kbd); }

int i, hh, mm, ss;   // globals for timer_handler
char clock[16] = {"00:00:00"};

int timer_task()    // code of timer_task
{
  while(1){
    printf("timer_task %d running\n", running->pid);
    ksleep((int)&timer);
    // use timer tick to update ss, mm, hh; then display wall clock
    clock[7]='0'+(ss%10); clock[6]='0'+(ss/10);
    clock[4]='0'+(mm%10); clock[3]='0'+(mm/10);
    clock[1]='0'+(hh%10); clock[0]='0'+(hh/10);
    for (i=0; i<8; i++){
        kpchar(clock[i], 0, 70+i);
    }
  }
}

int uart_task()    // code of uart_task
{
  char line[128];
  while(1){
    uprintf("uart_task %d sleep for line from UART\n", running->pid);
    ksleep((int)&uart);
```

```
    uprintf("uart_task %d running\n", running->pid);
    ugets(line);
    uprintf("line = %s\n", line);
  }
}

int kbd_task()        // code of kbd_task
{
  char line[128];
  while(1){
    printf("KBD task %d sleep for a line from KBD\n", running->pid);
    ksleep((int)&kbd);
    printf("KBD task %d running\n", running->pid);
    kgets(line);
    printf("line = %s\n", line);
  }
}

int main()
{
   fbuf_init();     // LCD driver
   uart_init();     // UART driver
   kbd_init();      // KBD driver
   printf("Welcome to Wanix in ARM\n");
   // initialize VIC interrupts: same as before
   timer_init();    // timer driver
   timer_start();
   kernel_init();   // initialize kernel and run P0
   printf("P0 create tasks\n");
   kfork((int)timer_task, 3);   // timer task
   kfork((int)uart_task,  2);   // uart task
   kfork((int)kbd_task,   1);   // kbd task
   while(1){ // P0 runs whenever no task is runnable
     if (readyQueue)
        tswitch();
   }
}
```

5.7.1 Demonstration of Event-Driven Embedded System Using Sleep/Wakeup

Figure 5.7 shows the output screens of running the example program C5.4, which demonstrates an event-driven multi-tasking system. As the Fig. 5.7 shows, the timer task displays a wall clock on the LCD on each second. The uart task prints a line to UART0 only when there is an input line from the UART0 port, and the kbd task prints a line to the LCD only when there is an input line from the keyboard. While these tasks are sleeping for their awaited events, the system is running the idle process P0, which is diverted to handles all the interrupts. As soon as a task is woken up and entered into the readyQueue, P0 switches process to run the newly awakened task.

```
task Uart0 2  starts
task 2  sleep for a line from UART0
rx interrupt: u
rx interrupt: a
rx interrupt: r
rx interrupt: t
rx interrupt:
task 2  0  running
uart     line = uart
task 2  sleep for a line from UART0
```

```
QEMU
Welcome to Wanix in ARM                          00 02 44
timer_init()
timer_start base=0x 101E2000
kernel_init()
running = 0
P0 kfork tasks
proc 0  kforked a child  1
proc 0  kforked a child  2
proc 0  kforked a child  3
readyQueue = [ 1  3 ]->[ 2  2 ]->[ 3  1 ]->NULL
timertask  1  start
task KBD    3  starts
task KBD    3  sleep for a line from KBD
kkbd interrupt: c=0x 6B  k
bkbd interrupt: c=0x 62  b
dkbd interrupt: c=0x 64  d
kbd interrupt: c=0x 0D <cr>
task KBD    3  running
line = kbd
task KBD    3  sleep for a line from KBD
```

Fig. 5.7 Event-driven multitasking system using sleep/wakeup

5.8 Resource Management Using Sleep/Wakeup

In addition to replacing busy-wait loops in device drivers, sleep/wakeup may also be used for general process synchronization. A typical usage of sleep/wakeup is for resource management. A resource is something that can be used by only one process at a time, e.g. a memory region for updating, a printer, etc. Each resource is represented by a res_status variable, which is 0 if the resource is FREE, and nonzero if it's BUSY. Resource management consists of the following functions

int acquire_resource(); // acquire a resource for exclusive use
int release_resource(); // release a resource after use

When a process needs a resource, it calls acquire_resource(), trying to get a resource for exclusive use. In acquire_resource(), the process tests res_status first. If res_status is 0, the process sets it to 1 and returns OK for success. Otherwise, it goes to sleep, waiting for the resource to become FREE. While the resource is BUSY, any other process calling acquire_resource() would go to sleep on the same event value also. When the process which holds the resource calls release_recource(), it clears res_status to 0 and issues wakeup(&res_status) to wakeup ALL processes that are waiting for the resource. Upon waking up, each process must try to acquire the resource again. This is because when an awakened process runs, the resource may no longer be available. The following code segment shows the resource management algorithm using sleep/wakeup.

```
                    int res_status = 0;          // resource initially FREE
--------------------------------------------------------------------------------
int acquire_resource()           |  int release_resource()
{                                |  {
    while(1){                    |
      int SR = int_off();        |     int SR = int_off();
      if (res_status==0){        |
         res_status = 1;         |     res_status = 0;
         break;                  |
      }                          |
      sleep(&res_status);        |     wakeup(&res_status);
    }                            |
    int_on(SR);                  |     int_on(SR);
    return OK;                   |     return OK;
}                                |  }
-----------------------------------------------------
```

5.8.1 Shortcomings of Sleep/Wakeup

Sleep and wakeup are simple tools for process synchronization, but they also have the following shortcomings.

. An event is just a value. It does not have any memory location to record the occurrence of an event. Process must go to sleep first before another process or an interrupt handler tries to wake it up. The sleep-first-wakeup-later order can always be achieved in a UP system, but not necessarily in MP systems. In a MP system, processes may run on different CPUs simultaneously (in parallel). It is impossible to guarantee the execution order of the processes. Therefore, sleep/wakeup are suitable only for UP systems.
. When used for resource management, if a process goes to sleep to wait for a resource, it must retry to get the resource again after waking up, and it may have to repeat the sleep-wakeup-retry cycles many times before succeeding (if ever). The repeated retry loops means poor efficiency due to excessive overhead in context switching.

5.9 Semaphores

A better mechanism for process synchronization is the semaphore, which does not have the above shortcomings of sleep/wakeup. A **(counting)** semaphore is a data structure

```
typedef struct semaphore{
   int spinlock;   // spin lock, needed only in MP systems
   int value;      // initial value of semaphore
   PROC *queue     // FIFO queue of blocked processes
}SEMAPHORE;
```

In the semaphore structure, the spinlock field is to ensure any operation on a semaphore can only be performed as an atomic operation by one process at a time, even if they may run in parallel on different CPUs. Spinlock is needed only for multiprocessor systems. For UP systems, it is not needed and can be omitted. The most well-known operations on semaphores are P and V, which are defined (for UP kernels) as follows.

```
--------------------------------------------------------------------
int P(struct semaphore *s)          | int V(struct semaphore *s)
{                                   | {
  int SR = int_off();               |   int SR = int_off();
  s->value--;                       |   s->value++;
  if (s->value < 0)                 |   if (s->value <= 0)
     block(s);                      |      signal(s);
  int_on(SR);                       |   int_on(SR);
}                                   | }
--------------------------------------------------------------------
int block(struct semaphore *s){     | int signal(struct semaphore *s){
    running->status = BLOCK;        | PROC *p = dequeue(&s->queue);
    enqueue(&s->queue, running);    |       p->status = READY;
    tswitch();                      |       enqueue(&readyQueue, p);
}                                   | }
--------------------------------------------------------------------
```

A **binary semaphore** (Dijkstra 1965) is a semaphore which can only take on two distinct values, 1 for FREE and 0 for OCCUPIED. P/V operations on binary semaphores are defined as

```
-------------------- P/V on binary semaphores --------------------
int P(struct semaphore *s)          | int V(struct semaphore *s)
{                                   | {
  int SR = int_off();               |   int SR = int_off();
  if (s->value == 1)                |   if (s->queue == 0)
     s->value = 0;                  |      s->value = 1;
  else                              |   else
     block(s);                      |      signal(s);
  int_on(SR);                       |   int_on(SR);
}                                   | }
--------------------------------------------------------------------
```

Binary semaphores may be regarded as a special case of counting semaphores. Since counting semaphores are more general, we shall not use, nor discuss, binary semaphores.

5.10 Applications of Semaphores

Semaphores are powerful synchronizing tools which can be used to solve all kinds of process synchronization problems in both UP and MP systems. The following lists the most common usage of semaphores. To simplify the notations, we shall denote s.value = n by s = n, and P(&s)/V(&s) by P(s)/V(s), respectively.

5.10.1 Semaphore Lock

A **critical region (CR)** is a sequence of operations on shared data objects which can only be executed by one process at a time. Semaphores with an initial value = 1 can be used as locks to protect CRs of long durations. Each CR is associated with a semaphore s = 1. Processes access the CR by using P/V as lock/unlock, as in

```
            struct semaphore s = 1;
Processes:  P(s);           // acquire semaphore to lock the CR
                            // CR protected by lock semaphore s
            V(s);           // release semaphore to unlock the CR
```

With the semaphore lock, the reader may verify that only one process can be inside the CR at any time.

5.10.2 Mutex lock

A **mutex** (Pthreads 2015) is a lock semaphore with an additional owner field, which identifies the current owner of the mutex lock. When a mutex is created, its owner filed is initialized to 0, i.e. no owner. When a process acquires a mutex by mutex_lock(), it becomes the owner. A locked mutex can only be unlocked by its owner. When a process unlocks a mutex, it clears the owner field to 0 if there are no processes waiting on the mutex. Otherwise, it unblocks a waiting process from the mutex queue, which becomes the new owner and the mutex remains locked. Extending P/V on semaphores to lock/unlock of mutex is trivial. We leave it as an exercise for the reader. A major difference between mutexes and semaphores is that, whereas mutexes are strictly for locking, semaphores can be used for both locking and process cooperation.

5.10.3 Resource Management using Semaphore

A semaphore with initial value n > 0 can be used to manage n identical resources. Each process tries to get a unique resource for exclusive use. This can be achieved as follows.

```
                       struct semaphore s = n;
Processes: P(s);
                           use a resource exclusively;
           V(s);
```

As long as s > 0, a process can succeed in P(s) to get a resource. When all the resources are in use, requesting processes will be blocked at P(s). When a resource is released by V(s), a blocked process, if any, will be allowed to continue to use a resource. At any time the following invariants hold.

s >= 0 : s = the number of resources still available;
s < 0 : |s| = number of processes waiting in s queue

5.10.4 Wait for Interrupts and Messages

A semaphore with initial value 0 is often used to convert an external event, such as hardware interrupt, arrival of messages, etc. to unblock a process that is waiting for the event. When a process waits for an event, it uses P(s) to block itself in the semaphore's waiting queue. When the awaited event occurs, another process or an interrupt handler uses V(s) to unblock a process from the semaphore queue, allowing it to continue.

5.10.5 Process Cooperation

Semaphores can also be used for process cooperation. The most widely cited cases involving process cooperation are the producer-consumer problem and the reader-writer problem (Silberschatz et al. 2009; Stallings 2011; Tanenbaum et al. 2006; Wang 2015).

5.10.5.1 Producer-Consumer Problem

A set of producer and consumer processes share a finite number of buffers. Each buffer contains a unique item at a time. Initially, all the buffers are empty. When a producer puts an item into an empty buffer, the buffer becomes full. When a consumer gets an item from a full buffer, the buffer becomes empty, etc. A producer must wait if there are no empty buffers. Similarly, a consumer must wait if there are no full buffers. Furthermore, waiting processes must be allowed to continue when their awaited events occur. Figure 5.8 shows a solution of the Producer-Consumer problem using semaphores.

In Fig. 5.8, processes use mutex semaphores to access the circular buffer as CRs. Producer and consumer processes cooperate with one another by the semaphores full and empty.

5.10.5.2 Reader-Writer Problem

A set of reader and writer processes share a common data object, e.g. a variable or a file. The requirements are: an active writer must exclude all others. However, readers should be able to read the data object concurrently if there is no active writer. Furthermore, both readers and writers should not wait indefinitely (starve). Figure 5.9 shows a solution of the Reader-Writer Problem using semaphores.

In Fig. 5.9, the semaphore rwsem enforces FIFO order of all incoming readers and writers, which prevents starvation. The (lock) semaphore rsem is for readers to update the nreader variable in a critical region. The first reader in a batch of readers locks the wsem to prevent any writer from writing while there are active readers. On the writer side, at most one writer can be either actively writing or waiting in wsem queue. In either case, new writers will be blocked in the rwsem queue. Assume that there is no writer blocked at rwsem. All new readers can pass through both P(rwsem) and P(rsem), allowing them to read the data concurrently. When the last reader finishes, it issues V(wsem) to allow any writer blocked at wsem to continue. When the writer finishes, it unlocks both wsem and rwsem. As soon as a writer waits at rwsem, all new comers will be blocked at rwsem also. This prevents readers from starving writers.

5.10.6 Advantages of Semaphores

As a process synchronization tool, semaphores have many advantages over sleep/wakeup.

(1). Semaphores combine a counter, testing the counter and making decision based on the testing outcome all in a single indivisible operation. The V operation unblocks only one waiting process, if any, from the semaphore queue. After passing through the P operation on a semaphore, a process is guaranteed to have a resource. It does not have to retry to get the resource again as in the case of using sleep and wakeup.
(2). The semaphore's value records the number of times an event has occurred. Unlike sleep/wakeup, which must obey the sleep-first-wakeup-later order, processes can execute P/V operations on semaphores in any order.

5.10.7 Cautions of Using Semaphores

Semaphores use a locking protocol. If a process can not acquire a semaphore in P(s), it becomes blocked in the semaphore queue, waiting for someone else to unblock it via a V(s) operation. Improper usage of semaphores may lead to problems. The most well-known problem is deadlock (Silberschatz et al. 2009; Tanenbaum et al. 2006). Deadlock is a condition in which a

```
        DATA buf[N]                          /* N buffer cells */
        int head = tail = 0;                 /* index to buffer cells */
        SEMAPHORE empty = N; full = 0 pmutex =1; cmutex = 1;

------- Producer --------------------- Consumer ------------
while(1){                             | while(1){
   produce an item;                   |
   P(empty);                          |     P(full);
     P(pmutex);                       |       P(cmutex);
       buf[head++] = item;            |       item = buf[tail++];
         head %= N;                   |       tail %= N;
     V(pmutex);                       |       V(cmutex);
   V(full);                           |     V(empty);
}                                     | }
-----------------------------------------------------------
```

Fig. 5.8 Producer-consumer problem solution

```
        SEMAPHORE rwsem = 1; wsem = 1; rsem = 1;
        Int nreader = 0    /* number of active Readers */
-------------------------------------------------------------
ReaderProcess                    |  WriterProcess
{                                |  {
  while(1){                      |     while(1){
    P(rwsem);                    |       P(rwsem);
    P(rsem);                     |       P(wsem);
      nreader++;                 |         /* write data */
      if (nreader==1)            |       V(wsem);
         P(wsem);                |       V(rwsem);
    V(rsem);                     |     }
    V(rwsem);                    |  }
      /* read data */            |
    P(rsem);                     |
      nreader--;                 |
      if (nreader==0)            |
         V(wsem);                |
    V(rsem);                     |
-------------------------------------------------------------
```

Fig. 5.9 Reader-writer problem solution

set of processes mutually wait for one another forever, so that none of the processes can proceed. In multitasking systems, deadlocks must not be allowed to occur. Methods of dealing with deadlocks include **deadlock prevention**, **deadlock avoidance**, and **deadlock detection and recovery**. Among the various methods, only deadlock prevention is practical and used in real operating systems. A simple but effective way to prevent deadlocks is to ensure that processes request different semaphores in a unidirectional order, so that cross or circular locking can never occur. The reader may consult (Wang 2015) for how to deal with deadlocks in general.

5.10.8 Use Semaphores in Embedded Systems

We demonstrate the use of semaphores in embedded systems by the following examples.

5.10.8.1 Device Drivers Using Semaphores

In the keyboard driver of Sect. 5.6.2.1, instead of using sleep/wakeup, we may use semaphore for synchronization between processes and the interrupt handler. To do this, we simply redefine the KBD driver's data variable as a semaphore with the initial value 0.

```
typedef volatile struct kbd{ // base = 0x10006000
  char *base;              // base address of KBD, as char *
  char buf[BUFSIZE];       // input buffer
  int head, tail;
  struct semaphore data; // data.value=0; data.queue=0;
}KBD;

KBD kbd;

int kgetc() // main program calls kgetc() to return a char
{
  char c;
  KBD *kp = &kbd;
```

```
    P(&kp->data);              // P on KBD's data semaphore
    lock();
      c = kp->buf[kp->tail++]; // get a c and update tail index
      kp->tail %= BUFSIZE;
    unlock();                  // enable IRQ interrupts
    return c;
}
```

(3). kbd_handler(): Rewrite KBD interrupt handler to unblock a process, if any. Since process cannot interfere with interrupt handler, there is no need to protect the data variables inside the interrupt handler.

```
kbd_handler()
{
  struct KBD *kp = &kbd;
  scode = *(kp->base+KDATA);     // read scan code in data register
  if (scode & 0x80)              // ignore key releases
     return;
  if (kp->data.value==BUFSIZE) // input buffer FULL
      return;                    // ignore current key
  c = unsh[scode];               // map scan code to ASCII
  kp->buf[kp->head++] = c;       // enter key into CIRCULAR buf[ ]
  kp->head %= BUFSIZE;
  V(&kp->data);
}
```

Note that the interrupt handler only issues V() to unblock waiting process but it should never block or wait. If the input buffer is full, it simply discards the current input key and returns. As can be seen, the logic of the new driver using semaphore is much clearer and the code size is also reduced significantly.

5.10.8.2 Event-Driven Embedded System Using Semaphore

The Example Program C5.4 uses sleep/wakeup for process synchronization. In the next example program, C5.5, we shall use P/V on semaphores for process synchronization. For the sake of brevity, we only show the modified KBD driver and the kbd process code. For clarity, the modifications are shown in bold faced lines.

```
struct kbd{
  char *base;
  char buf[BUFSIZE];  // #define BUFSIZE 128
  int head, tail;
  struct semaphore data, line;
} kbd;
kbd_init()
{
  struct kbd *kp = &kbd;
  kp->base = 0x10006000; // KBD base in Versatilepb
  kp->head = kp->tail = 0;
  kp->data.value = 0; kp->data.queue = 0;
  kp->line.value = 0; kp->line.queue = 0;
}
int kbd_handler()
{
```

```
  struct kbd *kp = &kbd;
  // same code as before: enter ASCII key into input buffer;
  V(&kp->data);
  if (c=='\r')     // return key: has an input line
     V(&kp->line);
}
int kgetc() // process call kgetc() to return a char
{
  char c;
  KBD *kp = &kbd;
  P(&kp->data);
  lock();                        // disable IRQ interrupts
   c = kp->buf[kp->tail++]; // get a c and update tail index
   kp->tail %= BUFSIZE;
  unlock();                      // enable IRQ interrupts
  return c;
}
int kgets(char *line)          // get a string
{
  char c;
  while((c= kgetc()) != '\r')
     *line++ = c;
  *line = 0;
}
int kbd_task()
{
  char line[128];
  struct kbd *kp = &kbd;
  while(1){
     P(&kp->line);             // wait for a line
     kgets(line);
     printf("line = %s\n", line);
  }
}
int main()
{  // initialization code SAME as before
   printf("P0 create tasks\n");
   kfork((int)kbd_task,  1);
   while(1){ // P0 runs whenever no task is runnable
     while(!readyQueue); // loop if readyQueue empty
     tswitch();
   }
}
```

Figure 5.10 shows the outputs of running the C5.5 program.

5.11 Other Synchronization Mechanisms

Many OS kernels use other mechanisms for process synchronization. These include

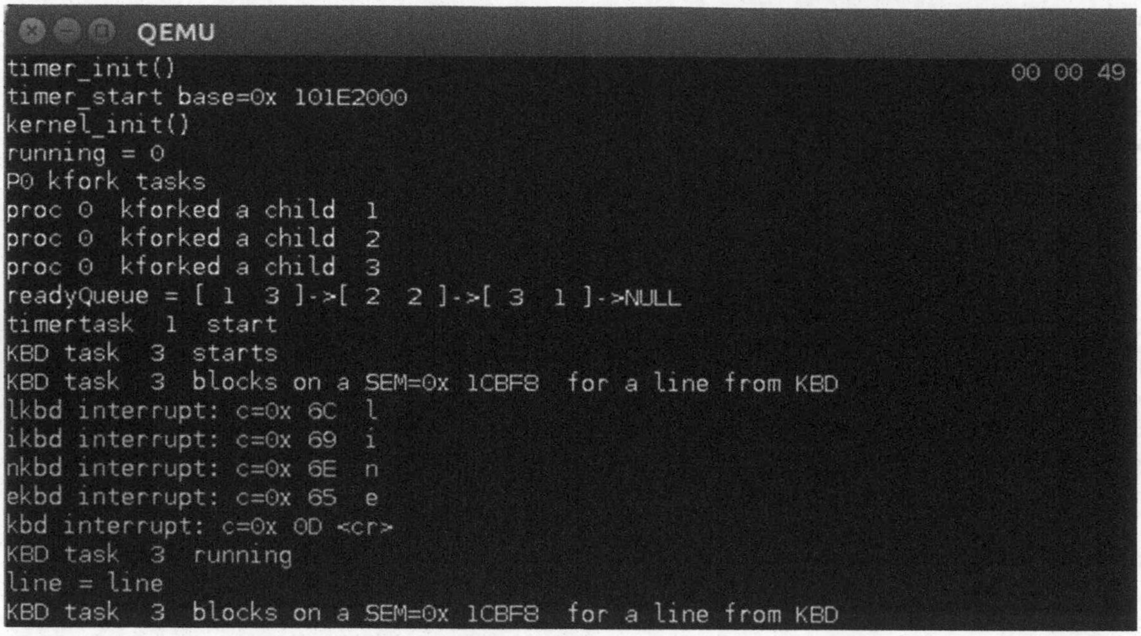

```
QEMU
timer_init()                                                    00 00 49
timer_start base=0x 101E2000
kernel_init()
running = 0
P0 kfork tasks
proc 0  kforked a child  1
proc 0  kforked a child  2
proc 0  kforked a child  3
readyQueue = [ 1  3 ]->[ 2  2 ]->[ 3  1 ]->NULL
timertask  1  start
KBD task  3  starts
KBD task  3  blocks on a SEM=0x 1CBF8  for a line from KBD
lkbd interrupt: c=0x 6C  l
ikbd interrupt: c=0x 69  i
nkbd interrupt: c=0x 6E  n
ekbd interrupt: c=0x 65  e
kbd interrupt: c=0x 0D <cr>
KBD task  3  running
line = line
KBD task  3  blocks on a SEM=0x 1CBF8  for a line from KBD
```

Fig. 5.10 Event-driven multitasking system using semaphores

5.11.1 Event Flags in OpenVMS

OpenVMS (formerly VAX/VMS) (OpenVMS 2014) uses event flags for process synchronization. In its simplest form, an event flag is a single bit, which is in the address spaces of many processes. Either by default or by explicit syscall, each event flag is associated with a specific set of processes. OpenVMS provides service functions for processes to manipulate their associated event flags by

```
set_event(b)  : set b to 1 and wakeup waiter(b) if any;
clear_event(b): clear b to 0;
test_event(b) : test b for 0 or 1;
wait_event(b) : wait until b is set;
```

Naturally, access to an event flag must be mutually exclusive. The differences between event flags and Unix events are:

. A Unix event is just a value, which does not have a memory location to record the occurrence of the event. A process must sleep on an event first before another process tries to wake it up later. In contrast, each event flag is a dedicated bit, which can record the occurrence of an event. Therefore, when using event flags the order of set_event and wait_event does not matter. Another difference is that Unix events are only available to processes in kernel mode, event flags in OpenVMS can be used by processes in user mode.
. Event flags in OpenVMS are in clusters of 32 bits each. A process may wait for a specific bit, any or all of the events in an event cluster. In Unix, a process can only sleep for a single event.
. As in Unix, wakeup(e) in OpenVMS also wakes up all waiters on an event.

5.11.2 Event Variables in MVS

IBM's MVS (2010) uses event variables for process synchronization. An event variable is a structure

```
struct event_variable{
        bit w;             // wait flag initial = 0
        bit p;             // post flag initial = 0
        struct proc *ptr;  // pointer to waiting PROC
} e1, e2,..., en;          // event variables
```

Each event variable e can be awaited by at most one process at a time. However, a process may wait for any number of event variables. When a process calls wait(e) to wait for an event, it does not wait if the event already occurred (post bit=1). Otherwise, it turns on the w bit and waits for the event. When an event occurs, another process uses post(e) to post the event by turning on the p bit. If the event's w bit is on, it unblocks the waiting process if all its awaited events have been posted.

5.11.3 ENQ/DEQ in MVS

In addition to event variables IBM's MVS (2010) also uses ENQ/DEQ for resource management. In their simplest form, ENQ(resource) allows a process to acquire the exclusive control of a resource. A resource can be specified in a variety of ways, such as a memory area, the contents of a memory area, etc. A process blocks if the resource is unavailable. Otherwise, it gains the exclusive control of the resource until it is released by a DEQ(resource) operation. Like event variables, a process may call ENQ(r1,r2,...rn) to wait for all or a subset of multiple resources.

5.12 High-Level Synchronization Constructs

Although P/V on semaphores are powerful synchronization tools, their usage in concurrent programs is scattered. Any misuse of P/V may lead to problems, such as deadlocks. To help remedy this problem, many high-level process synchronization mechanisms have been proposed.

5.12.1 Condition Variables

In Pthreads (Buttlar et al. 1996; Pthreads 2015), threads may use condition variables for synchronization. To use a condition variable, first create a mutex, m, for locking a CR containing shared variables, e.g. a counter. Then create a condition variable, con, associated with the mutex. When a thread wants to access the shared variable, it locks the mutex first. Then it checks the variable. If the counter value is not as expected, the thread may have to wait, as in

```
int count                          // shared variable of threads
pthread_mutex_lock(m);             // lock mutex first
  if (count is not as expected)
      pthread_cond_wait(con, m);   // wait in con and unlock mutex
pthread_mutex_unlock(m);           // unlock mutex
```

pthread_cond_wait(con, m) blocks the calling thread on the condition variable, which automatically and atomically unlocks the mutex m. While a thread is blocked on the condition variable, another thread may use pthread_cond_signal (con) to unblock a waiting thread, as in

```
pthread_lock(m);
   change the shared variable count;
   if (count reaches a certain value)
      pthread_cond_signal(con); // unblock a thread in con
pthread_unlock(m);
```

When an unblocked thread runs, the mutex m is automatically and atomically locked, allowing the unblocked thread to resume in the CR of the mutex m. In addition, a thread may use pthread_cond_broadcast(con) to unblock all threads that are waiting for the same condition variable, which is similar to wakeup in Unix. Thus, mutex is strictly for locking, condition variables may be used for threads cooperation.

5.12.2 Monitors

A monitor (Hoare 1974) is an Abstract Data Type (ADT), which includes shared data objects and all the procedures that operate on the shared data objects. Like an ADT in object-oriented programming (OOP) languages, instead of scattered codes in different processes, all codes which operate on the shared data objects are encapsulated inside a monitor. Unlike an ADT in OOP, a monitor is a CR which allows only one process to execute inside the monitor at a time. Processes can only access shared data objects of a monitor by calling monitor procedures, as in

MONITOR m.procedure(parameters);

The concurrent programming language compiler translates monitor procedure calls as entering the monitor CR, and provides run-time protection automatically. When a process finishes executing a monitor procedure, it exits the monitor, which automatically unlocks the monitor, allowing another process to enter the monitor. While executing inside a monitor, if a process becomes blocked, it automatically exits the monitor first. As usual, a blocked process will be eligible to enter the monitor again when it is SINGALed up by another process. Monitors are similar to condition variables but without an explicit mutex lock, which makes them somewhat more "abstract" than condition variables. The goal of monitor and other high-level synchronization constructs is to help users write "synchronization correct" concurrent programs. The idea is similar to that of using strong type-checking languages to help users write "syntactically correct" programs. These high-level synchronizing tools are used mostly in concurrent programming but rarely used in real operating systems.

5.13 Process Communication

Process communication refers to schemes or mechanisms that allow processes to exchange information. Process communication can be accomplished in many different ways, all of which depend of process synchronization.

5.13.1 Shared Memory

The simplest way for processes communication is through shared memory. In most embedded systems, all processes run in the same address space. It is both natural and easy to use shared memory for process communication. To ensure processes access the shared memory exclusively, we may use either locking semaphore or mutex to protect the shared memory as a critical region. If some processes only read but do not modify the shared memory, we may use the reader-writer algorithm to allow concurrent readers. When using shared memory for process communication, the mechanism only guarantees processes read/write shared memory in a controlled manner. It is entirely up to the user to define and interpret the meaning of the shared memory contents.

5.13.2 Pipes

Pipes are unidirectional inter-process communication channels for processes to exchange streams of data. A pipe has a read end and a write end. Data written to the write end of a pipe can be read from the read end of the pipe. Since their debut in the original Unix, pipes have been incorporated into almost all OS, with many variations. Some systems allow pipes to be bidirectional, in which data can be transmitted in both directions. Ordinary pipes are for related processes. Named pipes are FIFO communication channels between unrelated processes. Reading and writing pipes are usually synchronous and blocking. Some systems support non-blocking and asynchronous read/write operations on pipes. For simplicity, we shall consider a pipe as a finite-sized FIFO communication channel between a set of processes. Reader and writer processes of a pipe are synchronized in the following manner. When a reader reads from a pipe, if the pipe has data, the reader reads as much as it needs (up to the pipe size) and returns the number of bytes read. If the pipe has no data but still has writers, the reader waits for data. When a writer writes data to a pipe, it wakes up the waiting readers, allowing them to continue. If the pipe has no data and also no writer, the reader returns 0. Since readers wait for data if the pipe still has writers, the 0 return value means only one thing, namely the pipe has no data and also no writer. In that case, the reader can stop reading from the pipe. When a writer writes to a pipe, if the pipe has room, it writes as much as it needs to or until the pipe is full. If the pipe has no room but still has readers, the writer waits for room. When a reader reads data from the pipe to create more rooms, it wakes up the waiting writers, allowing them to continue. However, if a pipe has no more readers, the writer must detect this as a broken pipe error and aborts.

5.13.2.1 Pipes in Unix/Linux

In Unix/Linux, pipes are an integral part of the file system, just like I/O devices, which are treated as special files. Each process has three standard file streams; stdin for inputs, stdout for outputs and stderr for displaying error messages, which is usually associated with the same device as stdout. Each file stream is identified by a file descriptor of the process, which is 0 for stdin, 1 for stdout and 2 for stderr. Conceptually, a pipe is a two-ended FIFO file which connects the stdout of a writer process to the stdin of a reader process. This is done by replacing the file descriptor 1 of the writer process with the write-end of the pipe, and replacing the file descriptor 0 of the reader process with the read-end of the pipe. In addition, the pipe uses state variables to keep track of the status of the pipe, allowing it to detect abnormal conditions such no more writers and broken pipe, etc.

5.13.2.2 Pipes in Embedded Systems

Most embedded systems either do not support a file system or the file system may not be Unix-compatible. Therefore, processes in an embedded system may not have opened files and file descriptors. Despite this, we still can implement pipes for process communication in embedded systems. In principle, pipes are similar to the producer-consumer problem, except for the following differences.

. In the producer-consumer problem, a blocked producer process can only be signaled up by another consumer process, and vice versa. Pipes use state variables to keep track of the numbers of reader and writer processes. When a pipe writer detects the pipe has no more readers, it returns with a broken pipe error. When a reader detects the pipe has no more writers and also no data, it returns 0.
. The producer-consumer algorithm uses semaphores for synchronization. Semaphores are suitable for processes to write/read data of the same size. In contrast, pipe readers and writers do not have to read/write data of the same size. For example, writers may write lines but readers read chars, and vice versa.
. The V operation on a semaphore unblocks at most one waiting process. Although rare, a pipe may have multiple writers and readers at both ends. When a process at either end changes the pipe status, it should unblock all waiting processes on the other end. In this case, sleep/wakeup are more suitable than P/V on semaphores. For this reason, pipes are usually implemented using sleep/wakeup for synchronization.

In the following, we shall show how to implement a simplified pipe for process communication. The simplified pipe behaves as named pipes in Linux. It allows processes to write/read a sequence of bytes through the pipe, but it does not check or handle abnormal conditions, such as broken pipe. Full implementation of pipes as file streams will be shown later in Chap. 8 when we discuss general purpose embedded operating systems. The simplified pipe is implemented as follows.

(1). The pipe object: a pipe is a (global) data structure

```
typedef struct pipe{
    char  buf[PSIZE];       // circular data buffer
    int   head, tail;       // circular buf index
    int   data, room;       // number of data & room in pipe
    int   status;           // FREE or BUSY
}PIPE;
PIPE pipe[NPIPE];           // global PIPE objects
```

When the system starts, all the pipe objects are initialized to FREE.

(1). PIPE *create_pipe(): this creates a PIPE object in the (shared) address space of all the processes. It allocates a free PIPE object, initializes it and returns a pointer to the created PIPE object.
(2). Read/write pipe: For each pipe, the user must designate a process as either a writer or a reader, but not both. Writer processes call.

int write_pipe(PIPE *pipePtr, char buf[ ], int n);

to write n bytes from buf[ ] to the pipe. The return value is the number of bytes written to the pipe. Reader processes call

int read_pipe(PIPE *pipePtr, char buf[ ], int n);

which tries to read n bytes from the pipe. The return value is the actual number of bytes read. The following shows the pipe read/write algorithms, which use sleep/wakeup for process synchronization.

```
/*---------- Algorithm of pipe_read -------------*/
int read_pipe(PIPE *p, char *buf, int n)
{  int r = 0;
   if (n<=0)
      return 0;
   validate PIPE pointer p; // p->status must not be FREE
   while(n){
      while(p->data){
         *buf++ = p->buf[p->tail++] // read a byte to buf
         tail %= PSIZE;
         p->data--; p->room++; r++; n--;
         if (n==0)
            break;
      }
      wakeup(&p->room);     // wakeup writers
      if (r)                // if has read some data
         return r;
      // pipe has no data
      sleep(&p->data);      // sleep for data
   }
}

/*---------- Algorithm of write_pipe -----------*/
int write_pipe(PIPR *p, char *buf, int n)
{  int r = 0;
   if (n<=0)
      return 0;
   validate PIPE pointer p; // p->status must not be FREE
   while(n){
      while(p->room)
         p->buf[p->head++] = *buf++; // write a byte to pipe;
         p->head %= PSIZE;
         p->data++; p->room--; r++; n--;
         if (n==0)
            break;
      }
      wakeup(&p->data);     // wakeup readers, if any.
      if (n==0)
         return r;          // finished writing n bytes
      // still has data to write but pipe has no room
```

```
            sleep(&p->room);      // sleep for room
        }
    }
```

Note that when a process tries to read n bytes from a pipe, it may return less than n bytes. If the pipe has data, it reads either n bytes or the number of available bytes in the pipe, whichever is smaller. It waits only if the pipe has no data. Thus, each read returns at most PSIZE bytes.

(3). When a pipe is no longer needed, it may be freed by destroy_pipe(PIPE *pipePtr), which deallocates the PIPE object and wake up all the sleeping processes on the pipe.

5.13.2.3 Demonstration of Pipes

The sample system C5.6 demonstrates pipe in an embedded system with static processes.

When the system starts, the initialization code creates a pipe pointed by kpipe. When the initial process P0 runs, it creates two processes, P1 as the pipe writer, and P2 as the pipe reader. For demonstration purpose, we set the pipe's buffer size to a rather small value, PSIZE=16, so that if the writer tries to write more than 16 bytes, it will wait for rooms. After reading from the pipe, the reader wakes up the writer, allowing it to continue. In the demonstration program, P1 first gets a line from the UART0 port. Then it tries to write the line to the pipe. It waits for room if the pipe is full. P2 reads from the pipe and displays the bytes read. Although each time P2 tries to read 20 bytes, the actual number of bytes read is at most PSIZE. For the sake of brevity, we only show the t.c file of the sample program.

```
/*********** t.c file of Pipe Program C5.6 ***********/
PIPE *kpipe;              // global PIPE pointer
#include "queue.c"
#include "pv.c"
#include "kbd.c"
#include "uart.c"
#include "vid.c"
#include "exceptions.c"
#include "kernel.c"
#include "timer.c"
#include "pipe.c"         // pipe implementation

int pipe_writer()         // pipe writer task code
{
  struct uart *up = &uart[0];
  char line[128];
  while(1){
    uprintf("Enter a line for task1 to get : ");
    printf("task%d waits for line from UART0\n", running->pid);
    ugets(up, line);
    uprints(up, "\r\n");
    printf("task%d writes line=[%s] to pipe\n", running->pid, line);
    write_pipe(kpipe, line, strlen(line));
  }
}
int pipe_reader()         // pipe reader task code
{
  char line[128];
  int i, n;
  while(1){
    printf("task%d reading from pipe\n", running->pid);
    n = read_pipe(kpipe, line, 20);
    printf("task%d read n=%d bytes from pipe : [", running->pid, n);
```

```
    for (i=0; i<n; i++)
      kputc(line[i]);
    printf("]\n");
  }
}
int main()
{
    fbuf_init();
    kprintf("Welcome to Wanix in ARM\n");
    uart_init();
    kbd_init();
    pipe_init();              // initialize PIPEs
    kpipe = create_pipe(); // create global kpipe
    kernel_nit();
    kprintf("P0 kfork tasks\n");
    kfork((int)pipe_writer, 1);   // pipe writer process
    kfork((int)pipe_reader, 1);   // pipe reader process
    while(1){
      if (readyQueue)
         tswitch();
    }
}
```

Figure 5.11 shows the sample outputs of running the pipe program C5.6.

```
Enter a line for task1 to get : testing pipe
Enter a line for task1 to get :
QEMU
Welcome to Wanix in ARM                                  00:00:24
pipe_init()
creating pipe .....OK
timer_init()
kernel_init()
P0 kfork tasks: readyQueue = [ 1  1 ]->[ 2  1 ]->NULL
task 1  waits for line from UART0
task 2  reading from pipe
proc 2  sleep on event=0x 1E058
task 1  writes line=[testing pipe] to pipe
proc 1  wrote  8  bytes
proc 1  wakeup  2
proc 1  sleep on event=0x 1E05C
proc 2  wakeup  1
task 2  read n= 8  bytes from pipe : [testing ]
task 2  reading from pipe
proc 2  sleep on event=0x 1E058
proc 1  wrote  4  bytes
proc 1  wakeup  2
proc 1  finished writing  12  bytes
task 1  waits for line from UART0
task 2  read n= 4  bytes from pipe : [pipe]
task 2  reading from pipe
proc 2  sleep on event=0x 1E058
```

Fig. 5.11 Demonstration of pipe

5.13.3 Signals

Like interrupts to a CPU, signals are (software) interrupts to a process (Unix 1990), which diverts the process from its normal executions to do signal processing. In an ordinary OS, processes execute in one of two distinct modes; kernel mode or user mode. The CPU checks for pending interrupts at the end of each instruction, which is invisible to processes executing on the CPU. Similarly, a process checks for pending signals only in kernel mode, which is invisible to the process in user mode. In most embedded systems, all processes execute in the same address space, so they do not have a separate user mode. If we use signals for process communication, each process must check for pending signals explicitly in the process processing loop, which is equivalent to polling for events. Thus, for embedded systems with only a single address space, signals are unsuited to process communication.

5.13.4 Message Passing

Message passing allows processes to communicate by exchanging messages. Message passing has a wide range of applications. In operating systems, it is a general form of Inter-Process Communication (IPC) (Accetta et al. 1986). In computer networks, it is the basis of server-client oriented programming. In distributed computing, it is used for parallel processes to exchange data and synchronization. In operating system design, it is the basis of so called microkernel, etc. In this section, we shall show the design and implementation of several message passing schemes using semaphores.

The goal of message passing is to allow processes to communicate by exchanging messages. If processes have distinct (user mode) address spaces, they can not access each other's memory area directly. In that case, message passing must go through the kernel. If all processes only execute in the same address space of a kernel, message passing allows processes to exchange information in a controlled manner but hides the synchronization details from the processes. The contents of a message can be designed to suit the needs of the communicating processes. For simplicity, we shall assume that message contents are text strings of finite length, e.g. 128 bytes. To accommodate the transfer of messages, we assume that the kernel has a finite set of message buffers, which are defined as

```
typedef struct mbuf{
        struct mbuf *next;       // pointer to next mbuf
        int pid;                 // sender pid
        int priority;            // message priority
        char contents[128];      // message contents
}MBUF;
MBUF mbuf[NMBUF];                // NMBUF = number of mbufs
```

Initially, all message buffers are in a free mbufList. To send a message, a process must get a free mbuf first. After receiving a message, it releases the mbuf for reuse. Since the mbufList is accessed by many processes, it is a critical region (CR), which must be protected. So we define a semaphore mlock = 1 for processes to access the mbufList exclusively. The algorithm of get_mbuf() and put_mbuf() is

```
   MBUF *get_mbuf()  // return a free mbuf pointer or NULL if none
{
   P(mlock);
     MBUF *mp = dequeue(mbuflList); // return first mbuf pointer
   V(mlock);
   return mp;
}
int put_mbuf(MBUF *mp)          // free a used mbuf to mbuflist
{
   P(mlock);
     enqueue(mbufList)
   V(mlock);
}
```

Instead of a centralized message queue, we assume that each PROC has a private message queue, which contains mbufs delivered to, but not yet received by, the process. Initially, every PROC's mqueue is empty. The mqueue of each process is also a CR because it is accessed by all the sender processes as well as the process itself. So we define another semaphore PROC.mlock = 1 for protecting the process message queue.

5.13.4.1 Asynchronous Message Passing

In the asynchronous message passing scheme, both send and receive operations are non-blocking. If a process can not send or receive a message, it returns a failed status, in which case the process may retry the operation again later. Asynchronous communication is intended mainly for loosely-coupled systems, in which interprocess communication is infrequent, i.e. processes do not exchange messages on a planned or regular basis. For such systems, asynchronous message passing is more suitable due to its greater flexibility. The algorithms of asynchronous send-receive operations are as follows.

```
int a_send(char *msg, int pid) // send msg to target pid
{
  MBUF *mp;
  // validate target pid, e.g. proc[pid] must be a valid processs
  if (!(mp = get_mbuf()))      // try to get a free mbuf
     return -1;                // return -1 if no mbuf
  mp->pid = running->pid;      // running proc is the sender
  mp->priority = 1;    // assume SAME priority for all messages
  copy(mp->contents, msg);     // copy msg to mbuf
  // deliver mbuf to target proc's message queue
  P(proc[pid].mlock);          // enter CR
    enter mp into PROC[pid].mqueue by priority
  V(proc[pid].lock);           // exit CR
  V(proc[pid].message);    // V the target proc's messeage semaphore
  return 1;                    // return 1 for SUCCESS
}

int a_recv(char *msg)     // receive a msg from proc's own mqueue
{
  MBUF *mp;
  P(running->mlock);           // enter CR
  if (running->mqueue==0){     // check proc's mqueue
     V(running->mlock);        // release CR lock
     return -1;
  }
  mp = dequeue(running->mqueue); // remove first mbuf from mqueue
  V(running->mlock);           // release mlock
  copy(msg, mp->contents);     // copy contents to msg
  int sender=mp->pid;          // sender ID
  put_mbuf(mp);                // release mbuf as free
  return sender;
}
```

The above algorithms work under normal conditions. However, if all processes only send but never receive, or a malicious process repeatedly sends messages, the system may run out of free message buffers. When that happens, the message facility would come to a halt since no process can send anymore. One good thing about the asynchronous protocol is that there cannot be any deadlocks because it is non-blocking.

5.13.4.2 Synchronous Message Passing

In the synchronous message passing scheme, both send and receive operations are blocking. A sending process must "wait" if there is no free mbuf. Similarly, a receiving process must "wait" if there is no message in its message queue. In general, synchronous communication is more efficient than asynchronous communication. It is well suited to tightly-coupled systems in which processes exchange messages on a planned or regular basis. In such a system, processes can expect messages to come when they are needed, and the usage of message buffers is carefully planned. Therefore, processes can wait for messages or free message buffers rather than relying on retries. To support synchronous message passing, we define additional semaphores for process synchronization and redesign the send-receive algorithm as follows.

```
SEMAPHORE nmbuf = NMBUF; // number of free mbufs
SEMAPHORE PROC.nmsg = 0; // for proc to wait for messages

MBUF *get_mbuf()            // return a free mbuf pointer
{
   P(nmbuf);                // wait for free mbuf
   P(mlock);
     MBUF *mp = dequeue(mbufList)
   V(mlock);
   return mp;
}
int put_mbuf(MBUF *mp)      // free a used mbuf to freembuflist
{
   P(mlock);
     enqueue(mbufList, mp);
   V(mlock);
   V(nmbuf);
}

int s_send(char *msg, int pid)// synchronous send msg to target pid
{
   // validate target pid, e.g. proc[pid] must be a valid processs
   MBUF *mp = get_mbuf();       // BLOCKing: return mp must be valid
   mp->pid = running->pid;      // running proc is the sender
   copy(mp->contents, msg);     // copy msg from sender space to mbuf
   // deliver msg to target proc's mqueue
   P(proc[pid].mlock);          // enter CR
     enqueue(proc[pid].mqueue, mp);
   V(proc[pid].lock);           // exit CR
   V(proc[pid].nmsg);           // V the target proc's nmsg semaphore
}

int s_recv(char *msg) // synchronous receive from proc's own mqueue
{
   P(running->nmsg);            // wait for message
   P(running->mlock);           // lock PROC.mqueue
     MBUF *mp = dequeue(running->mqueue); // get a message
   V(running->mlock);           // release mlock
   copy(mp->contents, msg);     // copy contents to Umode
   put_mbuf(mp);                // free mbuf
}
```

The above s_send/s_recv algorithm is correct in terms of process synchronization, but there are other problems. Whenever a blocking protocol is used, there are chances of deadlock. Indeed, the s_send/s_recv algorithm may lead to the following deadlock situations.

(1). If processes only send but do not receive, all processes would eventually be blocked at P(nmbuf) when there are no more free mbufs.

(2). If no process sends but all try to receive, every process would be blocked at its own nmsg semaphore.

(3). A process Pi sends a message to another process Pj and waits for a reply from Pj, which does exactly the opposite. Then Pi and Pj would mutually wait for each other, which is the familiar cross-locked deadlock.

As for how to handle deadlocks in message passing, the reader may consult Chap. 6 of (Wang 2015), which also contains a server-client based message passing protocol.

5.13.4.3 Demonstration of Message Passing

The sample Program C5.7 demonstrates synchronous message passing.

(1). Type.h file : added MBUF type

```
struct semaphore{
  int value;
  struct proc *queue;
};
typedef struct mbuf{
  struct mbuf *next;
  int priority;
  int pid;
  char contents[128];
}MBUF;
typedef struct proc{
  // same as before, but added
  MBUF *mQueue;
  struct semaphore mQlock;
  struct semaphore nmsg;
  int    kstack[SSIZE];
}PROC;
```

(2). Message.c file

```
/******** message.c file ************/
#define NMBUF 10
struct semaphore nmbuf, mlock;
MBUF mbuf[NMBUF], *mbufList; // mbufs buffers and mbufList
int menqueue(MBUF **queue, MBUF *p){// enter p into queue by priority}
MBUF *mdequeue(MBUF **queue){// return first queue element}
int msg_init()
{
  int i;  MBUF *mp;
  printf("mesg_init()\n");
  mbufList = 0;
  for (i=0; i<NMBUF; i++)              // initialize mbufList
    menqueue(&mbufList, &mbuf[i]); // all priority=0, so use menqueue()
  nmbuf.value = NMBUF; nmbuf.queue = 0; // counting semaphore
  mlock.value = 1;     mlock.queue = 0; // lock semaphore
}
MBUF *get_mbuf()          // allocate a mbuf
{
  P(&nmbuf);
  P(&mlock);
    MBUF *mp = mdequeue(&mbufList);
  V(&mlock);
  return mp;
```

```
}
int put_mbuf(MBUF *mp) // release a mbuf
{
  P(&mlock);
    menqueue(&mbufList, mp);
  V(&mlock);
  V(&nmbuf);
}
int send(char *msg, int pid) // send msg to partet pid
{
  if (checkPid()<0)          // validate receiving pid
     return -1;
  PROC *p = &proc[pid];
  MBUF *mp = get_mbuf();
  mp->pid = running->pid;
  mp->priority = 1;
  strcpy(mp->contents, msg);
  P(&p->mQlock);
    menqueue(&p->mQueue, mp);
  V(&p->mQlock);
  V(&p->nmseg);
  return 0;
}
int recv(char *msg)          // recv msg from own msgqueue
{
  P(&running->nmsg);
  P(&running->mQlock);
    MBUF *mp = mdequeue(&running->mQueue);
  V(&running->mQlock);
  strcpy(msg, mp->contents);
  int sender = mp->pid;
  put_mbuf(mp);
  return sender;
}
```

(3). t.c file:

```
#include "type.h"
#include "message.c"
int sender() // send task code
{
  struct uart *up = &uart[0];
  char line[128];
  while(1){
    ugets(up, line);
    printf("task%d got a line=%s\n", running->pid, line);
    send(line, 4);
    printf("task%d send %s to pid=4\n", running->pid,line);
  }
}
int receiver() // receiver task code
{
  char line[128];
  int pid;
  while(1){
```

```
    printf("task%d try to receive msg\n", running->pid);
    pid = recv(line);
    printf("task%d received: [%s] from task%d\n",
            running->pid, line, pid);
  }
}
int main()
{
   msg_init();


   kprintf("P0 kfork tasks\n");
   kfork((int)sender, 1);       // sender process
   kfork((int)receiver, 1);     // receiver process
   while(1){
     if (readyQueue)
        tswitch();
   }
}
```

Figure 5.12 shows the sample outputs of running the C5.7 Program.

5.14 Uniprocessor (UP) Embedded System Kernel

An embedded system kernel comprises dynamic processes, all of which execute in the same address space of the kernel. The kernel provides functions for process management, such as process creation, synchronization, communication and termination. In this section, we shall show the design and implementation of uniprocessor (UP) embedded system kernels. There are two distinct types of kernels; non-preemptive and preemptive. In a non-preemptive kernel, each process runs until it gives up CPU voluntarily. In a preemptive kernel, a running process can be preempted either by priority or by time-slice.

5.14.1 Non-preemptive UP Kernel

A Uniprocessor (UP) kernel is non-preemptive if each process runs until it gives up the CPU voluntarily. While a process runs, it may be diverted to handle interrupts, but control always returns to the point of interruption in the same process at the end of interrupt processing. This implies that in a non-preemptive UP kernel only one process runs at a time. Therefore, there is no need to protect data objects in the kernel from the concurrent executions of processes. However, while a process runs, it may be diverted to execute an interrupt handler, which may interfere with the process if both try to access the same data object. To prevent interference from interrupt handlers, it suffices to disable interrupts when a process executes a piece of critical code. This simplifies the system design.

The Example Program C5.8 demonstrates the design and implementation of a non-preemptive kernel for uniprocessor embedded systems. We assume that the system hardware consists of two timers, which can be programmed to generate timer interrupts with different frequencies, a UART, a keyboard and a LCD display. The system software consists of a set of concurrent processes, all of which execute in the same address space but with different priorities. Process scheduling is by non-preemptive priority. Each process runs until it goes to sleep, blocks itself or terminates. Timer0 maintains the time-of-day and displays a wall-clock on the LCD. Since the task of displaying the wall clock is short, it is performed by the timer0 interrupt handler directly. Two periodic processes, timer_task1 and timer_task2, each of which calls the pause(t) function to suspend itself for a number of seconds. After registering a pause time in the PROC structure, the process changes status to PAUSE, enters itself into a pauseList and gives up the CPU. On each second, Timer2 decrements the pause time of every process in the pauseList by 1. When the time reaches 0, it makes the paused process ready to run again. Although this can be accomplished by the sleep-wakeup mechanism, it is intended to show that periodic tasks can be implemented in the

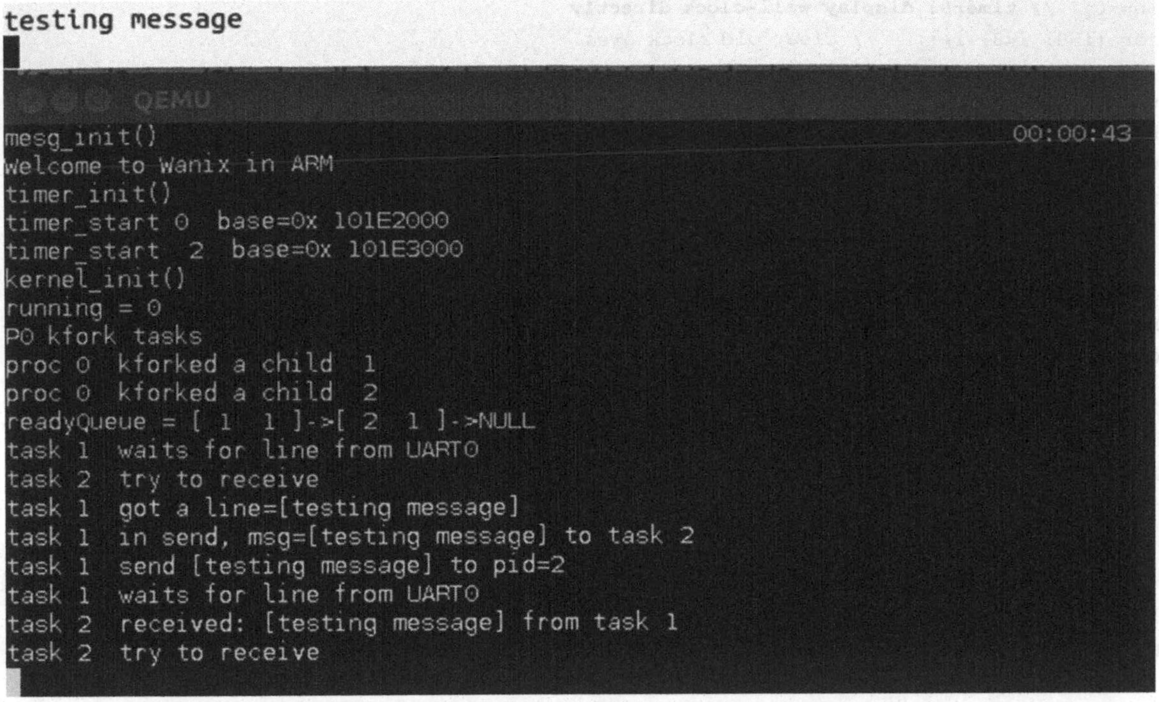

```
testing message

QEMU
mesg_init()                                                    00:00:43
Welcome to Wanix in ARM
timer_init()
timer_start 0  base=0x 101E2000
timer_start  2  base=0x 101E3000
kernel_init()
running = 0
P0 kfork tasks
proc 0  kforked a child  1
proc 0  kforked a child  2
readyQueue = [ 1  1 ]->[ 2  1 ]->NULL
task 1  waits for line from UART0
task 2  try to receive
task 1  got a line=[testing message]
task 1  in send, msg=[testing message] to task 2
task 1  send [testing message] to pid=2
task 1  waits for line from UART0
task 2  received: [testing message] from task 1
task 2  try to receive
```

Fig. 5.12 Demonstration of message passing

general framework of timer service functions. In addition, the system supports two sets of cooperative processes, which implement the producer-consumer problem to demonstrate process synchronization using semaphores. Each producer process tries to get an input line from UART0. Then it deposits the chars into a shared buffer, char pcbuffer[N] of size N bytes. Each consumer process tries to get a char from the pcbuff[N] and displays it to the LCD. Producer and consumer processes share the common data buffer as a pipe. For brevity, we only show the relevant code segments of the system.

```
(1). /********* timer.c file of C5.8 **********/
PROC *pauseList;           // a list of pausing processes
typedef struct timer{
  u32 *base;               // timer's base address;
  int tick, hh, mm, ss;    // per timer data area
  char clock[16];
}TIMER; TIMER timer[4];    // 4 timers;
void timer_init()
{
  printf("timer_init()\n");
  pauseList = 0;           // pauseList initially 0
  // initialize all 4 timers
}
void timer_handler(int n) // n=timer unit
{
    int i; TIMER *t = &timer[n];
    t->tick++; t->ss = t->tick;
    if (t->ss == 60){
      t->ss=0; tp->mm++;
      if (t->mm == 60){
        t->mm = 0; t->hh++;
      }
    }
```

```
    if (n==0){ // timer0: display wall-clock directly
       for (i=0; i<8; i++)    // clear old clock area
          unkpchar(t->clock[i], n, 70+i);
       t->clock[7]='0'+(t->ss%10); t->clock[6]='0'+(t->ss/10);
       t->clock[4]='0'+(t->mm%10); t->clock[3]='0'+(t->mm/10);
       t->clock[1]='0'+(t->hh%10); t->clock[0]='0'+(t->hh/10);
       for (i=0; i<8; i++)     // display new wall clock
          kpchar(t->clock[i], n, 70+i);
    }
    if (n==2){// timer2: process PAUSed PROCs in pauseList
       PROC *p, *tempList = 0;
       while ( p = dequeue(&pauseList) ){
         p->pause--;
         if (p->pause == 0){ // pause time expired
          p->status = READY;
            enqueue(&readyQueue, p);
         }
         else
          enqueue(&tempList, p);
       }
       pauseList = tempList;  // updated pauseList
    }
    timer_clearInterrupt(n);
}
int timer_start(int n)
{
  TIMER *tp = &timer[n];
  kprintf("timer_start %d base=%x\n", n, tp->base);
  *(tp->base+TCNTL) |= 0x80;  // set enable bit 7
}
int timer_clearInterrupt(int n) {
  TIMER *tp = &timer[n];
  *(tp->base+TINTCLR) = 0xFFFFFFFF;
}

typedef struct uart{
  char *base;          // base address; as char *
  u32 id;              // uart number 0-3
  char inbuf[BUFSIZE];
  int  inhead, intail;
  struct semaphore indata, uline;
  char outbuf[BUFSIZE];
  int  outhead, outtail;
  struct semaphore outroom;
  int txon;            // 1=TX interrupt is on
}UART; UART uart[4];  // 4 UART structures

int uart_init()
{
  int i; UART *up;
  for (i=0; i<4; i++){ // only use UART0
      up = &uart[i];
      up->base = (char *)(0x101f1000 + i*0x1000);
      *(up->base+0x2C) &= ~0x10; // disable FIFO
      *(up->base+0x38) |= 0x30;
      up->id = i;
```

```
      up->inhead = up->intail = 0;
      up->outhead = up->outtail = 0;
      up->txon = 0;
      up->indata.value = 0; up->indata.queue = 0;
      up->uline.value = 0;  up->uline.queue = 0;
      up->outroom.value = BUFSIZE; up->outroom.queue = 0;
   }
}
void uart_handler(UART *up)
{
  u8 mis = *(up->base + MIS);  // read MIS register
  if (mis & 0x10)  do_rx(up);
  if (mis & 0x20)  do_tx(up);
}
int do_rx(UART *up)               // UART rx interrupt handler
{
  char c;
  c = *(up->base+UDR);
  up->inbuf[up->inhead++] = c;
  up->inhead %= BUFSIZE;
  V(&up->indata);
  if (c=='\r'){
     V(&up->uline);
  }
}
int do_tx(UART *up)               // UART tx interrupt handler
{
  char c; u8 mask;
  if (up->outroom.value >= SBUFSIZE){ // outbuf[ ] empty
    // disable TX interrupt; return
    *(up->base+IMSC) = 0x10;          // mask out TX interrupt
    up->txon = 0;
    return;
  }
  c = up->outbuf[up->outtail++];
  up->outtail %= BUFSSIZE;
  *(up->base + UDR) = (int)c;
  V(&up->outroom);
}
int ugetc(UART *up)
{
  char c;
  P(&up->indata);
  lock();
    c = up->inbuf[up->intail++];
    up->intail %= BUFSIZE;
  unlock();
  return c;
}
int ugets(UART *up, char *s)
{
  while ((*s = (char)ugetc(up)) != '\r'){
    uputc(up, *s);
    s++;
  }
```

```
  *s = 0;
}
int uputc(UART *up, char c)
{
  if (up->txon){ // if TX is on => enter c into outbuf[]
     P(&up->outroom); // wait for room
     lock();
      up->outbuf[up->outhead++] = c;
      up->outhead %= 128;
     unlock();
     return;
  }
  int i = *(up->base+UFR);          // read FR
  while( *(up->base+UFR) & 0x20 ); // loop while FR=TXF
  *(up->base + UDR) = (int)c;       // write c to DR
  *(up->base+IMSC) |= 0x30;
  up->txon = 1;
}

(3). /**************** t.c file  of C5.8*****************/
#include "type.h"
#include "string.c"
#include "queue.c"
#include "pv.c"
#include "vid.c"
#include "kbd.c"
#include "uart.c"
#include "timer.c"
#include "exceptions.c"
#include "kernel.c"

// IRQ interrupts handler entry point
void irq_chandler()
{
    int vicstatus, sicstatus;
    // read VIC SIV status registers to find out which interrupt
    vicstatus = VIC_STATUS;
    sicstatus = SIC_STATUS;
    if (vicstatus & (1<<4))    // bit4: timer0
       timer_handler(0);
    if (vicstatus & (1<<5))    // bit5: timer2
       timer_handler(2);
    if (vicstatus & (1<<12))  // Bit12: uart0
       uart_handler(&uart[0]);
    if (vicstatus & (1<<13))  // bit13: uart1
       uart_handler(&uart[1]);
    if (vicstatus & (1<<31)){ // SIC interrupts=bit_31=>KBD at bit 3
       if (sicstatus & (1<<3))
          kbd_handler();
    }
}
int pause(int t)
{
  lock();     // disable IRQ interrupts
  running->pause = t;
  running->status = PAUSE;
```

```
  enqueue(&pauseList, running);
  tswitch();
  unlock();   // enable IRQ interrupts
}
int timer1_task()
{
  int t = 5;
  printf("timer1_task %d start\n", running->pid);
  while(1){
     pause(t);
     printf("proc%d run once in %d seconds\n", running->pid, t);
  }
}
int timer2_task()
{
  int t = 7;
  printf("timer2_task %d start\n", running->pid);
  while(1){
    pause(t);
    printf("proc%d run once in %d seconds\n", running->pid, t);
  }
}

/****** SHARED buffer of Producers and Consumers *****/
#define PRSIZE 16
char pcbuf[PRSIZE];
int head, tail;
struct semaphore full, empty, mutex;
int producer()
{
  char c, *cp;
  struct uart *up = &uart[0];
  char line[128];
  while(1){
    ugets(up, line);
    cp = line;
    while(*cp){
      printf("Producer %d P(empty=%d)\n", running->pid, empty.value);
      P(&empty);
      P(&mutex);
        pcbuf[head++] = *cp++;
        head %= PRSIZE;
      V(&mutex);
      printf("Producer %d V(full=%d)\n", running->pid, full.value);
      // show full.queue
      V(&full);
    }
  }
}
int consumer()
{
  char c;
  while(1){
    printf("Consumer %d P(full=%d)\n", running->pid, full.value);
    P(&full);
```

```
    P(&mutex);
      c = pcbuf[tail++];
      tail %= PRSIZE;
    V(&mutex);
    printf("Consumer %d V(empty=%d) ", running->pid, empty.value);
    // show empty.queue
    V(&empty);
  }
}
int main()
{
   fbuf_init();
   uart_init();
   kbd_init();
   printf("Welcome to Wanix in ARM\n");
   //Configure VIC for IRQ interrupts: same as before
   kernel_init();
   timer_init();
   timer_start(0);
   timer_start(2);
   /* initialize data buffer for producer-consumer */
   head = tail = 0;
   full.value = 0;        full.queue = 0;
   empty.value = PRSIZE; empty.queue = 0;
   mutex.value = 1;       mutex.queue = 0;
   printf("P0 kfork tasks\n");
   kfork((int)timer1_task, 1);
   kfork((int)timer2_task, 1);
   kfork((int)producer,    2);
   kfork((int)consumer,    2);
   kfork((int)producer,    3);
   kfork((int)consumer,    3);
   printQ(readyQueue);
   while(1){
     if (readyQueue)
        tswitch();
   }
}
```

5.14.2 Demonstration of Non-preemptive UP Kernel

Figure 5.13 shows the outputs of running the program C5.8, which demonstrate a non-preemptive UP kernel.

5.14.3 Preemptive UP Kernel

In a preemptive UP kernel, while a process runs, CPU can be taken away from it to run another process. Process preemption is triggered by events that require rescheduling of processes. For example, when a higher priority process becomes ready to run or, if using time-sliced process scheduling, when a process has exhausted its time quantum. The preemption policy can be either restrictive or nonrestrictive (fully preemptive). In **restrictive preemption**, while a process is executing a piece of critical code that can not be interfered by other processes, the kernel may disable interrupts or the process scheduler to prevent process switch, thus deferring process preemption until it is safe to do so. In **nonrestrictive preemption**, process

```
QEMU
P0 kfork tasks                                                    00:00:08
proc 0  kforked a child  1
proc 0  kforked a child  2
proc 0  kforked a child  3
proc 0  kforked a child  4
proc 0  kforked a child  5
proc 0  kforked a child  6
readyQueue = [ 5  3 ]->[ 6  3 ]->[ 3  2 ]->[ 4  2 ]->[ 1  1 ]->[ 2  1 ]->NULL
timer1_task  1  start
timer2_task  2  start
Producer  5  V(full=- 2 ) full.queue=[ 6 ]->[ 4 ]->NULL
Producer  5  V(full=- 1 ) full.queue=[ 4 ]->NULL
Consumer  6  V(empty=- 1 ) empty.queue=[ 5 ]->NULL
c = a
Producer  5  V(full=- 1 ) full.queue=[ 6 ]->NULL
Consumer  6  V(empty=- 1 ) empty.queue=[ 5 ]->NULL
c = b
Producer  5  V(full=- 1 ) full.queue=[ 6 ]->NULL
Consumer  6  V(empty=0 )
c = c
Consumer  4  V(empty= 1 )
c = d
proc 1  run once in  5  seconds
proc 2  run once in  7  seconds
```

Fig. 5.13 Demonstration of non-preemptive UP kernel

switch takes place immediately, regardless what the current running process is doing. This implies that, in a fully preemptive UP kernel, processes run logically in parallel. As a result, all shared kernel data structures must be protected as critical regions. This makes a fully preemptive UP kernel logically equivalent to a MP kernel since both must support the concurrent executions of multiple processes. The only difference between a MP kernel and a fully preemptive UP kernel is that, whereas processes in the former may run on different CPUs in parallel, processes in the latter can only run concurrently on the same CPU, but their logical behavior are the same. In the following, we shall only consider fully preemptive UP kernels. MP kernels will be covered in Chap. 9 on multiprocessor systems.

We demonstrate the design and implementation of a fully preemptive UP kernel by an example. The system hardware components are the same as in the sample program C5.8. The system software consists of a set of concurrent processes with different priorities, all of which execute in the same address of the kernel. Process scheduling policy is by fully preemptive priority. In order to support full preemption, we first identify the shared data structures in the kernel that must be protected. These include

(1). **PROC *freeList**: which is used for dynamic task creation and termination.
(2). **PROC *readyQueue**: which is used for process scheduling.
(3). **PROC *sleepList, *pauseList**: which are used for sleep/wakeup operations.

For each shared kernel data structure, we implement its access functions as critical regions, each protected by a mutex lock. In addition, we define the following global variables to control process preemption.

(4). **int swflag**: switch process flag, cleared to 0 when a process is scheduled to run, set to 1 whenever a reschedule event occurs, e.g. when a ready process is added to the readyQueue.
(5). **int intnest**: IRQ interrupts nesting counter, initially 0, increment by 1 when enter an IRQ handler, decrement by 1 when exit an IRQ handler. We assume IRQ interrupts are processed in IRQ handlers directly, i.e. not by pseudo interrupt processing tasks. Process switch may occur only at the end of interrupt processing. For nested interrupts, process switch is deferred until the end of all nested interrupts processing.

The fully preemptive UP kernel consists of the following components.

(1). ts.s file of C5.9:

```
/*************** ts.s file of Program C5.9 ***************/
    .set vectorAddr, 0x10140030
reset_handler:              // SAME AS BEFORE
irq_handler:
   sub   lr, lr, #4         // adjust lr
   stmfd sp!, {r0-r12, lr} // save context in IRQ mode stack
   mrs   r12, spsr          // copy spsr into r12
   stmfd sp!, {r12}         // save spsr in IRQ stack
   msr cpsr, #0x93          // to SVC mode
   ldr  r1, =vectorAddr     // read vectorAddr to ACK interrupt
   ldr  r0, [r1]
   msr cpsr, #0x13     // SVC mode with IRQ enabled
   bl  irq_chandler    // handle IRQ in SVC mode: return 1 for tswitch
   msr cpsr, #0x92
   ldr  r1, =vectorAddr // issue EOI
   str  r0, [r1]
   cmp r0, #0
   bgt do_switch
   ldmfd sp!, {r12}        // get spsr
   msr   spsr, r12         // restore spsr
   ldmfd sp!, {r0-r12, pc}^
do_switch:
   msr cpsr, #0x93
   bl  tswitch             // switch task in SVC mode
   msr cpsr, #0x92
   ldmfd sp!, {r12}        // get spsr
   msr   spsr, r12         // restore spsr
   ldmfd sp!, {r0-r12, pc}^
```

Explanations of the ts.s file of Program C5.9:

Reset_handler: As usual, reset_handler is the entry point. It sets SVC mode stack pointer to the high end of proc[0] and copies the vector table to address 0. Next, it changes to IRQ mode to set the IRQ mode stack pointer. Then it calls main() in SVC mode. During system operation, all processes run in SVC mode in the same address of the kernel.

Irq_handler: Process switch is usually triggered by interrupts, which may wake up sleeping processes, make a blocked process ready to run, etc. Thus, irq_handler is the most important piece of assembly code relevant to process preemption. So we only show the irq_handler code. As pointer out in Chap. 3, the ARM CPU can not handle nested interrupts in IRQ mode. To handle nested interrupts, interrupt processing must be done in a different privileged mode. In order to support process preemption due to interrupts, we choose to handle IRQ interrupts in SVC mode. The reader may consult Chap. 3 for how to handle nested interrupts. The irq_handler code interacts with irq_chnadler() in C to support preemption.

```
int irq_chandler()
{
  void *(*f)();             // f is a function pointer
  intnest++;                // interrupt nest counter++
  f =(void *)*((int *)(VIC_BASE_ADDR+0x30)); // read ISR address
  (*f)();                   // call the ISR function
  intnest--;                // interrupt nest counter--
  if (intnest==0 && swflag){ // if end of nested IRQs & swflag is set
     swflag = 0;
```

```
    return 1;           // return 1 to switch task
  }
  return 0;             // return 0 for no task switch
```

The algorithm of nested IRQ handler for full preemption is as follows.
/******* Algorithm of IRQ handler for full preemption ********/**

(1). Upon entry, adjust return lr; save working registers, lr and spsr in IRQ stack
(2). Change to SVC mode, read vector interrupt controller to ACK interrupt
(3). Call irq_chandler() in SVC mode with IRQ interrupts enabled

/************* irq_chandler() in C *****************/

(4). Increment interrupt nesting counter by 1;
(5). Call ISR in SVC mode; ISR may set the switch task flag by V()
(6). Decrement interrupt nesting counter by 1
(7). Return 1 to switch task if end of IRQs and swflag is set, else retrun 0

/************* back to irq_handler in assembly *******/

(8). Issue EOI for interrupt
(9). If irq_chandler() return 0: normal return; else switch task in SVC mode
(10). When the switched out task resumes, return via IRQ stack

(2). Modified kernel functions for preemption

For brevity, we only show the modified kernel functions in support of process preemption.

These include kwakeup, V on semaphore and mutex_unlock, all of which may make a sleeping or blocked process ready to run and change the readyQueue. In addition, kfork may also create a new process with a higher priority than the current running process. All these functions call reschedule(), which may switch process immediately or defer process switch until the end of IRQ interrupts processing.

```
int reschedule()
{
  int SR = int_off();
  if (readyQueue && readyQueue->priority >= running->priority){
    if (intnest==0){ // not in IRQ handler: preempt immediately
      printf("%d PREEMPT %d NOW\n", readyQueue->pid, running->pid);
      tswitch();
    }
    else{             // still in IRQ handler: defer preemption
      printf("%d DEFER PREEMPT %d\n", readyQueue->pid, running->pid);
      swflag = 1;     // set need to switch task flag
    }
  }
  int_on(SR);
}
int kwakeup(int event)
{
  PROC *p, *tmp=0;
  int SR = int_off();
  while((p = dequeue(&sleepList))!=0){
    if (p->event==event){
       p->status = READY;
```

```
            enqueue(&readyQueue, p);
        }
        else{
            enqueue(&tmp, p);
        }
    }
    sleepList = tmp;
    reschedule();  // if wake up any higher priority procs
    int_on(SR);
}
int V(struct semaphore *s)
{
    PROC *p; int cpsr;
    int SR = int_off();
    s->value++;
    if (s->value <= 0){
        p = dequeue(&s->queue);
        p->status = READY;
        enqueue(&readyQueue, p);
        printf("timer: V up task%d pri=%d; running pri=%d\n",
               p->pid, p->priority, running->priority);
        resechedule();
    }
    int_on(SR);
}
int mutex_unlock(MUTEX *s)
{
    PROC *p;
    int SR = int_off();
    printf("task%d unlocking mutex\n", running->pid);
    if (s->lock==0 || s->owner != running){ // unlock error
        int_on(SR); return -1;
    }
    // mutex is locked and running task is owner
    if (s->queue == 0){ // mutex has no waiter
        s->lock = 0;      // clear lock
        s->owner = 0;     // clear owner
    }
    else{ // mutex has waiters: unblock one as new owner
        p = dequeue(&s->queue);
        p->status = READY;
        s->owner = p;
        printf("%d mutex_unlock: new owner=%d\n", running->pid,p->pid);
        enqueue(&readyQueue, p);
        resechedule();
    }
    int_on(SR);
    return 0;
}
int kfork(int func, int priority)
{
    // create new task with priority as before
    mutex_lock(&readyQueuelock);
     enqueue(&readyQueue, p);
    mutex_unlock(&readyQueuelock);
```

```
  reschedule();
  return p-pid;
}
```

(3). t.c file:

```
/**************** t.c file of Program C5.9***************/
#include "type.h"
MUTEX *mp;              // global mutex
struct semaphore s1;    // global semaphore
#include "queue.c"
#include "pv.c"         // P/V on semaphores
#include "mutex.c"      // mutex functions
#include "kbd.c"
#include "uart.c"
#include "vid.c"
#include "exceptions.c"
#include "kernel.c"
#include "timer.c"  // timer drivers
int copy_vectors(){ // same as before }
int irq_chandler(){ // shown above }
int task3()
{
   printf("PROC%d running: ", running->pid);
   mutex_lock(mp);
    printf("PROC%d inside CR\n", running->pid);
   mutex_unlock(mp);
   printf("PROC%d exit\n", running->pid);
   kexit();
}
int task2()
{
   printf("PROC%d running\n", running->pid);
   kfork((int)task3, 3); // create P3 with priority=3
   printf("PROC%d exit:", running->pid);
   kexit();
}
int task1()
{
  printf("proc%d start\n", running->pid);
  while(1){
   P(&s1);  // task1 is Ved up by timer periodically
   mutex_lock(mp);
    kfork((int)task2, 2); // create P2 with priority=2
    printf("proc%d inside CR\n", running->pid);
   mutex_unlock(mp);
   printf("proc%d finished loop\n", running->pid);
  }
}
int main()
{
   fbuf_init();                // LCD driver
   uart_init();                // UARTs driver
   kbd_init();                 // KBD driver
```

```
    kprintf("Welcome to Wanix in ARM\n");
    // configure VIC for vectored nterrupts
    timer_init();
    timer_start(0);                  // timer0: wall clock
    timer_start(2);                  // timer2: timing events
    mp = mutex_create();             // create global mutex
    s1.value = 0;                    // initialize semaphore s1
    s1.queue = (PROC*)0;
    kernel_init();                   // initialize kernel and run P0
    kfork((int)task1, 1);            // create P1 with priority=1
    while(1){                        // P0 loop
      if (readyQueue)
         tswitch();
    }
}
```

5.14.4 Demonstration of Preemptive UP Kernel

The sample system C5.9 demonstrates fully preemptive process scheduling. When the system starts, it creates and runs the initial process P0, which has the lowest priority 0. P0 creates a new process P1 with priority=1. Since P1 has a higher priority than P0, it immediately preempts P0, which demonstrates direct preemption without any delay. When P1 runs in task1(), it first waits for a timer event by P(s1=0), which is Ved up by a timer periodically (every 4 seconds). While P1 waits on the semaphore, P0 resumes running. When the timer interrupt handler V up P1, it tries to preempt P0 by P1. Since task switch is not allowed inside interrupt handler, the preemption is deferred, which demonstrates preemption may be delayed by interrupts processing. As soon as interrupt processing ends, P1 will preempt P0 to become running again.

To illustrate process preemption due to blocking, P1 first locks the mutex mp. While holding the mutex lock, P1 creates a process P2 with a higher priority=2, which immediately preempts P1. We assume that P2 does not need the mutex. It creates a process P3 with a higher priority=3, which immediately preempts P2. In the task3() code, P3 tries to lock the same mutex mp, which is still held by P1. Thus, P3 gets blocked on the mutex, which switches to run P2. When P2 finishes, it calls kexit () to terminate, causing P1 to resume running. When P1 unlocks the mutex, it unblocks P3, which has a higher priority than P1, so it immediately preempts P1. After P3 terminates, P1 resumes running again and the cycle repeats. Figure 5.14 shows the sample outputs of running the program C5.9.

It is noted that, in a strict priority system, the current running process should always be the one with the highest priority. However, in the sample system C5.9, when the process P3, which has the highest priority, tries to lock the mutex that is already held by by P1, which has a lower priority, it becomes blocked on the mutex and switches to run the next runnable process. In the sample system, we assumed that the process P2 does not need the mutex, so it becomes the running process when P3 gets blocked on the mutex. In this case, the system is running P2, which does not have the highest priority. This violates the strict priority principle, resulting in what's known as a **priority inversion** (Lampson and Redell 1980), in which a low priority process may block a higher priority process. If P2 keeps on running or it switches to another process of the same or lower priority, process P3 would be blocked for an unknown amount of time, resulting in an **unbounded priority inversion**. Whereas simple priority inversion may be considered as natural whenever processes are allowed to compete for exclusive control of resources, unbounded priority inversion could be detrimental to systems with cirtical timing requirements. The sample program C5.9 actually implements a scheme called **priority inheritance**, which prevents unbounded priority inversion. We shall discuss priority inversion in more detail later in Chap. 10 on real-time systems.

```
next running = proc0
V up task 1  pri= 1 ; running pri=0                                   00:00:11
proc 1  DEFER PREEMPT 0  until end of IRQ
proc0 switch task: readyQueue=[ 1  1 ]->[0 0 ]->Nil
next running = proc 1
proc 1  kforked a new PROC 2 : proc 2  PREEMPT proc 1  NOW
proc 1  switch task: readyQueue=[ 2  2 ]->[ 1  1 ]->[0 0 ]->Nil
next running = proc 2
proc 2  kforked a new PROC 3 : proc 3  PREEMPT proc 2  NOW
proc 2  switch task: readyQueue=[ 3  3 ]->[ 2  2 ]->[ 1  1 ]->[0 0 ]->Nil
next running = proc 3
PROC 3  running: TASK 3  BLOCK ON MUTEX: RAISE PROC 1  PRIORITY to  3
proc 3  switch task: readyQueue=[ 1  3 ]->[ 2  2 ]->[0 0 ]->Nil
next running = proc 1
proc 1  inside CR:  1  mutex_unlock: new owner= 3 : restore task 1  pri to  1
proc 3  PREEMPT proc 1  IMMEDIATELY
proc 1  switch task: readyQueue=[ 3  3 ]->[ 2  2 ]->[ 1  1 ]->[0 0 ]->Nil
next running = proc 3
PROC 3  inside CR
proc 3  in kexit: freeList=[ 3 ]->[ 4 ]->[ 5 ]->[ 6 ]->[ 7 ]->[ 8 ]->Nil
proc 3  switch task: readyQueue=[ 2  2 ]->[ 1  1 ]->[0 0 ]->Nil
next running = proc 2
proc 2  running: PROC 2  finished
proc 2  in kexit: freeList=[ 2 ]->[ 3 ]->[ 4 ]->[ 5 ]->[ 6 ]->[ 7 ]->[ 8 ]->Nil
proc 2  switch task: readyQueue=[ 1  1 ]->[0 0 ]->Nil
next running = proc 1
proc 1  switch task: readyQueue=[0 0 ]->Nil
next running = proc0
```

Fig. 5.14 Demonstration of process preemption

5.15 Summary

This chapter covers process management. It introduces the process concept, the principle and technique of multitasking by context switching. It showed how to create processes dynamically and discussed the principles of process scheduling. It covered process synchronization and the various kinds of process synchronization mechanisms. It showed how to use process synchronization to implement event-driver embedded systems. It discussed the various kinds of process communication schemes, which include shared memory, pipes amd message passing. It shows how to integrate these concepts and techniques to implement a uniprocessor (UP) kernel that supports process management with both non-preemptive and preemptve process scheduling. The UP kernel will be the foundation for developing complete operating systems in later chapters.

List of Sample Programs

C5.1: Context switch
C5.2: Multitasking
C5.3. Dynamic Processes
C5.4. Event-driven multitasking system using sleep/wakeup
C5.5: Even-driven multitasking system using semaphores
C5.6: Pipe
C5.7. Message passing
C5.8: Non-preemptive UP kernel
C5.9: Preemptive UP kernel

Problems

1. In the example program C5.2, the tswitch function saves all the CPU registers in the process kstack and it restores all the save registers of the next process when it resumes. Since tswitch() is called as function, it is clearly unnecessary to save/restore R0. Assume that the tswitch function is implemented as

```
tswitch:
// disable IRQ interrupts
stmfd sp!, {r4-r12, lr}
LDR r0, =running        // r0=&running
LDR r1, [r0, #0]        // r1->runningPROC
str sp, [r1, #4]        // running->ksp = sp
bl   scheduler
LDR r0, =running
LDR r1, [r0, #0]        // r1->runningPROC
lDR sp, [r1, #4]
// enable IRQ interrupts
ldmfd sp!, {r4-r12, pc}
```

(1). Show how to initialize the kstack of a new process for it to start to execute the body() function.
(2). Assume that the body() function is written as
    int body(int dummy, int pid, int ppid){ }
where the parameters pid, ppid are the process id and the parent process id of the new process. Show how to modify the kfork() function to accomplish this.

2. Rewrite the UART driver in Chap. 3 by using sleep/wakeup to synchronize processes and the interrupt handler.
3. In the example program C5.3, all the processes are created with the same priority (so that they take turn to run).

   (1). What would happen if the processes are created with different priorities?
   (2). Implement a change_priority(int new_priority) function, which changes the running task's priority to new_priority. Switch process if the current running process no longer has the highest priority.

4. With dynamic processes, a process may terminate when it has completed its task. Implement a kexit() function for tasks to terminate.
5. In all the example programs, each PROC structure has a statically allocate 4KB kstack.

   (1). Implement a simple memory manager to allocate/deallocate memory dynamically. When the system starts, reserve a piece of memory, e.g. a 1MB area beginning at 4MB, as a free memory area. The function

char *malloc(int size)

allocates a piece of free memory of size bytes.When a memory area is no longer needed, it is released back to the free memory area by

void mfree(char *address, int size)

Design a data structure to represent the current available free memory. Then implement the malloc() and mfree() functions.

(2). Modify the kstack field of the PROC structure as an integer pointer

```
int *kstack;
```

and modify the kfork() function as

int kfork(int func, int priority, **int stack_size**)

which dynamically allocates a memory area of stack_size (in 1KB units) for the new process.

(3). When a process terminates, its stack area must be freed. How to implement this?

6. It is well known that an interrupt handler must never go to sleep, become blocked or wait. Explain why?
7. Misuse of semaphores may result in deadlocks. Survey the literatures to find out how to deal with deadlocks by deadlock prevention, deadlock avoidance, deadlock diction and recovery.
8. The pipe program C5.6 is similar to named pipes (FIFOs) in Linux. Read the Linux man page on fifo to learn how to use named pipes for inter-process communication.
9. Modify the example program C5.8 by adding another set of cooperative produced-consumer processes to the system. Let producers get lines from the KBD, process the chars and pipe them to the consumers, which output the chars to the second UART terminal.
10. Modify the sample program C5.9 to handle nested IRQ interrupts in SYS mode but still allows task preemption at the end of nested interrupt processing.
11. Assume that all processes have the same priority. Modify the sample program C5.9 to support process scheduling by time slice.

References

Accetta, M. et al., "Mach: A New Kernel Foundation for UNIX Development", Technical Conference - USENIX, 1986.
ARM toolchain: http://gnutoolchains.com/arm-eabi, 2016.
Bach, M. J., "The Design of the Unix operating system", Prentice Hall, 1990.
Buttlar, D, Farrell, J, Nichols, B., "PThreads Programming, A POSIX Standard for Better Multiprocessing", O'Reilly Media, 1996.
Dijkstra, E.W., "Co-operating Sequential Processes", in Programming Languages, Academic Press, 1965.
Hoare, C.A.R, "Monitors: An Operating System Structuring Concept", CACM, Vol. 17, 1974.
IBM MVS Programming Assembler Services Guide, Oz/OS V1R11.0, IBM, 2010.
Lampson, B; Redell, D. (June 1980). "Experience with processes and monitors in MESA". Communications of the ACM (CACM) **23** (2): 105–117, 1980.
OpenVMS: HP OpenVMS systems Documentation, http://www.hp.com/go/openvms/doc, 2014.
Pthreads: https://computing.llnl.gov/tutorials/pthreads/, 2015.
QEMU Emulators: "QEMU Emulator User Documentation", http://wiki.qemu.org/download/qemu-doc.htm, 2010.
Silberschatz, A., P.A. Galvin, P.A., Gagne, G, "Operating system concepts, 8th Edition", John Wiley & Sons, Inc. 2009.
Stallings, W. "Operating Systems: Internals and Design Principles (7th Edition)", Prentice Hall, 2011.
Tanenbaum, A. S., Woodhull, A. S., "Operating Systems, Design and Implementation, third Edition", Prentice Hall, 2006.
Versatilepb: Versatile Application Baseboard for ARM926EJ-S, ARM Information Center, 2016.
Wang, K.C., "Design and Implementation of the MTX Operating Systems, Springer International Publishing AG, 2015.

Memory Management in ARM

6

6.1 Process Address Spaces

After power-on or reset, the ARM processor starts to execute the reset handler code in Supervisor (SVC) mode. The reset handler first copies the vector table to address 0, initializes stacks of the various privileged modes for interrupts and exception processing, and enables IRQ interrupts. Then it executes the system control program, which creates and starts up processes or tasks. In the static process model, all the tasks run in SVC mode in the same address space of the system kernel. The main disadvantage of this scheme is the lack of memory protection. While executing in the same address space, tasks share the same global data objects and may interfere with one another. An ill-designed or misbehave task may corrupt the shared address space, causing other tasks to fail. For better system security and reliability, each task should run in a private address space, which is isolated and protected from other tasks. In the ARM architecture, tasks may run in the unprivileged User mode. It is very easy to switch the ARM CPU from a privileged mode to User mode. However, once in User mode, the only way to enter privileged mode is by one of the following means.

Exceptions: when an exception occurs, the CPU enters a corresponding privileged mode to handle the exception
Interrupts: an interrupt causes the CPU to enter either FIQ or IRQ mode
SWI: the SWI instruction causes the CPU to enter the Supervisor or SVC mode

In the ARM architecture, System mode is a separate privileged mode, which share the same set of CPU registers with User mode, but it is not the same system or kernel mode found in most other processors. To avoid confusion, we shall refer to the ARM SVC mode as the Kernel mode. SWI can be used to implement system calls, which allow a User mode process to enter Kernel mode, execute kernel functions and return to User mode with the desired results. In order to separate and protect the memory regions of individual tasks, it is necessary to enable the memory management hardware, which provides each task with a separate virtual address space. In this chapter, we shall cover the ARM memory management unit (MMU) and demonstrate virtual address mapping and memory protection by example programs.

6.2 Memory Management Unit (MMU) in ARM

The ARM Memory Management Unit (MMU) (ARM926EJ-S 2008) performs two primary functions: First, it translates virtual addresses into physical addresses. Second, it controls memory access by checking permissions. The MMU hardware which performs these functions consists of a Translation Lookaside Buffer (TLB) , access control logic and translation table walking logic. The ARM MMU supports memory accesses based on either Sections or Pages. Memory management by sections is a one-level paging scheme. The level-1 page table contains section descriptors, each of which specifies a 1 MB block of memory. Memory management by paging is a two-level paging scheme. The level-1 page table contains page table descriptors, each of which describes a level-2 page table. The level-2 page table contains page descriptors, each of which specifies a page frame in memory and access control bits. The ARM paging scheme supports two different page sizes. Small pages consist of 4 KB blocks of memory and large pages consist of 64 KB blocks of memory. Each page comprises 4 sub-pages. Access control can be extended to 1 KB sub-pages within small pages and to 16 KB sub-pages within large pages. The ARM MMU also supports the concept of domains. A domain is a memory area that can be defined with individual access rights. The Domain Access Control Register (DACR) specifies the access rights for up to 16 different

K.C. Wang, *Embedded and Real-Time Operating Systems*,
DOI 10.1007/978-3-319-51517-5_6

domains, 0–15. The accessibility of each domain is specified by a 2-bit permission, where 00 for no access, 01 for client mode, which checks the Access Permission (AP) bits of the domain or page table entries, and 11 for manager mode, which does not check the AP bits in the domain.

The TLB contains 64 translation entries in a cache. During most memory accesses, the TLB provides the translation information to the access control logic. If the TLB contains a translated entry for the virtual address, the access control logic determines whether access is permitted. If access is permitted, the MMU outputs the appropriate physical address corresponding to the virtual address. If access is not permitted, the MMU signals the CPU to abort. If the TLB does not contain a translated entry for the virtual address, the translation table walk hardware is invoked to retrieve the translation information from a translation table in physical memory. Once retrieved, the translation information is placed into the TLB, possibly overwriting an existing entry. The entry to be overwritten is chosen by cycling sequentially through the TLB locations. When the MMU is turned off, e.g. during reset, there is no address translation. In this case every virtual address is a physical address. Address translation takes effect only when the MMU is enabled.

6.3 MMU Registers

The ARM processor treats the MMU as a coprocessor. The MMU contains several 32-bit registers which control the operation of the MMU. Figure 6.1 shows the format of MMU registers. MMU registers can be accessed by using the MRC and MCR instructions.

The following is a brief description of the ARM MMU registers c0 to c10.

Register c0 is for access to the ID Register, Cache Type Register, and TCM Status Registers. Reading from this register returns the device ID, the cache type, or the TCM status depending on the value of Opcode_2 used.
Register c1 is the **Control Register**, which specifies the configuration of the MMU. In particular, setting the M bit (bit 0) enables the MMU and clearing the M bit disables the MMU. The V bit of c1 specifies whether the vector table is remapped during reset. The default vector table location is 0x00. It may be remapped to 0xFFFF0000 during reset.
Register c2 is the **Translation Table Base Register (TTBR)**. It holds the physical address of the first-level translation table, which must be on a 16kB boundary in main memory. Reading from c2 returns the pointer to the currently active first-level translation table. Writing to register c2 updates the pointer to the first-level translation table.
Register c3 is the **Domain Access Control Register.** It consists of 16 two-bit fields, each defines the access permissions for one of the sixteen Domains (D15-D0).
Register c4 is currently not used.

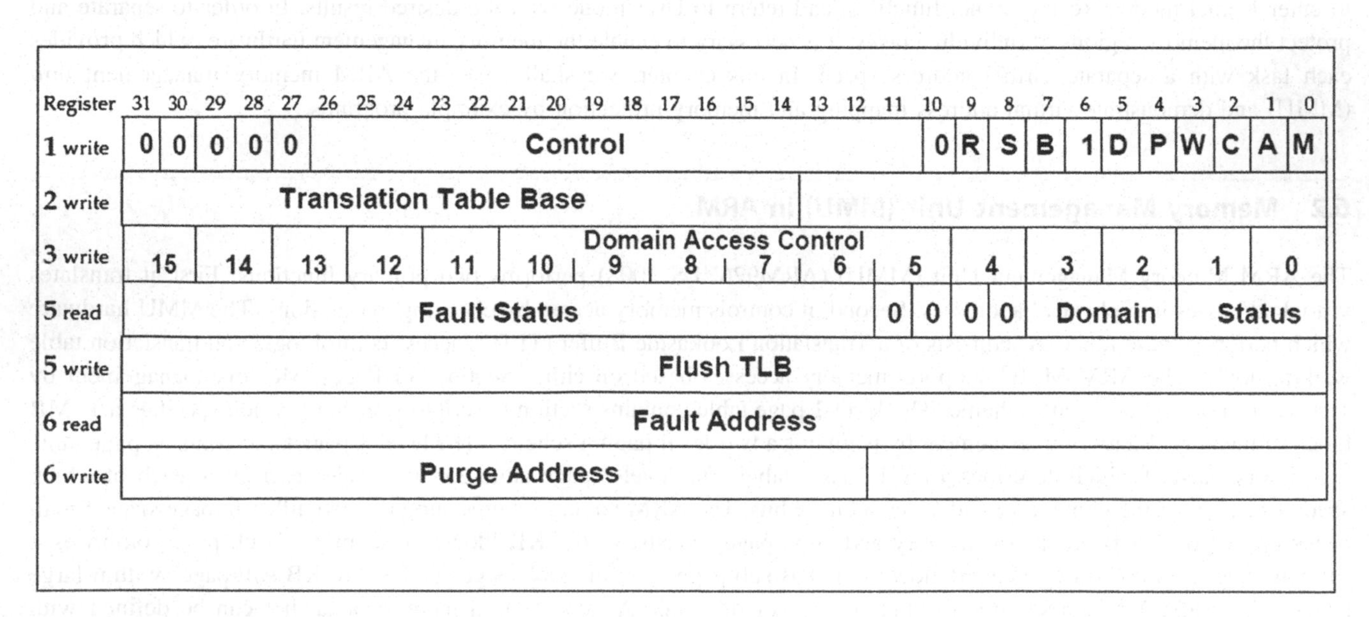

Fig. 6.1 ARM MMU Registers

Register c5 is the **Fault Status Register (FSR).** It indicates the domain and type of access being attempted when an abort occurred. Bits 7:4 specify which of the sixteen domains (D15–D0) was being accessed, and Bits 3:1 indicate the type of access being attempted. A write to this register flushes the TLB.
Register c6 accesses the **Fault Address Register (FAR).** It holds the virtual address of the access when a fault occurred. A write to this register causes the data written to be treated as an address and, if it is found in the TLB, the entry is marked as invalid. This operation is known as a TLB purge. The Fault Status Register and Fault Address Register are only updated for data faults, not for prefetch faults.
Register c7 controls the caches and the write buffer
Register c8 is the **TLB Operations Register**. It is used mainly to invalidate TLB entries. The TLB is divided into two parts: a set-associative part and a fully-associative part. The fully-associative part, also referred to as the lockdown part of the TLB, is used to store entries to be locked down. Entries held in the lockdown part of the TLB are preserved during an invalidate TLB operation. Entries can be removed from the lockdown TLB using an invalidate TLB single entry operation. The invalidate TLB operations invalidate all the unpreserved entries in the TLB. The invalidate TLB single entry operations invalidate any TLB entry corresponding to the virtual address.
Register c9 accesses the Cache Lockdown and TCM Region Register on some ARM boards equipped with TCM.
Register c10 is the **TLB Lockdown Register**. It controls the lockdown region in the TLB.

6.4 Accessing MMU Registers

The registers of CP15 can be accessed by MRC and MCR instructions in a privileged mode. The instruction format is shown in Fig. 6.2.

```
MCR{cond} p15,<Opcode_1>,<Rd>,<CRn>,<CRm>,<Opcode_2>
MCR{cond} p15,<Opcode_1>,<Rd>,<CRn>,<CRm>,<Opcode_2>
```

The CRn field specifies the coprocessor register to access. The CRm field and Opcode_2 fields specify a particular operation when addressing registers. The L bit distinguishes MRC (L = 1) and MCR (L = 0) instructions.

6.4.1 Enabling and Disabling the MMU

The MMU is enabled by writing the M bit, bit 0, of the CP15 Control Register c1. On reset, this bit is cleared to 0, disabling the MMU.

6.4.1.1 Enable MMU

Before enabling the MMU, the system must do the following:

1. Program all relevant CP15 registers. This includes setting up suitable translation tables in memory.
2. Disable and invalidate the Instruction Cache. The instruction cache can be enabled when enabling the MMU.

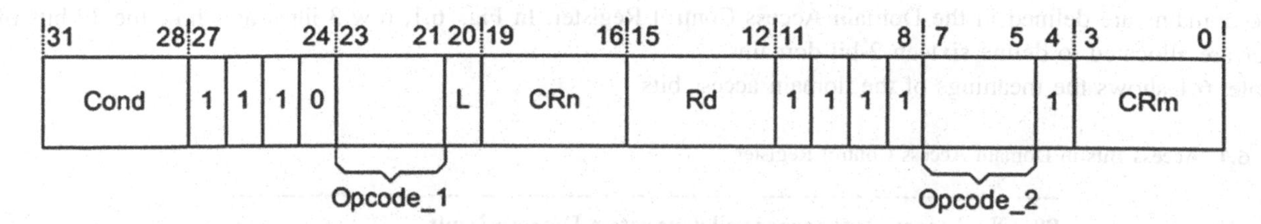

Fig. 6.2 MCR and MRC instruction format

To enable the MMU proceed as follows:

1. Program the Translation Table Base and Domain Access Control Registers.
2. Program first-level and second-level descriptor page tables as needed.
3. Enable the MMU by setting bit 0 in the CP15 Control Register c1.

6.4.1.2 Disable MMU

To disable the MMU proceed as follows:

(1). Clear bit 2 in the CP15 Control Register c1. The Data Cache must be disabled prior to, or at the same time as the MMU being disabled, by clearing bit 2 of the Control Register.

If the MMU is enabled, then disabled, and subsequently re-enabled, the contents of the TLBs are preserved. If the TLB entries are now invalid, they must be invalidated before the MMU is re-enabled.

(2). Clear bit 0 in the CP15 Control Register c1.

When the MMU is disabled, memory accesses are treated as follows:

- All data accesses are treated as Noncacheable. The value of the C bit, bit 2, of the CP15 Control Register c1 should be zero.
- All instruction accesses are treated as Cacheable if the I bit (bit 12) of the CP15 Control Register c1 is set to 1, and Noncacheable if the I bit is set to 0.
- All explicit accesses are Strongly Ordered. The value of the W bit, bit 3, of the CP15 Control Register c1 is ignored.
- No memory access permission checks are performed, and no aborts are generated by the MMU.
- The physical address for every access is equal to its Virtual Address. This is known as a flat address mapping.
- The FCSE PID Should Be Zero when the MMU is disabled. This is the reset value of the FCSE PID. If the MMU is to be disabled the FCSE PID must be cleared.
- All CP15 MMU and cache operations work as normal when the MMU is disabled.
- Instruction and data prefetch operations work as normal. However, the Data Cache cannot be enabled when the MMU is disabled. Therefore a data prefetch operation has no effect. Instruction prefetch operations have no effect if the Instruction Cache is disabled. No memory access permissions are performed and the address is flat mapped.
- Accesses to the TCMs work as normal if the TCMs are enabled.

6.4.2 Domain Access Control

Memory access is controlled primarily by domains. There are 16 domains, each defined by 2 bits in the Domain Access Control register. Each domain supports two kinds of users.

Clients:Clients use a domain
Managers:Managers control the behavior of the domain.

The domains are defined in the Domain Access Control Register. In Fig. 6.1, row 3 illustrates how the 32 bits of the register are allocated to define sixteen 2-bit domains.

Table 6.1 shows the meanings of the domain access bits.

Table 6.1 Access Bits in Domain Access Control Register

00: No Access. Any access will generate a Domain Fault.
01: Client. Check access by permission bits in Section or Page descriptor.
10: Reserved. Currently behaves as no access mode.
11: Manager. Access NOT checked against Section or Page permission bits.

Table 6.2 FSR Status Field Encoding

Priority	Source	Size	Status	Domain
High	Alignment	-	b00x1	Invalid
	External abort	First level	b1100	Invalid
		Second level	b1110	Valid
	Translation	Section Page	b0101	Invalid
			b0111	Valid
	Domain	Section Page	b1001	Valid
			b1011	Valid
	Permission	Section Page	b1101	Valid
			b1111	Valid
Low	External abort	Section or page	b10x0	Invalid

6.4.3 Translation Table Base Register

Register c2 is the Translation Table Base Register (TTBR), for the base address of the first-level translation table. Reading from c2 returns the pointer to the currently active first-level translation table in bits [31:14] and an Unpredictable value in bits [13:0]. Writing to register c2 updates the pointer to the first-level translation table from the value in bits [31:14] of the written value. Bits [13:0] Should Be Zero. The TTBR can be accessed by the following instructions.

MRC p15, < Rd > ,c2, c0, 0 read TTBR
MRC p15, < Rd > ,c2, c0, 0 write TTBR

The CRm and Opcode_2 fields are SBZ (Should-Be-Zero) when writing to c2.

6.4.4 Domain Access Control Register

Register c3 is the Domain Access Control Register consisting of 16 two-bit fields. Each two-bit field defines the access permissions for one of the 16 domains, D15–D0. Reading from c3 returns the value of the Domain Access Control Register. Writing to c3 writes the value of the Domain Access Control Register. The 2-bit Domain access control bits are defined as

Value	Meaning	Description
00	No access	Any access generates a domain fault
01	Client	Accesses are checked against the access permission bits in the section or page descriptor
10	Reserved	Currently behaves like the no access mode
11	Manager	Accesses are not checked against the access permission bits so a permission fault cannot be generated

The Domain Access Control Register can be accessed by the following instructions:

MRC p15, 0, < Rd > , c3, c0, 0 ; read domain access permissions
MCR p15, 0, < Rd > , c3, c0, 0 ; write domain access permissions

6.4.5 Fault Status Registers

Register c5 is the Fault Status Registers (FSRs). The FSRs contain the source of the last instruction or data fault. The instruction-side FSR is intended for debug purposes only. The FSR is updated for alignment faults, and external aborts that occur while the MMU is disabled. The FSR accessed is determined by the value of the Opcode_2 field:

Opcode_2 = 0 Data Fault Status Register (DFSR).
Opcode_2 = 1 Instruction Fault Status Register (IFSR).

The FSR can be accessed by the following instructions.

```
MRC p15, 0, < Rd > , c5, c0, 0 ;read DFSR
MCR p15, 0, < Rd > , c5, c0, 0 ;write DFSR
MRC p15, 0, < Rd > , c5, c0, 1 ;read IFSR
MCR p15, 0, < Rd > , c5, c0, 1 ;write IFSR
```

The format of the Fault Status Register (FSR) is

```
|31                    9| 8 |7 6 5 4 |3 2 1 0 |
_______________________________________________
|        UNP/SBZ        | 0 | Domain | Status |
_______________________________________________
```

The following describes the bit fields in the FSR.

Bits	Description
[31:9]	UNP/SBZP
[8]	Always reads as zero. Writes ignored
[7:4]	Specify the domain (D15–D0) being accessed when a data fault occurred
[3:0]	Type of fault generated. Table 6.2 shows the encodings of the status field in the FSR, and if the Domain field contains valid information

6.4.6 Fault Address Register

Register c6 is the Fault Address Register (FAR). It contains the Modified Virtual Address of the access being attempted when a Data Abort occurred. The FAR is only updated for Data Aborts, not for Prefetch Aborts. The FAR is updated for alignment faults, and external aborts that occur while the MMU is disabled. The FAR can be accessed by using the following instructions.

```
MRC p15, 0,  < Rd > , c6, c0, 0     ; read FAR
MCR p15, 0,  < Rd > , c6, c0, 0     ; write FAR
```

Writing c6 sets the FAR to the value of the data written. This is useful for a debugger to restore the value of the FAR to a previous state. The CRm and Opcode_2 fields are SBZ (Should Be Zero) when reading or writing CP15 c6.

6.5 Virtual Address Translations

The MMU translates virtual addresses generated by the CPU into physical addresses to access external memory, and also derives and checks the access permission. Translation information, which consists of both the address translation data and the access permission data, resides in a translation table located in physical memory. The MMU provides the logic needed to traverse the translation table, obtain the translated address, and check the access permission. The translation process consists of the following steps.

6.5.1 Translation Table Base

The Translation Table Base (TTB) Register points to the base of a translation table in physical memory which contains Section and/or Page descriptors.

6.5.2 Translation Table

The translation table is the level-one page table. It contains 4096 4-byte entries and it must be located on a 16 KB boundary in physical memory. Each entry is a descriptor, which specifies either a level-2 page table base or a section base. Figure 6.3 shows the format of the level-one page entries.

6.5.3 Level-One Descriptor

A level-one descriptor is either a Page Table Descriptor or a Section Descriptor, and its format varies accordingly. Figure 6.3 shows the format of level-one descriptors. The descriptor type is specified by the two least significant bits.

6.5.3.1 Page Table Descriptor

A page table descriptor (second row in Fig. 6.3) defines a level-2 page table. We shall discuss 2-level paging in Sect. 6.6.

6.5.3.2 Section Descriptor

A section descriptor (third row in Fig. 6.3) has a 12-bit base address, a 2-bit AP field, a 4-bit domain field, the C and B bits and a type identifier (b10). The bit fields are defined as follows.

Bits 31:20: base address of a 1 MB section in memory.
Bits 19:12: always 0.
Bits 11:10 (AP): access permissions of this section. Their interpretation depends on the S and R bits (bits 8–9 of MMU control register C1). The most commonly used AP and SR setting are as follows.

```
AP SR  -Supervisor-      —User –        ——————————— Note———————————————
00 xx   No Access        No Access
01 xx   Read/Write       No Access      Access allowed only in Supervisor mode
10 xx   Read/Write       Read Only      Writes in User mode cause permission fault
11 xx   Read/Write       Read/Write     Access permitted in both modes.
```

Bits 8:5: specify one of the sixteen possible domains (in the Domain Access Control Register) that form the primary access controls.
Bit 4: should be 1.
Bits 3:2 (C and B) control the cache and write buffer related functions as follows:
C—Cacheable: data at this address will be placed in the cache (if the cache is enabled).
B—Bufferable: data at this address will be written through the write buffer (if the write buffer is enabled).

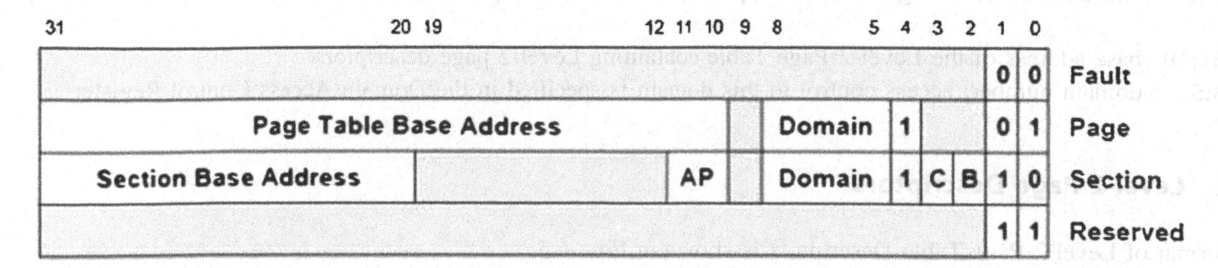

Fig. 6.3 Level-one Descriptors

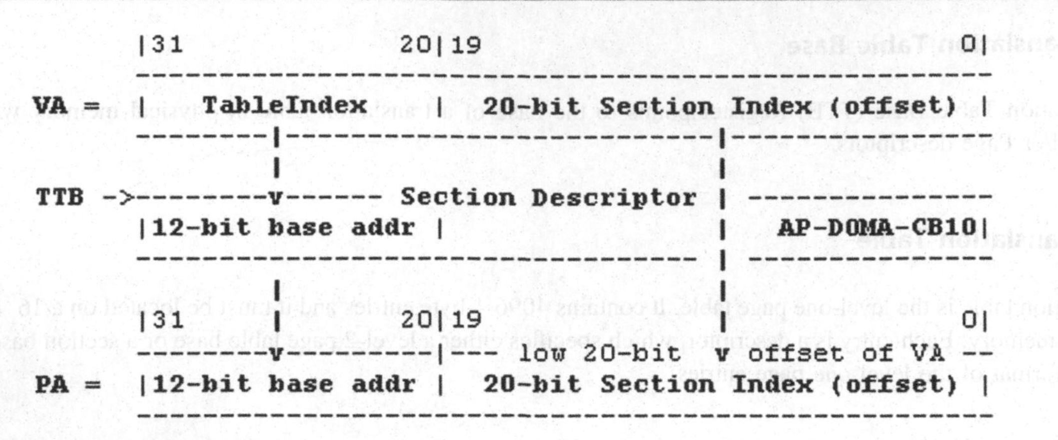

Fig. 6.4 Translation of Section References

6.6 Translation of Section References

In the ARM architecture, the simplest kind of paging scheme is by 1 MB sections, which uses only a level-one page table. So we discuss memory management by sections first. When the MMU translates a virtual address to a physical address, it consults the page tables. The translation process is commonly referred to as a page table walk. When using sections, the translation consists of the following steps, which are depicted in Fig. 6.4.

(1). A Virtual address (VA) comprises a 12-bit Table index and a 20-bit Section index, which is the offset in the section. The MMU uses the 12-bit Table index to access a section descriptor in the translation table pointed by the TTBR.
(2). The section descriptor contains a 12-bit base address, which points to a 1 MB section in memory, a (2-bit) AP field and a (4-bit) domain number. First, it checks the domain access permissions in the Domain Access Control register. Then, it checks the AP bits for accessibility to the Section.
(3). If the permission checking passes, it uses the 12-bit section base address and the 20-bit Section index to generate the physical address as

```
(32-bit)PA  =  ((12-bit)Section_base_address  << 20)  +  (20-bit)Section_index
```

6.7 Translation of Page References

When using 2-level paging, the page tables consist of the following.

6.7.1 Level-1 Page Table

The level-1 page table contains Page Table descriptors (Second row in Fig. 6.3). The contents of a Page Table descriptor are

Bits 31:10: base address of the Level-2 Page Table containing Level-2 page descriptors.
Bits 8:5 : domain number; access control to this domain is specified in the Domain Access Control Register.

6.7.2 Level-2 Page Descriptors

The format of Level-2 Page Table Descriptors is shown in Fig. 6.5.

In a Level-2 page descriptor, the two least significant bits indicate the page size and validity. Other bits are interpreted as follows.

31 … 20 19 … 16	15 … 12	11 10	9 8	7 6	5 4	3	2	1	0	
								0	0	Fault
Large Page Base Address		ap3	ap2	ap1	ap0	C	B	0	1	Large Page
Small Page Base Address		ap3	ap2	ap1	ap0	C	B	1	0	Small Page
								1	1	Reserved

Fig. 6.5 Page table entry (Level Two descriptor)

Bits 31:16 (large pages) or bits **31:12** (small pages) contain the physical address of the page frame in memory. Large page size is 64 KB and small page size is 4 KB.

Bits 11:4 specify the access permissions (ap3–ap0) of the four sub-pages. This allows for finer access control within a page, but it is rarely used in practice.

Bit 3 C—Cacheable: indicates that data at this address will be placed in the IDC (if the cache is enabled).

Bit 2 B—Bufferable: indicates that data at this address will be written through the write buffer (if the write buffer is enabled).

6.7.3 Translation of Small Page References

Page translation involves one additional step beyond that of a section translation: the Level-1 descriptor is a Page Table descriptor, which points to the Level-2 page table containing Level-2 page descriptors. Each Level-2 page descriptor points to a page frame in physical memory. Translation of small page references consists of the following steps, which are depicted in Fig. 6.6.

(1). A Virtual Address VA comprises a 12-bit Level-1 Table Index, an 8-bit Level-2 Table Index and a 12-bit Page Index, which is the byte offset within the page.
(2). Use the 12-bit Level-1 Table Index to access a Level-1 descriptor in the translation table pointed by the translation table base register (TTBR).
(3). Check the domain access permission in the Level-1 descriptor as follows: 00 = abort, 01 = check AP in level-2 page table, 11 = do not check AP of page table.
(4). The leading 22-bits of Level-1 descriptor specifies the (physical) address of a Level-2 Page Table containing 256 page entries. Use the 8-bit Level-2 Table Index to access a Level-2 page descriptor in the Level-2 Page table.

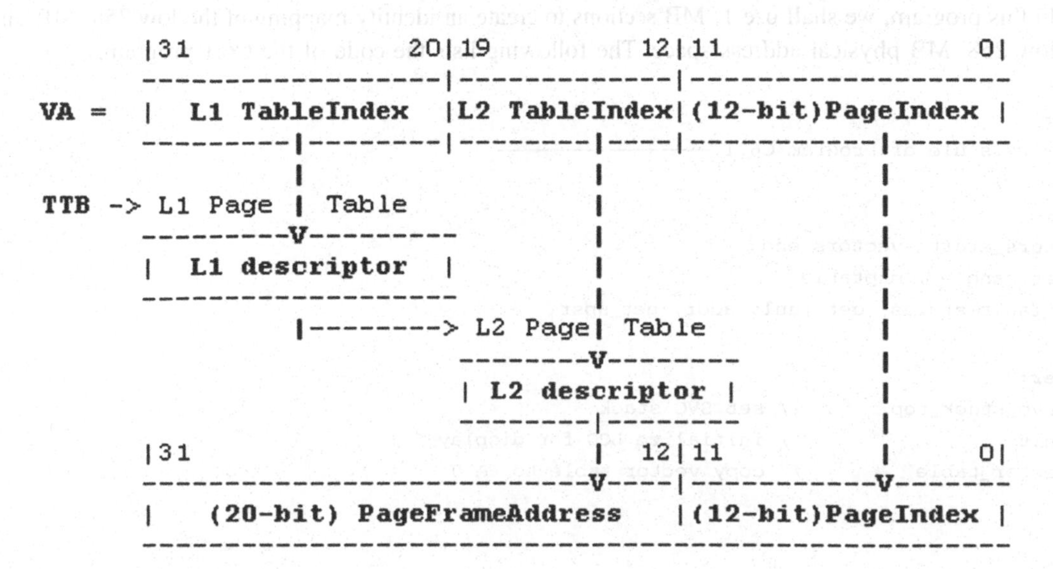

Fig. 6.6 Translation of Small Pages

(5). If domain access bits = 01, check page access permission AP bits (ap3-ap0) in the Level-2 descriptor.
(6). The leading 20 bits of Level-2 page descriptor specifies the PageFrameAddress of the page frame in physical memory. If the access permission checking passes, the generated Physical Address (PA) is

```
(32-bit)PA  =  ((20-bit)PageFrameAddress  << 12)   +   (12-bit)PageIndex
```

6.7.4 Translation of Large Page References

Translation of large page references is similar to that of translating small pages, except for the following differences.

(1). For Large pages, a **VA = [12-bit L1 Index| 8-bit L2 Index| 16-bit Page Index]**.
(2). Since the upper four bits of the Page Index and low order four bits of the Level-2 Page Table index overlap, each page table entry for a Large Page must be duplicated 16 times in consecutive memory locations in the Level-2 Page Table. This is a rather peculiar property of the ARM paging tables for large pages. Since large pages are rarely used in practice, we shall only consider small pages of 4 KB pages size.

6.8 Memory Management Example Programs

This section presents several programming examples which illustrate how to configure the ARM MMU for memory management.

6.8.1 One-Level Paging Using 1 MB Sections

In the first example program, denoted by C6.1, we shall use 1 MB sections to map the VA space to PA space. The program consists of the following components: A ts.s file in assembly code and a t.c file in C, which are (cross) compile-linked to a binary executable t.bin. When running on the emulated Versatilepb board under QEMU, it will be loaded to 0x10000 and runs from there. The program supports the following I/O devices: A LCD for display, a keyboard for inputs, a UART for serial port I/O and also a timer. Since the object here is to demonstrate memory management, we shall focus on how to set up the MMU for virtual address space. The ARM Versatilepb board supports 256 MB RAM and a 2 MB I/O space beginning at 256 MB. In this program, we shall use 1 MB sections to create an identity mapping of the low 258 MB virtual address space to the low 258 MB physical address space. The following lists the code of the C6.1 program.

1. ts.s file:

```
//———————— ts.s file of Program C6.1 ————————
        .text
.code 32
.global vectors_start, vectors_end
.global reset_handler, mkptable
.global get_fault_status, get_fault_addr, get_spsr

reset_handler:
  LDR sp, =svc_stack_top        // set SVC stack
  BL fbuf_init                  // initialize LCD for display
  BL copy_vector_table          // copy vector table to PA 0
```

```
//(m1): build level-1 page table using 1MB sections in C code
  BL mkptable

// (m2): set TTB register to 0x4000
  mov r0, #0x4000
  mcr p15, 0, r0, c2, c0, 0    // set TTB register
  mcr p15, 0, r0, c8, c7, 0    // flush TLB

//(m3): set domain0 01=client(check permission) 11=master(no check)
  mov r0,#1                     // 01 for client mode
  mcr p15, 0, r0, c3, c0, 0


//(m4): enable MMU
  mrc p15, 0, r0, c1, c0, 0    // get c1 into r0
  orr r0, r0, #0x00000001      // set bit0 to 1
  mcr p15, 0, r0, c1, c0, 0    // write to c1
  nop                           // time to allow MMU to finish
  nop
  nop
  mrc p15, 0, r2, c2, c0, 0    // read TLB base reg c2 into r2
  mov r2, r2


// go in ABT mode to set ABT stack
  MSR cpsr, #0x97
  LDR sp, =abt_stack_top
// go in UND mode to set UND stack
  MSR cpsr, #0x9B
  LDR sp, =und_stack_top
// go in IRQ mode to set IRQ stack and enable IRQ interrupts
  MSR cpsr, #0x92             // write to cspr, so in IRQ mode now
  LDR sp, =irq_stack_top      // set IRQ stack poiner
// go back in SVC mode
  MSR cpsr, #0x13             // SVC mode with IRQ enabled
// call main() in SVC mode
  BL main
  B .
swi_handler:                  // dummy swi_handler, not used yet
data_handler:
  sub lr, lr, #4
  stmfd sp!, {r0-r12, lr}
  bl data_chandler            // call data_chandler() in C
  ldmfd sp!, {r0-r12, pc}^
irq_handler:
  sub lr, lr, #4
  stmfd sp!, {r0-r12, lr}
  bl irq_chandler
  ldmfd sp!, {r0-r12, pc}^
vectors_start:
  LDR PC, reset_handler_addr
```

```
  LDR PC, undef_handler_addr
  LDR PC, swi_handler_addr
  LDR PC, prefetch_abort_handler_addr
  LDR PC, data_abort_handler_addr
  B .
  LDR PC, irq_handler_addr
  LDR PC, fiq_handler_addr
  reset_handler_addr:           .word reset_handler
  undef_handler_addr:           .word undef_handler
  swi_handler_addr:             .word swi_handler
  prefetch_abort_handler_addr:  .word prefetch_abort_handler
  data_abort_handler_addr:      .word data_handler
  irq_handler_addr:             .word irq_handler
  fiq_handler_addr:             .word fiq_handler
vectors_end:
get_fault_status:               // read and return MMU reg 5
  MRC p15,0,r0,c5,c0,0          // read DFSR c5
  mov pc, lr
get_fault_addr:                 // read and return MMU reg 6
  MRC p15,0,r0,c6,c0,0          // read DFAR
  mov pc, lr
get_spsr:                       // get SPSR
  mrs r0, spsr
  mov pc, lr
// ----------- end of ts.s file of Program C6.1 ------------
```

Explanations of ts.s file: Since the program is compile-linked without using virtual addresses, the program code can be executed directly during reset when the MMU is off. Therefore, we may call functions in the C code when the program starts up. Upon entry to reset_handler, it first sets the SVC mode stack pointer. Then it calls functions in C to initialize the LCD for display and copy the vector table to address 0. Then it sets up the page table and enables the MMU for VA to PA address translation. The steps are marked as (m1) to (m4), which are explained in more detail below.

(m1): It calls mkptable() in C to set up the level-1 page table using 1 MB sections. The emulated Versatilepb board under QEMU supports 256 MB RAM and a 2 MB I/O space at 256 MB. The level-1 page table is set up to create an identity mapping of the low 258 MB VA to PA, which includes 256 MB RAM and the 2 MB I/O space. The attributes of each section descriptor is set to 0x412 for AP = 01(client), domain = 0000 CB = 00 (D cache and W buffer disabled) and type = 01 for 1 MB sections.
(m2): It sets the Translation Table Base register (TTBR) to point at the page table.
(m3): It sets the access bits of domain 0 to 01 (client) to ensure that the domain can only be accessed in privileged mode. Alternatively, we may also set the domain access bits to 11 (manager mode) to allow access in any mode without domain permission checking.
(m4): It enables the MMU for address translation.

After these steps, every virtual address (VA) is mapped to a physical address (PA). In this case, both addresses are the same due to the identity mapping. The remaining parts of the ts.s code do the following. The program runs in SVC mode but it may enter IRQ mode to handle IRQ interrupts. It may also enter data abort mode to handle data abort exceptions. So, it initializes stack pointers for the various modes. Then it calls main() in SVC mode with IRQ interrupts enabled.

(2). The t.c file:

```
/************** t.c file of Program C6.1 **************/
#include "type.h"
#include "string.c"
#include "uart.c"
#include "kbd.c"
```

```
#include "timer.c"
#include "vid.c"
#include "exceptions.c"

/* set up MMU using 1 MB sections to ID map VA to PA */
// Versatilepb: 256 MB RAM  +  2 MB I/O sections at 256 MB
/************** L1 section descriptor ****************
 |31          20|19      12|--|9|8765|4|32|10|
 | section addr |          |AP|0|DOMN|1|CB|10|
 |              |000000000|01|0|0000|1|00|10|  =  0x412
 |                        KRW  dom0          |
*****************************************************/
int mkptable()   // build level-1 pgtable using 1 MB sections
{
  int i, pentry, *ptable;
  printf("1. build level-1 pgtable at 16 KB\n");
  ptable  =  (int *)0x4000;
  for (i = 0; i < 4096; i ++){ // zero out the pgtable
    ptable[i]  =  0;
  }
  printf("2. fill 258 entries of pgtable to ID map 258 MB VA to PA\n");
  pentry  =  0x412;  // AP = 01,domain = 0000, CB = 00, type = 02 for section
  for (i = 0; i < 258; i ++){ // 258 level-1 page table entries
    ptable[i]  =  pentry;
    pentry  += 0x100000;
  }
  printf("3. finished building level-1 page table\n");
  printf("4. return to set TTB, domain and enable MMU\n");
}
int data_chandler() //data abort handler
{
  u32 fault_status, fault_addr, domain, status;
  int spsr  =  get_spsr();
  printf("data_abort exception in ");
  if ((spsr & 0x1F) ==0x13)
     printf("SVC mode\n");
  fault_status   =  get_fault_status();
  fault_addr     =  get_fault_addr();
  domain  =  (fault_status & 0xF0) >> 4;
  status  =  fault_status & 0xF;
  printf("status   =  %x: domain = %x status = %x (0x5 = Trans Invalid)\n",
         fault_status, domain, status);
  printf("VA addr  =  %x\n", fault_addr);
}
int copy_vector_table(){
    u32 *vectors_src  =  &vectors_start;
    u32 *vectors_dst  =  (u32 *)0;
    while(vectors_src  <  &vectors_end)
       *vectors_dst ++  =  *vectors_src ++;
}
int irq_chandler() // IRQ interrupts handler
{
   // read VIC, SIC status registers to find out which interrupt
   int vicstatus  =  VIC_STATUS;
   int sicstatus  =  SIC_STATUS;
   if (vicstatus & (1 << 4))
```

```
        timer0_handler();
    if (vicstatus & (1 << 12))
        uart0_handler();
    if (vicstatus & (1 << 31) && sicstatus & (1 << 3))
        kbd_handler();
}
int main()
{
    int i, *p;
    char line[128];
    kbd_init();
    uart_init();
    VIC_INTENABLE |=  (1 << 4);  // timer0 at 4
    VIC_INTENABLE |=  (1 << 12); // UART0 at 12
    VIC_INTENABLE  =  1 << 31;   // SIC to VIC's IRQ31
    UART0_IMSC  =  1 << 4;       // enable UART RX interrupts
    SIC_ENSET  =  1 << 3;        // KBD int = 3 on SIC
    SIC_PICENSET  =  1 << 3;     // KBD int = 3 on SIC
    kbd->control  =  1 << 4;
    timer_init();   timer_start(0);
    printf("test MMU protection: try to access VA = 0x00200000\n");
    p  =  (int *)0x002000000; *p  =  123;
    printf("test MMU protection: try to access VA = 0x02000000\n");
    p  =  (int *)0x020000000; *p  =  123;
    printf("test MMU protection: try to access VA = 0x20000000\n");
    p  =  (int *)0x20000000;  *p  =  123;
    while(1){
      printf("main running Input a line: ");
      kgets(line);
      printf("  line  =  %s\n", line);
    }
}
```

The t.c file contains the main function of the program. It first initializes the device drivers and IRQ interrupt handlers. Then it demonstrates memory protection by trying to access invalid virtual addresses, which generate data_abort exceptions. In the data abort handler, data_chandler(), it reads the MMU's data fault register c5 and fault address register c6 to display the reason of the exception (domain invalid) as well as the VA that caused the exception. It is noted that when a data abort exception occurs, PC-8 points to the instruction that caused the exception. In the data abort handler, if we adjust the link register by −8, it would return to the same bad instruction again, resulting in an infinite loop. For this reason, we adjust the return PC by −4 to allow the execution to continue.

(4). **Linker and mk scripts files t.ld**: This is a standard linker script. It defines the program's entry point and allocates memory areas as privileged mode stacks

(5) **Compile-link command**: This is a sh script used to (cross) compile-link the.s and.c files. The starting virtual address of the program is **0x10000**.

```
arm-none-eabi-as -mcpu = arm926ej-s -g ts.s -o ts.o
arm-none-eabi-gcc -c -mcpu = arm926ej-s -g t.c -o t.o
arm-none-eabi-ld -T t.ld ts.o t.o -Ttext = 0x10000 -o t.elf
arm-none-eabi-objcopy -O binary t.elf t.bin
```

(6). **Run the program under QEMU**

```
qemu-system-arm -M versatilepb -m 256 M -kernel t.bin -serial mon:stdio
```

(7). **Sample Outputs:** Fig. 6.7 shows the sample outputs of running the C6.1 program. As the figure shows, trying to access any invalid VA generates a data abort exception.

6.8.2 Two-Level Paging Using 4 KB Pages

The second MMU example program, C6.2, uses 2-level paging. It consists of the following components.

1. ts.s file: This is identical to the t.s file of Program C6.1. During startup, it calls mkptable() in C to set up the two-level page tables. Then it sets the TTB and domain access permission bits and enables the MMU. Then it call main() in SVC mode.
2. t.c file: This is the same as the t.c file of Program C6.1, except for the modified mkptable() function. Instead of building a level-1 page table using 1 MB sections, it builds a level-1 page table and its associated level-2 page tables for 2-level paging. For the sake of brevity, we only show the modified mkptable() function.

```
int mkptable()     // build 2-level pgtables for 2-level paging
{
  int i, j, pentry, *ptable, *pgtable, paddr;
  printf("Welcome to Wanix in ARM\n");
  ptable  =  (int *)0x4000;
  pentry  =  0x412;
  printf("1. build level-1 pgtable at 16 KB to map 258 MB VA to PA\n");
  for (i = 0; i < 4096; i ++){  // zero out 4096 entries
      ptable[i]  =  0;
  }
  printf("2. fill 258 entries in level-1 pgdir with 258 pgtables\n");
  for (i = 0; i < 258; i ++){   // ASSUME 256 MB RAM  +  2 MB I/O space at 256 MB
```

```
QEMU
Welcome to Wanix in ARM                                              00:00:15
1. build level-1 pgtable at 16KB
2. fill low 258 entries of pgtable to ID map 258MB VA to PA
3. finished building level-1 page table
4. return to assembly code to set TTB, domain and enable MMU
kbd init() uart init() timer_init() timer_start
test MMU protection: try to access VA=0x00200000
test MMU protection: try to access VA=0x02000000
data_abort exception in SVC mode
status  = 0x 5 : domain=0x0  status=0x 5  (0x5=Trans Invalid)
VA addr = 0x 2000000
test MMU protection: try to access VA=0x20000000
data_abort exception in SVC mode
status  = 0x 5 : domain=0x0  status=0x 5  (0x5=Trans Invalid)
VA addr = 0x 20000000
main running Input a line: test  line = test
main running Input a line: 
```

Fig. 6.7 Memory Management using 1 MB Sections

```
      ptable[i]  =  (0x500000  +  i*1024) | 0x11; // domain = 0,CB = 00,type = 01
  }
  printf("3. build 258 level-2 pgtables at 5 MB\n");
  for (i = 0; i < 258; i ++){          // 258 page tables
    pgtable  =  (u32 *)((u32)0x500000  +  (u32)i*1024);
    paddr  =  i*0x100000 | 0x55E;   // all APs = 01|01|01|01|CB = 11|type = 10
    for (j = 0; j < 256; j ++){         // 256 entries, each points to 4 KB PA
      pgtable[j]  =  paddr  +  j*4096; // inc by 4 KB
    }
  }
  printf("4. finished building Two-level Page tables\n");
  printf("5. return to assembly to set TTB, domain and enable MMU\n");
}
```

Figure 6.8 shows the sample outputs of running the C6.2 program, which demonstrates two-level paging. When the program starts, it first builds the two-level page tables in 5 steps. Then it tests memory protection by trying to access some VA locations. As the figure shows, attempts to access VA = 0x00200000 (2 MB) and VA = 0x02000000 (16 MB), do not cause any data-abort exception because both are within the 258 MB VA space of the kernel. However, for VA = 0xA0000000, it generates a data_abort exception because the VA is outside of the 258 MB VA space of the kernel.

6.8.3 One-Level Paging with High VA Space

The third MMU program, C6.3, uses 1 MB sections to map the virtual address space to **0x80000000 (2 GB)**. The program will be loaded to the physical address 0x10000 by QEMU. It is compile-linked with the starting virtual address **0x80010000**. Since the program is compiled with virtual addresses, we can not call any function in the program's C code before setting up the page table and enabling MMU to map VA to PA. For this reason, the initial page table must be built in assembly code during reset while the MMU is off. When the program starts, we first set up an initial page table to ID map the lowest 1 MB VA to PA. This is because the vector table is located at the physical address 0 and the entry points of the exception handlers are located within 4 KB from the vector table. In addition to ID map the low 1 MB, we also fill the page table entries 2048–2295 to map the virtual address space VA = (0x80000000, 0x80000000 + 258 MB) to the low 258 MB PA. Next, we enable the MMU to start VA to PA address translation. Then we call main() in C using its VA at 0x80000000 + main. Since the entire program resides in the lowest 1 MB physical memory, we may also call main() using its PA. The following list the assembly code.

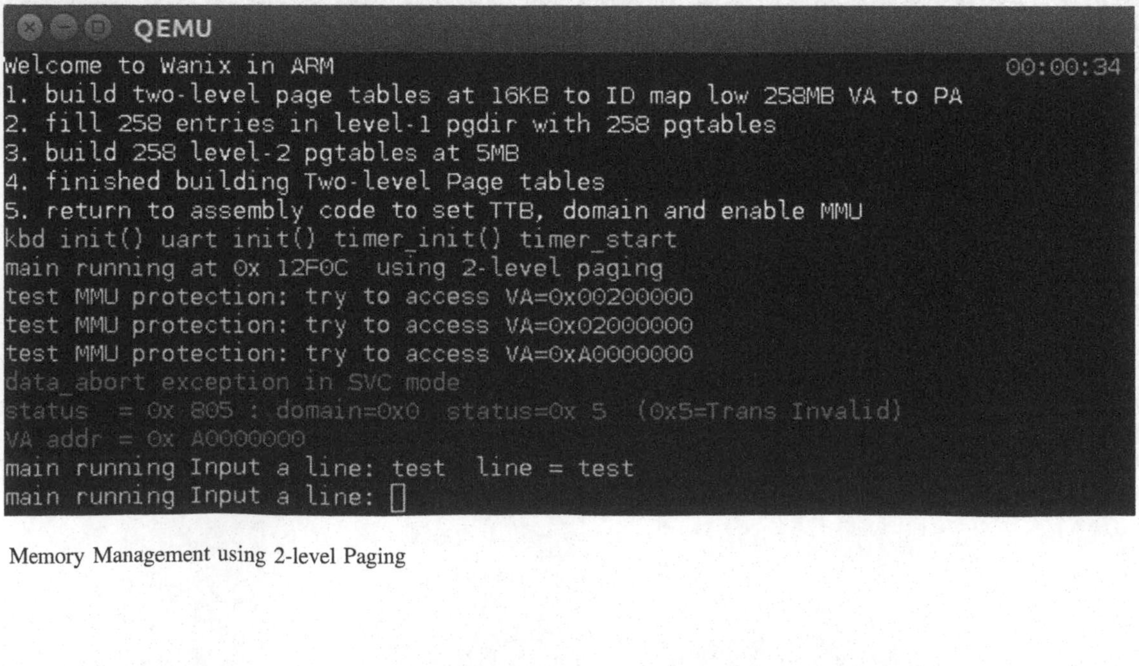

```
QEMU
Welcome to Wanix in ARM                                          00:00:34
1. build two-level page tables at 16KB to ID map low 258MB VA to PA
2. fill 258 entries in level-1 pgdir with 258 pgtables
3. build 258 level-2 pgtables at 5MB
4. finished building Two-level Page tables
5. return to assembly code to set TTB, domain and enable MMU
kbd init() uart init() timer_init() timer_start
main running at 0x 12F0C  using 2-level paging
test MMU protection: try to access VA=0x00200000
test MMU protection: try to access VA=0x02000000
test MMU protection: try to access VA=0xA0000000
data_abort exception in SVC mode
status  = 0x 805 : domain=0x0  status=0x 5  (0x5=Trans Invalid)
VA addr = 0x  A0000000
main running Input a line: test  line = test
main running Input a line:
```

Fig. 6.8 Memory Management using 2-level Paging

1. ts.s file:

```
//  ---------------------- ts.s file of Program C6.3 ----------------------
          .text
.code 32
.global reset_handler, vectors_start, vectors_end
.global get_fault_status, get_fault_addr
reset_handler:             // entry point
// Versatilepb: 256 MB RAM, 2 1 MB I/O sections at 256 MB
// clear ptable at 0x4000 (16 KB) to 0
  mov r0, #0x4000          // ptable at 0x4000 = 16 KB
  mov r1, #4096            // 4096 entries
  mov r2, #0               // fill all with 0
1:
  str  r2, [r0], #4        // store r3 to [r0]; inc r0 by 4
  subs r1, r1, #1          // r1-; set condition flags
  bgt 1b                   // loop r1 = 4096 times
//(m1): ptable[0] ID map low 1 MB VA to PA
//      ptable[2048-2295] map VA = [2 GB,2 GB + 258 MB] to low 258 MB PA
  mov r0, #0x4000
  mov r1, r0
  add r1, r1, #(2048*4)  // entry 2048 in ptable[ ]
  mov r2, #256           // r2 = 256
  add r2, r2, #2         // r2 = 258 entries
  mov r3, #0x100000      // r3 = 1 M increments
  mov r4, #0x400         // r4 = AP = 01 (KRW, user no) AP = 11: both KU r/w
  orr r4, r4, #0x12      // r4  =  0x412 (OR 0xC12 if AP = 11)
  str r4, [r0]           // ptable[0]  =  0x412
// ptable[2048-2257] map to low 258 MB PA
2:
  str r4, [r1], #4       // store r4 to [r1]; inc r1 by 4
  add r4, r4, r3         // inc r4 by 1 M
  subs r2,r2, #1         // r2-
  bgt 2b                 // loop r2 = 258 times

//(m2): set TTB register pointing at pgtable at 0x4000
  mov r0, #0x4000
  mcr p15, 0, r0, c2, c0, 0  // set TTBR with PA = 0x4000
  mcr p15, 0, r0, c8, c7, 0  // flush TLB

//(m3): set domain0: 01 = client(check permission)11 = manager(no check)
  mov r0, #0x1                  // b01 for CLIENT
  mcr p15, 0, r0, c3, c0, 0     // write to domain REG c3

//(m4): enable MMU
  mov r0, #0x1
  mrc p15, 0, r0, c1, c0, 0     // read control REG to r0
  orr r0, r0, #0x00000001       // set bit0 of r0 to 1
  mcr p15, 0, r0, c1, c0, 0     // write to control REG c1  ==> MMU on
  nop
  nop
  nop
  mrc p15, 0, r2, c2, c0, 0    // read TLB base reg c2 into r2
  mov r2, r2                   // time to allow MMU to finish
// set SVC stack to HIGH END of svc_stack[ ]
  LDR r0,  = svc_stack        // r0 points svc_stack[]
```

```
  ADD r1, r0, #4096          // r1 - >  high end of svc_stack[]
  MOV sp, r1
// set IRQ stack and enable IRQ interrupts
  MSR cpsr, #0x92       // write to cspr
  ldr sp,  = irq_stack // u32 irq_stack[1024] in t.c
  add sp, sp, #4096     // ensure it's a VA from 2 GB
// set ABT stack */
  MSR cpsr, #0x97
  LDR sp,  = abt_stack_top
// go back to SVC mode
  MSR cpsr, #0x93       // SVC mode with IRQ off
  BL  copy_vector_table  // copy vector table to PA 0
  MSR cpsr, #0x13       // SVC mode with IRQ on
  BL main               // call main() in C
  B .
data_handler:
  sublr, lr, #4
  stmfd sp!, {r0-r12, lr}  // save regs in abt_stack
  bldata_chandler          // call data_chandler in C
  ldmfd sp!, {r0-r12, pc}^ // pop abt_stack and return
irq_handler:               // IRQ interrupts entry point
  sublr, lr, #
  stmfd sp!, {r0-r12, lr}  // save regs in irq_stack
  bl  irq_chandler         // call irq_chandler() in C
  ldmfd sp!, {r0-r12, pc}^ // pop regs and return
getsp:                     // return current sp
   mov r0, sp
   mov pc, lr
svc_entry:                     // dummy SVC entry, not used yet
vectors_start:                 // vector table
  LDR PC, reset_handler_addr
  LDR PC, undef_handler_addr
  LDR PC, svc_handler_addr
  LDR PC, prefetch_abort_handler_addr
  LDR PC, data_abort_handler_addr
  B .
  LDR PC, irq_handler_addr
  LDR PC, fiq_handler_addr
  reset_handler_addr:          .word reset_handler
  undef_handler_addr:          .word undef_handler
  svc_handler_addr:            .word svc_entry
  prefetch_abort_handler_addr: .word prefetch_abort_handler
  data_abort_handler_addr:     .word data_handler
  irq_handler_addr:            .word irq_handler
  fiq_handler_addr:            .word fiq_handler
vectors_end:
// other utility functions: not shown
//—————————— end of ts.s file ——————————
```

As before, we focus on the code that sets up the MMU, which are labeled as (m1) to (m4).

(m1): Upon entry, it first sets up a level-1 page table at 0x4000 (16 KB) using 1 MB sections. Entry 0 of the page table, ptable [0], is used to ID map the lowest 1 MB of VA to PA, which is required by the vector table. Then it fills the page table entries ptable [2048] to ptable [2048 + 258] with section descriptors, which map VA = (0x80000000, 0x80000000

+ 258 MB) to the low 258 MB PA. The attributes field of each section descriptor is set to AP = 01 (client), domain = 0000, CB = 00 (D cache and W buffer off) and type = 10 (section).
(m2): After setting up the level-1 page table, it sets the TTBR to the page table at 0x4000.
(m3): It sets the access bits of domain0 to 01 for client mode.
(m4): Then it enables the MMU for VA to PA translation.

With MMU enabled, it can now call functions in C, which are compiled with virtual addresses starting at 0x80010000. It sets up the stacks for the various modes, copies the vector table to address 0 and calls main() in SVC mode.

2. The t.c file:

```
/************* t.c file of Program C6.3 ************/
#include "type.h"
#include "string.c"
#include "uart.c"
#include "kbd.c"
#include "timer.c"
#include "vid.c"
#include "exceptions.c"
int irq_chandler()      { // same as in Program C6.2 }
int copy_vector_table(){ // same as in Program C6.2 }
extern int reset_handler();
int svc_stack[1024], irq_stack[1024]; // SVC and IRQ stacks in VA
int g; // global variable to show VA
int main()
{
   int a, sp, *p;
   char line[128];
   fbuf_init();
   kbd_init();
   uart_init();
   timer_init(); timer_start(0);
   // enable device IRQ interrupts: same as before
   printf("Welcome to WANIX in Arm\n");
   printf("Demonstration of one-level sections VA = 0x80000000(2G)\n");
   printf("main running at VA = %x using level-1 1 MB sections\n", main);
   printf("reset_handler  =  %x\n", reset_handler);
   printf("data_chandler  =  %x\n", data_chandler);
   printf("SVC stack pointer   =  %x\n", getsp());
   printf("global variable g at %x\n", &g);
   printf("local  variable a at %x\n", &a);
   printf("test MMU protection: try to access VA = 0x80200000\n");
   p  =  (int *)(0x80200000);  a  =  *p;
   printf("test MMU protection: try to access VA = 0x200000\n");
   p  =  (int *)0x200000;      a  =  *p;
   printf("test MMU protection: try to access VA = 0xA0000000\n");
   p  =  (int *)0xA0000000;    a  =  *p;
   while(1){
     printf("enter a line at %x: ", line);
     kgets(line);
     printf("  line = %s\n", line);
   }
}
```

Explanations of t.c code: In main(), we display the VA of some functions and variables to show that they are in the virtual address range above 2 GB (0x80000000). Then we verify the memory protection mechanism of the MMU by trying to access some invalid VA, which would generate data abort exceptions. In the data_abort_handler(), we read and display the MMU's fault_status and fault_addr registers to show the reason as well as the invalid VA that caused the exception.

3. **Starting Virtual Address**: In order to use virtual addresses starting from 0x80000000, the compile-link commands are modified as

```
arm-none-eabi-as -mcpu = arm926ej-s ts.s -o ts.o
arm-none-eabi-gcc -c -mcpu = arm926ej-s t.c -o t.o
arm-none-eabi-ld -T t.ld vector.o ts.o t.o -Ttext = 0x80010000 -o t.elf
```

4. **Other Modifications**: With a starting VA = 0x80000000, the base addresses of all the I/O devices must be changed to virtual addresses. These are done by a VA(x) macro

```
#define VA(x) (0x80000000 + (u32)x)
```

which adds 0x80000000 to their base addresses in the memory map.

5. **Sample Outputs of Program C6.3**: Figure 6.9 shows the sample outputs of running the C6.3 program. As the figure shows, the VA range is above 0x80000000, and any attempt to access an invalid VA generates a data_abort exception.

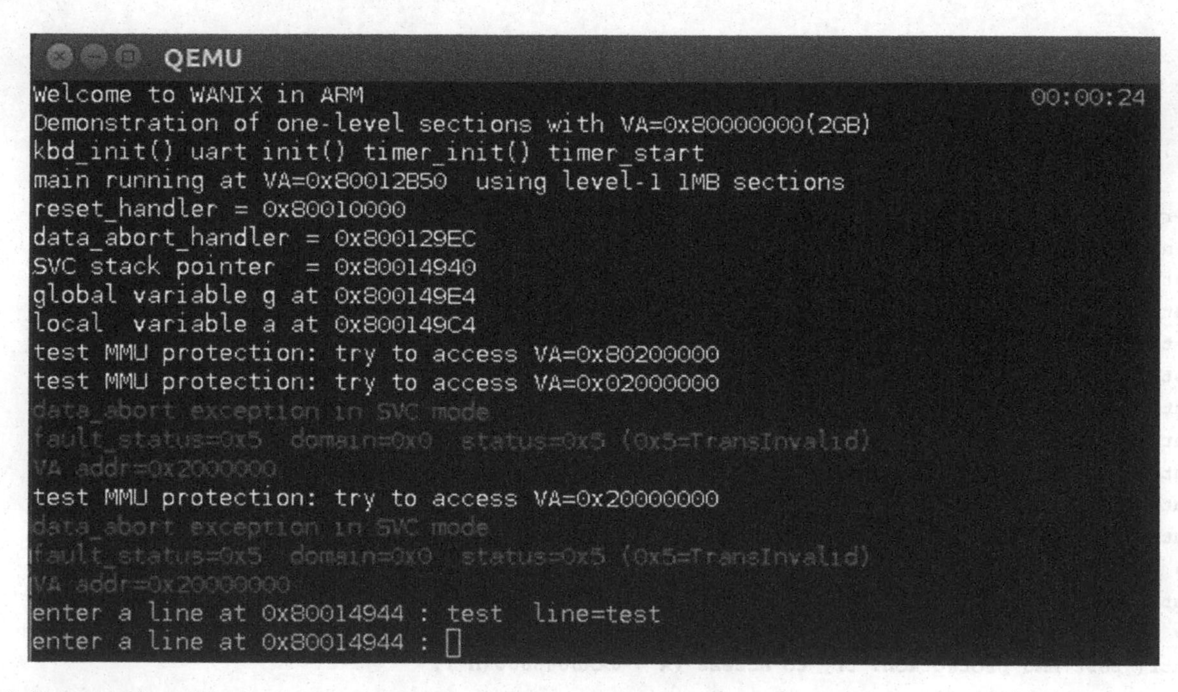

Fig. 6.9 Demonstration of One-level Paging with High VA Space

6.8.4 Two-Level Paging with High VA Space

The sample program C6.4 uses 2-level paging with virtual address space beginning at 0x80000000 (2 GB). Since the program will be loaded at the physical address 0x10000 and runs from there, it is compiled-linked with the starting VA = 0x80010000. Similar to the Program C6.3, we must set up the page tables and enable the MMU in assembly code when the program starts. Since building page tables in assembly is very tedious, we shall do it in two separate steps. In the first step, we set up an initial one-level page table using 1 MB sections to map in the VA = (2 GB, 2 GB + 258 MB) range exactly the same as in Program C6.3. After enabling the MMU for address translation, we call a function in C to build a new level-1 page table (pgdir) at 32 KB and its associated level-2 page tables at 5 MB using 4 KB small pages. Then we switch TTB to the new level-1 page table and flush the TLB, thereby switching the MMU from one-level paging to two-level paging. The following show the code of example program C6.4.

(1). **ts.s file**: The ts.s file is the same as that of Program C6.3, except for the added switchPgdir function, which is list below.

```
switchPgdir: // switch pgdir to new pgdir passed in r0
  mcr p15, 0, r0, c2, c0, 0     // set TTBase to C2
  mov r1, #0
  mcr p15, 0, r1, c8, c7, 0     // flush TLB
  mcr p15, 0, r1, c7, c10, 0    // flush I and D Caches
  mrc p15, 0, r2, c2, c0, 0     // read TLB base reg C2
  // set domain0: 01 = client(check permission) 11 = manager(no check)
  mov r0, #0x1                  // b01 for CLIENT
  mcr p15, 0, r0, c3, c0, 0     // write to domain reg C3
  mov pc, lr                    // return
```

(2). **t.c file**: The t.c file is identical to that of Program C6.3, except for the added mkPtable() function, which creates a new level-1 page table (pgdir) at 32 KB and its associated level-2 page tables at 5 MB. Then it switches TTB to the new pgdir to let the MMU use 2-level paging. The choices of the new pgdir at 32 KB and the level-2 page tables at 5 MB are quite arbitrary. They can be built anywhere is physical memory.

```
int mkPtable()
{
   int i, j, paddr, *pgdir, *pgtable
   printf("1. build two-level page tables at 32 KB\n");
   pgdir  =  (int *)VA(0x8000);   // 0x80000000
   for (i = 0; i < 4096; i ++){  // zero out 4096 entries
       pgdir[i]  =  0;
   }
   // build NEW pgtables at 5 MB
   printf("2. fill 258 entries in level-1 pgdir with 258 pgtables\n");
   for (i = 0; i < 258; i ++){  // ASSUME 256 MB RAM; 2 I/O sections
       pgdir[i + 2048]  =  (int)(0x500000  +  i*1024) | 0x11;
       // descriptor attribute = 0x11: DOMAIN = 0,CB = 00,type = 01
   }
   printf("3. build 258 level-2 pgtables at 5 MB\n");
   for (i = 0; i < 258; i ++){
       pgtable  =  (int *)(VA(0x500000)  +  (int)i*1024);
       paddr  =  i*0x100000 | 0x55E; // APs = 01|01|01|01|CB = 11|type = 10
       for (j = 0; j < 256; j ++){ // 256 entries, each points to 4 KB PA
```

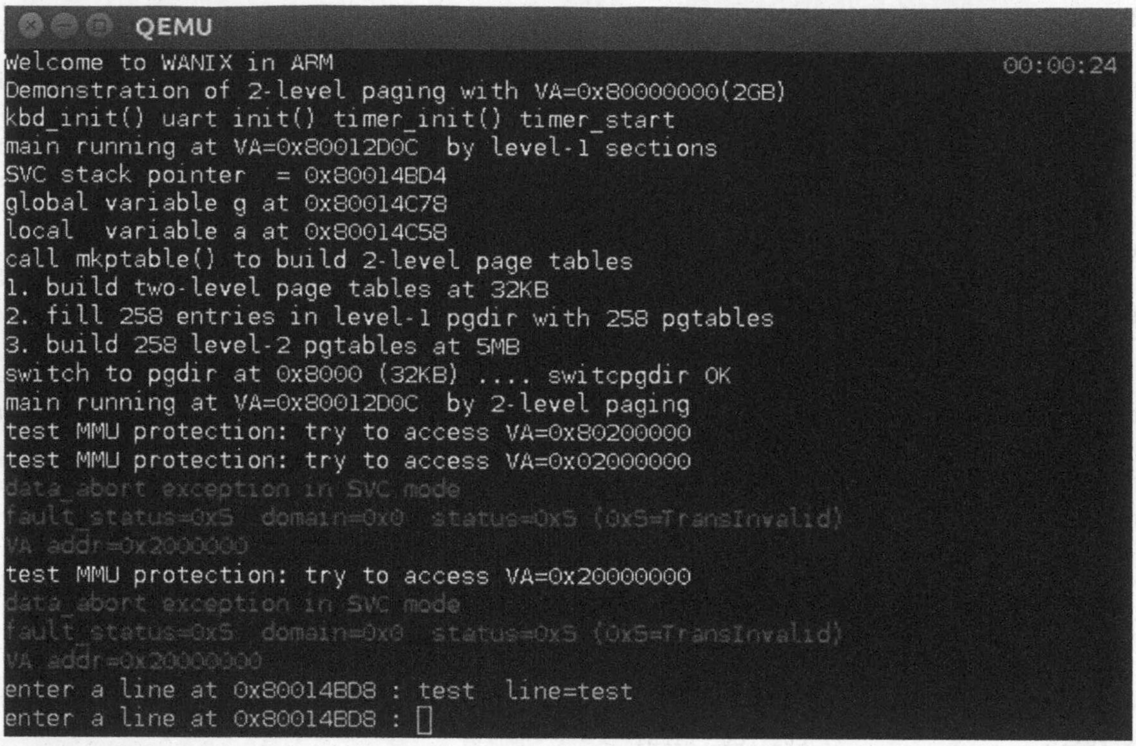

```
QEMU
Welcome to WANIX in ARM                                        00:00:24
Demonstration of 2-level paging with VA=0x80000000(2GB)
kbd_init() uart init() timer_init() timer_start
main running at VA=0x80012D0C  by level-1 sections
SVC stack pointer   = 0x80014BD4
global variable g at 0x80014C78
local  variable a at 0x80014C58
call mkptable() to build 2-level page tables
1. build two-level page tables at 32KB
2. fill 258 entries in level-1 pgdir with 258 pgtables
3. build 258 level-2 pgtables at 5MB
switch to pgdir at 0x8000 (32KB) .... switcpgdir OK
main running at VA=0x80012D0C  by 2-level paging
test MMU protection: try to access VA=0x80200000
test MMU protection: try to access VA=0x02000000
data_abort exception in SVC mode
fault_status=0x5  domain=0x0  status=0x5 (0x5=TransInvalid)
VA addr=0x2000000
test MMU protection: try to access VA=0x20000000
data_abort exception in SVC mode
fault_status=0x5  domain=0x0  status=0x5 (0x5=TransInvalid)
VA addr=0x20000000
enter a line at 0x80014BD8 : test  line=test
enter a line at 0x80014BD8 : 
```

Fig. 6.10 Demonstration of 2-level Paging with High VA

```
            pgtable[j]  =  paddr  +  j*4096; // inc by 4 KB
        }
    }
    pgdir[0]   =  pgdir[2048]; // pgdir[0] and pgdir[2048]- > low 1 MB PA
    printf("4. switch to pgdir at 0x8000 (32 KB) .... ");
    switchPgdir(0x8000);
    printf("switchPgdir OK\n");
}
```

(3). **Sample Outputs of Program C6.4**: Fig. 6.10 shows the sample outputs of running the C6.4 program.

6.9 Summary

This chapter covers the ARM memory management unit (MMU) and virtual address space mappings. It covers the ARM MMU in detail and shows how to configure the MMU for virtual address mapping using both one-level and two-level paging. In addition, it also shows the distinction between low VA space and high VA space mappings. Rather than only discussing the principles, it demonstrates the various kinds of virtual address mappings by complete working example programs.

List of Sample Programs

C6.1: One-level paging using 1 MB sections with VA mapped low
C6.2: Two-level paging using 4 KB pages with VA mapped low

C6.3: One-level paging using 1 MB sections with VA mapped high
C6.4: Two-level paging using 4 KB pages with VA mapped high

Problems

1. In the example C6.1, the level-1 page table is built by the mkptable() function in C.
(1). Why is it possible to build the page table in C?
(2). Implement the mkptable() function in assembly code.
2. The example program C6.2 implements 2-level paging using 4 KB small pages. Modify it to implement 2-level paging using 64 KB large pages.
3. In the example program C6.3, which maps the VA space to 2 GB, the page table is built in assembly code, rather than in C, when the system starts.
(1). Why is it necessary to build the page table in assembly code?
(2).
4. In the example program C6.4, which uses 2-level paging to map VA to 2 GB, the page tables are built in two stages, all in C. Alternatively, the page tables can also be built in a single step, all in assembly code.
(1). Try to build the page tables in assembly code in one step. Compare the amount of programming efforts needed in both approaches.
(2). The mkPtable() function of C6.4 contains the lines of code

```
pgdir = (int *)VA(0x8000);  // pgdir at 32 KB
```

which sets the level-1 page table at 32 KB of physical memory, and

```
pgdir[i + 2048] = (int)(0x500000 + i*1024) | 0x11;
```

which fills the level-1 page descriptors with page frames beginning at 5 MB of physical memory. While the first line of code uses VA, the second line of code uses PA.
Why the difference?

Reference

ARM MMU: ARM926EJ-S, ARM946E-S Technical Reference Manuals, ARM Information Center 2008.

User Mode Process and System Calls

7

7.1 User Mode Processes

In Chap. 5, we developed a simple uniprocessor kernel for process management. The simple kernel supports dynamic process creation, process synchronization and process communication. It can be used as a model for many simple embedded systems. A simple embedded system comprises a fixed number of processes, all of which execute in the same address space of the kernel. The system can be implemented as event-driven, with processes as execution entities. Events can be interrupts from hardware devices, process cooperation through semaphores or messages from other processes. The disadvantage of this kind of systems is the lack of memory protection. An ill-designed or malfunctioning process may corrupt the shared address space, causing other processes to fail. For both reliability and security reasons, each process should run in a private virtual address space that is isolated and protected from other processes. In order to support processes with virtual address spaces, it is necessary to use the memory management hardware to provide both virtual address mapping and memory protection. In Chap. 6, we discussed the ARM Memory Management Unit (MMU) (ARM MMU 2008) in detail and showed how to configure the MMU for virtual address mappings. In this chapter, we shall extend the simple kernel to support user mode processes. In the extended kernel, each process may execute in two different modes, kernel mode and user mode. While in kernel mode, all processes execute in the same address space of the kernel, which is non-preemptive. While in user mode, each process executes in a private virtual address space and is preemptable. User mode processes may enter kernel through exceptions, interrupts and system calls. System call is a mechanism which allows user mode processes to enter kernel mode to execute kernel functions. After executing a system call function in kernel, the process returns to user mode (except exit which never returns) with the desired results. For simplicity, we shall ignore exceptions first and focus on developing a kernel to support user mode processes and system calls.

7.2 Virtual Address Space Mapping

When an embedded system boots up, the system kernel is usually loaded to the low end of physical memory, e.g. to the physical address 0 or 16 KB as in the case of the ARM Versatilepb VM under QEMU. When the kernel starts, it first configures the Memory Management Unit (MMU) to enable virtual address translation. Each process may run in two different modes, Kernel mode and User mode, each with a distinct virtual address space. With 32-bit addressing, the total VA space range is 4 GB. We may divide the 4 GB VA space evenly into two equal halves and assign each mode a VA space range of 2 GB. There are two ways to create the Kernel and User mode VA spaces. In the **Kernel Mapped Low (KML)** scheme, the Kernel mode VA space is mapped to low virtual addresses and User mode VA space is mapped to high virtual addresses. In this case, Kernel VA to PA mapping is usually one-to-one or identity mapping, so that every VA is the same as PA. User mode VA space is mapped to high virtual address range of 0x80000000 (2 GB) and above. In the **Kernel Mapped High (KMH)** scheme, the VA address mapping is reversed. In this case, the kernel mode VA space is mapped to high virtual addresses and User mode VA space is mapped to low virtual addresses. From a memory protection point of view, there is no difference between the two mapping schemes. However, from a programming point of view, there may be some significant differences. For example, in the KML scheme, the kernel can be compile-linked with real addresses. When the kernel starts, it can execute in real address mode directly without configuring the MMU for address translation first. In contrast, in the KMH scheme, the kernel must be compile-linked with virtual addresses. When the kernel starts, it can not

K.C. Wang, *Embedded and Real-Time Operating Systems*,
DOI 10.1007/978-3-319-51517-5_7

execute any code that uses virtual addresses directly. In this case, it must configure the MMU to use virtual address first. The ARM architecture does not support the notion of floating vector tables, which allows the vector table to be remapped to any physical memory. On some ARM machines, the vector table can only be remapped to 0xFFFF0000 during booting. Without vector remapping, the vector table must be located at the physical address 0. In the vector table, the Branch or LDR instructions have an address range limit of 4 KB. This implies that both the vector table and the exception handler entry points must all reside in the lowest 4 KB of physical memory. For these reasons, we shall mainly use the **KML** scheme because it is more natural and simpler. However, we shall also show how to use the **KMH** scheme and demonstrate its differences from the KML scheme by sample systems.

7.3 User Mode Process

From now on, we shall assume that a process may execute in two different modes; Kernel mode (SVC mode in ARM) and User mode. For the sake of brevity, we shall simply refer to them as **Kmode** and **Umode**, respectively. Each mode has its own VA space. When configuring the MMU for virtual address mapping, we shall use the KML scheme so that the VA space of Kmode is from 0 to the amount of physical memory, and the VA space of Umode is from 0x80000000 (2 GB) to the size of Umode image.

7.3.1 User Mode Image

First, we show how to develop User mode process images. A user mode program consists of an assembly file, us.s, and a set of C files, which are shown and explained below.

```
(1). /* us.s file in ARM assembly */
   .global entryPoint, main, syscall, getcsr, getAddr
   .text
   .code 32
   . global _exit // syscall(99, 0, 0 ,0) to terminate
// upon entry, r0 points to cmdline string in Umode stack
entryPoint:
   bl main
   bl _exit
syscall:       // syscall(a,b,c,d) : a,b,c,d are passed in r0-r3
   swi #0
   mov pc, lr
get_cpsr:
   mrs r0, cpsr
   mov pc, lr
```

Explanation of us.s file: us.s is the entry point of all Umode programs. As will be shown shortly, prior to entering us.s in User mode, the kernel has already set up the execution environment of the program, including a Umode stack. So, upon entry it simply calls main(). If main() returns, it calls _exit(), which issues a syscall(99, 0, 0, 0) to terminate. Umode processes may enter kernel to execute kernel functions via system calls, which is

```
int r  = syscall(int a, int b, int c, int d)
```

When issuing a system call, the first parameter a is the system call number, b, c, d are parameters to the kernel function, and r is the return value. In ARM based systems, system call, or **syscall** for short, is implemented by the SWI instruction, which causes the CPU to enter the privileged Supervisor (SVC) mode. Therefore, processes in kernel run in SVC mode. The function get_cpsr() returns the current status register of CPU. It is used to verify that the process is indeed executing in User mode (mode = 0x10).

```
(2).  /*********** ucode.c file ***********/
#include "string.c"    // string functions
#include "uio.c"       // uprintf(), etc.
int umenu()            // show a command menu
{
  uprintf("-------------------------\n");
  uprintf("getpid getppid ps chname \n");
  uprintf("-------------------------\n");
}
// syscalls to kernel
int getpid()         { return syscall(0,0,0,0); }
int getppid()        { return syscall(1,0,0,0); }
int ps()             { return syscall(2,0,0,0); }
int chname(char *s){ return syscall(3,s,0,0); }

// User mode command functions, each issues a syscall
int ugetpid()
{
  int pid = getpid();
  uprintf("pid = %d\n", pid);
}
int ugetppid()
{
  int ppid = getppid();
  uprintf("ppid = %d\n", ppid);
}
int uchname()
{
  char s[32];
  uprintf("input a name string : ");
  ugetline(s); uprintf("\n");
  chname(s);
}
int ups(){ ps(); }
/** BASIC Umode I/O are syscalls to kernel **/
int ugetc()       {  return syscall(90,0,0,0); }
int uputc(char c){  return syscall(91,c,0,0); }
```

Explanation of ucode.c file: ucode.c contains system call interface functions. When a user mode program runs, it first displays some startup information, such as the CPU mode and starting virtual address. Then, it displays a menu and asks for a user command. For demonstration purpose, each user command issues a system call to the kernel. Each syscall is assigned a number for identification, which corresponds to a function in kernel. The syscall number is entirely up to the system designer's choice. Since User mode programs run in Umode address spaces, they can not access the I/O space in kernel directly. Therefore, basic I/O in Umode, such as ugetc() and uputc(), are also system calls. Since all user mode programs rely on system calls, the same ucode.c file can be shared by all user mode programs. In a real system, system call interface is usually pre-compiled as part of the link library, which is used by the linker to develop all user mode programs.

(3). **u1.c file**: This is the main body of the Umode program. It may be used as a template to develop other Umode programs.

```
/********* u1.c file *************/
#include "ucode.c" // all user mode program share the same ucode.c
int main()
{
  int i, pid, ppid, mode;
```

```
   char line[64];
   mode = get_cpsr() & 0x1F); // get CPSR to determine CPU mode
   printf("CPU mode=%x\n", mode);      // show CPU mode
   pid  = getpid();
   ppid = getppid();
   while(1){
      printf("This is process %d in Umode at %x: parent=%d\n",
              pid, &entryPoint, ppid);
      umenu();
      uprintf("input a command : ");
      ugetline(line); uprintf("\n");
      if (!strcmp(line, "getpid"))
         ugetpid();
      if (!strcmp(line, "getppid"))
         ugetppid();
      if (!strcmp(line, "ps"))
         ups();
      if (!strcmp(line, "chname"))
         uchname();
   }
}
```

Explanation of u1.c file: u1.c is the main body of a Umode program. After displaying the CPU mode to verify it is indeed executing in Umode, it issues system calls to get its pid and ppid of the parent process. Then it executes an infinite loop. First, it shows the process ID and the starting virtual address of the Umode image. Then it displays a menu. To begin with, the menu only includes four commands: getpid, getppid, ps, and chname. As we continue to expand the kernel, we shall add more user commands later. Each user command invokes an interface function in ucode.c, which issues a system call to execute a syscall function in kernel. For example, the ps command causes the process to execute kps() in kernel, which prints all the PROC status information. Each process is initialized with a name string in the PROC.name field. The chname command changes the name string of the current running process. After changing name, the user may use the ps command to verify the results.

(4). **mku script**: The mku sh script is used to generate the binary image u1.o

```
arm-none-eabi-as -mcpu=arm926ej-s us.s -o us.o
arm-none-eabi-gcc -c -mcpu=arm926ej-s -o $1.o $1.c
arm-none-eabi-ld -T u.ld us.o $1.o -Ttext=0x80000000 -o $1.elf
arm-none-eabi-objcopy -O binary $1.elf $1
arm-none-eabi-objcopy -O elf32-littlearm -B arm $1 $1.o
cp -av $1.o ../
```

Explanation of the mku script file: The mku script generates a binary executable image file. First, it (cross) compile-link us.s and u1.c into an ELF file with the starting virtual address **0x800000000 (2 GB)**. Then it uses objcopy to convert the ELF file into a raw binary image file. Before developing a loader to load program images, we shall use the binary image as a raw data section in the kernel image. This is done in the linker script t.ld file.

```
#----------------- link script t.ld ----------------------
ENTRY(reset_handler)
SECTIONS
{
  . = 0x10000;
  .text : { ts.o t.o }
  .data : { ts.o(.data) t.o (.data) }
```

```
  .bss   : { *(.bss) }
  . = ALIGN(8);
  . = . + 0x1000; /* 4kB of irq stack space */
  irq_stack_top = .;
  . = . + 0x1000; /* 4kB of abt stack memory */
  abt_stack_top = .;
  . = . + 0x1000; /* 4kB of und stack memory */
  und_stack_top = .;
  . = ALIGN(1024);
 .data : { /* include u1.o as a RAW data section */
      u1.o
  }
}
```

7.4 System Kernel Supporting User Mode Processes

The system kernel consists of the following components: interrupt handlers, device drivers, I/O and queue manipulation functions, and process management functions. Most of the kernel components, e.g. interrupt handlers, device drivers and basic process management functions, etc. are already covered in previous Chapters. In the following, we shall focus on the new features of the kernel. For the sake of clarity, in addition to the section titles we shall also use sequence numbers (in parentheses) to show the kernel code segments.

7.4.1 PROC Structure

(1). **The PROC structure (in type.h file)**

```
#define NPROC      9
#define FREE       0
#define READY      1
#define SLEEP      2
#define BLOCK      3
#define ZOMBIE     4
#define SSIZE   1024

typedef struct proc{        // byte offsets:
  struct proc *next;        // 0
  int     *ksp;             // 4 : saved Kmode sp when not running
  int     *usp;             // 8 : Umode sp at time of syscall
  int     *upc;             // 12: Umode pc at time of syscall
  int     *ucpsr;           // 16: Umode cpsr
  int     *pgdir;           // level-1 page table pointer
  int     status;           // FREE|READY|SLEEP|BLOCK|ZOMBIE, etc.
  int     priority;
  int     pid;
  int     ppid;             // parent pid
  struct proc *parent;      // parent PROC pointer
  int     event;            // event to sleep on
  int     exitCode;         // exit code
```

```
  char    name[64];        // name field
  int     kstack[SSIZE]; // Kmode stack
}PROC;
```

Each process is represented by a PROC structure. The new fields in the PROC structure in support of Umode operations are **usp, upc, ucpsr**: When a process enters kernel via syscall, it saves the Umode sp, lr and cpsr in the PROC structure for return to Umode later.
pgdir: Each process has a level-1 page table (pgdir) pointed by PROC.pgdir. The pgdir and its associated page tables define the virtual address spaces of the process in both Kernel and User modes.
ppid and parent PROC pointer: parent process pid and pointer to parent PROC
exitCode: for process to terminate with an exitCode value
name: process name string used to demonstrate system call.
(2) **ts.s file**: The kernel's assembly code consists of five parts, which are highlighted in the code listing shown below.

```
// ********** ts.s file ***************
        .text
        .code 32
.global reset_handler, vectors_start, vectors_end
.global proc, procsize
.global tswitch, scheduler, running, goUmode, switchPgdir
.global int_off, int_on, lock, unlock, get_cpsr
// Part 1: reset_handler
reset_handler:
  // set SVC stack to HIGH END of proc[0].kstack[]
  LDR r0, =proc        // r0 points to proc's
  LDR r1, =procsize  // r1 -> procsize
  LDR r2, [r1, #0]    // r2 =  procsize
  ADD r0, r0, r2      // r0 -> high end of proc[0]
  MOV sp, r0
  // go to IRQ mode to set IRQ stack
  MSR cpsr, #0x92
  LDR sp, =irq_stack_top  // set IRQ stack
  // go to ABT mode to set ABT stack
  MSR cpsr, #0x97
  LDR sp, =abt_stack_top  // set abt stack
  // go to UND mode to set UND stack
  MSR cpsr, #0x9B
  LDR sp, =und_stack_top  // set UND stack
  // go back in SVC mode, set SPSR to User mode with IRQ on
  MSR cpsr, #0x93         // SVC mode
  MSR spsr, #0x10         // set SPSR to Umode with IRQ enabled
  BL copy_vectors         // copy vector table to address 0
  BL mkPtable             // create page table in C
  // initialize MMU control register C1
  // bit12=1: EnIcache; bits9-8=RS=11; bit2=1: EnDcache;
  // all other bits of C1=0 for default
  LDR r0, regC1                 // load r0 with 0x1304
  MCR p15, 0, r0, c1, c0, 0   // write to MMU control reg C1
  // set page table base register, flush TLB
  LDR r0, MTABLE                // pgdir at 0x4000 (16KB)
  MCR p15, 0, r0, c2, c0, 0   // set Page table Base register C2
  MCR p15, 0, r0, c8, c7, 0   // flush TLB
  // set domain0: 01=client(check permission),11=manager(no check)
```

```
  MOV r0, #0x1                    // b01 for CLIENT
  MCR p15, 0, r0, c3, c0, 0
  // enable MMU: turn on bit0 of control reg C1
  MRC p15, 0, r0, c1, c0, 0
  ORR r0, r0, #0x00000001         // set bit0 to 1
  MCR p15, 0, r0, c1, c0, 0       // write to c1 to enable MMU
  nop                             // time to allow MMU to finish
  nop
  nop
  BL  main                        // call main() in t.c
  B .                             // if main() returns, loop here
MTABLE:   .word 0x4000        // MTABLE at 16KB
regC1:    .word 0x1304        // P15 control register c1 setting

// part 2: IRQ and exception handler entry points
irq_handler:                    // IRQ interrupts entry point
  SUB  lr, lr, #4               // adjust link register lr
  STMFD sp!,{r0-r12, lr}        // save context in IRQ stack
  BL  irq_chandler              // call irq_handler() in C
  LDMFD sp!,{r0-r12, pc}^       // pop IRQ stack, restore SPSR
Data_handler:                   // data abort handler
  SUBlr, lr, #4
  STMFD sp!, {r0-r12, lr}
  BL  data_chandler
  LDMFD sp!, {r0-r12, pc}^

// Part 3: task context switching, switch pgdir
tswitch: // tswitch() in Kmode
  mrs   r0, cpsr                 // IRQ OFF
  orr   r0, #0x80
  msr   cpsr, r0
  STMFD sp!, {r0-r12, lr}        // save context
  LDR r0, =running               // r0=&running
  LDR r1, [r0, #0]               // r1->runningPROC
  STR sp, [r1, #4]               // running->ksp = sp
BL scheduler     // call scheduler() in C
  LDR r0, =running
  LDR r1, [r0, #0]               // r1->runningPROC
  LDR sp, [r1, #4]               // sp = running->ksp
  mrs   r0, cpsr                 // IRQ ON
  bic   r0, #0x80
  msr   cpsr, r0
  LDMFD sp!, {r0-r12, pc}        // all in Kmode
switchPgdir:  // switch to new PROC's pgdir during task switch
  // r0 contains PA of new PROC's pgdir
  MCR p15, 0, r0, c2, c0, 0  // set TTB in C2
  MOV r1, #0
  MCR p15, 0, r1, c8, c7, 0  // flush TLB
  MCR p15, 0, r1, c7, c10, 0 // flush Cache
  MRC p15, 0, r2, c2, c0, 0
  // set domain AP bits to CLIENT mode: check AP bits
  MOV r0, #0x5                   // 0101: |domain1|domain0=CLIENT
  MCR p15, 0, r0, c3, c0, 0
  MOV pc, lr                     // return
```

```
// Part 4: SVC entry, system call routing and return to User mode
svc_entry: // r0-r3 contain syscall params, do not disturb   stmfd sp!, {r0-r12, lr}
   ldr r4, =running    // r4=&running
   ldr r5, [r4, #0]    // r5 -> PROC of running process
   mov r6, spsr
   str r6, [r5, #16]   // save Umode SR in PROC.ucpsr at offset 16
// go to SYS mode;
   mrs r6, cpsr        // r6 = SVC mode cpsr
   mov r7, r6          // save a copy in r7
   orr r6, r6, #0x1F   // r6 = SYS mode
   msr cpsr, r6        // change cpsr to SYS mode
// now in SYS mode, save Umode sp, pc into running PROC
   str sp, [r5, #8]    // save usp into proc.usp at offset 8
   str lr, [r5, #12]   // save upc into proc.upc at offset 12
// go back to SVC mode
   msr cpsr, r7
// replace saved lr in kstack with Umode PC at syscall
   mov r6, sp
   add r6, r6, #52     // entry 13 => offset=13*4 = 52
   ldr r7, [r5, #12]   // Umode LR at syscall, NOT at swi
   str r7, [r6]        // replace saved LR in kstack
// enable interrupts
   mrs r6, cpsr
   bic r6, r6, #0xC0   // I and F bits=0 enable IRQ,FIQ
   msr cpsr, r6
   bl  svc_handler     // call svc_handler() in C
goUmode:
   ldr r4, =running    // r4=&running
   ldr r5, [r4, #0]    // r5 -> PROC of running
   ldr r6, [r5, #16]   // saved spsr
   msr spsr, r6        // restore spsr of THIS process
// go to SYS mode
   mrs r6, cpsr        // r6 = SVC mode cpsr
   mov r7, r6          // save a copy in r7
   orr r6, r6, #0x1F   // r6 = SYS mode
   msr cpsr, r6        // change to SYS mode
// now in SYS mode
   ldr sp, [r5, #8]    // restore Umode sp from PROC.usp
// go back to SVC mode
   msr cpsr, r3
//pop regs from kstack AND restore Umode cpsr
   ldmfd sp!, {r0-r12, pc}^ // pop kstack AND spsr: back to Umode

// Part 5: Utility functions
// IRQ interrupts enable/disable functions
int_off:            // SR = int_off()
  MRS r0, cpsr
  MOV r1, r0
  ORR r1, r1, #0x80
  MSR cpsr, r1      // return value in r0 = original cpsr
  MOV pc, lr
int_on:             // int_on(SR);  SR in r0
  MSR cpsr, r0
  MOV pc, lr
lock:
```

```
  MRS r0, cpsr
  ORR r0, r0, #0x80
  MSR cpsr, r0
  MOV pc, lr
unlock:
  MRS r0, cpsr
  BIC r0, r0, #0x80
  MSR cpsr, r0
  MOV pc, lr
get_cpsr:               // return cpsr to verify MODE
  MRS r0, cpsr
  MOV pc, lr
// vector table: copied to PA 0 in reset_handler
vectors_start:
  LDR PC, reset_handler_addr
  LDR PC, undef_handler_addr
  LDR PC, svc_handler_addr
  LDR PC, prefetch_abort_handler_addr
  LDR PC, data_abort_handler_addr
  B .
  LDR PC, irq_handler_addr
  LDR PC, fiq_handler_addr
  reset_handler_addr:          .word reset_handler
  undef_handler_addr:          .word undef_handler
  svc_handler_addr:            .word svc_entry
  prefetch_abort_handler_addr: .word prefetch_abort_handler
  data_abort_handler_addr:     .word data_handler
  irq_handler_addr:            .word irq_handler
  fiq_handler_addr:            .word fiq_handler
vectors_end:
```

7.4.2 Reset Handler

The reset_handler consists of three steps.

7.4.2.1 Exception and IRQ Stacks

Step 1: Set up stacks: Assume NPROC=9 PROCs, each PROC has a Kmode stack in the PROC structure. The system starts in SVC mode. Reset_handler initializes the SVC mode stack pointer to the high end of proc[0], thus making proc[0].kstack as the initial stack. It also sets spsr to User mode, making the CPU ready to return to User mode when it exits SVC mode. Then it initializes the stack pointers of other privileged modes, e.g. IRQ, data_abort, undef_abort, etc. Each privileged mode (except FIQ mode, which is not used) has a separate 4 KB stack area (defined in linker script t.ld) for interrupt and exception processing.

7.4.2.2 Copy Vector Table

Step 2: Copy_vector_table: During reset, the memory management MMU is off and the vector table remap enable bit (V bit in the MMU control register c0) is 0, meaning that the vector table is not remapped to 0xFFFF0000. At this moment, every address is a physical address. Reset_handler copies the vector table to the physical address 0 as required by the ARM CPU's vector hardware.

7.4.2.3 Create Kernel Mode Page Table

Step 3: Create Kmode Page Table: After initializing the stack pointers of the various privileged modes, reset_handler calls mkPtable() to set up the Kernel mode page table. To begin with, we shall use simple one-level paging with 1 MB sections to create an identity mapping of VA to PA. Assuming 256 MB physical memory plus 2 MB I/O space (of the ARM Versatilepb VM) at 256 MB, the mkPtable() function in C is

```
#define PTABLE 0x4000
int mkPtable() // create ID mapped Ptable with 1MB sections
{
   int i;
   int *ut = (int *)PTABLE; // PTABLE at 16KB of PA
   int entry = 0 | 0x41E;   // |AP=01|domain=0000|1|CB=11|type=10|
   for (i=0; i<258; i++){   // assume 256MB RAM + 2MB I/O space
       ut[i] = entry;
       entry += 0x100000;   // section size=1MB bytes
   }
}
```

The attributes of the page table entries are set to 0x41E for AP = 01, domain = 0000, CB = 11 and type = 10 (Section). Alternatively, the CB bits can be set to 00 to disable instruction and data caching and write buffering. The entire Kmode space is treated as domain 0 with permission bits = 01 for R|W in privileged modes but no access in User mode. Then it enables the MMU for VA to PA address translation. After these, every virtual address VA is mapped to a physical address PA by the MMU hardware. In this case, the VA and PA addresses are the same due to the identity mapping. Alternative virtual address mapping schemes will be discussed later. Then it calls main() in C to continue the kernel initialization.

7.4.2.4 Process Context Switching Function

Part 2: Process Context Switching: tswitch() is for switching process in Kernel. When a process gives up CPU, it calls tswitch(), in which it saves CPU registers in the process kstack, saves the stack pointer into PROC.ksp and call scheduler() in C. In scheduler(), the process enters the readyQueue by priority if it is still READY to run. Then it picks the highest priority process from the readyQueue as the next running process. If the next running process is different from the current process, it calls switchPgdir() to switch page table to that of the next running process. SwitchPgdir() also flushes the TLB, invalidates the instruction and data caches and flushes the write buffer to prevent the CPU from using TLB entries belonging to an old process context. In order to speed up the address translation process, the ARM MMU supports many advanced options, such as lock-down instruction and data cache, invalidating selected TLB and cache entries, etc. In order to keep the system simple, we shall not use these advanced MMU features.

7.4.2.5 System Call Entry and Exit

Part 3: System Call entry and exit: User mode processes use syscall(a, b, c, d) to execute system call functions in kernel. In syscall(), it issues a SWI to enter SVC mode, which is routed to SVC handler via the SWI vector.

7.4.2.6 SVC Handler

SVC handler: svc_entry is is the system call entry point. System call parameters a, b, c, d are passed in registers r0–r3. Upon entry, the process first saves all the CPU registers in the process Kmode stack (PROC.kstack). In addition, it also saves Umode sp, lr and cpsr into PROC's usp, upc and ucpsr fields, respectively. In order to access Umode registers, it temporarily switches the CPU to System mode, which shares the same set of registers with User mode. Then, it replaces the saved lr in kstack with the Umode upc. This is because the saved lr points to the SWI instruction in Umode, not the upc at the time of system call. Then, it enables IRQ interrupts and calls svc_handler() in C, which actually handles the system call. When the process exits kernel, it executes goUmode() to return to Umode. In the goUmode code, it fisrt restores Umode sp and cpsr from the saved usp and cpsr in PROC. Then it returns to Umode by

```
ldmfd sp!, {r0-r12, pc}^
```

The reader may wonder why is it necessary to save and restore Umode sp and cpsr during system calls. The problem is as follows. When a process enters Kernel, it may not return to User mode immediately. For example, the process may become suspended in Kernel and switches to another process. When the new process returns from Kernel to User mode, the CPU's usp and spsr are that of the suspended process, not that of the current process.

It is noted that in most stack-oriented architecture, saving and restoring User mode stack pointer and status registers is automatic during system calls. For example, in the Intel x86 CPU (Intel 1990, 1992), the INT instruction is similar to the ARM SWI instruction, both cause the process to enter Kernel mode. The major difference is that when the Intel x86 CPU executes the INT instruction, it automatically stacks User mode [uss, usp], uflags, [ucs, upc], which are equivalent to the User mode SP, CPSR, LR of ARM CPU. When the Intel x86 CPU exits Kernel mode by IRET (which is similar to the ^ operation involving PC in ARM), it restores all the saved Umode registers from the Kernel mode stack. In contrast, the ARM processor does not stack any Umode registers automatically when it enters a privileged mode. The system programmer must do the save and restore operations manually.

7.4.2.7 Exception Handlers

(4) **Exception Handlers**: For the time being, we only handle data_abort exceptions, which are used to demonstrate the memory protection capability of the MMU. All other exception handlers are while(1) loops. The following shows the algorithm of the data_abort exception handler.

```
/*** exceptions.c file: only show data_abort_handler ***/
int data_chandler()
{
// read MMU registers C5=fault_status, C6=fault_address;
// print fault_address and fault_status;
}
```

7.4.3 Kernel Code

(5) **Kernel.c file**: This file defines the kernel data structures and implements kernel functions. When the system starts, reset_handler calls main(), which calls kernel_init() to initialize the kernel. First, it initializes the free PROC list and the readyQueue. Then it creates the initial process P0, which runs only in Kmode with the lowest priority 0. Then it sets up page tables for the PROCs. When setting up the process page tables, we shall assume that the kernel's VA space is mapped low, from 0 to the amount of available physical memory. The User mode VA space is mapped high, from 0x80000000 (2 GB) to 2 GB + Umode image size. Each PROC has a unique pid (1 to NPROC-1) and a level-1 page table pointer pgdir. The process page tables are constructed in the physical memory area of 6 MB. Each page table requires 4096 * 4 = 16 K bytes space. The 1 MB area from 6 to 7 MB has enough space for 64 PROC page tables. Each process (except P0) has a page table at 6 MB + (pid − 1) * 16 KB. In each page table the low 2048 entries define the process kernel mode address space, which are identical for all processes since they share the same address space in Kmode. The high 2048 entries of the page table define the process Umode address space, which are filled in only when the process is created.

In the following, we shall assume that each process (except P0) has a 1 MB Umode image at 8 MB + (pid − 1) * 1 MB, e.g. P1 at 8 MB, P2 at 9 MB, etc. This assumption is not critical. If desired, the reader may assume different Umode image sizes. With a 1 MB Umode image size, each page table only needs one entry for User mode VA space, i.e. entry 2048 points to the 1 MB Umode area of the process. Recall that we have designated the Kernel mode memory area as domain 0. We shall assign all User mode memory areas to domain 1. Accordingly, we set the Umode page entry attributes to 0xC3E (AP = 11, domain = 0001, CB = 11 and type = 10). The access permission (AP) bits of domain0 are set to 01 to allow access from Kmode but not from Umode. However, the AP bits of either the Umode page table descriptor or domain1 (of the domain access control register) must be set to 11 to allow access from Umode. Since the AP bits of domain1 are set to 01 in switchPgdir(), the AP bits of Umode page descriptors must be set to 11. The following lists the kernel.c file code.

```
/********* kernel.c file ***********/
PROC proc[NPROC], *freeList, *readyQueue, *sleepList, *running;
int procsize = sizeof(PROC);
char *pname[NPROC]={"sun", "mercury", "venus", "earth", "mars",
                    "jupiter", "saturn", "uranus", "neptune"};
int kernel_init()
{
  int i, j;
  PROC *p; char *cp;
  int *MTABLE, *mtable, paddr;
  printf("kernel_init()\n");
  for (i=0; i<NPROC; i++){
    p = &proc[i];
    p->pid = i;
    p->status = FREE;
    p->priority = 0;
    p->ppid = 0;
    p->parent = 0;
    strcpy(p->name, pname[i]);
    p->next = p + 1;
    p->pgdir = (int *)(0x600000 + p->pid*0x4000);
  }
  proc[NPROC-1].next = 0;
  freeList = &proc[0];          // all PROCs are in freeList
  readyQueue = 0;
  sleepList = 0;
  // create and run P0
  running = get_proc(&freeList);
  running->status = READY;
  printList(freeList); printQ(readyQueue);
  printf("building pgdirs at 6MB\n");
  // create pgdir's for ALL PROCs at 6MB; MTABLE at 16KB in ts.s
  MTABLE = (int *)0x4000;       // Mtable at 0x4000
  mtable = (int *)0x600000;     // mtables begin at 6MB
  // Each pgdir MUST be at a 16K boundary ==>
  // 1MB at 6MB has space for 64 pgdirs of 64 PROCs
  for (i=0; i<64; i++){         // for 64 PROC mtables
    for (j=0; j<2048; j++){
      mtable[j] = MTABLE[j]; // copy low 2048 entries of MTABLE
    }
    mtable += 4096;             // advance mtable to next 16KB
  }
  mtable = (int *)0x600000;     // PROC mtables begin at 6MB
  for (i=0; i<64; i++){
    for (j=2048; j<4096; j++){ // zero out high 2048 entries
      mtable[j] = 0;
    }
    if (i) // exclude P0, page attribute=0xC3E:AP=11,domain=1
      mtable[2048]=(0x800000 + (i-1)*0x100000) | 0xC3E;
    mtable += 4096;
  }
}
```

```
int scheduler()
{
  PROC *old = running;
  if (running->status==READY){
    enqueue(&readyQueue, running);
  }
  running = dequeue(&readyQueue);
  if (running != old){
    // switch to new running's pgdir; flush TLB and I&D caches
    switchPgdir((u32)running->pgdir);
  }
}
```

(6) System call handler and kernel functions

```
/*** system call functions in Kernel ***/
int kgetpid() { return running->pid;  }
int kgetppid(){ return running->ppid; }
int kchname(char *s){ fetch *s from Umode; strcpy(running->name, s); }
char *pstatus[]={"FREE","READY","SLEEP","BLOCK","ZOMBIE"};
int kps()
{
  int i; PROC *p;
  for (i=0; i<NPROC; i++){
    p = &proc[i];
    printf("proc[%d]: pid=%d ppid=%d", i, p->pid, p->ppid);
    if (p==running)
      printf("%s ", "RUNNING");
    else
      printf("%s", pstatus[p->status]);
    printf(" name=%s\n", p->name);
  }
}
/********* syscall handler svc.c file ***********/
int svc_handler(volatile int a, int b, int c, int d)
{
  int r = -1; // default BAD return value
  switch(a){
    case 0:  r = kgetpid();            break;
    case 1:  r = kgetppid();           break;
    case 2:  r = kps();                break;
    case 3:  r = kchname((char *)b); break;
    case 90: r = kgetc() & 0x7F;      break;
    case 91: r = kputc(b);             break;
    default: printf("invalid syscall %d\n", a);
  }
  running->kstack[SSIZE-14] = r; // saved r0 in kstack = r
}
```

The function svc_handler() is essentially a system call router. System call parameters (a, b, c, d) are passed in registers r0–r3, which are the same in all CPU modes. Based on the system call number a, the call is routed to a corresponding kernel function. The kernel.c file implements all the system call functions in kernel. At this moment, our purpose is to demonstrate the mechanism and control flow of system calls. Exactly what the system call functions do is unimportant. So we only implement four very simple system calls: getpid, getppid, ps and chname. Each function returns a value r, which is loaded to the saved r0 in kstack as the return value to User mode.

(7) **t.c file**: This is the main body of the system kernel. Since most of the system components, such as interrupts, device drivers, etc. are already explained in previous chapters, we shall only focus on the new features, which are highlighted in bold faced lines. Before entering main(), the kernel mode page table is already set up in ts.s by the mkPtable() function and the MMU enabled for address translation. Because of the identity mapping of virtual address to physical address in kernel mode, no changes are needed in the kernel code. When main() starts, it copies the u1 program image to 8 MB. This is because we assume that the User mode image of the process P1 is at the physical address 8 MB but it runs in the virtual address space of 0x80000000 to 0x80100000 (2G to 2G + 1 MB) in User mode. Then it calls kfork() to create the process P1 and switches process to run P1.

```
/************ t.c file of C7.1 *************/
#include "type.h"
#include "string.c"
#include "uart.c"
#include "kbd.c"
#include "timer.c"
#include "vid.c"
#include "exceptions.c"
#include "queue.c"
#include "kernel.c"
#include "svc.c"
int copy_vectors(){ // copy vectors to 0; same as before }
int mkPtable()    { // create Kmode page table at 0x4000;  }
int irq_chandler(){ // IRQ interrupts handler entry point; }
int main()           // entered with interrupts OFF
{
   int i, usize;
   char line[128], *cp, *cq;
   /* enable VIC and SIC device interrupts */
   VIC_INTENABLE |= (1<<4);        // timer0 at 4
   VIC_INTENABLE |= (1<<12);       // UART0 at 12
   VIC_INTENABLE |= (1<<13);       // UART1 at 13
   VIC_INTENABLE = 1<<31;          // SIC to VIC's IRQ31
   fbuf_init();                    // LCD display
   kbd_init();                     // KBD driver
   uart_init();                    // UART driver
   timer_init(); timer_start(0); // timer
   kernel_init();      // initialize kernel, create and run P0
   unlock();           // enable IRQ interrupts
   printf("Welcome to Wanix in ARM\n");
   kfork();            // create P1 with Umode image=u1 at 8MB
   //code demo MMU protection: try to access invalid VA=>data_abort
   printf("P0 switch to P1 : enter a line : \n");
   kgets(line);        // enter a line to continue
   while(1){           // P0 code
     while(readyQueue==0); // loop if readyQueue empty
     tswitch();        // switch to run any ready process
   }
}
```

7.4.3.1 Create Process with User Mode Image

(8) **fork.c file**: This file implements the kfork() function, which creates a child process with a User mode image u1 in the User mode area of the new process. In addition, it also ensures that the new process can execute its Umode image in User mode when it runs.

```
#define UIMAGE_SIZE 0x100000
PROC *kfork()
{
  extern char _binary_u1_start, _binary_u1_end;
  int usize;
  char *cp, *cq;
  PROC *p = get_proc(&freeList);
  if (p==0){
     printf("kfork failed\n");
     return (PROC *)0;
  }
  p->ppid = running->pid;
  p->parent = running;
  p->status = READY;
  p->priority = 1;        // all NEW procs priority=1
  // clear all "saved" regs in kstack to 0
  for (i=1; i<29; i++)  // all 28 cells = 0
      p->kstack[SSIZE-i] = 0;
  // set kstack to resume to goUmode in ts.s
  p->kstack[SSIZE-15] = (int)goUmode;
  // let saved ksp point to kstack[SSIZE-28]
  p->ksp = &(p->kstack[SSIZE-28]);
  // load Umode image to Umode memory
  cp = (char *)&_binary_u1_start;
  usize = &_binary_u1_end - &_binary_u1_start;
  cq = (char *)(0x800000 + (p->pid-1)*UIMAGE_SZIE);
  memcpy(cq, cp, usize);
  // return to VA 0 in Umode image;
  p->kstack[SSIZE-1] = (int)VA(0);
  p->usp = VA(UIMAGE_SIZE); // empty Umode stack pointer
  p->ucpsr = 0x10;           // CPU.spsr=Umode
  enqueue(&readyQueue, p);
  printf("proc %d kforked a child %d\n", running->pid, p->pid);
  printQ(readyQueue);
  return p;
}
```

Explanation of kfork(): kfork() creates a new process with a User mode image and enters it into the readyQueue. When the new process begins to run, it first resumes in kernel. Then it returns to User mode to execute the Umode image. The current kfork() function is the same as in Chap. 5 except for the User mode image part. Since this part is crucial, we shall explain it in more detail. In order for a process to execute its Umode image, we may ask the question: how did the process wind up in the readyQueue? The sequence of events must be as follows.

(1).
It did a system call from Umode by SWI #0, which causes it to enter kernel to execute SVC handler (in ts.s), in which it uses STMFD sp!, {r0–r12, lr} to save Umode registers into the (empty) kstack, which becomes

```
---------------------------------------------------------------
|uLR|ur12 ur11 ur10 ur9 ur8 ur7 ur6 ur5 ur4 ur3 ur2 ur1|ur0|
---------------------------------------------------------------
 -1  -2   -3   -4  -5  -6  -7  -8  -9  -10 -11 -12 -13 -14
```

where the prefix u denotes Umode registers and −i means SSIZE-i. It also saves Umode sp and cpsr into PROC.usp and PROC.ucpsr. Then, it called tswitch() to give up CPU, in which it again uses STMFD sp!, {r0–r12, lr} to save Kmode registers into kstack. This adds one more frame to kstack, which becomes

```
                                                          ksp
-------goUmode-------------------------------------------|---
|KLR kr12 kr11 kr10 kr9 kr8 kr7 kr6 kr5 kr4 kr3 kr2 kr1 kr0|
--------------------------------------------------------------
 -15  -16  -17  -18  -19 -20 -21 -22 -23 -24 -25 -26 -27 -28
```

where the prefix k denotes Kmode registers. In the PROC kstack,

kLR = where the process called tswitch() and that's where it shall resume to,

uLR = where the process did the system call, and that's where it shall return to when it goes back to Umode.

Since the process never really ran before, all other "saved" CPU registers do not matter, so they can all be set to 0. Accordingly, we initialize the new process kstack as follows.

1. Clear all "saved" registers in kstack to 0

```
for (i=1; i<29; i++){ p->kstack[SSIZE-i] = 0; }
```

2. Set saved ksp to kstack[SSIZE-28]

```
p->ksp = &(p->kstack[SSIZE-28]);
```

3. Set kLR = goUmode, so that p will resume to goUmode (in ts.s)

```
p->kstack[SSIZE-15] = (int)goUmode;
```

4. Set uLR to VA(0), so that p will execute from VA=0 in Umode

```
p->kstack[SSIZE-1] = VA(0); // beginning of Umode image
```

5. Set new process usp to point at ustack TOP and ucpsr to Umode

```
p->usp = (int *)VA(UIAMGE_SIZE); // high end of Umode image
p->ucpsr = (int *)0x10;          // Umode status register
```

7.4.3.2 Execution of User Mode Image

With this setup, when the new process begins to run, it first resumes to goUmode (in ts.s), in which it sets Umode sp=PROC.usp, cpsr=PROC.ucpsr. Then it executes

```
ldmfd sp!, {r0-r12, pc}^
```

which causes it to execute from uLR = VA(0) in Umode, i.e. from the beginning of the Umode image with the stack pointer pointing at the high end of the Umode image. Upon entry to us.s, it calls main() in C. To verify that the process is indeed executing in Umode, we get the CPU's cpsr register and show the current mode, which should be 0x10. To test memory protection by the MMU, we try to access VAs outside of the process 1 MB VA range, e.g. 0x80200000, as well as

VA in kernel space, e.g. 0x4000. In either case, the MMU should detect the error and generate a data abort exception. In the data_abort handler, we read the MMU's fault_status and fault_address registers to show the cause of the exception as well as the VA address that caused the exception.When the data_abort handler finishes, we let it return to PC-4, i.e. skip over the bad instruction that caused the data_abort exception, allowing the process continue. In a real system, when a Umode process commits memory access exceptions, it is a very serious matter, which usually causes the process to terminate. As for how to deal with exceptions caused by Umode processes in general, the reader may consult Chap. 9 of Wang (2015) on signal processing.

7.4.4 Kernel Compile-Link Script

(2). **mk script**: generate kernel image and run it under QEMU

```
(cd USER; mku u1) # create binary executable u1.o
arm-none-eabi-as -mcpu=arm926ej-s ts.s -o ts.o
arm-none-eabi-gcc -c -mcpu=arm926ej-s t.c -o t.o
arm-none-eabi-ld -T t.ld ts.o t.o u1.o -Ttext=0x10000 -o t.elf
arm-none-eabi-objcopy -O binary t.elf t.bin
qemu-system-arm -M versatilepb -m 256M -kernel t.bin
```

In the linker script, **u1.o** is used as a raw binary data section in the kernel image. For each raw data section, the linker exports its symbolic addresses, such as

```
_binary_u1_start, _binary_u1_end, _binary_u1_size
```

which can be used to access the raw data section in the loaded kernel image.

7.4.5 Demonstration of Kernel with User Mode Process

Figure 7.1 shows the output screen of running the C7.1 program. When a process begins to execute the Umode image, it first gets the CPU's cpsr to verify it is indeed executing in User mode (mode = 0x10). Then it tries to access some VA outside of its VA space. As the figure show, each invalid VA generates a data abort exception. After testing the MMU for memory protection, it issues syscalls to get its pid and ppid. Then, it shows a menu and asks for a command to execute. Each command issues a syscall, which causes the process to enter kernel to execute the corresponding syscall function in kernel. Then it returns to Umode and prompts for a command to execute again. The reader may run the program and enter commands to test system calls in the sample system.

7.5 Embedded System with User Mode Processes

Based on the example program C7.1, we propose two different models for embedded systems to support multiple User mode processes.

7.5.1 Processes in the Same Domain

Instead of a single Umode image, we may create many Umode images, denoted by u1, u2, …, un. Modify the linker script to include all the Umode images as separate raw data sections in the kernel image. When the system starts, create n processes,

```
QEMU
                    Welcome to WANIX in Arm                          00:00:18
LCD display initialized : fbuf = 0x100000
kbd_init(): uart_init(): timer_init() timer_start: kernel_init()
freeList     = [1 ]->[2 ]->[3 ]->[4 ]->[5 ]->[6 ]->[7 ]->[8 ]->NULL
building pgdirs at 6MB
proc 0  kforked a child 1 : readyQueue = [1 1 ]->NULL
test memory protection? (y|n): y
try to access 2GB: should cause data_abort
data_abort exception in SVC mode
domain=0x0  status=0x5  addr=0x80000000
P0 switch to P1 : enter a line :
proc 0  in scheduler
readyQueue = [1 1 ]->[0 0 ]->NULL
next running = 1
switch to proc 1  pgdir at 0x604000  pgdir[2048] = 0x800C3E
test memory protection? [y|n]: y
try 0x80200000 : data_abort exception in USER mode
domain=0x0  status=0x5  addr=0x80200000
try 0x1000 : data_abort exception in USER mode
domain=0x0  status=0xD  addr=0x1000
This is process #1  in Umode at 0x800000  parent=0
------------------------
getpid getppid ps chname
------------------------
input a command :
```

Fig. 7.1 Demonstration of user mode process and system calls

P1, P2, …, Pn, each executes a corresponding image in User mode. Modify kfork() to kfork(int i), which creates process Pi and loads the image ui to the memory area of process Pi. On ARM based systems, use the simplest memory management scheme by allocating each process a 1 MB Umode image area by process PID, e.g. P1 at 8 MB, P2 at 9 MB, etc. Some of the processes can be periodic while others can be event-driven. All the processes run in the same virtual address space of [2 GB to 2 GB + 1 MB] but each has a separate physical memory area, which is isolated from other processes and protected by the MMU hardware. We demonstrate such a system by an example.

7.5.2 Demonstration of Processes in the Same Domain

In this example system, we create 4 Umode images, denoted by u1 to u4. All Umode images are compile-linked with the same starting virtual address 0x80000000. They executes the same ubody(int i) function. Each process calls ubody(pid) with a unique process ID number for identification. When setting up the process page tables, the kernel space is assigned the domain number 0. All Umode spaces are assigned the domain number 1. When a process begins execution in Umode, it allows the user to test memory protection by trying to access invalid virtual addresses, which would generate memory protection faults. In the data abort exception handler, it displays the MMU's fault_status and fault_addr registers to show the exception cause as well as the faulting virtual address. Then each process executes an infinite loop, in which it prompts for a command and executes the command. Each command invokes a system call interface, which issues a system call, causing the process to execute the system call function in kernel. To demonstrate additional capabilities of the system, we add the following commands:

```
switch  :   enter kernel to switch process;
sleep       : enter kernel to sleep on process pid;
wakeup:     enter kernel to wake up sleeping process by pid;
```

If desired, we may also use sleep/wakeup to implement event-driven processes, as well as for process cooperation. To support and test the added User mode commands, simply add them to the command and syscall interfaces.

```
#include "string.c"
#include "uio.c"
int ubody(int id)
{
  char line[64]; int *va;
  int pid = getpid();
  int ppid = getppid();
  int PA = getPA();
  printf("test memory protection\n");
  printf("try VA=0x80200000 : ");
  va = (int *)0x80200000; *va = 123;
  printf("try VA=0x1000 : ");
  va = (int *)0x1000;      *va = 456;
  while(1){
    printf("Process #%d in Umode at %x parent=%d\n", pid, PA, ppid);
    umenu();
    printf("input a command : ");
    ugetline(line); uprintf("\n");
    if (!strcmp(line, "switch"))     // only show the added commands
       uswitch();
    if (!strcmp(line, "sleep"))
       usleep();
    if (!strcmp(line, "wakeup"))
       uwakeup();
  }
}
/******* u1.c file ******/ // u2.c, u3.c u4.c are similar
#include "ucode.c"
main(){ ubody(1); }
```

The following shows the (condensed) kernel files. First, the syscall routing table, svc_handler(), is modified to support the added syscalls.

```
int svc_handler(volatile int a, int b, int c, int d)
{
  int r;
  switch(a){
     case 0: r = kgetpid();              break;
     case 1: r = kgetppid();             break;
     case 2: r = kps();                  break;
     case 3: r = kchname((char *)b);     break;
     case 4: r = ktswitch();             break;
     case 5: r = ksleep(running->pid); break;
     case 6: r = kwakeup((int)b);        break;
     case 92:r = kgetPA(); // return runnig->pgdir[2048]&0xFFF00000;
  }
  running->kstack[SSIZE-14] = r;
}
```

In the kernel t.c file, after system initialization it creates 4 processes, each with a different Umode image. In kfork(pid), it uses the new process pid to load the corresponding image into the process memory area. The loading addresses are P1 at

8 MB, P2 at 9 MB, P3 at 10 MB and P4 at 11 MB. Process page tables are set up in the same way as in Program C7.1. Each process has a page table in the 6 MB area. In the process page tables, entry 2048 (VA = 0x80000000) points to the process Umode image area in physical memory.

```
/*********** kernel t.c file ************/
#define VA(x) (0x80000000 + (u32)x)
int main()
{  // initialize kernel as before
   for (int i=1; i<=4, i++){ // create P1 to P4
       kfork(i);
   }
   printf("P0 switch to P1\n");
   tswitch();  // switch to run P1 ==> never return again
}
PROC *kfork(int i) // i = 1 to 4
{
  // creat new PROC *p with pid=i as before
  u32 *addr = (char *)(0x800000 + (p->pid-1)*0x100000);
  // copy Umode image ui of p to addr
  p->usp = (int *)VA(0x100000);  // high end of 1MB VA
  p->kstack[SSIZE-1] = VA(0);    // starting VA=0
  enqueue(&readyQueue, p);
  return p;
}
```

Figure 7.2 shows the outputs of running the C7.2 program. It shows that the switch command switches the running process from P1 to P2. The reader may run the system and enter other User mode commands to test the system.

7.5.3 Processes with Individual Domains

In the first system model of C7.2, each process has its own page table. Switching process requires switching page table, which in turn requires flushing the TLB and I&D caches. In the next system model, the system supports $n < 16$ user mode processes with only one page table. In the page table, the first 2048 entries define the kernel mode virtual address space, which is ID mapped to the available physical memory (258 MB). As before, the kernel space is designated as domain 0. Assume that all user mode images are 1 MB in size. Entry 2048 + pid of the page table maps to the 1 MB physical memory of process Pi. The pgdir entry attributes are set to 0xC1E | (pid ≪ 5), so that each user image area is in a unique domain numbered 1 to pid. When switching process to Pi, instead of switching page tables, it calls

```
set_domain( (1 << 2*pid) | 0x01);
```

which sets the access bits of domain0 and domainPid to b01 and clears the access bits of all other domains to 0, making them inaccessible. This way, each process runs only in its own virtual address space, which is protected from other processes. Naturally, processes in kernel mode can still access all the memory because it is running in the privileged SVC mode. The limitation of this model is that the system can only support 15 user mode processes. Another drawback is that each user mode image must be compile-linked with a different starting virtual address that matches its page table index. The sample system C7.3 implements processes with individual domains.

7.5.4 Demonstration of Processes with Individual Domains

Figure 7.3 shows the outputs of running the C7.3 program. As the figure shows, all processes share the same page table at 0x4000 but each process has a different entry in the page table. The figure also shows that each process can only access its

```
QEMU
domain=0x0  status=0x5  addr=0x80000000                                   00:00:51
P0 switch to P1 : enter a line :
proc 0  in scheduler
readyQueue = [1 1 ]->[2 1 ]->[3 1 ]->[4 1 ]->[0 0 ]->NULL
next running = 1
switch to proc 1  pgdir at 0x604000  pgdir[2048] = 0x800C3E
test memory protection? [y|n]: y
try 0x80200000 : data_abort exception in USER mode
domain=0x0  status=0x5  addr=0x80200000
try 0x1000 : data_abort exception in USER mode
domain=0x0  status=0xD  addr=0x1000
This is process #1  in Umode at 0x800000  parent=0
-------------------------------------------
getpid getppid ps chname switch sleep wakeup
-------------------------------------------
input a command : switch
proc 1  in scheduler
readyQueue = [2 1 ]->[3 1 ]->[4 1 ]->[1 1 ]->[0 0 ]->NULL
next running = 2
switch to proc 2  pgdir at 0x608000  pgdir[2048] = 0x900C3E
This is process #2  in Umode at 0x900000  parent=0
-------------------------------------------
getpid getppid ps chname switch sleep wakeup
-------------------------------------------
input a command :
```

Fig. 7.2 Outputs of sample system C7.2

own VA space. Any attempt to access a VA outside of its VA space will generate a data_abort exception due to invalid domains.

7.6 RAM Disk

In the previous programming examples, User mode images are included as raw data sections in the kernel image. When the kernel image boots up, it relies on the symbolic addresses generated by the linker to load (copy) the various User mode images to their memory locations. This scheme works well if the number of User mode images is small. It can be very tedious when the number of User mode images becomes large, which also increases the kernel image size. If we intend to run a large number of User mode processes with different images, a better way to manage the user mode images is needed. In this section, we shall show how to use a ramdisk file system to manage user mode images. First, we create a virtual ramdisk and formt it as a file system. Then we generate User mode images as executable ELF files in the ramdisk file system. When booting up the system, we also load the ramdisk image to make it accessible to the kernel. When the kernel starts, we move the ramdisk image to a known memory area, e.g. at 4 MB, and use it as a RAMdisk in memory. There are two ways to make the ramdisk image accessible to the kernel.

(1).
Include ramdisk image as a raw data section: Convert the ramdisk image to binary and include it as a raw data section in the kernel image, similar to the individual Umode images before.

(2).
As an initial ramdisk image: Run QEMU with the **–initrd ramdisk** option, as in

```
qemu-system-arm -M versatilpb -m 256M -kernel t.bin -initrd ramdisk
```

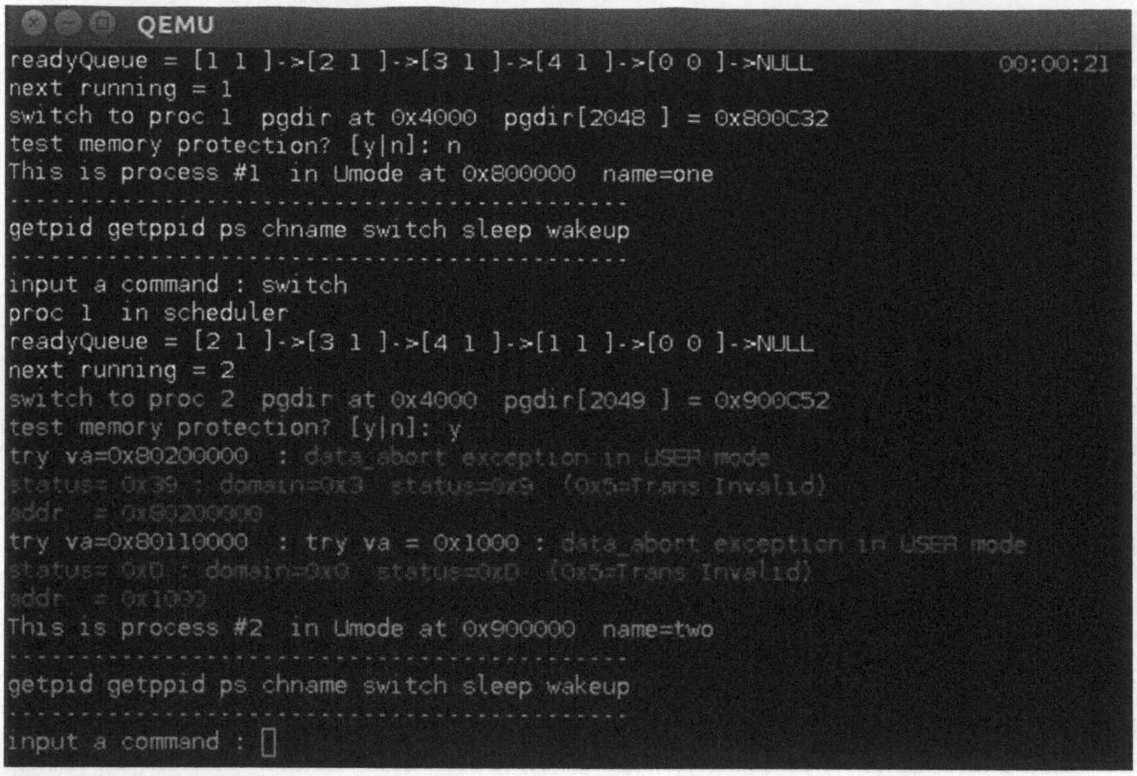

```
QEMU
readyQueue = [1 1 ]->[2 1 ]->[3 1 ]->[4 1 ]->[0 0 ]->NULL                00:00:21
next running = 1
switch to proc 1  pgdir at 0x4000  pgdir[2048 ] = 0x800C32
test memory protection? [y|n]: n
This is process #1  in Umode at 0x800000  name=one
---------------------------------------------
getpid getppid ps chname switch sleep wakeup
---------------------------------------------
input a command : switch
proc 1  in scheduler
readyQueue = [2 1 ]->[3 1 ]->[4 1 ]->[1 1 ]->[0 0 ]->NULL
next running = 2
switch to proc 2  pgdir at 0x4000  pgdir[2049 ] = 0x900C52
test memory protection? [y|n]: y
try va=0x80200000  : data_abort exception in USER mode
status= 0x39 : domain=0x3  status=0x9  (0x5=Trans Invalid)
addr  = 0x80200000
try va=0x80110000  : try va = 0x1000 : data_abort exception in USER mode
status= 0xD : domain=0x0  status=0xD  (0x5=Trans Invalid)
addr  = 0x1000
This is process #2  in Umode at 0x900000  name=two
---------------------------------------------
getpid getppid ps chname switch sleep wakeup
---------------------------------------------
input a command :
```

Fig. 7.3 Outputs of sample system C7.3

QEMU will load the kernel image to 0x10000 (64 KB) and the initial ramdisk image to 0x4000000 (64 MB). Although the QEMU documents state that the initial ramdisk image will be loaded to 0x800000 (8 MB), it actually depends on the memory size of the virtual machine. As a general rule, the VM's memory size should be a power of 2. The ramdisk image loading address is the memory size divided by 2, with an upper limit of 128 MB. The following lists some of the commonly used VM memory sizes and loading addresses of the initial ramdisk image.

VM memory size (MB)	Ramdisk loading address (MB)
16	8
32	16
64	32
128	64
512	128 (upper limit)

A simple way to find out the loading address of the ramdisk image is to dump a string, e.g. "ramdisk begin" to the beginning of the ramdisk image. When the kernel starts, scan each 1 MB memory area to detect the string. Once the loading address is known, we can move the ramdisk image to a memory location and use it as a RAMdisk. In order to access Umode images in the RAMdisk file system, we add a RAMdisk driver to read-write RAMdisk blocks. Then we develop an ELF file loader to load the image files into process memory areas. There are several popular file systems used in embedded systems. Most early embedded systems use the Microsoft FAT file system. Since we use Linux as the development platform, we shall use a Linux compatible file system in order to avoid any unnecessary file conversions. For this reason, we shall use the EXT2 file system, which is totally Linux compatible. The reader may consult (EXT2 2001; Card et al. 1995; Cao et al. 2007) for EXT2 file system specifications. In the following, we show how to create an EXT2 file system image and use it as a ramdisk.

7.6.1 Creating RAM Disk Image

(1). Under Linux, run the following commands (or as a sh script). For small EXT2 file systems, it is better to use 1 KB file block size.

```
dd if=/dev/zero of=ramdisk bs=1024 count=1024 # create ramdisk file
mke2fs -b 1024 ramdisk 1024 # format it as a 1MB EXT2 FS
mount  -o loop ramdisk  /mnt            # mount it as a loop device
mkdir  /mnt/bin                         # create a /bin directory
umount  /mnt                            # umount it
```

(2). Compile-link User mode source files in a USER directory as ELF executables and copy them into /bin directory of ramdisk, as shown by the following mku script.

```
arm-none-eabi-as -mcpu=arm926ej-s us.s -o us.o
arm-none-eabi-gcc -c -mcpu=arm926ej-s -o $1.o $1.c
arm-none-eabi-ld -T u.ld us.o $1.o -Ttext=0x80000000 -o $1
mount -o loop ../ramdisk /mnt
cp $1 /mnt/bin
umount /mnt
```

(3). Run the following sh script to generate User mode images and convert ramdisk into an object file.

```
(cd USER; mku u1; mku u2) # assume files are in a USER directory
arm-none-eabi-objcopy -I binary -O elf32-littlearm -B arm \
                      ramdisk ramdisk.o
```

(4). Compile-link the kernel files to include ramdisk.o as a raw data section in the kernel image.
(5). In the t.c file, copy ramdisk.o to a memory location and use it as a RAMdisk.

```
extern char _binary_ramdisk_start, _binary_ramdisk_end;
char *RAMdisk = (char *)0x400000; // global at 4MB
int main()
{
   char *cp, *cq = RAMdisk;
   int size;
   cp = (char *)&_binary_ramdisk_start;
   size = &_binary_ramdisk_end - &_binary_ramdisk_start;
   // copy ramdisk image to RAMdisk at 4MB
   memcpy(cq, cp, size);
   fbuf_init();
   kbd_init();
   uart_init();
   // enable SIC VIC for device interrupts as before
   timer_init(); timer_start(0);
   kernel_init();      // kernel init
   printf("RAMdisk start=%x size=%x\n", RAMdisk, size);
   kfork("/bin/u1"); // Create P1 with Umode image="/bin/u1"
   printf("P0 switch to P1\n");
   tswitch();          // switch to run P1; never return
}
```

Alternatively, we may also rely on QEMU to load the ramdisk image by the –initrd option directly, as in

```
qemu-system-arm –M versatilpb –m 128M –kernel t.bin -initrd ramdisk
```

In that case, we may either move the ramdisk image from its loading address (64 MB) to 4 MB or use the ramdisk loading address directly. The following shows the RAMdisk block I/O functions, which are essentially memory copy functions.

```
/******* RAMdisk I/O driver **********/
#define BLKSZIE 1024
int get_block(u32 blk, char *buf) // read from a 1KB block
{
  char *cp = RAMdisk + blk*BLKSIZE;
  memcpy(buf, cp, BLKSIZE);
}
int put_block(u32 blk, char *buf) // write to a 1KB block
{
  char *cp = RAMdisk + blk*BLKSIZE;
  memcpy(cp, buf, BLKSIZE);
}
```

7.6.2 Process Image File Loader

An image loader consists of two parts. The first part is to locate the image file and check whether it is executable. The second part is to actually load the image file's executable contents into memory. We explain each part in more detail.

Part 1 of image loader: In an EXT2 file system, each file is represented by a unique INODE data structure, which contains all the information of the file. Each INODE has an inode number (ino), which is the position (counting from 1) in the INODE table. To find a file amounts to finding its INODE. The algorithm is as follows.

(1). Read in the Superblock (block 1) Check the magic number (0xEF53) to verify it's indeed an EXT2 FS.
(2). Read in the group descriptor block (block 2) to access the group 0 descriptor. From the group descriptor's bg_inode_table entry, find the INODEs begin block number, call it the InodesBeginBlock.
(3). Read in InodeBeginBlock to get the INODE of root directory /, which is second inode (ino = 2) in the INODE Table
(4). Tokenize the pathname into component strings and let the number of components be n. For example, if pathname = /a/b/c, the component strings are "a", "b", "c", with n = 3. Denote the components by name[0], name[1], ..., name[n − 1].
(5). Start from the root INODE in (3), search for name[0] in its data block(s). Each data block of a DIR INODE contains dir_entry structures of the form

```
[ino rlen nlen NAME] [ino rlen nlen NAME] ......
```

where NAME is a sequence of nlen chars (without a terminating NULL). For each data block, read the block into memory and use a dir_entry *dp to point at the loaded data block. Then use nlen to extract NAME as a string and compare it with name[0]. If they do not match, step to the next dir_entry by

```
dp = (dir_entry *)((char *)dp + dp->rlen);
```

and continue the search. If name[0] exists, we can find its dir_entry and hence its inode number.

(6). Use the inode number (ino) to compute the disk block containing the INODE and its offset in that block by the Mailman's algorithm (Wang 2015).

$$\text{blk} = (\text{ino} - 1)/(\text{BLKSIZE}/\text{sizeof}(\text{INODE})) + \text{InodesBeginBlock};$$
$$\text{offset} = (\text{ino} - 1)\%(\text{BLKSIZE}/\text{sizeof}(\text{INODE}));$$

Then read in the INODE of /a, from which we can determine whether it's a DIR. If /a is not a DIR, there can't be /a/b, so the search fails. If /a is a DIR and there are more components to search, continue for the next component name[1]. The problem now becomes: search for name[1] in the INODE of /a, which is exactly the same as that of Step (5).

(7). Since Steps 5–6 will be repeated n times, it's better to write a search function

```
u32 search(INODE *inodePtr, char *name)
{
   // search for name in the data blocks of this INODE
   // if found, return its ino; else return 0
}
```

Then all we have to do is to call search() n times, as sketched below.

```
Assume: n,  name[0], ..., name[n-1] are globals
INODE *ip points at INODE of /
for (i=0; i<n; i++){
    ino = search(ip, name[i])
    if (!ino){ // can't find name[i], return 0;}
    use ino to read in INODE and let ip point to INODE
}
```

If the search loop ends successfully, ip must point at the INODE of pathname. Then we can check its file type and file header (if necessary) to ensure it is executable.

Part 2 of image loader: From the file's INODE, we know its size and file blocks. The second part of the loader loads the file's executable contents into memory. This step depends on the executable file type.

(1). A **flat binary executable file** is a single piece of binary code, which is loaded in its entirety for direct execution. This can be done by converting all ELF files to binary files first. In this case, loading file contents is the same as loading file blocks.

(2). **ELF executable file format**: An ELF executable file begins with an elf-header, followed by one or more program section headers, which are defined as ELF header and ELF program section header structures.

```
struct elfhdr {    // ELF File header
  u32 magic;       // ELF_MAGIC 0x464C457F
  u8  elf[12];
  u16 type;
  u16 machine;
  u32 version;
  u32 entry;
  u32 phoffset;    // byte offset of program header
  u32 shoffset;    // byte offset of sections
  u32 flags;
  u16 ehsize;      // elf header size
```

```
  u16 phentsize; // program header size
  u16 phnum;      // number of program section headers
  u16 shentsize;
  u16 shnum;
  u16 shstrndx;
};
struct proghdr {  // ELF Program section header
  u32 type;       // 1 = loadable image
  u32 offset;     // byte offset of program section
  u32 vaddr;      // virtual address
  u32 paddr;      // physical address
  u32 filesize;   // number of bytes of program section
  u32 memsize;    // load memory size
  u32 flags;      // R|W|Ex flags
  u32 align;      // alignment
};
```

The reader may consult (ELF 1995) for the ELF file format. For help information, the reader may also use the (Linux) readelf command to view ELF file contents. For example,

readelf –eSt file.elf

displays the headers (e), section headers (S) and section details (t) of an ELF file. For ELF executable files, the loader must load the various sections of an ELF file to their specified virtual addresses. In addition, each loaded section should be marked with appropriate R|W|Ex attributes for protection. For example, the code section pages should be marked for RO (read-only), data section pages should be marked for RW, etc. For generality, our image loader can load either binary or ELF executable files. The loader's algorithm is as follows. The reader may consult the ELF loader code in the loadelf.c fie for implementation details.

```
/*********** Loader Algorithm ***********/
find file's INODE; return 0 if fails;
read elf-header to check whether it's an ELF file;
if (!ELF){ // assume flat BINARY file
    determine file size;
     load file blocks to process image area by file size;
     return 1;                // for SUCCESS
}
/*************** ELF file ***************/
locate program header(s);
for each program header do{
     get section's offset, loading address and size;
     load section to virtual address until size;
     set section's R|W|Ex attributes in loaded pages;
}
     return 1;                // for SUCCESS
```

7.7 Process Management

7.7.1 Process Creation

Most embedded systems are designed for specific application environments. A typical embedded system comprises a fixed number of processes. Each process is an independent execution unit, which does not have any relation or interactions with other processes. For such systems, the current kfork() function, which creates a process to execute a specific function is adequate. In order to support dynamic user mode processes, we shall extend the kernel to impose a parent-child relation between processes. When the kernel starts, it runs the initial process P0, which is handcrafted or created by brute force. Thereafter, every other process is created by

```
int newpid = kfork(char *filename,  int priority);
```

which creates a new process with a specified priority to execute a Umode image filename. When creating a new process, the creator is the parent and the newly created process is the child. In the PROC structure, the field ppid records the parent process pid, and the parent pointer points to the parent PROC. Thus, the processes form a family tree with P0 as the root.

7.7.2 Process Termination

In a multitasking system with dynamic processes, a process may terminate or die, which is a common term of process termination. A process may terminate in two possible ways:

Normal termination: The process has completed its task, which may not be needed again for a long time. To conserve system resources, such as PROC structures and memory, the process calls kexit(int exitValue) in kernel to terminate itself, which is the case we are discussing here.

Abnormal termination: The process terminates due to an exception, which renders the process impossible to continue. In this case it calls kexit(value) with a unique value that identifies the exception.

In either case, when a process terminates, it eventually calls kexit() in kernel. The general algorithm of kexit() is as follows.

```
/*************** Algorithm of kexit ******************/
 kexit(int exitValue)
 {
      1. erase process user-mode context, e.g. close file descriptors,
         release resources, deallocate user-mode image memory, etc.
      2. dispose of children processes, if any.
      3. record exitValue in PROC.exitCode for parent to get.
      4. become a ZOMBIE (but do not free the PROC)
      5. wakeup parent and, if needed, also the INIT process P1.
      6. switch process to give up CPU
 }
```

So far, our system model does not yet support a file system. Due to the simple memory allocation scheme, each process runs in a dedicated 1 MB memory area, deallocation of user-mode image is also trivial. When a process terminates, its user mode memory area will be left unused until a process with the same pid is created again. So we begin by discussing Step 2 of kexit(). Since each process is an independent execution entity, it may terminate at any time. If a process with children terminates first, all the children of the process would have no parent anymore, i.e. they become orphans. The question is then: what to do with such orphans? In human society, they would be sent to grandma's house. But what if grandma already died? Following this reasoning, it immediately becomes clear that there must be a process which should not terminate if there are other processes still existing. Otherwise, the parent-child process relation would soon break down. In all Unix-like systems, the process P1, which is also known as the INIT process, is chosen to play this role. When a process dies, it sends all the orphaned children, dead or alive, to P1, i.e. become P1's children. Following suit, we shall also designate P1 as such a process. Thus, P1 should not die if there are other processes still existing. The remaining problem is how to implement Step 2 efficiently. In order for a dying process to dispose of children, the process must be able to determine whether it has any child and, if it has children, find all the children quickly. If the number of processes is small, both questions can be answered effectively by searching all the PROC structures. For example, to determine whether a process has any child, simply search the PROCs for any one that is not FREE and its ppid matches the process pid. If the number of processes is large, e.g. in the order of hundreds, this simple search scheme would be too slow. For this reason, most large OS kernels keep track of process relations by maintaining a process family tree.

7.7.3 Process Family Tree

Typically, the process family tree is implemented as a binary tree by a pair of child and sibling pointers in each PROC, as in

```
struct proc *child, *sibling, *parent;
```

where child points to the first child of a process and sibling points to a list of other children of the same parent. For convenience, each PROC also uses a parent pointer pointing to its parent. With a process tree, it is much easier to find the children of a process. First, follow the child pointer to the first child PROC. Then follow the sibling pointers to traverse the sibling PROCs. To send all children to P1, simply detach the children list and append it to the children list of P1 (and change their ppid and parent pointer also). Because of the small number of PROCs in all the sample systems of this book, we do not implement the process tree. This is left as a programming exercise. In either case, it should be fairly easy to implement Step 2 of kexit().

Each PROC has a 2-byte exitCode field, which records the process exit status. In Linux, the high byte of exitCode is the exitValue and the low byte is the exception number that caused the process to terminate. Since a process can only die once, only one of the bytes has meaning. After recording exitValue in PROC.exitCode, the process changes its status to ZOMBIE but does not free the PROC. Then the process calls kwakeup() to wake up its parent and also P1 if it has sent any orphans to P1. The final act of a dying process is to call tswitch() for the last time. After these, the process is essentially dead but still has a dead body in the form of a ZOMBIE PROC, which will be buried (set FREE) by the parent process through the wait operation.

7.7.4 Wait for Child Process Termination

At any time, a process may call the kernel function

```
int pid = kwait(int *status)
```

to wait for a ZOMBIE child. If successful, the returned pid is the ZOMBIE child pid and status contains the exitCode of the ZOMBIE child. In addition, kwait() also releases the ZOMBIE PROC back to the freeList, allowing it to be reused for another process. The algorithm of kwait is

```
int kwait(int *status)
{
    if (caller has no child)
       return -1 for error;
    while(1){          // caller has children
       search for a (any) ZOMBIE child;
       if (found a ZOMBIE child){
          get ZOMBIE child pid;
          copy ZOMBIE child exitCode to *status;
          release child PROC to freeList as FREE;
          return ZOMBIE child pid;
       }
       ksleep(running);  // sleep on its PROC address
    }
}
```

In the kwait algorithm, the process returns −1 for error if it has no child. Otherwise, it searches for a (any) ZOMBIE child. If it finds a ZOMBIE child, it collects the ZOMBIE child pid and exitCode, releases the ZOMBIE PROC to freeList and returns the ZOMBIE child pid. Otherwise, it goes to sleep on its own PROC address, waiting for a child to terminate. Correspondingly, when a process terminates, it must issue

```
kwakeup(parent); // parent is a pointer to parent PROC
```

to wake up the parent. When the parent process wakes up in kwait(), it will find a dead child when it executes the while loop again. Note that each kwait() call handles only one ZOMBIE child, if any. If a process has many children, it may have to call kwait() multiple times to dispose of all dead children. Alternatively, a process may terminate first without waiting for any dead child. When a process dies, all of its children become children of P1. As we shall see later, in a real system P1 executes an infinite loop, in which it repeatedly waits for dead children, including adopted orphans. Instead of sleep/wakeup, we may also use semaphore to implement the kwait()/kexit() functions. Variations to the kwait() operation include waitpid and waitid of Linux (Linux Man Page 2016), which allows a process to wait for a specific child by pid with many options.

7.7.5 Fork-Exec in Unix/Linux

When a process executes, it may need to save information generated by the execution. A good example is a log, which records important events occurred during execution. The log can be used to trace process executions for debugging in case something went wrong. A process may also need inputs to control its paths of execution. Saving and retrieving information require the support of a file system. An operating system kernel usually provides basic file system support to allow processes to do file operations. In such systems, the execution environment of each process includes both its execution context and its ability to access files. The current kfork() mechanism can only create processes to execute different images, but it does not provide any means for file operations. In order to support the latter, we need an alternative way to create and run processes. In Unix/Linux, the system call

```
int pid = fork();
```

creates a child process with a Umode image identical to that of the parent. In addition, it also passes all opened file descriptors to the child, allowing the child to inherit the same file operation environment of the parent. If successful, fork() returns the child process pid. Otherwise, it returns −1. When the child process runs, it returns to its own Umode image and the returned pid is 0. This allows us to write User mode programs as

```
int pid = fork();       // fork a child process
if (pid){
            // parent executes this part;
}
else{   // child  executes this part;
}
```

The code segment uses the returned pid to differentiate between the parent and child processes. Upon return from fork(), the child process usually uses the system call

```
int r = exec(char *filename, char *para-list);
```

to change its execution image to a different file, passing as parameters para-list to the new image when execution starts. If successful, exec() merely replaces the original Umode image with a new image. It is still the same process but with a different Umode image. This allows a process to execute different programs. Fork and exec may be called the bread and butter of Unix/Linux because almost every operation depends on fork-exec. For example, when a user enters a command line of the form

```
cmdLine = "cmd arg1 arg2 .... argn"
```

the sh process forks a child and waits for the child to terminate. The child process uses exec to change its image to the cmd file, passing arg1 to argn as parameters to the new image. When the child process terminates, it wakes up the parent sh, which prompts for another command, etc. Note that fork-exec creates a process to execute a new image in two steps. The main advantages of the fork-exec paradigm are twofold. First, fork creates a child with an identical image. This eliminates the need for passing information across different address spaces between the parent and the child. Second, before executing the new image, the child can examine the command line parameters to alter its execution environment to suit its own needs. For example, the child process may redirect its standard input (stdin) and output (stdout) to different files.

7.7.6 Implementation of Fork

The implementation of fork-exec is rather simple. The algorithm of fork() is

```
/********************** Algorith of fork() ***************************/
```

1. get a PROC for the child and initialize it, e.g. ppid = parent pid, priority=1, etc.
2. copy parent Umode image to child, so that their Umode images are identical;
3. copy (part of) parent kstack to child kstack; Ensure that the child return to the same virtual address as the parent but in its own Umode image;
4. copy parent usp and spsr to child;
5. mark child PROC READY and enter it into readyQueue;
6. return child pid;

The following shows the fork() code that implement the fork algorithm

```
.
int fork()
{
  int i;
  char *PA, *CA;
  PROC *p = get_proc(&freeList);
  if (p==0){ printf("fork failed\n"); return -1; }
  p->ppid = running->pid;
  p->parent = running;
  p->status = READY;
  p->priority = 1;
  PA = (char *)running->pgdir[2048] & 0xFFFF0000; // parent Umode PA
  CA = (char *)p->pgdir[2048] & 0xFFFF0000;       // child  Umode PA
  memcpy(CA, PA, 0x100000); // copy 1MB Umode image
  for (i=1; i <= 14; i++){  // copy bottom 14 entries of kstack
     p->kstack[SSIZE-i] = running->kstack[SSIZE-i];
  }
  p->kstack[SSIZE - 14] = 0;                // child return pid = 0
  p->kstack[SSIZE-15] = (int)goUmode;       // child resumes to goUmode
  p->ksp = &(p->kstack[SSIZE-28]);          // child saved ksp
  p->usp = running->usp;                    // same usp as parent
  p->ucpsr = running->ucpsr;                // same spsr as parent
  enqueue(&readyQueue, p);
  return p->pid;
}
```

We explain the fork() code in mode detail. When the parent executes fork() in kernel, it has saved Umode registers in kstack by stmfd sp!, {r0–r12, LR}, and replaced the saved LR with the proper return address to Umode. Therefore, its kstack bottom contains

```
  -1   -2   -3   -4  -5  -6  -7  -8  -9  -10 -11 -12 -13 -14
 --------------------------------------------------------------
 |uLR ur12 ur11 ur10 ur9 ur8 ur7 ur6 ur5 ur4 ur3 ur2 ur1 ur0=0|
---------------------------------------------------------------
```

which are copied to the bottom of the child's kstack. These 14 entries will be used by the child to return to Umode when it executes ldmfs sp!, {r0–12, pc}^ in goUmode. The copied LR at entry −1 allows the child to return to the same VA as the parent, i.e. to the same pid = fork() syscall. In order for the child to return pid = 0, the saved r0 at entry −14 must be set to 0. In order for the child to resume in kernel, we append a RESUME stack frame to the child's kstack for it to resume when it is scheduled to run. The added stack frame must be consistent with the RESUME part of tswitch().The added kstack frame is shown below, and the child saved ksp points to entry −28.

```
   -15   -16  -17  -18  -19 -20 -21 -22 -23 -24 -25 -26 -27 -28
 ------------------------------------------------------------------
 |goUmode kr12 kr11 kr10 kr9 kr8 kr7 kr6 kr5 kr4 kr3 kr2 kr1 kr0 |
----|----------------------------------------------------------|---
   klr                                                        ksp
```

Since the child resumes running in kernel, all the "saved" Kmode registers do not matter, except the resume klr at entry −15, which is set to goUmode. When the child runs, it uses the RESUME kstack frame to execute goUmode directly. Then it executes goUmode with the copied syscall stack frame, causing it to return to the same VA as the parent but in its own memory area with a 0 return value.

7.7.7 Implementation of Exec

Exec allows a process to replace its Umode image with a different executable file. We assume that the parameter to exec(char *cmdlie) is a command line of the form

```
cmdline = "cmd arg1 arg2 ... argn";
```

where cmd is the file name of an executable program. If cmd starts with a /, it is an absolute pathname. Otherwise it is a file in the default /bin directory. If exec succeeds, the process returns to Umode to execute the new file, with the individual token strings as command line parameters. Corresponding to the command line, the cmd program can be written as

```
main(int argc, char *argv[]){      }
```

where argc = n + 1 and argv is a null terminated array of string pointers, each points to a token string of the form

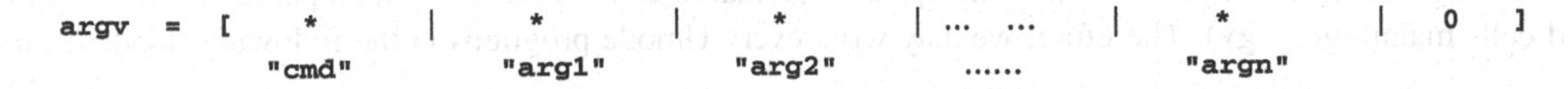

The following shows the algorithm and implementation of the exec operation.

```
/***************** Algorithm of exec ******************/
```

1. fetch cmdline from Umode space;
2. tokenize cmdline to get cmd filename;
3. check cmd file exists and is executable; return −1 if fails;
4. load cmd file into process Umode image area;

5. copy cmdline to high end of usatck, e.g. to **x = high end−128**;
6. reinitialize syscall kstack frame to return to VA = 0 in Umode;
7. **return x**;

The following exec() code implements the above exec algorithm.

```
int exec(char *cmdline) // cmdline=VA in Uspace
{
  int i, upa, usp;
  char *cp, kline[128], file[32], filename[32];
  PROC *p = running;
  strcpy(kline, cmdline); // fetch cmdline into kernel space
  // get first token of kline as filename
  cp = kline; i = 0;
  while(*cp != ' '){
    filename[i] = *cp;
    i++; cp++;
  }
  filename[i] = 0;
  file[0] = 0;
  if (filename[0] != '/') // if filename relative
     strcpy(file, "/bin/"); // prefix with /bin/
  strcat(file, filename);
upa = p->pgdir[2048] & 0xFFFF0000; // PA of Umode image
// loader return 0 if file non-exist or non-executable
if (!loadelf(file, p))
    return -1;
// copy cmdline to high end of Ustack in Umode image
  usp = upa + 0x100000 - 128;  // assume cmdline len < 128
  strcpy((char *)usp, kline);
  p->usp = (int *)VA(0x100000 - 128);
  // fix syscall frame in kstack to return to VA=0 of new image
  for (i=2, i<14; i++)      // clear Umode regs r1-r12
    p->kstack[SSIZE - i] = 0;
  p->kstack[SSIZE-1]    = (int)VA(0);   // return uLR = VA(0)
  return (int)p->usp; // will replace saved r0 in kstack
}
```

When execution begins from us.s in Umode, r0 contains p->usp, which points to the original cmdline in the Umode stack. Instead of calling main() directly, it calls a C startup function main0(char *cmdline), which parses the cmdline into argc and argv, and calls main(argc, argv). Therefore, we may write every Umode programs in the following standard form as usual.

```
#include "ucode.c"
main(int argc, char *argv[ ]){..............}
```

The following lists the code segments of main0(), which plays the same role as the standard C startup file crt0.c.

```
int argc; char *argv[32]; // assume at most 32 tokens in cmdline
int parseArg(char *line)
{
  char *cp = line; argc = 0;
  while (*cp != 0){
     while (*cp == ' ') *cp++ = 0;   // skip over blanks
     if (*cp != 0)                   // token start
```

```
        argv[argc++] = cp;           // pointed by argv[ ]
    while (*cp != ' ' && *cp != 0) // scan token chars
        cp++;
    if (*cp != 0)
        *cp = 0;                     // end of token
    else
        break;                       // end of line
    cp++;                            // continue scan
  }
  argv[argc] = 0;                    // argv[argc]=0
}
main0(char *cmdline)
{
  uprintf("main0: cmdline = %s\n", cmdline);
  parseArg(cmdline);
  main(argc, argv);
}
```

7.7.8 Demonstration of Fork-Exec

We demonstrate a system that supports dynamic processes with fork, exec, wait and exit functions by the sample system C7.4. Since all the system components have already been covered and explained before, we only show the t.c file containing the main() function. In order to demonstrate exec, we need a different Umode image file. The u2.c file is identical to u1.c, except that it displays in German (just for fun).

```
/****************** t.c file of Program C7.4 ******************/
#include "uart.c"
#include "kbd.c"
#include "timer.c"
#include "vid.c"
#include "exceptions.c"
#include "queue.c"
#include "svc.c"
#include "kernel.c"
#include "wait.c"

#include "fork.c"      // kfork() and fork()
#include "exec.c"      // exec()#include "disk.c"       // RAMdisk driver
#include "loadelf.c"   // ELF file loader

extern char _binary_ramdisk_start, _binary_ramdisk_end;
int main()
{
   char *cp, *cq;
   int dsize;
   cp = disk = &_binary_ramdisk_start;
   dsize = &_binary_ramdisk_end - &_binary_ramdisk_start;
   cq = (char *)0x400000;
   memcpy(cq, cp, dsize);
   fbuf_init();
   printf("Welcome to WANIX in Arm\n");
   kbd_init();
   uart_init();
```

```
    timer_init();
    timer_start(0);
    kernel_init();
    unlock();            // enable IRQ interrupts
    printf("RAMdisk start=%x size=%x\n", disk, dsize);
    kfork("/bin/u1");    // cteate P1 to execute u1 image
    while(1){            // P0 as the idle process
      if (readyQueue)
        tswitch();
    }
}
```

Figure 7.4a shows the outputs of running the fork and switch commands. As the figure shows, P1 runs at the physical address PA = 0x800000 (8 MB). When it forks a child P2 at PA = 0x900000 (9 MB), it copies both the Umode image and the kstack to the child. Then it return to Umode with the child pid = 2. The switch command causes P1 to enter kernel to switch process to P2, which returns to Umode with child pid = 0, indicating that it is the forked child. The reader may test the exit and wait operations as follows.

(1) While P2 runs, enter the command wait. P2 will issue a wait syscall to execute kwait() in kernel. Since P2 does not have any child, it returns -1 for no child error.
(2) While P2 runs, enter the exit command and enter an exit value, e.g. 1234. P2 will issue an exit(1234) syscall to terminate in kernel. In kexit(), it records the exitValue in its PROC.exitCode, becomes a ZOMBIE and tries to wakeup its parent. Since P1 is not yet in the wait condition, the wakeup call of P2 will have no effect. After becoming a ZOMBIE, P2 is no longer runnable, so it switches process, causing P1 to resume running. While P1 runs, enter the wait command. P1 will enter kernel to executes kwait(). Since P2 has already terminated, P1 will find the ZOMBIE child P2 without going to sleep. It frees the ZOMBIE PROC and returns the dead child pid = 2, as well as its exit value.
(3) Alternatively, the parent P1 may wait first and the child P2 exits later. In that case, P1 will go to sleep in kwait() until it is woken up by P2 when the latter (or any child) terminates. Thus, the orders of parent-wait and child-exit do not matter.

Figure 7.4b shows the outputs of executing the exec command with command line parameters. As the figure shows, for the command line

u2 one two three

the process changes its Umode image to the u2 file. When execution of the new image starts, the command line parameters are passed in as argv[ ] strings with argc = 4.

7.7.9 Simple sh for Command Execution

With fork-exec, we can standardize the execution of user commands by a simple sh. First, we precompile main0.c as crt0.o and put it into the link library as the C startup code of all Umode programs. Then we write Umode programs in C as

```
/********** filename.c file ***************/
#include "ucode.c"  // user commands and syscall interface
main(int argc, char *argv[ ])
{  // C code of Umode program }
```

Then we implement a rudimentary sh for command execution as follows.

```
/******************** sh.c file ***********************/
#include "ucode.c"  // user commands and syscall interface
main(int argc, char *argv[ ])
{
```

(a)

```
QEMU
input a command : fork                                              00:00:44
running usp=0x800FFF50  linkR=0x80000A2C
FORK: child  2  uimage at 0x900000
copy Umode image from 0x800000  to 0x900000
copy kernel mode stack
FIX UP child resume PC to 0x80000A2C
KERNEL: proc 1  forked a child 2
readyQueue = [2 1 ]->[0 0 ]->NULL
parent 1  forked a child 2
This is process 1  in Umode at 0x800000  parent=0
-----------------------------------------
ps chname kfork switch wait exit fork exec
-----------------------------------------
input a command : switch
1  in ktswitch()
proc 1  in scheduler
readyQueue = [2 1 ]->[1 1 ]->[0 0 ]->NULL
next running = 2
switch to proc 2  pgdir at 0x608000  pgdir[2048] = 0x900C12
child 2  return from fork(), pid=0
This is process 2  in Umode at 0x900000  parent=1
-----------------------------------------
ps chname kfork switch wait exit fork exec
-----------------------------------------
input a command : 
```

(b)

```
QEMU
next running = 2                                                    00:03:35
switch to proc 2  pgdir at 0x608000  pgdir[2048] = 0x900C12
child 2  return from fork(), pid=0
This is process 2  in Umode at 0x900000  parent=1
-----------------------------------------
ps chname kfork switch wait exit fork exec
-----------------------------------------
input a command : exec
enter a command string : u2 one two three
line=u2 one two three
EXEC: proc 2  cmdline=0x800FFF10
EXEC: proc 2  kline = u2 one two three
load file /bin/u2 to 0x900000
usp=0x9FFF80  p->usp = 0x800FFF80  kexec exit
main0: s = u2 one two three
argc=4
argv[0 ]=u2
argv[1 ]=one
argv[2 ]=two
argv[3 ]=three
DAS IST PROZESS 2  IM USER-MODUS at 0x900000  ELTERN=1
-----------------------------------------
ps chname kfork switch wait exit fork exec
-----------------------------------------
BEFEHL : 
```

Fig. 7.4 a Demonstration of fork and b demonstration of exec

```
    int pid, status;
    while(1){
       display executable commands in /bin directory
       prompt for a command line cmdline = "cmd a1 a2 .... an"
       if (!strcmp(cmd,"exit"))
           exit(0);
       // fork a child process to execute the cmd line
       pid = fork();
       if (pid)                 // parent sh waits for child to die
           pid = wait(&status);
       else                     // child exec cmdline
           exec(cmdline);       // exec("cmd a1 a2 ... an");
    }
}
```

Then compile all Umode programs as binary executables in the /bin directory and run sh when the system starts. This can be improved further by changing P1's Umode image to an init.c file. These would make the system to have similar capability as Unix/Linux in terms of process management and command executions.

```
/******* init.c file : initial Umode image of P1 ********/
main( )
{
  int sh, pid, status;
  sh = fork();
  if (sh){                      // P1 runs in a while(1) loop
     while(1){
        pid = wait(&status);  // wait for ANY child to die
        if (pid==sh){         // if sh died, fork another one
           sh = fork();
           continue;
        }
        printf("P1: I just buried an orphan %d\n", pid);
     }
  }
  else
     exec("sh");               // child of P1 runs sh
}
```

7.7.10 vfork

In all Unix-like systems, the standard way of creating processes to run different programs is by the fork-exec paradigm. The main drawback of the paradigm is that it must copy the parent process image, which is time-consuming. In most Unix-like systems, the usual behaviors of parent and child processes are as follows.

```
if (fork())             // parent fork() a child process
    wait(&status);      // parent waits for child to terminate
else
    exec(filename);     // child executes a new image file
```

After creating a child, the parent waits for the child to terminate. When the child runs, it changes the Umode image to a new file. In this case, copying image in fork() would be a waste since the child process abandons the copied image immediately. For this reason, most Unix-like systems support a vfork operation, which create a child process without copying the parent image. Instead, the child process is created to share the same image with the parent. When the child does exec to change image, it only detaches itself from the shared image without destroying it. If every child process behaves this way, the scheme would work fine. But what if users do not obey this rule and allow the child to modify the shared image? It would alter the shared image, causing problems to both processes. To prevent this, the system must rely on memory protection. In systems with memory protection hardware, the shared image can be marked as read-only so that processes sharing the same image can only execute but not modify it. If either process tries to modify the shared image, the image must be split into separate images. In ARM based systems, we can also implement vfork by the following algorithm.

```
/********************* Algorithm of vfork ***********************/
```

1. create a child process ready to run in Kmode, return −1 if fails;
2. copy a section of parent's ustack from parent.usp all the way back to where it called pid = vfork(), e.g. the bottom 1024 entries; set child usp = parent usp - 1204;
3. let child pgdir = parent pgdir, so that they share the same page table;
4. mark child as vforked; return child pid;

For simplicity, in the vfork algorithm we do not mark the shared page table entries READ-ONLY. Corresponding to vfork, the exec function must be modified to account for possible shared images. The following shows the modified exec algorithm

```
/******************** Modified exec algorithm *********************/
```

1. fetch cmdline from (possibly shared) Umode image;
2. **if caller is vforked: switch to caller's page table and switchPgdir**;
3. load file to Umode image;
4. copy cmdline to ustack top and set usp;
5. modify syscall stack frame to return to VA = 0 in Umode
6. turn off vforked flag; return usp;

In the modified exec algorithm, all the steps are the same as before, except step 2, which switches to the caller's page table, detaching it from the parent image. The following lists the code segments of kvfork() and (modified) kexec() functions, which support vfork.

```
int kvfork()
{
  int i, cusp, pusp;
  PROC *p = get_proc(&freeList);
  if (p==0){ printf("vfork failed\n"); return -1;  }
  p->ppid = running->pid;
  p->parent = running;
  p->status = READY;
  p->priority = 1;
  p->vforked = 1;                  // add vforked entry in PROC
  p->pgdir = running->pgdir; // share parent's pgdir
  for (i=1; i <= 14; i++){
     p->kstack[SSIZE-i] = running->kstack[SSIZE-i];
  }
  for (i=15; i<=28; i++){      // zero out Umode registers
```

```
    p->kstack[SSIZE-i] = 0;
  }
  p->kstack[SSIZE - 14] = 0; // child return pid = 0
  p->kstack[SSIZE-15] = (int)goUmode; // resume to goUmode
  p->ksp = &(p->kstack[SSIZE-28]);
  p->ucpsr = running->ucpsr;
  pusp = (int)running->usp;
  cusp = pusp - 1024;          // child ustack: 1024 bytes down
  p->usp = (int *)cusp;
  memcpy((char *)cusp, (char *)pusp, 128); // 128 entries enough
  enqueue(&readyQueue, p);
  printf("proc %d vforked a child %d\n",running->pid, p->pid);
  return p->pid;
}

int kexec(char *cmdline)        // cmdline=VA in Umode space
{
  // fetch cmdline and get cmd filename: SAME as before
  if (p->vforked){
     p->pgdir = (int *)(0x600000 + p->pid*0x4000);
     printf("%d is VFORKED: switchPgdir to %x",p->pid, p->pgdir);
     switchPgdir(p->pgdir);
     p->vforked = 0;
  }
  // load cmd file, set up kstack; return to VA=0: SAME as before
}
```

7.7.11 Demonstration of vfork

The sample system C7.5 implements vfork. To demonstrate vfork, we add a vfork command to Umode programs. The vfork command calls the uvfork() function, which issues a syscall to execute kvfork() in kernel.

```
int uvfork()
{
  int ppid, pid, status;
  ppid = getpid();
  pid = syscall(11, 0,0,0);  // vfork() syscall# = 11
  if (pid){
    printf("vfork parent %d return child pid=%d ", ppid, pid);
    printf("wait for child to terminate\n");
    pid = wait(&status);
    printf("vfork parent: dead child=%d, status=%x\n", pid, status);
  }
  else{
    printf("vforked child: mypid=%d ", getpid());
    printf("vforked child: exec(\"u2 test vfork\")\n");
    syscall(10, "u2 test vfork",0,0);
  }
}
```

In uvfork(), the process issues a syscall to create a child by vfork(). Then it waits for the vforked child to terminate. The vforked child issues an exec syscall to change image to a different program. When the child exits, it wakes up the parent,

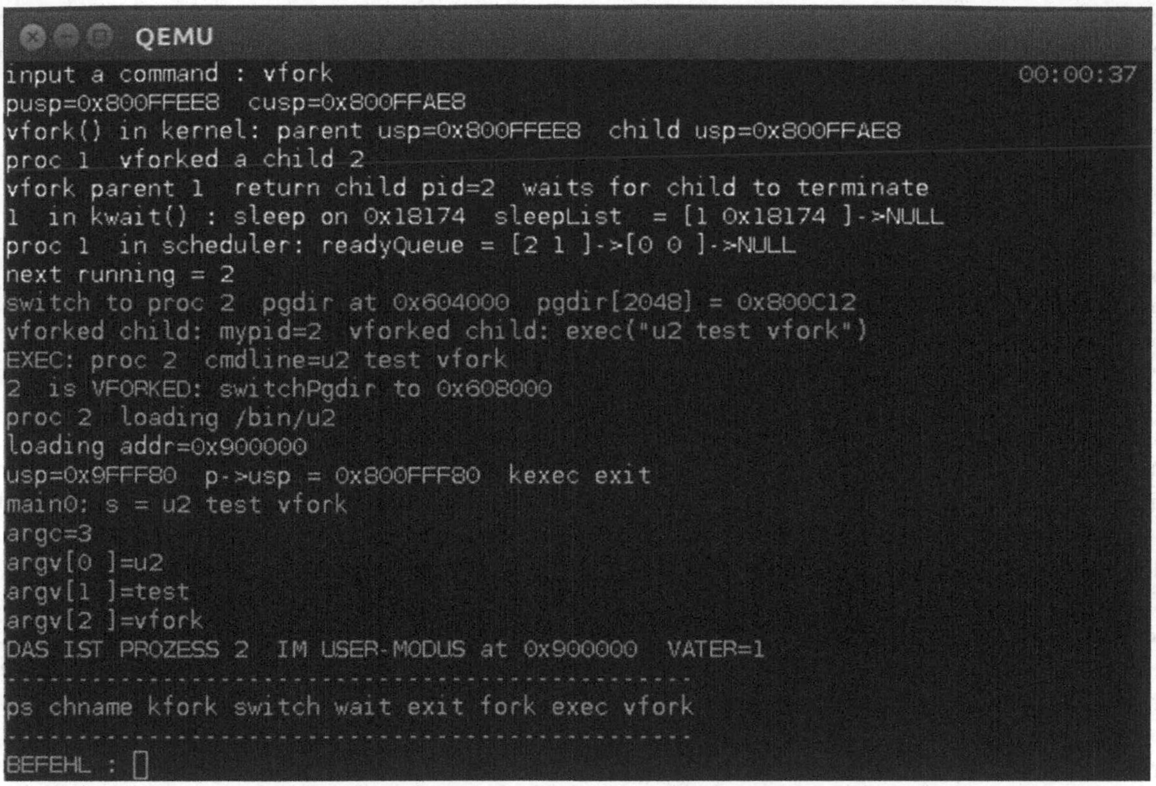

```
QEMU
input a command : vfork                                                    00:00:37
pusp=0x800FFEE8  cusp=0x800FFAE8
vfork() in kernel: parent usp=0x800FFEE8  child usp=0x800FFAE8
proc 1  vforked a child 2
vfork parent 1  return child pid=2  waits for child to terminate
1 in kwait() : sleep on 0x18174  sleepList  = [1 0x18174 ]->NULL
proc 1  in scheduler: readyQueue = [2 1 ]->[0 0 ]->NULL
next running = 2
switch to proc 2  pgdir at 0x604000  pgdir[2048] = 0x800C12
vforked child: mypid=2  vforked child: exec("u2 test vfork")
EXEC: proc 2  cmdline=u2 test vfork
2 is VFORKED: switchPgdir to 0x608000
proc 2  loading /bin/u2
loading addr=0x900000
usp=0x9FFF80  p->usp = 0x800FFF80  kexec exit
main0: s = u2 test vfork
argc=3
argv[0 ]=u2
argv[1 ]=test
argv[2 ]=vfork
DAS IST PROZESS 2  IM USER-MODUS at 0x900000  VATER=1
----------------------------------------------
ps chname kfork switch wait exit fork exec vfork
----------------------------------------------
BEFEHL : 
```

Fig. 7.5 Demonstration of vfork

which would never know that the child was executing in its Umode image before. Figure 7.5 shows the outputs of running the samples system C7.5.

7.8 Threads

In the process model, each process is an independent execution unit in a unique Umode address space. In general, the Umode address spaces of processes are all distinct and separate. The mechanism of vfork() allows a process to create a child process which temporarily shares the same address space with the parent, but they eventually diverge. The same technique can be used to create separate execution entities in the same address space of a process. Such execution entities in the same address space of a process are called **light-weighted processes**, which are more commonly known as **threads** (Silberschatz et al. 2009). The reader may consult (Posix 1C 1995; Buttlar et al. 1996; Pthreads 2015), for more information on threads and threads programming. Threads have many advantages over processes. For a detailed analysis of the advantages of threads and their applications, the reader may consult Chap. 5 of Wang (2015). In this section, we shall demonstrate the technique of extending vfork() to create threads.

7.8.1 Thread Creation

(1) **Thread PROC structures:**

As an independent execution entity, each thread needs a PROC structure. Since threads of the same process execute in the same address space, they share many things in common, such the pgdir, opened file descriptors, etc. It suffices to maintain only one copy of such shared information for all threads in the same process. Rather than drastically altering the PROC structure, we shall add a few fields to the PROC structure and use it for both processes and threads. The modified PROC structure is

```
#define NPROC   10
#define NTHREAD 16
typedef struct proc{
   // same as before, but add
   int type;            // PROCESS|THREAD type
   int tcount;          // number of threads in a process
   int kstack[SSIZE]
}PROC;
PROC proc[NPROC+NTHREAD],*freeList,*tfreeList,*readyQueue;
```

During system initialization, we put the first NPROC PROCs into freeList as before, but put the remaining NTHREAD PROCs into a separate tfreeList. When creating a process by fork() or vfork(), we allocate a free PROC from the freeList and the type is PROCESS. When creating a thread, we allocate a PROC from tfreeList and the type is THREAD. The advantage of this design is that it keeps the needed modifications to a minimum. For instance, the system falls back the pure process model if NTHREAD = 0. With threads, each process may be regarded as a container of threads. When creating a process, it is created as the main thread of the process. With only the main thread, there is virtually no difference between a process and thread. However, the main thread may create other threads in the same process. For simplicity, we shall assume that only the main thread may create other threads and the total number of threads in a process is limited to TMAX.

(2) **Thread creation**: The system call

```
int thread(void *fn,  int *ustack,  int *ptr);
```

creates a thread in the address space of a process to execute a function, fn(ptr), using the ustack area as its execution stack. The algorithm of thread() is similar to vfork(). Instead of temporarily sharing Umode stack with the parent, each thread has a dedicated Umode stack and the function is executed with a specified parameter (which can be a pointer to a complex data structure). The following shows the code segment of kthread() in kernel.

```
// create a thread to execute fn(ptr); return thread's pid
int kthread(int fn, int *stack, int *ptr)
{
  int i, uaddr, tcount;
  tcount = running->tcount;
  if (running->type!=PROCESS || tcount >= TMAX){
     printf("non-process OR max tcount %d reached\n", tcount);
     return -1;
  }
  p = (PROC *)get_proc(&tfreeList); // get a thread PROC
  if (p == 0){ printf("\nno more THREAD PROC  "); return -1; }
  p->status = READY;
  p->ppid = running->pid;
  p->parent = running;
  p->priority  = 1;
  p->event = 0;
  p->exitCode = 0;
  p->pgdir = running->pgdir;  // same pgdir as parent
  p->type = THREAD;
  p->tcount = 0;              // tcount not used by threads
  p->ucpsr = running->ucpsr;
```

```
    p->usp = stack + 1024;            // high end of stack
    for (i=1; i<29; i++)              // zero out "saved" registers
       p->kstack[SSIZE-i] = 0;
    p->kstack[SSIZE-1] = fn;                // uLR = fn
    p->kstack[SSIZE-14] = (int)ptr;         // saved r0 for fn(ptr)
    p->kstack[SSIZE-15] = (int)goUmode; // resume to goUmode
    p->ksp = &p->kstack[SSIZE-28];          // aved ksp
    enqueue(&readyQueue, p);
    running->tcount++;                      // inc caller tcount
    return(p->pid);
}
```

When a thread starts to run, it first resumes to goUmode. Then it follows the syscall stack frame to return to Umode to execute fn(ptr) as if it was invoked by a function call. When control enters fn(), it uses

stmfd sp!, {fp, lr}

to save the return link register in Umode stack. When the function finishes, it returns by

ldmfs sp!, {fp, pc}

In order for the fn() function to return gracefully when it finishes, the initial Umode lr register must contain a proper return address. In the scheduler() function, when the next running PROC is a thread, we load the Umode lr register with the value of VA(4). At the virtual address 4 of every Umode image is an exit(0) syscall (in ts.s), which allows the thread to terminate normally. Each thread may run statically, i.e. in an infinite loop, and never terminate. If needed, it uses either sleep or semaphore to suspend itself. A dynamic thread terminates when it has completed the designated task. As usual, the parent (main) thread may use the wait system call to dispose of terminated children threads. For each terminated child, it decrements its tcount value by 1. We demonstrate a system which supports threads by an example.

7.8.2 Demonstration of Threads

The sample system C7.6 implements support for threads. In the ucode.c file, we add a thread command and thread related syscalls to the kernel. The thread command calls uthread(), in which it issues the thread syscall to create N $\leq$ 4 threads. All the threads executes the same function fn(ptr), but each has its own stack and a different ptr parameter. Then the process waits for all the threads to finish. Each thread prints its pid, the parameter value and the physical address of the process image. When all the threads have terminated, the main process continues.

```
int a[4] = {1,2,3,4};   // parameters to threads
int stack[4][1024];     // threads stacks
int fn(int *ptr)        // thread function
{
  int pid = getpid();
  int pa  = getPA();    // get PA of process memroy
printf("thread %d in %x ptr=%x *ptr=%d\n", pid, pa, ptr, *ptr);
}
int uthread()
{
  int i, status, pid, cid, N = 2;
  pid = getpid();
  for (i=0; i<N; i++){
    printf("proc %d create thread %d\n", pid, i);
       syscall(12, fn, stack[i], &a[i]);
  }
  printf("proc %d wait for %d threads to finish\n", pid, N);
```

```
QEMU
-----------------------------------------------------------         00:00:37
ps chname kfork switch wait exit fork exec vfork thread
-----------------------------------------------------------
input a command : thread
kernel thread(): fn=0x80000C18   stack=0x8000166C  ptr=0x800015D8  *ptr=1
readyQueue = [10 1 ]->[0 0 ]->NULL
proc 1  created a thread 10  in PA=0x800000  tcount=2
kernel thread(): fn=0x80000C18   stack=0x8000266C  ptr=0x800015DC  *ptr=2
readyQueue = [10 1 ]->[11 1 ]->[0 0 ]->NULL
proc 1  created a thread 11  in PA=0x800000  tcount=3
proc 1  wait for 2  child threads to finish
proc 1  in scheduler: readyQueue = [10 1 ]->[11 1 ]->[0 0 ]->NULL
this is thread 10  in 0x800000  ptr=0x800015D8 ; *ptr=1
10  in kexit, value=128
proc 10  in scheduler: readyQueue = [11 1 ]->[1 1 ]->[0 0 ]->NULL
this is thread 11  in 0x800000  ptr=0x800015DC ; *ptr=2
11  in kexit, value=128
proc 11  in scheduler: readyQueue = [1 1 ]->[0 0 ]->NULL
proc 1 : dead child=10  status=0x80
proc 1 : dead child=11  status=0x80
This is process 1  in Umode at 0x800000  parent=0
-----------------------------------------------------------
ps chname kfork switch wait exit fork exec vfork thread
-----------------------------------------------------------
input a command : 
```

Fig. 7.6 Demonstration of threads

```
  for (i=0; i<N; i++){
      cid = wait(&status);
      printf("proc %d: dead child=%d status=%x\n",pid, cid, status);
  }
}
```

Figure 7.6 shows the outputs of running the C7.6 program. As the figure shows, all the threads execute in the same address space (PA = 0x800000) of the parent process P1. It also shows that the parent process P1 waits for all the children threads to terminate by the usual wait for child termination operation.

7.8.3 Threads Synchronization

Whenever multiple execution entities share the same address space, they may access and modify shared (global) data objects. Without proper synchronization, they may corrupt the shared data objects, causing problems to all execution entities. While it's very easy to create many processes or threads, to synchronize their executions to ensure the integrity of shared data objects is no easy task. The standard tools for threads synchronization in Pthreads are mutex and condition variables (Pthreads 2015). We shall cover threads synchronization in Chap. 9 when we discuss multiprocessor systems.

7.9 Embedded System with Two-Level Paging

This section shows how to configure the ARM MMU for two-level paged virtual memory to support User mode processes. First, we show the system memory map, which specifies the planned usage of the system memory space.

```
 --Range--      ----------- Usage ------------
  0 -64K        Vectors, initial pgidr/pgtable
 64K-2MB        System kernel
 2MB-4MB        LCD display frame buffer
 4MB-5MB        RAMdisk file system
 5MB-6MB        Kernel level-2 page tables
 6MB-7MB        Process level-1 pgdirs
 7MB-8MB        Process Umode level-2 pgtables
 8MB-256MB      Free RAM for proc Umode images
256MB-257MB     Memory mapped I/O space
----------------------------------------------
```

When the system starts, we first set up a one-level paging using 1 MB sections as before. While in this simple paging environment, we initialize the kernel data structures, create and run the initial process P0 in Kmode. In kernel_init(), we set up a new level-1 page table (pgdir), denoted by ktable, at 32 KB and its associated level-2 page tables (pgtables) at 5 MB to create an identity mapping of VA to PA. Then we create 64 pgdirs at 6 MB for other processes, each has a pgdir at 6 MB + (pid − 1) * 16 KB. Since the Kmode address spaces of all the processes are identical, their pgdirs are simply copied from ktable. Then we change pgdir to the ktable at 32 KB, which switches the MMU to 2-level paging. The following shows the makePageTable() function code in C.

```
int *mtable = (int *)0x4000;     // initial pgdir at 16KB
int *ktable = (int *)0x8000;     // new 2-level pgdir at 32KB
int makePageTable()
{
  int i;
  for (i=0; i<4096; i++){  // zero out 4096 entries of ktable[]
      ktable[i] = 0;
  }
  for (i=0; i<258; i++){   // ASSUME 256MB PA + 2MB I/O space
      ktable[i] = (0x500000 + i*1024) | 0x11; // DOMAIN 0,type=01
  }
  // build Kmode level-2 pgtables at 5MB
  for (i=0; i<258; i++){
      pgtable = (int *)(0x500000 + i*1024);
      paddr = i*0x100000 | 0x55E;  // APs=01|01|01|01|CB=11|type=10
      for (j=0; j<256; j++){    // 256 entries, each points to 4KB PA
          pgtable[j] = paddr + j*4096;
      }
  }
// build 64 level-1 pgdirs for other PROCs at 6MB
ktable = (int *)0x600000;    // build 64 proc's pgdir at 6M
  for (i=0; i<64; i++){          // 512KB area in 6MB
      ktable = (int *)(0x600000 + i*0x4000); // each pgdir 16KB
      for (j=0; j<4096; j++){ // copy ktable[ ]
          ktable[j] = mtable[j];
    }
  }
  running->pgdir = ktable;  // change P0's pgdir to ktable
  switchPgdir((int)ktable); // switch to 2-level pgdir
}
```

In the level-1 page table (pgdir) of each process, the high 2048 entries are initially 0's. These entries define the Umode VA space of each process, which will be filled in when the process is created in kfork() or fork(). This part depends on the memory allocation scheme for process Umode images, which may be either static or dynamic.

7.9.1 Static 2-Level Paging

In static paging, each process is allocated a fixed size memory area of PSIZE for its Umode image. For simplicity, we may assume that PSIZE is a multiple of 1 MB. We allocate each process Umode image at 8 MB + (pid − 1) * PSIZE. For example, if PSZIE = 4 MB, then P1 is at 8 MB, P2 is at 12 MB, etc. Then we set up the process page tables to access the Umode image as pages. The following shows the code segments in kfork(), fork() and kexec().

```
#define PSIZE 0x400000
PROC *fork1()  // common code of creating a new proc *p
{
  int i, j;
  int *pgtable, npgdir;
  PROC *p = get_proc(&freeList);
  if (p==0){
    kprintf("fork1() failed\n");
    return (PROC *)0;
  }
  p->ppid = running->pid;
  p->parent = running;
  p->status = READY;
  p->priority = 1;
  p->tcount = 1;
  p->paddr = (int *)(0x800000 + (p->pid-1)*PSIZE);
  p->psize = PSIZE;
  p->pgdir = (int *)(0x600000 + (p->pid-1)*0x4000);
  // fill in high pgdir entries and construct Umode pgtables
  npgdir = p->size/0x100000; // no. of Umode pgdir entries and pgtables
  for (i=0; i<npgdir; i++){
      pgtable = (int *)(0x700000 + (pid-1)*1024 + i*1024);
      p->pgdir[2048+i] = (int)pgtable | 0x31; // DOMAIN=1,type=01
      for (j=0; j<256; j++){
          pgtable[j] = (int)(p->paddr + j*1024) | 0xFFE; // APs=11
      }
  }
  return p;
}
PROC *kfork(char *filename)
{
  int i; char *cp;
  PROC *p = fork1();
  if (p==0){ printf("kfork failed\n"); return 0; }
  for (i=1; i<29; i++){  // all 28 cells in kstack[] = 0
  p->kstack[SSIZE-i] = 0;
  }
  p->kstack[SSIZE-15] = (int)goUmode;  // in dec reg=address ORDER !!!
  cp = (char *)p->paddr + PSIZE - 128;
  strcpy(cp, istring);
  p->usp   = (int *)VA(0x200000 - 128); // usp at VA(2MB-128)
  p->ucpsr = (int *)0x10;
  p->upc   = (int *)VA(0);
  p->kstack[SSIZE-14] = (int)p->usp;     // saved r0
  p->kstack[SSIZE-1] = VA(0);
  enqueue(&readyQueue, p);
  return p;
```

```
}
int fork()   // fork a child proc with identical Umode image
{
  int i, PA, CA;
  int *pgtable;
  PROC *p = fork1();
  if (p==0){ printf("fork failed\n"); return -1; }
  PA = (int)running->paddr;
  CA = (int)p->paddr;
  memcpy((char *)CA, (char *)PA, running->psize);  // copy image
  p->usp = running->usp;    // both should be VA in their sections
  p->ucpsr = running->ucpsr;
  for (i=1; i <= 14; i++){
     p->kstack[SSIZE-i] = running->kstack[SSIZE-i];
  }
  for (i=15; i<=28; i++){
     p->kstack[SSIZE-i] = 0;
  }
  p->kstack[SSIZE-14] = 0; // child return pid=0
  p->kstack[SSIZE-15] = (int)goUmode;
  p->ksp = &(p->kstack[SSIZE-28]);
  enqueue(&readyQueue, p);
  return p->pid;
}
int kexec(char *cmdline) // cmdline=VA in Uspace
{
  int i, j, upa, usp;
  char *cp, kline[128], file[32], filename[32];
  PROC *p = running;
  int *pgtable;
  // fetch cmdline, get cmd file name: SAME AS BEFORE
  if (p->vforked){ // create its own Umode pgtables
     p->paddr = (int *)(0x800000 + (p->pid-1)*PSIZE);
     p->psize = PSIZE;
     p->pgdir = (int *)(0x600000 + (p->pid-1)*0x4000);
     npgdir = p->psize/0x100000;
     for (i=0; i<npgdir; i++){
        pgtable = (int *)(0x700000 + (p->pid-1)*1024 + i*1024);
        p->pgdir[2048+i] = (int)((int)pgtable | 0x31);
        for (j=0; j<256; j++){
            pgtable[j] = (int)(p->paddr + j*1024) | 0x55E;
        }
     }
     p->vforked = 0;          // turn off vfored flag
     switchPgdir(p->pgdir);   // switch to its own pgdir
  }
  loadelf(file, p);
  // copy kline to Ustack and return to VA=0: SAME AS BEFORE
}
```

7.9.2 Demonstration of Static 2-Level Paging

Figure 7.7 shows the outputs of the sample system C7.7, which uses 2-level static paging.

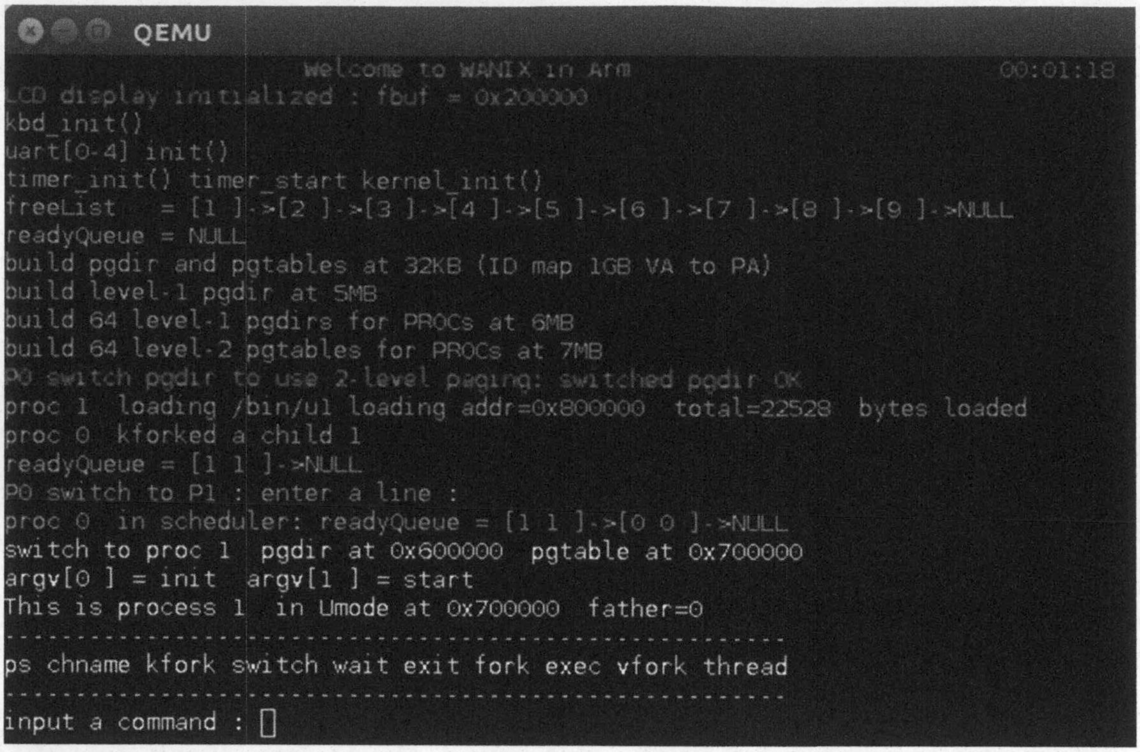

```
QEMU
                    welcome to WANIX in Arm                          00:01:18
LCD display initialized : fbuf = 0x200000
kbd_init()
uart[0-4] init()
timer_init() timer_start kernel_init()
freeList    = [1 ]->[2 ]->[3 ]->[4 ]->[5 ]->[6 ]->[7 ]->[8 ]->[9 ]->NULL
readyQueue = NULL
build pgdir and pgtables at 32KB (ID map 1GB VA to PA)
build level-1 pgdir at 5MB
build 64 level-1 pgdirs for PROCs at 6MB
build 64 level-2 pgtables for PROCs at 7MB
P0 switch pgdir to use 2-level paging: switched pgdir OK
proc 1  loading /bin/u1 loading addr=0x800000  total=22528  bytes loaded
proc 0  kforked a child 1
readyQueue = [1 1 ]->NULL
P0 switch to P1 : enter a line :
proc 0  in scheduler: readyQueue = [1 1 ]->[0 0 ]->NULL
switch to proc 1  pgdir at 0x600000  pgtable at 0x700000
argv[0 ] = init  argv[1 ] = start
This is process 1  in Umode at 0x700000  father=0
-------------------------------------------------------
ps chname kfork switch wait exit fork exec vfork thread
-------------------------------------------------------
input a command :
```

Fig. 7.7 Demonstration of static 2-level paging

7.9.3 Dynamic 2-Level Paging

In dynamic paging, the Umode memory area of each process consists of page frames that are allocated dynamically. In order to support dynamic paging, the system must manage available memory in the form of free page frames. This can be done by either a bitmap or a link list. If the number of page frames is large, it is more efficient to manage them by a link list. When the system starts, we construct a free page link list, pfreeList, which threads all the free page frames in a linked list. When a page frame is needed, we allocate a page frame from pfreeList. When a page frame is no longer needed, we release it back to pfreeList for reuse. The following code segments show the free page list management functions.

```
int *pfreeList;            // free page frame list
int *palloc()              // allocate a free page frame
{
  int *p = pfreeList;
  if (p){
     pfreeList = (int *)(*p);
     *p = 0;
  }
  return p;
}
void pdealloc(int p)   // release a page frame
{
  u32 *a = (u32 *)((int)p & 0xFFFFF000); // frame's PA
  *a = (int)pfreeList;
  pfreeList = a;
}
```

```
int *free_page_list(int *startva, int *endva) // build pfreeList
{
  int *p = startva;
  while(p < (int *)(endva-1024)){
    *p = (int)(p + 1024);
     p += 1024;
  }
  *p = 0;
  return startva;
}
pfreeList = free_page_list((int *)8MB, (int *)256MB);
```

7.9.3.1 Modifications to Kernel for Dynamic Paging

When using dynamic paging, the Umode image of each process is no longer a single piece of contiguous memory. Instead, it consists of dynamically allocated page frames, which may not be contiguous. We must modify the kernel code that manages process images to accommodate these changes. The following shows the modified kernel code segments to suit the dynamic paging scheme.

(1). fork1(): fork1() is the base code of both kfork() and fork(). It creates a new process p and sets up its Umode image. Since the level-1 page table (pgdir) must be a single piece of 16 KB aligned memory, we can not build it by allocating page frames because the page frames may not be contiguous or aligned at a 16 KB boundary. So we still build the proc pgdirs at 6 MB as before. We only need to modify fork1() to construct the Umode image by allocating page frames.

```
// Assume: Umode image size in MB
#define PSIZE 0x400000
PROC *fork1()
{
  int i, j, npgdir, npgtable;
  int *pgtable;
  // create a new proc *p: SAME AS BEFORE
  p->pgdir = (int *)(0x600000 + (p->pid-1)*0x4000); // same as before
  // copy entries from ktable at 32KB
  for (i=0; i<4096; i++){
      p->pgdir[i] = ktable[i];
  }
  p->psize = 0x400000;          // Uimage size in MB
  npgdir = p->psize/0x100000;   // no. of pgdir entries and pgtables
  npgtable = npgdir;
  // fill in pgdir entries and construct Umode pgtables
  for (i=0; i<npgdir; i++){
    pgtable = (int *)palloc(); // allocate a page but only use 1KB
    p->pgdir[2048+i] = (int)pgtable | 0x31; // pgdir entry
    for (j=0; j<256; j++){                 // pgtable
        pgtable[j] = (int)((int)palloc() | 0xFFE);
    }
  }
  return p;
}
```

(2). **kfork()**: kfork() creates a new process with initial command line parameters passed in the Umode stack at the high end of the virtual address space. Since the caller's pgdir is different from that of the new process, we can not use VA(PSIZE) to access the (high end of) Umode stack of the new process. Instead, we must use the last allocated page frame to access its Umode stack.

(3). **fork()**: When copying image, we must copy the parent page frames to that of the child process, as shown by the following code segment.

```
// copy parent's page frames to child' page frames
void copyimage(PROC *parent, PROC *child)
{
  int i, j;
  int *ppgtable, *cpgtable, *cpa, *ppa;
  int npgdir = parent->psize/0x100000;
  for (i=0; i < npgdir; i++){ // each image has npgdir page tables
    ppgtable = (int *)(parent->pgdir[i+2048] & 0xFFFFFC00);
    cpgtable = (int *) (child->pgdir[i+2048] & 0xFFFFFC00);
    for (j=0; j<256; j++){ // copy page table frames
      ppa = (int *)(ppgtable[j] & 0xFFFFF000);
      cpa = (int *)(cpgtable[j] & 0xFFFFF000);
      memcpy((char *)cpa, (char *)ppa, 4096); // copy 4KB page frame
    }
  }
}
```

(4). **kexec()**: We assume that kexec() uses the same Umode image area of a process. In this case, no changes are needed, except for vforked process, which must create its own pgdir entries, allocate page frames and switch to its own pgdir.

(5). **loader**: the image loader must be modified to load the image file into page frames of the process. The modified (pseudo) loader code is

```
// locate ELF section header as before; ph->pgogram header
for (int i=1, ph=aph; i <= phnum; ph++, i++){
  if (ph->type != 1) break;
  lseek(fd, (long)ph->offset, 0); // set offset to ph->offset
  pn = PA(ph->vaddr)/0x1000; // convert vaddr to page number
  count = 0;
  addr = (char *)(pgtable[pn] & 0xFFFF000); // loading address
  while(count < ph->memsize){
     read(fd, dbuf, BLKSIZE);        // read by 1KB BLKSIZE
     memcpy(addr, dbuf, BLKSIZE);    // copy to addr
     addr  += BLKSIZE;
     count += BLKSIZE;
  }
}
```

(6). **kexit()**: When a process terminates, we release its page frames used as Umode pgdir entries and pgtables back to the pfreeList for reuse.

7.9.4 Demonstration of 2-Level Dynamic Paging

Figure 7.8 shows the output of running the C7.8 program, which uses dynamic 2-level paging. As the figure shows, the pgdir entries and pgtables of process P1 are all dynamically allocated.

7.10 KMH Memory Mapping

The sample systems C7.1 to C7.8 use the Kernel Mapped Low (KML) memory mapping scheme, in which the kernel mode space is mapped to low virtual addresses and the user mode space is mapped to high virtual addresses. The mapping scheme can be reversed. In the Kernel Mapped High (KMH) scheme, the kernel mode VA space is mapped high, e.g. to 2 GB (0x80000000) and user mode VA space is mapped low. In this section, we shall demonstrate the KMH memory mapping scheme, compare it with the KML scheme and discuss the differences between the two mapping schemes.

7.10.1 KMH Using One-Level Static Paging

(1). **High VA in kernel image**: Assume that the kernel VA space is mapped to 2 GB. In order to let the kernel use high virtual addresses, the link command must be modified to generate high virtual address. For the kernel image, the modified link command is

```
arm-none-eabi-ld -T t.ld ts.o t.o mtx.o -Ttext=0x80010000 -o t.elf
```

The kernel image is loaded at the physical address 0x10000 but its VA is 0x800100000. For user mode images, the link command is changed to

```
arm-none-eabi-ld -T u.ld us.o u1.o -Ttext=0x100000 -o u1.elf
```

Note that the starting virtual address of user mode images is not 0 but 0x100000 (1 MB). We shall explain and justify this later.

```
QEMU
timer_init() timer_start                                        00:02:39
kernel_init()
freeList   = [1 ]->[2 ]->[3 ]->[4 ]->[5 ]->[6 ]->[7 ]->[8 ]->[9 ]->NULL
readyQueue = NULL
build pgdir and pgtables at 32KB (ID map 512MB VA to PA)
build Kmode level-2 pgtables at 5MB
build 64 level-1 pgdirs for PROCs at 6MB
switch pgdir to use 2-level paging : switched pgdir OK
build pfreeList: start=0x800000  end=0x10000000  : 63487  4KB entries
disk start=0x400000  size=0x100000
pgtable=0x800000  pgtable[0 ]=0x80155E  pgtable[1 ]=0x80255E
pgtable=0x901000  pgtable[0 ]=0x90255E  pgtable[1 ]=0x90355E
pgtable=0xA02000  pgtable[0 ]=0xA0355E  pgtable[1 ]=0xA0455E
pgtable=0xB03000  pgtable[0 ]=0xB0455E  pgtable[1 ]=0xB0555E
proc 0  kforked a child 1  at 0x801000
readyQueue = [1 1 ]->NULL
P0 switch to P1 : enter a line :
proc 0  in scheduler: readyQueue = [1 1 ]->[0 0 ]->NULL
switch to proc 1  pgdir at 0x600000
CPU mode=0x10  argc=2  argv[0 ] = init argv[1 ] = start
This is process 1  in Umode at 0x801000  father=0
-----------------------------------------------------
ps chname kfork switch wait exit fork exec vfork thread
-----------------------------------------------------
input a command :
```

Fig. 7.8 Demonstration of dynamic 2-level paging

(2). **VA to PA conversion**: The kernel code uses VA but page table entries and I/O device base addresses must use PA. For convenience, we define the macros

```
#define VA(x) ((x) + 0x80000000)
#define PA(x) ((x) - 0x80000000)
```

for conversions between VA and PA.

(3). **VA for I/O Base Addresses**: The base addresses of I/O devices use PA, which must be remapped to VA. For the Versatilepb board, I/O devices are located in the 2 MB I/O space beginning at PA 256 MB. The I/O device base addresses must be mapped to VA by the VA(x) macro.

(4). **Initial Page Table**: Since the kernel code is compile-linked with VA, we can not execute the kernel's C code before configuring the MMU for address translation. The initial page table must be constructed in assembly code when the system starts.

(5). **The ts.s file**: The reset_handler sets up the initial page table and enables the MMU for address translation. Then it sets up the privileged mode stacks, copies the vector table to 0 and call main() in C. For the sake of brevity, we shall only show the code segments that are relevant to memory mapping. The assembly code shown below sets up an initial one-level page table at PA = 0x4000 (16 KB) using 1 MB sections. In the one-level page table, the 0th entry points to the lowest 1 MB of physical memory, which creates an identity mapping of the lowest 1 MB memory space. This is because the vector table and the exception handler entry addresses are all within the lowest 4 KB of physical memory. In the page table, entries 2048 to 2048 + 258 map VA = (2 GB, 2 GB + 258 MB) to the low 258 physical memory, which includes the 2 MB I/O space at 256 MB. The kernel mode space is assigned the domain 0, with the access bits set to AP = 01 (client mode) to prevent User mode process from accessing kernel's VA space. Then it sets the TLB base register and enables the MMU for address translation. After these, the system is running with VA in the range of (2 GB to 2 GB + 258 MB).

```
/************** ts.s file **************/
reset_handler:
 adr r4, MTABLE    // r4 point to MTABLE at 0x4000=16KB
 ldr r5, [r4]      // r5 = 0x4000
 mov r0, r5        // r0 = MTABLE contents=0x4000 at 16KB

// clear MTABLE 4096 entries to 0
  mov r1, #4096             // 4096 entries
  mov r3, #0                // r3=0
1:
  str r3, [r0], #4          // store r3 to [r0]; inc r0 by 4
  subs r1, r1, #1           // r1--
  bgt 1b                    // loop r1=4096 times

// Assume: Versatilepb has 256MB RAM + 2MB I/O at 256MB,
// entry 0 of MTABLE must point to lowest 1MB PA due to vector table
  mov r6, #(0x1 << 10)     // r6=AP=01 (KRW, user no) AP=11: both KU r/w
  orr r6, r6, #0x12         // r6 = 0x412 OR 0xC12 if AP=11: used 0xC12
  mov r3, r6                // r3 contains 0x412
  mov r0, r5                // r0 = 0x4000
  str r3, [r0]              // fill MTABLE[0] with 0x00000412

//******** map kernel 258MB VA to 2GB ************
  mov r0, r5        // r0 = 0x4000 again
  add r0, r0,#(2048*4)      // add 8K for 2048th entry
  mov r2, #0x100000         // r2 = 1MB increments
  mov r1, #256              // 256 entries
  add r1, r1, #2            // add 2 more entries for 2MB I/O space
  mov r3, r6                // r3 = 0x00000412
3:
```

```
    str r3, [r0], #4           // store r3 to [r0]; inc r0 by 4
    add r3, r3, r2             // inc r3 by 1M
    subs r1,r1, #1             // r1--
    bgt 3b                     // loop r1=258 times

// set TLB base register
    mov r0, r5                     // r0 = 0x4000
    mcr p15, 0, r0, c2, c0, 0   // set TTBase with PHYSICAL address 0x4000
    mcr p15, 0, r0, c8, c7, 0   // flush TLB
// set domain0: 01=client(check permission) 11=master(no check)
    mov r0, #0x1                   // 01 for client mode
    mcr p15, 0, r0, c3, c0, 0   // write to domain REG c3
// enable MMU
    mrc p15, 0, r0, c1, c0, 0   // read control REG to r0
    orr r0, r0, #0x00000001     // set bit0 of r0 to 1
    mcr p15, 0, r0, c1, c0, 0   // write to control REG c1 ==> MMU on
    nop
    nop
    nop
    mrc p15, 0, r2, c2, c0, 0   // read TLB base reg c2 into r2
    mov r2, r2
// set up privileged mode stack pointers, then call main() in C
MTABLE: .word 0x4000              // initial page table at 16KB

switchPgdir: // switch pgdir to new PROC's pgdir; passed in r0
    mcr p15, 0, r0, c2, c0, 0   // set TTBase to C2
    mov r1, #0
    mcr p15, 0, r1, c8, c7, 0   // flush TLB
    mcr p15, 0, r1, c7, c7, 0   // flush I and D Cache
    mrc p15, 0, r2, c2, c0, 0   // read TLB base reg C2
    // set domain: all 01=client(check permission) 11=master(no check)
    mov r0, #0x5                   // 0101=client for both domain 0 and 1
    mcr p15, 0, r0, c3, c0, 0   // write to domain reg C3
    mov pc, lr                     // return
```

(5). **Kernel.c file**: We only show the memory mapping part of kernel_init(). It creates 64 page directories (level-1 page tables) at 6 MB for 64 PROCs. Each proc's pgdir is at 6 MB + pid * 16 KB. Since the kernel mode VA spaces of all the processes are the same, their kernel mode pgdir entries are all identical. For simplicity, we assume that every process, except P0, has a Umode image of size 1 MB, which is statically allocated at the physical address (pid+7) MB, i.e. P1 at 8 MB, P2 at 9 MB, etc. In each process pgdir, entry 1 defines the process Umode image. Corresponding to this, every Umode image is compile-linked with the starting VA=0x100000 (1 MB). During task switch, if the current process differs from the next process, it calls switchPdgir() to switch pgdir to that of the next process.

```
/****************** kernel_init() function *******************/
kernel_init()
{
  int i, j, *mtable, *MTABLE;
  // initialize kernel data structures, create P0: SAME as before
  printf("building pgdirs at 5MB\n");
  // create pgdir's for ALL PROCs at 5MB;
  MTABLE = (int *)0x80004000;      // initial Mtable at PA=0x4000
  mtable = (int *)0x80600000;      // mtables begin at 6MB
  for (j=0; j<4096; j++)           // clear mtable entries to 0
      mtable[j] = 0;
```

```
    // pgdir MUST be at 16K boundary; 6M-7MB has space for 64 pgdirs
  for (i=0; i<64; i++){      // for 64 PROC mtables
     mtable[0] = MTABLE[0]; // entry 0 for low 1MB for vectors
     for (j=2048; j<4096; j++){  // last 2048 entries copy from MTABLE
         mtable[j] = MTABLE[j];
   }
   // ASSUME: every Umode image size=1MB => only need one entry
   if (i){ // excluding P0 which does not have Umode image
        mtable[1]=(0x800000 + (i-1)*0x100000)|0xC12; // Umode=entry #1
     }
     mtable += 4096;              // advance mtable to next 16KB
  }
  printf("switch to P0's pgdir : ");
  switchPgdir((int)0x600000);      // parameter must be a PA
}

int scheduler()
{
  PROC *old=running;
  printf("proc %d in scheduler: ", running->pid);
  if (running->status==READY)
     enqueue(&readyQueue, running);
  running = dequeue(&readyQueue);
  printf("next running = %d\n", running->pid);
  if (running != old)
     switchPgdir(PA(int)running->pgdir & 0xFFFFF000);
}
```

(6). **Use VA in Kernel Functions**: All kernel functions, such as kfork, fork, image file loader and kexec, must use VA.

(7). **Demonstration of KMH Memory Mapping**: Fig. 7.9 shows the output of running the program C7.9, which demonstrate the KMH address mapping using 1 MB sections.

The reader may run the system C7.9 and fork other processes. It should show that each process runs in a different PA area but all at the same VA = 0x100000.Variations of the one-level paging scheme are left as programming projects in the Problems section.

7.10.2 KMH Using Two-Level Static Paging

In this section, we shall demonstrate the KMH memory mapping scheme using two-level paging. This is accomplished in three steps.

Step 1: When the system starts, we first set up an initial one-level page table and enable the MMU exactly the same as in program C7.9. While in this simple paging environment, we can execute the kernel code using high virtual addresses.
Step 2: In kernel_init(), we build a two-level page pgdir at 32 KB (0x8000). The 0th entry of the level-1 pgdir is for ID map the lowest 1 MB memory. We build its level-2 page table at 48 KB (0xC000) to create an ID mapming of the lowest 1 MB memory. Assume 256 MB RAM plus 2 MB I/O space at 256 MB. We build 258 level-2 page tables at 5 MB. Then we build 64 level-1 pgdirs at 6 MB. Each proc has a pgdir at 5 MB + pid * 16 KB. The level-2 page tables of each pgdir are the same. Then we switch pgdir to 6 MB to use two-level paging.

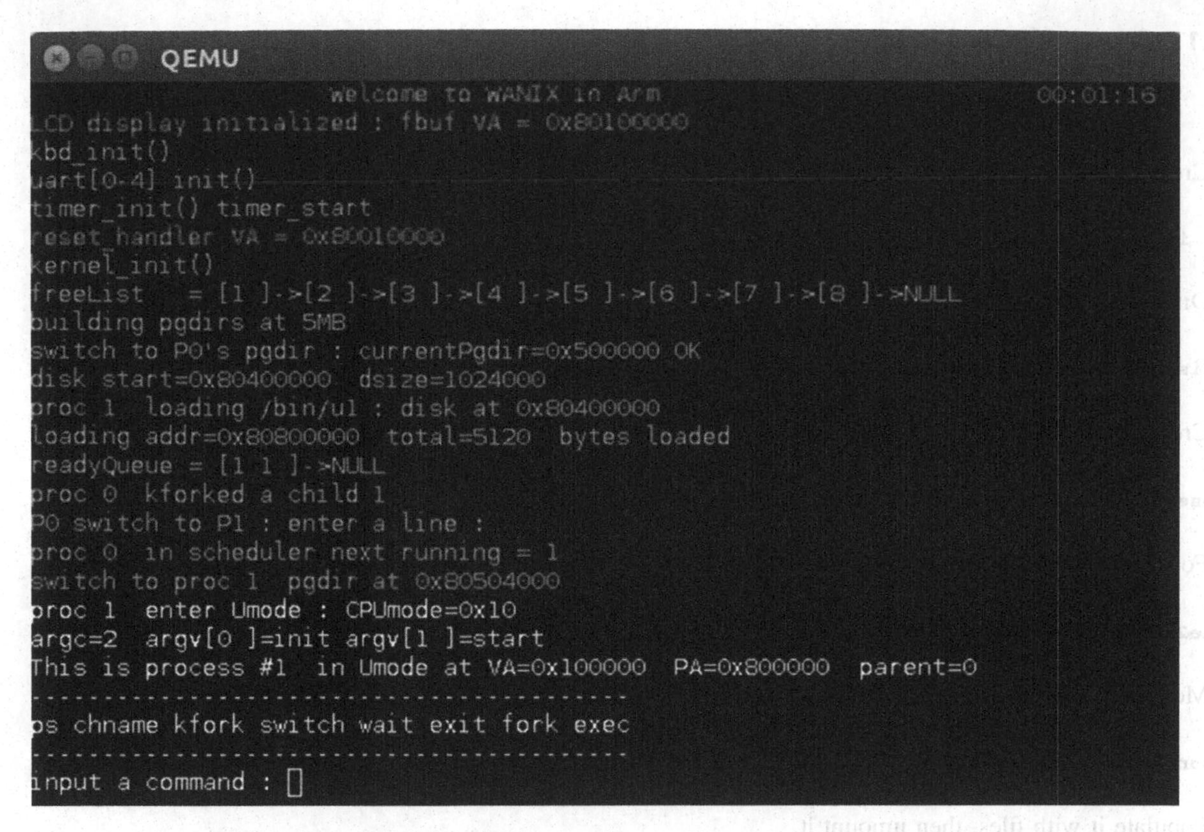

Fig. 7.9 Demonstration of KMH using one-level static paging

Step 3: Assume that each process Umode image is of size = USIZE MB, which is statically allocated at 7 MB + pid * 1 MB. When creating a new process in fork1(), we build its Umode level-1 pgdir entries and level-2 page tables at 7 MB + pid*1 KB

Figure 7.10 shows the outputs of running the sample program C7.10, which demonstrates the KMH mapping scheme using two-level static paging.

7.10.3 KMH Using Two-Level Dynamic Paging

It is fairly easy to extend the two-level KMH static paging to dynamic paging. This is left as an exercise in the Problems section.

7.11 Embedded System Supporting File Systems

So far, we have used a RAMdisk as a file system. Each process image is loaded as an executable file from the RAMdisk. During execution, processes may save information by writing to the RAMdisk. Since the RAMdisk is held in volatile memory, all the RAMdisk contents will vanish when the system power is turned off. In this section, we shall use a SD card (SDC) as a persistent storage device to support file systems. Like a hard disk, a SDC can be divided into partitions. The following shows how to create a flat disk image with only one partition. The resulting disk image can be used as a virtual SDC in most virtual machines that support SD cards.

7.11.1 Create SD Card Image

(1). Create a flat disk image file of 4096 1 KB blocks.

```
dd if=/dev/zero of=disk.img bs=1024 count=4096
```

(2). Divide disk.img into partitions. The simplest way is to create a single partition.

```
fdisk -H 16 -S 512 disk.img  # enter n, then press enter keys
```

(3). Create a loop device for disk.img.

```
losetup -o 1048576 --sizelimit 4193792 /dev/loop1 disk.img
```

(4). Format the partition as an EXT2 file system.

```
mke2fs -b 1024 disk.img 3072
```

(5). Mount the loop device.

```
mount /dev/loop1 /mnt
```

(6). Populate it with files, then umount it.

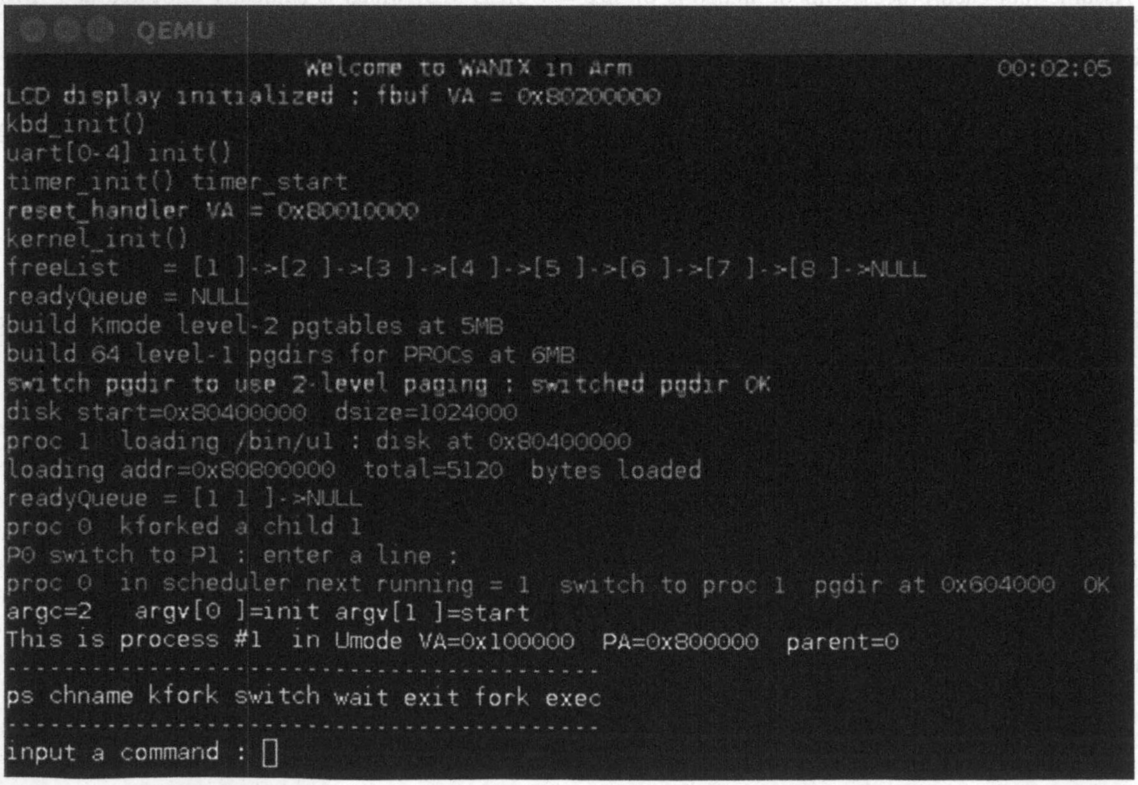

```
QEMU
                    Welcome to WANIX in Arm                         00:02:05
LCD display initialized : fbuf VA = 0x80200000
kbd_init()
uart[0-4] init()
timer_init() timer_start
reset_handler VA = 0x80010000
kernel_init()
freeList    = [1 ]->[2 ]->[3 ]->[4 ]->[5 ]->[6 ]->[7 ]->[8 ]->NULL
readyQueue = NULL
build Kmode level-2 pgtables at 5MB
build 64 level-1 pgdirs for PROCs at 6MB
switch pgdir to use 2-level paging : switched pgdir OK
disk start=0x80400000  dsize=1024000
proc 1  loading /bin/u1 : disk at 0x80400000
loading addr=0x80800000  total=5120  bytes loaded
readyQueue = [1 1 ]->NULL
proc 0  kforked a child 1
P0 switch to P1 : enter a line :
proc 0  in scheduler next running = 1  switch to proc 1  pgdir at 0x604000  OK
argc=2   argv[0 ]=init argv[1 ]=start
This is process #1  in Umode VA=0x100000  PA=0x800000  parent=0
------------------------------------------
ps chname kfork switch wait exit fork exec
------------------------------------------
input a command :
```

Fig. 7.10 Demonstration of KMH using 2-level static paging

```
mkdir /mnt/boot; umount /mnt
```

On a disk image, sectors count form 0. Step (2) creates a partition with first sector = 2048 (a default value by fdisk) and last sector = (4096 * 2 − 1) = 8191. In Step (3), the start offset and size limit are both in bytes = sector * sector_size (512). In Setp (4), the file system size is 4096 − 1024=3072 (1 KB) blocks. Since the file system size is less than 4096 blocks, it requires only one blocks group, which simplifies both the file system traversal algorithm and the management of inodes and disk blocks.

7.11.2 Format SD Card Partitions as File Systems

In the following sections, we shall use a SDC disk image with multiple partitions. The disk image is created as follows.

(1). **Run the sh script mkdisk with a filename**

```
# sh script mkdisk: Run as mkdisk diskname

  dd if=/dev/zero of=$1 bs=1024 count=21024     # about 2MB in size
  dd if=MBR of=$1 bs=512 count=1 conv=notrunc # MBR contains a ptable
  BEGIN=$(expr 2048 \* 512)
  for I in 1 2 3 4; do
     losetup -d /dev/loop$I    # delete exiting loop device
     END=$(expr $BEGIN + 5110000)
     losetup -o $BEGIN -sizelimt $END /dev/loop$I $1
     mount /dev/loop$I /mnt
     (cd /mnt; mkdir bin boot dev etc tmp user)
     umount /mnt
     BEGIN=$(expr $BEGIN + 5120000)
  done
```

(2). The MBR file is a MBR image containing a partition table created by fdisk. Rather than using fdisk to partition the new disk image manually again, we simply dump the MBR file to the MBR sector of the disk image. The resulting disk image has 4 partitions.

Partition	Start_sector	End_sector	Size (1 KB blocks)
1	2048	10,247	5000
2	12,048	22,047	5000
3	22,048	32,047	5000
4	32,048	42,047	5000

The partition type does not matter but it should be set to 0x90 to avoid confusions with other operating sytems, such as Linux, which uses partition types 0x82–0x83.

7.11.3 Create Loop Devices for SD Card Partitions

(3). The mkdisk script creates loop devices for the partitions, formats each partition as an EXT2 file system and populates it with some directories. After creating the loop devices, each partition can be accessed by mounting its corresponding loop device, as in

```
mount /dev/loopN MOUNT_POINT # N=1 to 4
```

(4). Modify the compile-link script by copying User mode images to the /bin directory of the SDC partitions.

7.12 Embedded System with SDC File System

In this section, we shall show an embedded system, denoted by C7.11, which supports dynamic processes with a SDC as the mass storage device. All User mode images are ELF executables in the /bin directory of a (EXT2) file system on the SDC. The system uses 2-level dynamic paging. The kernel VA space is from 0 to 258 MB. User mode VA space is from 2 to 2 GB + Umode image size. To simplify the discussion, we shall assume that every Umode image size is a multiple of 1 MB, e.g. 4 MB. When creating a process, its level-2 page table and page frames are allocated dynamically. When a process terminates, its page table and page frames are released for reuse. The system consists of the following components.

(1). **ts.s file**: The ts.s file is the same as that of Program C7.8.
(2). **Kernel.c file**: The kernel.c file is also the same as that of Program C7.8.

7.12.1 SD Card Driver Using Semaphore

(3). **sdc.c file**: This file contains the SDC driver. The SDC driver is interrupt-driven but it also supports polling for the following reason. When the system starts, only the initial process P0 is running. After initializing the SDC driver, it calls mbr() to display the SDC partitions. Since there is no other process yet, P0 can not go to sleep or become blocked. So it uses polling to read SDC blocks. Similarly, it also uses polling to load the Umode image of P1 from the SDC. After creating P1 and switching to run P1, processes and the SDC driver use semaphore for synchronization. In a real system, CPU is much faster than I/O devices. After issuing an I/O operation to a device, a process usually has plenty of time to suspend itself to wait for device interrupt. In this case, we may use sleep/wakeup to synchronize process and interrupt handler. However, an emulated virtual machine may not obey this timing order. It is observed that on the emulated ARM Versatilepb under QEMU, after a process issuing an I/O operation to the SDC, the SDC interrupt handler always finishes first before the process suspends itself. This makes the sleep/wakeup mechanism unsuitable for synchronization between the process and SDC interrupt handler. For this reason, it uses semaphore for synchronization. In the SDC driver, we define a semaphore s with the initial value 0. After issuing an I/O operation, the process uses P(s) to block itself, waiting for SDC interrupts. When the SDC interrupt handler completes the data transfer, it uses V(s) to unblock the process. Since the order of P and V on semaphores does not matter, using semaphores prevents any race conditions between processes and the interrupt handler. The following lists the SDC driver code.

```
#include "sdc.h"
#define FBLK_SIZE 1024
int partition, bsector;
typedef struct semaphore{ // uniprocessor, no need for spinlock
  int value;
  PROC *queue;
}SEMAPHORE;

SEMAPHORE s; // semaphore for SDC driver synchronization
int P(SEMAPHORE *s)
{
  int sr = int_off();
  s->value--;
  if (s->value < 0){
```

```
    running->status = BLOCK;
    enqueue(&s->queue, running);
    tswitch();
  }
  int_on(sr);
}
int V(SEMAPHORE *s)
{
  PROC *p;
  int sr = int_off();
  s->value++;
  if (s->value <= 0){
    p = dequeue(&s->queue);
    p->status = READY;
    enqueue(&readyQueue, p);
  }
  int_on(sr);
}
struct partition {       // partition table in MBR
      u8 drive;          /* 0x80 - active */
      u8 head;           /* starting head */
      u8 sector;         /* starting sector */
      u8 cylinder;       /* starting cylinder */
      u8 sys_type;       /* partition type */
      u8 end_head;       /* end head */
      u8 end_sector;     /* end sector */
      u8 end_cylinder;   /* end cylinder */
      int start_sector;  /* starting sector counting from 0 */
      int nr_sectors;    /* nr of sectors in partition */
};
int mbr(int partition)
{
  int i;
  char buf[FBLK_SIZE];
  struct partition *p;
  GD *gp;   // EXT2 Group Descriptor pointer
  bsector = 0;
  printf("read MBR to show partition table\n");
  get_block(0, buf);
  p = (struct partition *)&buf[0x1bE];
  printf("P# start  size\n");
  for (i=1; i<=4; i++){ // ASSUME: only 4 prime partitions
    printf("%d %d %d\n", i, p->start_sector, p->nr_sectors);
    if (i==partition)
      bsector = p->start_sector;
    p++;
  }
  printf("partition=%d bsector=%d ", partition, bsector);
  get_block(2, buf);
  gp = (GD *)buf;
  iblk = gp->bg_inode_table;
  bmap = gp->bg_block_bitmap;
  imap = gp->bg_inode_bitmap;
  printf("bmap=%d imap=%d iblk=%d ", bmap, imap, iblk);
  printf("MBR done\n");
```

```
}
// shared variables between process and interrupt handler
volatile char *rxbuf, *txbuf;
volatile int  rxcount, txcount, rxdone, txdone;

int sdc_handler()
{
  u32 status, status_err, *up;
  int i;
  // read status register to find out TXempty or RxAvail
  status = *(u32 *)(base + STATUS);
  if (status & (1<<17)){ // RxFull: read 16 u32 at a time;
     //printf("RX interrupt: ");
     up = (u32 *)rxbuf;
     status_err = status & (DCRCFAIL | DTIMEOUT | RXOVERR);
     if (!status_err && rxcount) {
        //printf("R%d ", rxcount);
        for (i = 0; i < 16; i++)
           *(up + i) = *(u32 *)(base + FIFO);
        up += 16;
        rxcount -= 64;
        rxbuf += 64;
     status = *(u32 *)(base + STATUS); // clear Rx interrupt
     }
     if (rxcount == 0){
         do_command(12, 0, MMC_RSP_R1); // stop transmission
        if (hasP1)
           V(&s);        // by semaphore
        else
           rxdone = 1;  // by polling
     }
  }
  else if (status & (1<<18)){ // TXempty: write 16 u32 at a time
    //printf("TX interrupt: ");
    up = (u32 *)txbuf;
    status_err = status & (DCRCFAIL | DTIMEOUT);
    if (!status_err && txcount) {
      //printf("W%d ", txcount);
       for (i = 0; i < 16; i++)
           *(u32 *)(base + FIFO) = *(up + i);
       up += 16;
       txcount -= 64;
       txbuf += 64;              // advance txbuf for next write
       status = *(u32 *)(base + STATUS); // clear Tx interrupt
    }
    if (txcount == 0){
       do_command(12, 0, MMC_RSP_R1); // stop transmission
       if (hasP1)
          V(&s);       // by semaphore
       else
          txdone = 1; // by polling
    }
  }
  //printf("write to clear register\n");
  *(u32 *)(base + STATUS_CLEAR) = 0xFFFFFFFF;
```

```
  // printf("SDC interrupt handler done\n");
}

int delay(){ int i; for (i=0; i<1000; i++); }

int do_command(int cmd, int arg, int resp)
{
  *(u32 *)(base + ARGUMENT) = (u32)arg;
  *(u32 *)(base + COMMAND)  = 0x400 | (resp<<6) | cmd;
  delay();
}

int sdc_init()
{
  u32 RCA = (u32)0x45670000; // QEMU's hard-coded RCA
  base    = (u32)0x10005000; // PL180 base address
  printf("sdc_init : ");
  *(u32 *)(base + POWER) = (u32)0xBF; // power on
  *(u32 *)(base + CLOCK) = (u32)0xC6; // default CLK

  // send init command sequence
  do_command(0,  0,   MMC_RSP_NONE);// idle state
  do_command(55, 0,   MMC_RSP_R1);  // ready state
  do_command(41, 1,   MMC_RSP_R3);  // argument must not be zero
  do_command(2,  0,   MMC_RSP_R2);  // ask card CID
  do_command(3,  RCA, MMC_RSP_R1);  // assign RCA
  do_command(7,  RCA, MMC_RSP_R1);  // transfer state: must use RCA
  do_command(16, 512, MMC_RSP_R1);  // set data block length

  // set interrupt MASK0 registers bits = RxFULL(17)|TxEmpty(18)
  *(u32 *)(base + MASK0) = (1<<17)|(1<<18);
  // initialize semaphore s
  s.value = 0; s.queue = 0;
}

int get_block(int blk, char *buf)
{
  u32 cmd, arg;
  rxbuf = buf; rxcount = FBLK_SIZE;
  rxdone = 0;
  *(u32 *)(base + DATATIMER) = 0xFFFF0000;
  // write data_len to datalength reg
  *(u32 *)(base + DATALENGTH) = FBLK_SIZE;
  // 0x93=|9|0011|=|9|DMA=0,0=BLOCK,1=Host<-Card,1=Enable
  //  *(u32 *)(base + DATACTRL) = 0x93;
  cmd = 18;          // CMD17 = READ single sector
  arg = ((bsector + blk*2)*512);
  do_command(cmd, arg, MMC_RSP_R1);
  // 0x93=|9|0011|=|9|DMA=0,0=BLOCK,1=Host<-Card,1=Enable
  *(u32 *)(base + DATACTRL) = 0x93;
  if (hasP1)
    P(&s);                   // by semaphore
  else
    while(rxdone == 0); // by polling
}

int put_block(int blk, char *buf)
{
```

```
  u32 cmd, arg;
  txbuf = buf; txcount = FBLK_SIZE;
  txdone = 0;
  *(u32 *)(base + DATATIMER) = 0xFFFF0000;
  *(u32 *)(base + DATALENGTH) = FBLK_SIZE;
  cmd = 25;       // CMD24 = Write single sector
  arg = (u32)((bsector + blk*2)*512);
  do_command(cmd, arg, MMC_RSP_R1);
  // write 0x91=|9|0001|=|9|DMA=0,BLOCK=0,0=Host->Card, Enable
  *(u32 *)(base + DATACTRL) = 0x91; // Host->card
  if (hasP1)
    P(&s);                    // by semaphore
  else
    while(txdone == 0); // by polling
}
```

7.12.2 System Kernel Using SD File System

(4). **kernel.c file**: The kernel.c file is the same as that of Program C6.x.
(5). **t.c file**:

```
/**************** t.c file **********************/
#include "type.h"
#include "string.c"
#define FBLK_SZIE 1024
#include "uart.c"
#include "kbd.c"
#include "timer.c"
#include "vid.c"
#include "exceptions.c"
#include "queue.c"
#include "kernel.c"
#include "wait.c"
#include "fork.c"
#include "exec.c"
#include "svc.c"
#include "loadelf.c"
#include "thread.c"
#include "sdc.c"

int copy_vector_table(){ // SAME as Program C6.x }
int mkPtable()          { // same as Program C6.x }
int irq_chandler(){// same as before but ADD SDC interrupt handler }

extern int hasP1, partition, bsector; // define in sdc.c file

int main()
{
   fbuf_init();
```

```
    printf("Welcome to WANIX in Arm\n");
    printf("LCD display initialized : fbuf = %x\n", fb);
    kbd_init();
    uart_init();
    // code to enable VIC, SIC and device interrupts, including SDC
    kernel_init();
    hasP1 = 0;          // only P0 running at this moment
    partition = 2;      // choose a partition number
    sdc_init();         // initialize SDC driver
    mbr(partition);     // show partition table and set bsector
    kfork("/bin/u1"); // create P1 with Umode image
    hasP1 = 1;
    printf("P0 switch to P1\n");
    while(1){ // P0 switch process whenever readyQueue not empty
      while (readyQueue == 0);
      tswitch();
    }
}
```

7.12.3 Demonstration of SDC File System

Figure 7.11 shows the sample outputs of running the sample system C7.11.

7.13 Boot Kernel Image from SDC

Usually, an ARM based system has an onboard booter implemented in firmware. When such an ARM system starts, the onboard firmware booter first loads a stage-2 booter, e.g. Das Uboot (UBOOT 2016), from a (FAT) partition of a flash memory or a SDC and executes it. The stage-2 booter then boots up a real operating system, such as Linux, from a different partition. The emulated ARM Verstilepb virtual machine is an exception. When the emulated Versatilepb VM starts, QEMU loads a specified kernel image to 0x10000 and transfers control to the loaded kernel image directly, bypassing the usual booting phase of most other real or virtual machines. In fact, when the Vesatilepb VM starts, QEMU simply loads a specified image file and executes the loaded image. It does not know, nor care about, whether the image is an OS kernel or just a piece of executable code. The loaded image could be a booter, which can be used to boot up a real OS kernel from a storage device. In this section, we shall develop a booter for the emulated Versatilepb VM to boot up a system kernel from SDC partitions. In this scheme, each partition of the SDC is a (EXT2) file system. The system kernel image is a file in the /boot directory of a SDC partition. When the system starts, QEMU loads a booter to 0x10000 and executes it first. The booter may ask for a partition to boot, or it may simply boot from a default partition. Then it loads a system kernel image from the /boot directory in the SDC partition and transfer control to the kernel image, causing the OS kernel to startup. The advantages of this scheme are two-fold. First, the system kernel can be loaded to any memory location and runs from there, making it no longer confined to 0x10000 as dedicated by QEMU. This would make the system more compatible with other real or virtual machines that require a booting phase. Second, the booter can collect information from the user and pass them as booting parameters to the kernel. If desired, the booter may also set up an appropriate execution environment prior to transferring control to the kernel, which simplifies the kernel's startup code. For instance, if the system kernel is compiled with virtual addresses, the booter can set up the MMU first to allow the kernel to start up by using virtual addresses directly. The following shows the organization of such a system. It consists of a booter, which is loaded by QEMU to 0x10000 when the emulated Versatilepb VM starts. The booter then boots up a system kernel from a SDC partition and starts up the kernel. First, we show the components of the booter program.

```
QEMU
build pgdir and pgtables at 32KB (ID map 512MB VA to PA)                    00:00:45
build Kmode level-2 pgtables at 5MB
build 64 level-1 pgdirs for PROCs at 6MB
switch pgdir to use 2-level paging : switched pgdir OK
build pfreeList: start=0x800000  end=0x10000000  : 63487  4KB entries
read MBR to show partition table
P# start  size
1  2048  10000
2  12048  10000
3  22048  10000
4  32048  10000
partition=2  bsector=12048  bmap=18  imap=19  iblk=20  MBR done
allocate pgtables for proc 1
ptables=0x800000  0x901000  0xA02000  0xB03000
proc 0  kforked a child 1  at 0x801000
readyQueue = [1 1 ]->NULL
P0 switch to P1 : enter a line :
proc 0  in scheduler: readyQueue = [1 1 ]->[0 0 ]->NULL
switch to proc 1  pgdir at 0x600000
CPU mode=0x10  argc=2  argv[0 ] = init argv[1 ] = start
This is process 1  in Umode at VA=0x80000000  PA=0x801000  parent=0
-------------------------------------------------------
ps chname kfork switch wait exit fork exec vfork thread
-------------------------------------------------------
input a command : 
```

Fig. 7.11 Sample outputs of system with SDC

7.13.1 SDC Booter Program

(1). **booter's ts.s file**: This is the entry point of the booter program. It initializes a UART for serial port I/O during booting. In order to keep the booter code simple, it does not use interrupts. So the UART driver uses polling for serial port I/O. The booter loads a kernel image from a SDC partition to 1 MB. Then it jumps to there to start up the kernel.

```
/************* booter's ts.s file ******************/
  .global reset_handler, main
reset_handler:
  /* initialize SVC stack */
  ldr sp, =svc_stack_top
  BL uart_init
  BL main
  mov pc, #0x100000  // jump to loaded kernel at 1MB
```

(2). **Booter's t.c file**: This is the main function of the booter program.

```
#include "type.h"
#include "string.c"
#include "uart.c"
#include "sdc.c"
#include "boot.c"
int main()
{
```

```
  printf("Welcome to ARM EXT2 Booter\n");
  sdc_init();
  mbr();        // read and display partition table
  boot();       // boot kernel image from a partition
  printf("BACK FROM booter: enter a key\n");
  ugetc();
}
```

(3). **Booter's sd.c file**: This file implements the SDC driver of the booter. It provides a

```
getblk(int blk, char *address)
```

function, which loads a (1 KB) block from the SDC into the specified memory address. In order to keep the booter simple, the booter's SDC driver uses polling for block I/O.

(4). **The boot.c file**: This file implements the SDC booter. For the emulated ARM Versatilepb VM, the booter is a separate image. It is loaded to 0x10000 by QEMU and starts to execute from there. It then boots up a kernel image in the /boot directory of a SDC partition.The function mbr() displays the partition table of the SDC and prompts for a partition number to boot. It writes the partition and the start sector number to 0x200000 (2 MB) for the kernel to get. Then it calls boot(), which locates a kernel image file in the /boot directory and loads the kernel image to 0x100000 (1 MB).

```
/****************** booter's boot.c file ********************/
int bmap, imap, iblk, blk, offset;
struct partition {
      u8 drive; /* 0x80 - active */
      u8 head; /* starting head */
      u8 sector; /* starting sector */
      u8 cylinder;        /* starting cylinder */
      u8 sys_type; /* partition type */
      u8 end_head; /* end head */
      u8 end_sector; /* end sector */
      u8 end_cylinder; /* end cylinder */
     u32 start_sector;    /* starting sector counting from 0 */
     u32 nr_sectors;      /* nr of sectors in partition */
};
char buf[1024], buf1[1024], buf2[1024];
int mbr()
{
  int i, pno, bno;
  int *partition = (int *)0x200000;
  int *sector =    (int *)0x200004;
  char line[8], c;
  struct partition *p;
  GD     *gp;
  bsector = 0;
  printf("read MBR to show partition table\n");
  get_block(0, buf);
  p = (struct partition *)&buf[0x1bE];
  printf("P# start  size\n");
  for (i=1; i<=4; i++){
    printf("%d %d %d\n", i, p->start_sector, p->nr_sectors);
    p++;
  }
  printf("enter partition number [1-4]:");
  c = ugetc();
```

```
  pno = c - '0';
  if (pno < 1 || partition > 4)
     pno = 2; // use default partition 2
  p = (struct partition *)&buf[0x1bE];
  for (i=1; i<=4; i++){
    if (i==pno){
      bno = p->start_sector;  break;
    }
    p++;
  }
  printf("partition=%d bsector = %d\n", partition, bsector);
  *partition = pno; *bsector = bno;
  get_block(2, buf); // read group descriptor block
  gp = (GD *)buf;     // access group 0 despriptor
  iblk = gp->bg_inode_table;
  bmap = gp->bg_block_bitmap;
  imap = gp->bg_inode_bitmap;
  printf("bmap=%d imap=%d iblk=%d\n", bmap, imap, iblk);
  printf("MBR done\n");
}
int search(INODE *ip, char *name)
{
   int i;
   char c, *cp;
   DIR  *dp;
   for (i=0; i<12; i++){
       if (ip->i_block[i]){
          get_block(ip->i_block[i], buf2);
          dp = (DIR *)buf2;
          cp = buf2;
          while (cp < &buf2[1024]){
              c = dp->name[dp->name_len];  // save last byte
              dp->name[dp->name_len] = 0;
              printf("%s ", dp->name);
              if ( strcmp(dp->name, name) == 0 ){
                 printf("found %s\n", name);
                 return(dp->inode);
              }
              dp->name[dp->name_len] = c; // restore that last byte
              cp += dp->rec_len;
              dp = (DIR *)cp;
          }
       }
   }
   printf("serach failed\n");
   return 0;
}
boot()
{
  int   i, ino, blk, iblk, count;
  char  *cp, *name[2],*location;
  u32   *up;
  GD    *gp;
  INODE *ip;
  DIR   *dp;
```

```
  name[0] = "boot";
  name[1] = "kernel";
  mbr();
  /* read blk#2 to get group descriptor 0 */
  get_block(2, buf1);
  gp = (GD *)buf1;
  iblk = (u16)gp->bg_inode_table;
  getblk(iblk, buf1);         // read first inode block block
  ip = (INODE *)buf1 + 1;     // ip->root inode #2
  /* serach for system name */
  for (i=0; i<2; i++){
      ino = search(ip, name[i]) - 1;
      if (ino < 0)
         return 0;
      get_block(iblk+(ino/8), buf1); // read inode block of ino
      ip = (INODE *)buf1 + (ino % 8);
  }
  /* read indirect block into b2 */
  if (ip->i_block[12])          // only if has indirect blocks
     get_block(ip->i_block[12], buf2);
  location = (char *)0x100000;
  count = 0;
  for (i=0; i<12; i++){
      get_block(ip->i_block[i], location);
      uputc('*');
      location += 1024;
      count++;
  }
  if (ip->i_block[12]){ // only if file has indirect blocks
     up = (u32 *)buf2;
     while(*up){
      get_block(*up, location);
       uputc('.');
       location += 1024;
       up++; count++;
     }
  }
  printf("loading done\n", count);
}
```

(5). **Kernel and User Mode Images**: Kernel and User mode images are generated by the following sh scripts files, which create image files and copies them to the SD partitions. Note that the kernel's starting VA is at **0x100000 (1 MB)** and the starting VA of Umode images is at **0x80000000 (2 GB)**.

```
# mkkernel script file

  arm-none-eabi-as -mcpu=arm926ej-s ts.s -o ts.o
  arm-none-eabi-gcc -c -mcpu=arm926ej-s t.c -o t.o
  arm-none-eabi-ld -T t.ld ts.o t.o -Ttext=0x100000 -o kernel.elf
  arm-none-eabi-objcopy -O binary kernel.elf kernel
  for I in 1 2 3 4
  do
     mount /dev/loop$I /mnt  # assume SDC partitions are loop devices
     cp -av kernel /mnt/boot/
     umount /mnt
```

```
done

# mku script file: mku u1; mku u2, etc.
  arm-none-eabi-as -mcpu=arm926ej-s us.s -o us.o
  arm-none-eabi-gcc -c -mcpu=arm926ej-s -o $1.o $1.c
  arm-none-eabi-ld -T u.ld us.o $1.o -Ttext=0x80000000 -o $1
  for I in 1 2 3 4
  do
     mount /dev/loop$I /mnt
     cp -av $1 /mnt/bin/
     umount /mnt
done
```

7.13.2 Demonstration of Booting Kernel from SDC

The sample system C7.12 demonstrates booting an OS kernel from SDC.

Figure 7.12 shows the UART screen of the booter. It displays the SDC partition table and asks for a partition to boot from. It locates the kernel image file, /boot/kernel, in the SDC partition and loads the kernel image to 0x100000 (1 MB). Then it sends the CPU to execute the loaded kernel code. When the kernel starts, it uses 2-level static paging. Since the kernel is compile-linked with real addresses, it can execute all the code directly when it starts up. In this case, there is no need for the booter to build any page table for the kernel. The page tables will be built by the kernel itself when it starts up.

Figure 7.13 shows the startup screen of the kernel. In kernel_init(), it initializes the kernel data structures, builds the 2-level page tables for the processes and switches pgdir to use 2-level static paging.

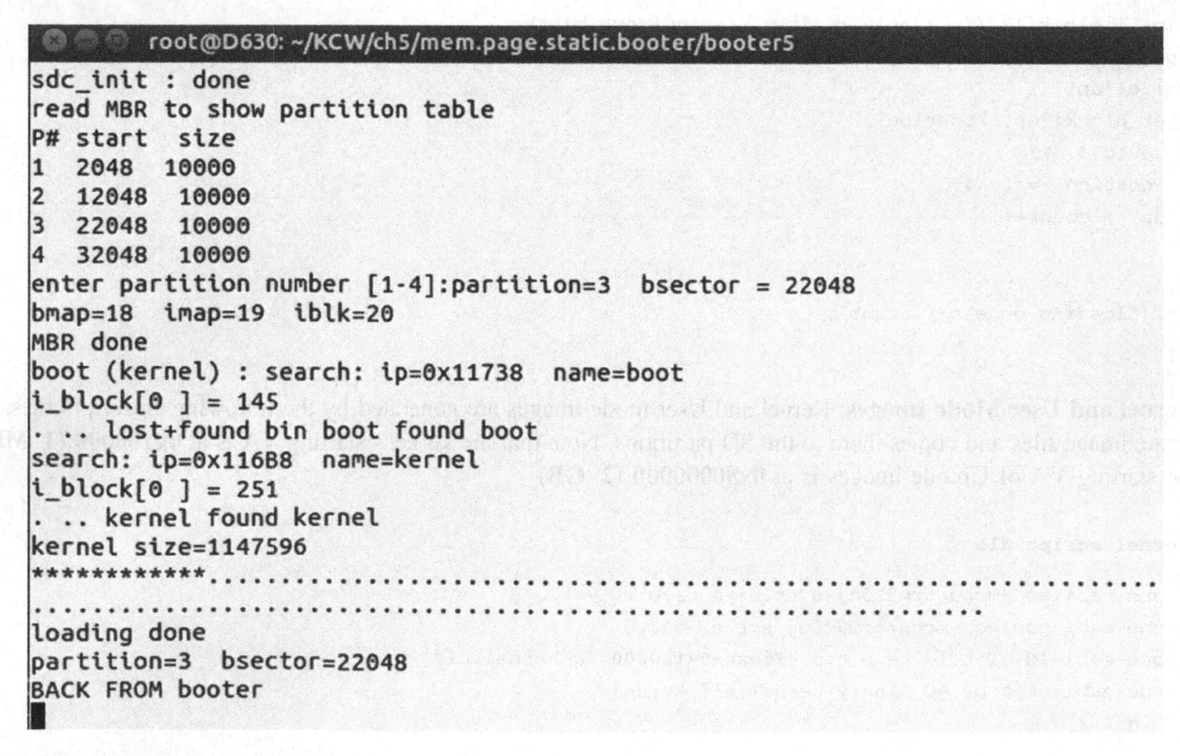

root@D630: ~/KCW/ch5/mem.page.static.booter/booter5
sdc_init : done
read MBR to show partition table
P# start size
1 2048 10000
2 12048 10000
3 22048 10000
4 32048 10000
enter partition number [1-4]:partition=3 bsector = 22048
bmap=18 imap=19 iblk=20
MBR done
boot (kernel) : search: ip=0x11738 name=boot
i_block[0] = 145
. .. lost+found bin boot found boot
search: ip=0x116B8 name=kernel
i_block[0] = 251
. .. kernel found kernel
kernel size=1147596
loading done
partition=3 bsector=22048
BACK FROM booter

Fig. 7.12 Demonstration of SDC booter

```
QEMU
                    Welcome to WANIX in Arm                         00:01:13
LCD display initialized : fbuf = 0x200000
partition=3  bsector=22048
sdc_init : done
kbd_init()
uart[0-4] init()
timer_init() timer_start
kernel_init()
freeList    = [1 ]->[2 ]->[3 ]->[4 ]->[5 ]->[6 ]->[7 ]->[8 ]->[9 ]->NULL
readyQueue = NULL
build pgdir and pgtables at 32KB (ID map 512MB VA to PA)
build Kmode level-2 pgtables at 5MB
build 64 level-1 pgdirs for PROCs at 6MB
switch pgdir to use 2-level paging : switched pgdir OK
proc 0  kforked a child 1
readyQueue = [1 1 ]->NULL
P0 switch to P1 : enter a line :
```

Fig. 7.13 Demonstration of booting OS Kernel from SDC

7.13.3 Booting Kernel from SDC with Dynamic Paging

The sample system C7.13 demonstrates booting a kernel that uses 2-level dynamic paging. The booter part is the same as before. Instead of static paging, the kernel uses 2-level dynamic paging. Figure 7.14 shows the startup screen of the kernel. As the figure shows, all the page table entries of P1 are dynamically allocated page frames.

```
QEMU
                    Welcome to WANIX in Arm                         00:00:52
LCD display initialized : fbuf = 0x200000
sdc_init : done
kbd_init()
uart[0-4] init()
timer_init() timer_start
kernel_init()
freeList    = [1 ]->[2 ]->[3 ]->[4 ]->[5 ]->[6 ]->[7 ]->[8 ]->[9 ]->NULL
readyQueue = NULL
build pgdir and pgtables at 32KB (ID map 512MB VA to PA)
build Kmode level-2 pgtables at 5MB
switch pgdir to use 2-level paging : switched pgdir OK
build pfreeList: start=0x800000  end=0x10000000  : 63487  4KB entries
MBR: partition=2  bsector=12048
bmap=18  imap=19  iblk=20
MBR done
pgtable=0x800000  pgtable[0 ]=0x80155E  pgtable[1 ]=0x80255E
pgtable=0x901000  pgtable[0 ]=0x90255E  pgtable[1 ]=0x90355E
pgtable=0xA02000  pgtable[0 ]=0xA0355E  pgtable[1 ]=0xA0455E
pgtable=0xB03000  pgtable[0 ]=0xB0455E  pgtable[1 ]=0xB0555E
proc 0  kforked a child 1  at 0x801000
readyQueue = [1 1 ]->NULL
P0 switch to P1 : enter a line :
```

Fig. 7.14 Booting OS Kernel from SDC with dynamic paging

7.13.4 Two-Stage Booting

In many ARM based systems, booting up an OS kernel consists of two stages. When an ARM system starts, the system's onboard boot-loader in firmware loads a booter from a storage device, such as a flash memory or a SD card, and transfers control to the loaded booter. The booter then loads an operating system (OS) kernel from a bootable device and transfers control to the OS kernel, causing the OS kernel to start up. In this case, the system's onboard boot-loader is a stage-1 booter, which loads and executes a stage-2 booter, which is designed to boot up a specific OS kernel image. The stage-2 booter can be installed on a SDC for booting up different kernel images. Some ARM boards require that the stage-2 booter be installed in a DOS partition, but the bootable kernel image may reside in a different partition. The principle and technique of installing a booter to SDC is the same as that of installing a booter to a regular hard disk or USB drive. The reader may consult Chap. 3 on Booting Operating Systems of Wang (2015) for more details. In this section, we demonstrate a two-stage booter for the emulated ARM Versatilepb VM. First, we show the code segments of the stage-1 booter.

7.13.4.1 Stage-1 Booter

(1). **ts.s file of stage-1 booter**: initialize UART0 for serial I/O, call main() to load stage-2 booter to 2 MB. Then jump to stage-2 booter.

```
.text
.code 32
reset_handler:
  ldr sp, =svc_stack_top    // set SVC stack
  bl  uart_init             // initialize UART0 for serial I/O
  bl  main                  // call main() in C
mov pc, #0x200000          // jump to stage-2 booter at 2MB
```

(2). **boot() function of stage-1 booter**: load stage-2 booter from SDC to 2 MB and jump to execute the stage-2 booter at 2 MB.

```
int boot1()                 // boot1() in boot.c file
{
  int i;
  char *p = (char *)0x200000;
  for (i=1; i<10; i++){     // load 10 blocks from SDC
    getblk(i, p);
    printf("%c", '.');
    p += 1024;
  }
  printf("\nloading done: jump to stage-2 booter\n");
}
```

(3). **t.c file of stage-1 booter.**

```
#include "type.h"
#include "string.c"
#include "uart.c"
#include "sdc.c"
#include "boot.c"
int main()
{
   printf("Stage-1 Booter\n");
```

```
    sdc_init();              // initialize SDC driver
    boot1();                 // load stage-2 booter to 2MB
}
```

7.13.4.2 Stage-2 Booter

The stage-2 booter is the same as the booter in Sect. 7.13.1. It is installed to the front part of a SDC, which will be loaded for execution by the stage-1 booter. The stage-2 booter size is about 8 KB. On a SDC with partitions, partition 1 begins from the sector 2048, so blocks 1 to 1023 are free spaces on the SDC. The stage-2 booter is installed to blocks 1 to 8 of the SDC by the following dd command.

```
dd if=booter2.bin of=../sdc bs=1024 count=8 seek=1 conv=notrunc
```

7.13.5 Demonstration of Two-Stage Booting

The sample system C7.14 demonstrates two-stage booting. The system is run from the stage-1 booter directory by

```
qemu-system-arm -M versatilepb -m 512M -sd ../sdc -kernel booter1.bin \
-serial mon:stdio
```

Figure 7.15 shows the screen of the 2-stage booters.
Figure 7.16 shows the kernel startup screen after booting up by the 2-stage booters.

```
Stage-1 Booter
sdc_init : done
load stage 2 booter from SDC to 2MB
........
loading done: transfer to stage-2 booter
Stage-2 Booter
sdc_init : done
read MBR to show partition table
P# start size
1  2048   10000
2  12048  10000
3  22048  10000
4  32048  10000
enter partition number [1-4] : 2
partition=2  bsector = 12048
bmap=18  imap=19  iblk=20  MBR done
boot (wanix) : search for boot
. .. lost+found bin boot found boot
search for wanix
. .. wanix found wanix
LOCATION=0x100000
************.....................................................
.................................................................
loading done
partition=2  bsector=12048
loading OS kernel complete: start up OS kernel
```

Fig. 7.15 Screen of two-stage booters

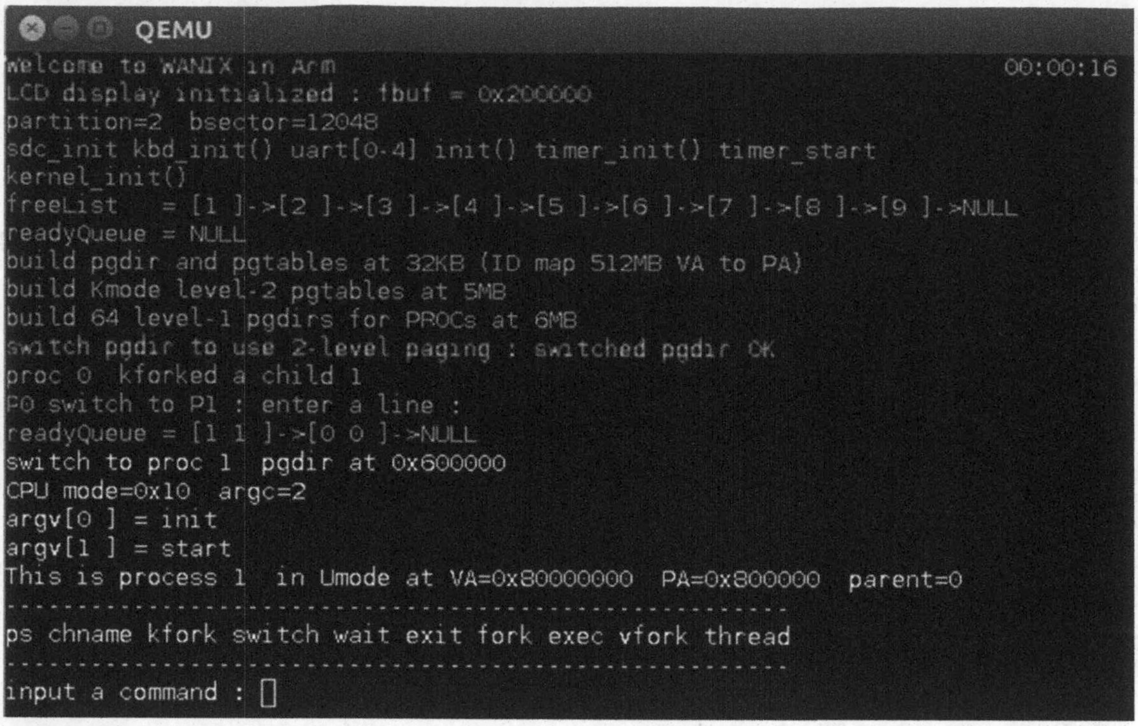

```
QEMU
welcome to WANIX in Arm                                            00:00:16
LCD display initialized : fbuf = 0x200000
partition=2  bsector=12048
sdc_init kbd_init() uart[0-4] init() timer_init() timer_start
kernel_init()
freeList    = [1 ]->[2 ]->[3 ]->[4 ]->[5 ]->[6 ]->[7 ]->[8 ]->[9 ]->NULL
readyQueue = NULL
build pgdir and pgtables at 32KB (ID map 512MB VA to PA)
build Kmode level-2 pgtables at 5MB
build 64 level-1 pgdirs for PROCs at 6MB
switch pgdir to use 2-level paging : switched pgdir OK
proc 0  kforked a child 1
P0 switch to P1 : enter a line :
readyQueue = [1 1 ]->[0 0 ]->NULL
switch to proc 1  pgdir at 0x600000
CPU mode=0x10  argc=2
argv[0 ] = init
argv[1 ] = start
This is process 1  in Umode at VA=0x80000000  PA=0x800000  parent=0
----------------------------------------------------------
ps chname kfork switch wait exit fork exec vfork thread
----------------------------------------------------------
input a command : 
```

Fig. 7.16 Demonstration of 2-stage booters

7.14 Summary

This chapter covers process management, which allows us to create and run processes dynamically in embedded systems. In order to keep the systems simple, it only shows the basic process management functions, which include process creation, process termination, process synchronization and wait for child process termination. Throughout the chapter, it shows how to use memory management to provide each process with a private User mode virtual address space that is isolated from other processes and protected by the MMU hardware. The memory management schemes use both one-level sections and two-level static and dynamic paging. In addition, it also discussed the advanced concepts and techniques of vfork and threads. Lastly, it showed how to use SD cards for storing both kernel and user mode image files in a SDC file system and how to boot up system kernel from SDC partitions. With this background, we are ready to show the design and implementation of general purpose operating system for embedded systems.

List of Sample Programs

C7.1: Kernel and User mode
C7.2: Tasks in the same domain
C7.3: Tasks with individual domain
C7.4: Fork-exec
C7.5: Vfork
C7.6: Threads
C7.7: 2-level static paging KML
C7.8: 2-level dynamic paging KML
C7.9: 1-level static paging KMH
C7.10: 2-level static paging KMH
C7.11: Paging with SDC
C7.12: Booting OS kernel from SDC
C7.13: Booting kernel with 2-level dynamic paging
C7.14: 2-stage booting

Problems

1. In the example program C7.1, both tswitch() and svc_handler() use

```
stmfd sp!, {r0-r12, lr}
```

save all CPU registers, which may be unnecessary since the generated code of most ARM C compilers preserve registers r4–r12 during function calls. Assume that both tswitch() and svc_handler() use

```
stmfd sp!, {r0-r3, lr}
```

to save CPU registers. Rewrite tswitch(), svc_handler and kfork() of the system. Verify that the modified system works.
2. Modify the sample system C7.1 to use 4 MB Umage image size.
3. In the example program C7.1, it builds the level-1 page tables of all (64) PROCs statically in the memory area of 6 MB. Modify it to build the process level-1 page table dynamically, i.e. only when a process is created.
4. In the example program C7.1, each process runs in a different Umode area but the Umode stack pointer of every process is initialized as
Explain why and how does it work?

```
#define UIAMGE_SIZE 0x100000
p->usp = (int *)VA(UIMAGE_SIZE);
```

5. In the example program C7.1, it assumes that the VA spaces of both Kmode and Umode are 2 GB. Modify it for 1 GB Kmode VA space and 3 GB Umode VA space.
6. For the sample system C7.4, implement process family tree and use it in kexit() and kwait().
7. For the sample system C7.4, modify the kexit() function to implement the following policy.
   (1). A terminating process must dispose of its ZOMBIE children, if any, first.
   (2). A process can not terminate until all the children processes have terminated.
   Discuss the advantages and disadvantages of these schemes.
8. In all the example programs, each PROC structure has a statically allocate 4 KB kstack.
   (1). Implement a simple memory manager to allocate/deallocate memory dynamically. When the system starts, reserve a 1 MB area, e.g. beginning at 4 MB, as a free memory area. The function

```
char *malloc(int size)
```

allocates a piece of free memory of size in 1 KB bytes. When a memory area is no longer needed, it is released back to the free memory area by

```
void mfree(char *address, int size)
```

Design a data structure to represent the current available free memory. Then implement the malloc() and mfree() functions.
   (2). Modify the kstack field of the PROC structure as an integer pointer

```
int *kstack;
```

and modify the kfork() function as

```
int kfork(int func, int priority, int stack_size)
```

which dynamically allocates a memory area of stack_size for the new task.
   (3). When a task terminates, its stack area must be (eventually) freed. How to implement this? If you think you may simply release the stack area in kexit(), think carefully again.

9. Modify the example program C7.5 to support large Umode image sizes, e.g. 4 MB.
10. In the example program C7.5, assume that the Umode image size in not a multiple of 1 MB, e.g. 1.5 MB. Show how to set up process page tables to suit the new image size.
11. Modify the example program C7.10 to use 2-level static paging.
12. Modify the example program C7.10 to map kernel VA space to [2 GB, 2 GB+512 MB].
13. Modify the example program C7.11 to use two-level dynamic paging.

References

ARM MMU: ARM926EJ-S, ARM946E-S Technical Reference Manuals, ARM Information Center 2008
Buttlar, D, Farrell, J, Nichols, B., "PThreads Programming, A POSIX Standard for Better Multiprocessing", O'Reilly Media, 1996
Card, R., Theodore Ts'o, T., Stephen Tweedie, S., "Design and Implementation of the Second Extended Filesystem", web.mit.edu/tytso/www/linux/ext2intro.html, 1995
Cao, M., Bhattacharya, S, Tso, T., "Ext4: The Next Generation of Ext2/3 File system", IBM Linux Technology Center, 2007.
ELF: Tool Interface Standard (TIS) Executable and Linking Format (ELF) Specification Version 1.2, 1995
EXT2: www.nongnu.org/ext2-doc/ext2.html, 2001
Intel 64 and IA-32 Architectures Software Developer's Manual, Volume 3, 1992
Intel i486 Processor Programmer's Reference Manual, 1990
Linux Man pages: https://www.kernel.org/doc/man-pages, 2016
Pthreads: https://computing.llnl.gov/tutorials/pthreads, 2015
POSIX.1C, Threads extensions, IEEE Std 1003.1c, 1995
Silberschatz, A., P.A. Galvin, P.A., Gagne, G, "Operating system concepts, 8th Edition", John Wiley & Sons, Inc. 2009
UBOOT, Das U-BOOT, http://www.denx.de/wiki/U-BootUboot, 2016
Wang, K.C., "Design and Implementation of the MTX Operating System", Springer Publishing International AG, 2015

8 General Purpose Embedded Operating Systems

8.1 General Purpose Operating Systems

A General Purpose Operating System (**GPOS**) is a complete OS that supports process management, memory management, I/O devices, file systems and user interface. In a GPOS, processes are created dynamically to perform user commands. For security, each process runs in a private address space that is isolated from other processes and protected by the memory management hardware. When a process completes a specific task, it terminates and releases all its resources to the system for reuse. A GPOS should support a variety of I/O devices, such as keyboard and display for user interface, and mass storage devices. A GPOS must support a file system for saving and retrieving both executable programs and application data. It should also provide a user interface for users to access and use the system conveniently.

8.2 Embedded General Purpose Operating Systems

In the early days, embedded systems were relative simple. An embedded system usually consists of a microcontroller, which is used to monitor a few sensors and generate signals to control a few actuators, such as to turn on LEDs or activates relays to control external devices. For this reason, the control programs of early embedded systems were also very simple. They were written in the form of either a super-loop or event-driven program structure. However, as computing power and demand for multi-functional systems increase, embedded systems have undergone a tremendous leap in both applications and complexity. In order to cope with the ever increasing demands for extra functionalities and the resulting system complexity, traditional approaches to embedded OS design are no longer adequate. Modern embedded systems need more powerful software. Currently, many mobile devices are in fact high-powered computing machines capable of running full-fledged operating systems. A good example is smart phones, which use the ARM core with gig bytes internal memory and multi-gig bytes micro SD card for storage, and run adapted versions of Linux, such as (Android 2016). The current trend in embedded OS design is clearly moving in the direction of developing multi-functional operating systems suitable for the mobile environment. In this chapter, we shall discuss the design and implementation of general purpose operating systems for embedded systems.

8.3 Porting Existing GPOS to Embedded Systems

Instead of designing and implementing a GPOS for embedded systems from scratch, a popular approach to embedded GPOS is to port existing OS to embedded systems. Examples of this approach include porting Linux, FreeBSD, NetBSD and Windows to embedded systems. Among these, porting Linux to embedded systems is especially a common practice. For example, Android (2016) is an OS based on the Linux kernel. It is designed primarily for touch screen mobile devices, such as smart phones and tablets. The ARM based Raspberry PI single board computer runs an adapted version of Debian Linux, called Raspbian (Raspberry PI-2 2016). Similarly, there are also widely publicized works which port FreeBSD (2016) and NetBSD (Sevy 2016) to ARM based systems.

When porting a GPOS to embedded systems, there are two kinds of porting. The first kind can be classified as a procedural oriented porting. In this case, the GPOS kernel is already adapted to the intended platform, such as ARM based

K.C. Wang, *Embedded and Real-Time Operating Systems*,
DOI 10.1007/978-3-319-51517-5_8

systems. The porting work is concerned primarily with how to configure the header files (.h files) and directories in the source code tree of the original GPOS, so that it will compile-link to a new kernel for the target machine architecture. In fact, most reported work of porting Linux to ARM based systems fall into this category. The second kind of porting is to adapt a GPOS designed for a specific architecture, e.g. the Intel x86, to a different architecture, such as the ARM. In this case, the porting work usually requires redesign and, in many cases, completely different implementations of the key components in the original OS kernel to suit the new architecture. Obviously, the second kind of porting is much harder and challenging than the procedural oriented porting since it requires a detailed knowledge of the architectural differences, as well as a complete understanding of operating system internals. In this book, we shall not consider the procedural oriented porting. Instead, we shall show how to develop an embedded GPOS for the ARM architecture from ground zero.

8.4 Develop an Embedded GPOS for ARM

PMTX (Wang 2015) is a small Unix-like GPOS originally designed for Intel x86 based PCs. It runs on uniprocessor PCs in 32-bit protected mode using dynamic paging. It supports process management, memory management, device drivers, a Linux-compatible EXT2 file system and a command-line based user interface. Most ARM processors have only a single core. In this chapter, we shall focus on how to adapt PMTX to single CPU ARM based systems. Multicore CPUs and multiprocessor systems will be covered later in Chap. 9. For ease of reference, we shall denote the resulting system as EOS, for Embedded Operating System.

8.5 Organization of EOS

8.5.1 Hardware Platform

EOS should be able to run on any ARM based system that supports suitable I/O devices. Since most readers may not have a real ARM based hardware system, we shall use the emulated ARM Versatilepb VM (ARM Versatilepb 2016) under QEMU as the platform for implementation and testing. The emulated Versatilepb VM supports the following I/O devices.

(1). SDC: EOS uses a SDC as the primary mass storage device. The SDC is a virtual disk, which is created as follows.

```
dd if=/dev/zero of=$1 bs=4096 count=33280 # 512+32768 4KB blocks
fdisk disk         # create partition 1 = [2048 to 266239] sectors
losetup -o $(expr 2048 \* 512) --sizelimit $(expr 266239 \* 512) \
          /dev/loop1 $1
mke2fs -b 4096 /dev/loop1 32768 # mke2fs with 32K 4KB blocks
mount /dev/loop1 /mnt              # mount as loop device
 (cd /mnt; mkdir bin boot dev etc user) # populate with DIRs
umount /mnt
```

For simplicity, the virtual SDC has only one partition, which begins from the (fdisk default) sector 2048. After creating the virtual SDC, we set up a loop device for the SDC partition and format it as an EXT2 file system with 4 KB block size and one blocks-group. The single blocks-group on the SDC image simplifies both the file system traversal and the inodes and disk blocks management algorithms. Then we mount the loop device and populate it with DIRs and files, making it ready for use. The resulting file system size is 128 MB, which should be big enough for most applications. For larger file systems, the SDC can be created with multiple blocks-groups, or multiple partitions. The following diagram shows the SDC contents.

```
---------|------------------Partition 1 -------------------------|
|M|booter|super|gd  |. . . |bmap|imap|inodes      |data blocks  |
|-----------------------------------------------------------------
            |< ---------------- EXT2 FS ---------------------- >|
            |-- bin : binary executable command files
```

```
|- boot : bootable kernel images
|-- dev : special files (I/O devices)
|-- etc : passwd file
|-- user: user home directories
```

On the SDC, the MBR sector (0) contains the partition table and the beginning part of a booter. The remaining part of the booter is installed in sectors 2 to booter_size, assuming that the booter size is no more than 2046 sectors or 1023 KB (The actual booter size is less than 10 KB). The booter is designed to boot up a kernel image from an EXT2 file system in a SDC partition. When the EOS kernel boots up, it mounts the SDC partition as the root file system and runs on the SDC partition.

(2). LCD: the LCD is the primary display device. The LCD and the keyboard play the role of the system console.
(3). Keyboard: this is the keyboard device of the Versatilepb VM. It is the input device for both the console and UART serial terminals.
(4). UARTs: these are the (4) UARTs of the Versatilepb VM. They are used as serial terminals for users to login. Although it is highly unlikely that an embedded system will have multiple users, our purpose is to show that the EOS system is capable of supporting multiple users at the same time.
(5). Timer: the VersatilepbVM has four timers. EOS uses timer0 to provide a time base for process scheduling, timer service functions, as well as general timing events, such as to maintain Time-of-Day (TOD) in the form of a wall clock.

8.5.2 EOS Source File Tree

The source files of EOS are organized as a file tree.

```
EOS
|- BOOTER     : stage-1 and stage-2 booters
|- type.h, include.h, mk scripts
|- kernel     : kernel source files
|- fs         : file system files
|- driver     : device driver files
|- USER       :  commands and user mode programs
```

BOOTER	:	this directory contains the source code of stage-1 and stage-2 booters
type.h	:	EOS kernel data structure types, system parameters and constants
include.h	:	constants and function prototypes
mk	:	sh scripts to recompile EOS and install bootable image to SDC partition

8.5.3 EOS Kernel Files

```
------------- Kernel: Process Management Part -------------
type.h      : kernel data structure types, such as PROC, resources, etc.
ts.s        : reset handler, tswitch, interrupt mask, interrupt handler entry/exit code, etc.
eoslib      : kernel library functions; memory and string operations.
------------- Kernel files -------------
queue.c     : enqueue, dequue, printQueue and list operation functions
wait.c      : ksleep, kwakeup, kwait, kexit functions
loader.c    : ELF executable image file loader
mem.c       : page tables and page frame management functions
fork.c      : kfork, fork, vfork functions
exec.c      : kexec function
```

```
threads.c      : threads and mutex functions
signal.c       : signals and signal processing
except.c       : data_abort, prefetch_abort and undef exception handlers
pipe.c         : pipe creation and pipe read/write functions
mes.c          : send/recv message functions
syscall.c      : simple system call functions
svc.c          : syscall routing table
kernel.c       : kernel initialization
t.c            : main entry, initialization, parts of process scheduler
--------------- Device Drivers ---------------------------------
lcd.c          : console display driver
pv.c           : semaphore operations
timer.c        : timer and timer service functions
kbd.c          : console keyboard driver
uart.c         : UART serial ports driver
sd.c           : SDC driver
--------------- File system -----------------------------------
fs             : implementation of an EXT2 file system
buffer.c       : block device (SDC) I/O buffering
---------------------------------------------------------------
```

EOS is implemented mostly in C, with less than 2% of assembly code. The total number of line count in the EOS kernel is approximately 14,000.

8.5.4 Capabilities of EOS

The EOS kernel consists of process management, memory management, device drivers and a complete file system. It supports dynamic process creation and termination. It allows process to change execution images to execute different programs. Each process runs in a private virtual address space in User mode. Memory management is by two-level dynamic paging. Process scheduling is by both time-slice and dynamic process priority. It supports a complete EXT2 file system that is totally Linux compatible. It uses block device I/O buffering between the file system and the SDC driver to improve efficiency and performance. It supports multiple user logins from the console and serial terminals. The user interface sh supports executions of simple commands with I/O redirections, as well as multiple commands connected by pipes. It unifies exception handling with signal processing, and it allows users to install signal catchers to handle exceptions in User mode.

8.5.5 Startup Sequence of EOS

The startup sequence of EOS is as follows. First, we list the logical order of the startup sequence. Then we explain each step in detail.

(1). Booting the EOS kernel
(2). Execute reset_handler to initialize the system
(3). Configure vectored interrupts and device drivers
(4). kernel_init: initialize kernel data structures, create and run the initial process P0
(5). Construct pgdir and pgtables for processes to use two-level dynamic paging
(6). Initialize file system and mount the root file system
(7). Create the INIT process P1; switch process to run P1
(8). P1 forks login processes on console and serial terminals to allow user logins.
(9). When a user login, the login process executes the command interpreter sh.
(10). User enters commands for sh to execute.
(11). When a user logout, the INIT process forks another login process on the terminal.

(1). SDC Booting: An ARM based hardware system usually has an onboard boot-loader implemented in firmware. When an ARM based system starts, the onboard boot-loader loads and executes a stage-1 booter form either a flash device or, in many cases, a FAT partition on a SDC. The stage-1 booter loads a kernel image and transfers control to the kernel image. For EOS on the ARM Versatilpb VM, the booting sequence is similar. First, we develop a stage-1 booter as a standalone program. Then we design a stage-2 booter to boot up EOS kernel image from an EXT2 partition. On the SDC, partition 1 begins from the sector 2048. The first 2046 sectors are free, which are not used by the file system. The stage-2 booter size is less than 10 KB. It is installed in sectors 2 to 20 of the SDC. When the ARM Versatilepb VM starts, QEMU loads the stage-1 booter to 0x10000 (64 KB) and executes it first. The stage-1 booter loads the stage-2 booter from the SDC to 2 MB and transfers control to it. The stage-2 booter loads the EOS kernel image file (/boot/kernel) to 1 MB and jumps to 1 MB to execute the kernel's startup code. During booting, both stage-1 and stage-2 booters use a UART port for user interface and a simple SDC driver to load SDC blocks. In order to keep the booters simple, both the UART and SDC drivers use polling for I/O. The reader may consult the source code in the booter1 and booter2 directories for details. It also shows how to install the stage-2 booter to the SDC.

8.5.6 Process Management in EOS

In the EOS kernel, each process or thread is represented by a PROC structure which consists of three parts.

```
. fields for process management
. pointer to a per-process resource structure
. kernel mode stack pointer to a dynamically allocated 4KB page as kstack
```

8.5.6.1 PROC and Resource Structures

In EOS, the PROC structure is defined as

```
typedef struct proc{
  struct proc *next;      // next proc pointer
  int     *ksp;           // at 4
  int     *usp;           // at 8 : Umode usp at syscall
  int     *upc;           // at 12: upc at syscall
  int     *ucpsr;         // at 16: Umode cpsr at syscall
  int     status;         // process status
  int     priority;       // scheduling priority
  int     pid;            // process ID
  int     ppid;           // parent process ID
  int     event;          // sleep event
  int     exitCode;       // exit code
  int     vforked;        // whether the proc is VFROKED
  int     time;           // time-slice
  int     cpu;            // CPU time ticks used in ONE second
  int     type;           // PROCESS or THREAD
  int     pause;          // seconds to pause
  struct proc *parent;    // parent PROC pointer
  struct proc *proc;      // process ptr of threads in PROC
  struct pres *res;       // per-process resource pointer
  struct semaphore *sem; // ptr to semaphore proc BLOCKed on
  int     *kstack;        // pointer to Kmode stack
}PROC;
```

In the PROC structure, the next field is used to link the PROCs in various link lists or queues. The ksp field is the saved kernel mode stack pointer of the process. When a process gives up CPU, it saves CPU registers in kstack and saves the stack pointer in ksp. When a process regains CPU, it resumes running from the stack frame pointed by ksp

The fields usp, upc and ucpsr are for saving the Umode sp, pc and cpsr during syscall and IRQ interrupt processing. This is because the ARM processor does not stack the Umode sp and cpsr automatically during SWI (system calls) and IRQ (interrupts) exceptions. Since both system calls and interrupts may trigger process switch, we must save the process Umode context manually. In addition to CPU registers, which are saved in the SVC or IRQ stack, we also save Umode sp and cpsr in the PROC structure. The fields pid, ppid, priority and status are obvious. In most large OS, each process is assigned a unique pid from a range of pid numbers. In EOS, we simply use the PROC index as the process pid, which simplifies the kernel code and also makes it easier for discussion. When a process terminates, it must wakeup the parent process if the latter is waiting for the child to terminate. In the PROC structure, the parent pointer points to the parent PROC. This allows the dying process to find its parent quickly. The event field is the event value when a process goes to sleep. The exitValue field is the exit status of a process. If a process terminates normally by an exit(value) syscall, the low byte of exitValue is the exit value. If it terminates abnormally by a signal, the high byte is the signal number. This allows the parent process to extract the exit status of a ZOMBIE child to determine whether it terminated normally or abnormally. The time field is the maximal timeslice of a process, and cpu is its CPU usage time. The timeslice determines how long can a process run, and the CPU usage time is used to compute the process scheduling priority. The pause field is for a process to sleep for a number of seconds. In EOS, process and thread PROCs are identical. The type field identifies whether a PROC is a PROCESS or THREAD. EOS is a uniprocessor (UP) system, in which only one process may run in kernel mode at a time. For process synchronization, it uses sleep/wakeup in process management and implementation of pipes, but it uses semaphores in device drivers and file system. When a process becomes blocked on a semaphore, the sem field points to the semaphore. This allows the kernel to unblock a process from a semaphore queue, if necessary. For example, when a process waits for inputs from a serial port, it is blocked in the serial port driver's input semaphore queue. A kill signal or an interrupt key should let the process continue. The sem pointer simplifies the unblocking operation. Each PROC has a res pointer pointing to a resource structure, which is

```
typedef struct pres{
  int      uid;
  int      gid;
  u32      paddress, psize;  // image size in KB
  u32      *pgdir;           // per proc level-1 page table pointer
  u32      *new_pgdir;       // new_pgdir during exec with new size
  MINODE *cwd;               // CWD
  char     name[32];         // executing program name
  char     tty[32];          // opened terminal /dev/ttyXX
  int      tcount;           // threads count in process
  u32      signal;           // 31 signals=bits 1 to 31
  int      sig[NSIG];        // 31 signal handlers
  OFT      *fd[NFD];         // open file descriptors
  struct semaphore mlock;    // message passing
  struct semaphore message;
  struct mbuf     *mqueue;
} PRES;
```

The PRES structure contains process specific information. It includes the process uid, gid, level-1 page table (pgdir) and image size, current working directory, terminal special file name, executing program name, signal and signal handlers, message queue and opened file descriptors, etc. In EOS, both PROC and PRES structures are statically allocated. If desired, they may be constructed dynamically. Processes and threads are independent execution units. Each process executes in a unique address space. All threads in a process execute in the same address space of the process. During system initialization, each PROCESS PROC is assigned a unique PRES structure pointed by the res pointer. A process is also the main thread of the process. When creating a new thread, its proc pointer points to the process PROC and its res pointer points to the same PRES structure of the process. Thus, all threads in a process share the same resources, such as opened file descriptors, signals and messages, etc. Some OS kernels allow individual threads to open files, which are private to the threads. In that case, each

PROC structure must have its own file descriptor array. Similarly for signals and messages, etc. In the PROC structure, kstack is a pointer to the process/thread kernel mode stack. In EOS, PROCs are managed as follows.

Free process and thread PROCs are maintained in separate free lists for allocation and deallocation. In EOS, which is a UP system, there is only one readyQueue for process scheduling. The kernel mode stack of the initial process P0 is statically allocated at 8 KB (0x2000). The kernel mode stack of every other PROC is dynamically allocated a (4 KB) page frame only when needed. When a process terminates, it becomes a ZOMBIE but retains its PROC structure, pgdir and the kstack, which are eventually deallocated by the parent process in kwait().

8.5.7 Assembly Code of EOS

The ts.s File: ts.s is the only kernel file in ARM assembly code. It consists of several logically separate parts. For easy of discussion and reference, we shall identified them as ts.s.1 to ts.s.5. In the following, we shall list the ts.s code and explain the functions of the various parts.

```
//--------------------- ts.s file --------------------------
   .text
.code 32
.global reset_handler
.global vectors_start, vectors_end
.global proc, procsize
.global tswitch, scheduler, running, goUmode
.global switchPgdir, mkPtable, get_cpsr, get_spsr
.global irq_tswitch, setulr
.global copy_vector, copyistack, irq_handler, vectorInt_init
.global int_on, int_off, lock, unlock
.global get_fault_status, get_fault_addr, get_spsr
```

8.5.7.1 Reset Handler

```
// -------------------- ts.s.1. ----------------------------
reset_handler:
// set SVC stack to HIGH END of proc[0].kstack[]
  ldr r0, =proc        // r0 points to proc's
  ldr r1, =procsize    // r1 -> procsize
  ldr r2,[r1, #0]      // r2 = procsize
  add r0, r0, r2       // r0 -> high end of proc[0]
  sub r0, #4           // r0 ->proc[0].kstack
  mov r1, #0x2000      // r1 = 8KB
  str r1, [r0]         // proc[0].kstack at 8KB
  mov sp, r1
  mov r4, r0           // r4 is a copy of r0, points PROC0's kstack top

// go in IRQ mode to set IRQ stack
  msr cpsr, #0xD2      // IRQ mode with IRQ and FIQ interrupts off
  ldr sp, =irq_stack // 4KB area defined in linker script t.ld
// go in FIQ mode to set FIQ stack
  msr cpsr, #0xD1
  ldr sp, =fiq_stack // set FIQ mode sp
// go in ABT mode to set ABT stack
  msr cpsr, #0xD7
  ldr sp, =abt_stack  // set ABT mode stack
// go in UND mode to set UND stack
```

```
  msr cpsr, #0xDB
  ldr sp, =und_stack // set UND mode stack
// go back in SVC mode
  msr cpsr, #0xD3
// set SVC mode spsr to USER mode with IRQ on
  msr spsr, #0x10     // write to previous mode spsr
```

ts.s.1 is the reset_handler, which begins execution in SVC mode with interrupts off and MMU disabled. First, it initializes proc[0]'s kstack pointer to 8 KB (0x2000) and sets the SVC mode stack pointer to the high end of proc[0].kstack. This makes proc[0]'s kstack the initial execution stack. Then it initializes the stack pointers of other privileged modes for exceptions processing. In order to run processes later in User mode, it sets the SPSR to User mode. Then it continues to execute the second part of the assembly code. In a real ARM based system, FIQ interrupt is usually reserved for urgent event, such as power failure, which can be used to trigger the OS kernel to save system information into non-volatile storage device for recovery later. Since most emulated ARM VMs do not have such a provision, EOS uses only IRQ interrupts but not the FIQ interrupt.

8.5.7.2 Initial Page Table

```
//---------------- ts.s.2 ---------------------------
// copy vector table to address 0
  bl copy_vector
// create initial pgdir and pgtable at 16KB
  bl mkPtable          // create pgdir and pgtable in C
  ldr r0, mtable
  mcr p15, 0, r0, c2, c0, 0   // set TTBR
  mcr p15, 0, r0, c8, c7, 0   // flush TLB
// set DOMAIN 0,1 : 01=CLIENT mode(check permission)
  mov r0,   #0x5              // b0101 for CLIENT
  mcr p15, 0, r0, c3, c0, 0
// enable MMU
  mrc p15, 0, r0, c1, c0, 0
  orr r0, r0, #0x00000001     // set bit0
  mcr p15, 0, r0, c1, c0, 0   // write to c1
  nop
  nop
  nop
  mrc p15, 0, r2, c2, c0
  mov r2, r2
// enable IRQ interrupts, then call main() in C
  mrs r0, cpsr
  bic r0, r0, #0xC0
  mrs cpsr, r0
  BL main                     // call main() directly
  B . // main() never return; in case it does, just loop here
mtable:  .word 0x4000         // initial pgdir at 16KB
```

ts.s.2: the second part of the assembly code performs three functions. First, it copies the vector table to address 0. Then it constructs an initial one-level page table to create an identity mapping of the low 258 MB VA to PA, which includes 256 MB RAM plus 2 MB I/O space beginning at 256 MB. The EOS kernel uses the KML memory mapping scheme, in which the kernel space is mapped to low VA addresses. The initial page table is built at 0x4000 (16 KB) by the mkPtable() function (in t.c file). It will be the page table of the initial process P0, which runs only in Kernel mode. After setting up the initial page table, it configures and enables the MMU for VA to PA address translation. Then it calls main() to continue kernel initialization in C.

8.5.7.3 System Call Entry and Exit

```
/******************** ts.s.3 ******************************/
// SVC (SWI) handler entry point
svc_entry: // syscall parameters are in r0-r3: do not touch
   stmfd sp!, {r0-r12, lr}
// access running PROC
   ldr r5, =running     // r5 = &running
   ldr r6, [r5, #0]     // r6 -> PROC of running
   mrs r7, spsr         // get spsr, which is Umode cpsr
   str r7, [r6, #16]    // save spsr into running->ucpsr
// to SYS mode to access Umode usp, upc
   mrs r7, cpsr         // r7 = SVC mode cpsr
   mov r8, r7           // save a copy in r8
   orr r7, r7, #0x1F    // r7 = SYS mode
   msr cpsr, r7         // change cpsr to SYS mode
// now in SYS mode, sp and lr same as User mode
   str sp, [r6, #8]     // save usp into running->usp
   str lr, [r6, #12]    // save upc into running->upc
// change back to SVC mode
   msr cpsr, r8
// save kmode sp into running->ksp at offest 4;
// used in fork() to copy parent's kstack to child's kstack
   str sp, [r6, #4]     // running->ksp = sp
// enable IRQ interrupts
   mrs r7, cpsr
   bic r7, r7, #0xC0    // I and F bits=0 enable IRQ,FIQ
   msr cpsr, r7
   bl svc_handler       // call SVC handler in C
// replace saved r0 on stack with the return value from svc_handler()
   add sp, sp, #4       // effectively pop saved r0 off stack
   stmfd sp!,{r0}       // push r as the saved r0 to Umode

goUmode:
// disable IRQ interrupts
   mrs r7, cpsr
   orr r7, r7, #0xC0    // I and F bits=1 mask out IRQ,FIQ
   msr cpsr, r7         // write to cpsr
   bl  kpsig            // handle outstanding signals
   bl  reschedule       // reschedule process
// access running PROC
   ldr r5, =running     // r5 = &running
   ldr r6, [r5, #0]     // r6 -> PROC of running
 // goto SYS mode to access user mode usp
   mrs r2, cpsr         // r2 = SVC mode cpsr
   mov r3, r2           // save a copy in r3
   orr r2, r2, #0x1F    // r2 = SYS mode
   msr cpsr, r2         // change to SYS mode
   ldr sp, [r6, #8]     // restore usp from running->usp
   msr cpsr, r3         // back to SVC mode
// replace pc in kstack with p->upc
   mov r3, sp
   add r3, r3, #52      // offset = 13*4 bytes from sp
   ldr r4, [r6, #12]
   str r4, [r3]
```

```
// return to running proc in Umode
   ldmfd sp!, {r0-r12, pc}^
```

8.5.7.4 IRQ Handler

```
// IRQ handler entry point
irq_handler:              // IRQ entry point
   sub lr, lr, #4
   stmfd sp!, {r0-r12, lr}  // save all Umode regs in IRQ stack

// may switch task at end of IRQ processing; save Umode info
   mrs r0, spsr
   and r0, #0x1F
   cmp r0, #0x10        // check whether was in Umode
   bne noUmode          // no need to save Umode context if NOT in Umode
// access running PROC
   ldr r5, =running     // r5=&running
   ldr r6, [r5, #0]     // r6 -> PROC of running
   mrs r7, spsr
   str r7, [r6, #16]
// to SYS mode to access Umode usp=r13 and cpsr
   mrs r7, cpsr         // r7 = SVC mode cpsr
   mov r8, r7           // save a copy of cpsr in r8
   orr r7, r7, #0x1F    // r7 = SYS mode
   msr cpsr, r7         // change cpsr to SYS mode
// now in SYS mode, r13 same as User mode sp r14=user mode lr
   str sp, [r6, #8]     // save usp into proc.usp at offset 8
   str lr, [r6, #12]    // save upc into proc.upc at offset 12
// change back to IRQ mode
   msr cpsr, r8
noUmode:
   bl irq_chandler      // call irq_handler() in C in SVC mode
// check mode
   mrs r0, spsr
   and r0, #0x1F
   cmp r0, #0x10        // check if in Umode
   bne kiret
// proc was in Umode when IRQ interrupt: handle signal, may tswitch
   bl kpsig
   bl reschedule        // re-schedule: may tswitch

// CURRENT running PROC return to Umode
   ldr r5, =running     // r5=&running
   ldr r6, [r5, #0]     // r6 -> PROC of running
// restore Umode.[sp,pc,cpsr] from saved PROC.[usp,upc,ucpsr]
   ldr r7, [r6, #16]    // r7 = saved Umode cpsr
// restore spsr to saved Umode cpsr
   msr spsr, r7
// go to SYS mode to access user mode sp
   mrs r7, cpsr         // r7 = SVC mode cpsr
   mov r8, r7           // save a copy of cpsr in r8
   orr r7, r7, #0x1F    // r7 = SYS mode
   msr cpsr, r7         // change cpsr to SYS mode
```

```
// now in SYS mode; restore Umode usp
   ldr sp, [r6, #8]     //  set usp in Umode = running->usp
// back to IRQ mode
   msr cpsr, r8        // go back to IRQ mode
kiret:
   ldmfd sp!, {r0-r12, pc}^ // return to Umode
```

ts.s.3: The third part of the assembly code contains the entry points of SWI (SVC) and IRQ exception handlers. Both the SVC and IRQ handlers are quite unique due to the different operating modes of the ARM processor architecture. So we shall explain them in more detail.

System Call Entry: svc_entry is the entry point of SWI exception handler, which is used for system calls to the EOS kernel. Upon entry, it first saves the process (Umode) context in the process Kmode (SVC mode) stack. System call parameters (a,b,c,d) are passed in registers r0–r3, which should not be altered. So the code only uses registers r4–r10. First, it lets r6 point to the process PROC structure. Then it saves the current spsr, which is the Umode cpsr, into PROC.ucpsr. Then it changes to SYS mode to access Umode registers. It saves the Umode sp and pc into PROC.usp and PROC.upc, respectively. Thus, during a system call, the process Umode context is saved as follows.

Umode registers [r0 – r12, r14] saved in PROC.kstack
Umode [sp, pc, cpsr] saved in PROC.[usp, upc, ucpsr]

In addition, it also saves Kmode sp in PROC.ksp, which is used to copy the parent kstack to child during fork(). Then it enables IRQ interrupts and calls svc_chandler() to process the system call. Each syscall (except kexit) returns a value, which replaces the saved r0 in kstack as the return value back to Umode.

System Call Exit: goUmode is the syscall exit code. It lets the current running process, which may or may not be the original process that did the syscall, return to Umode. First, it disables IRQ interrupts to ensure that the entire goUmode code is executed in a critical section. Then it lets the current running process check and handle any outstanding signals. Signal handling in the ARM architecture is also quite unique, which will be explained later. If the process survives the signal, it calls reschedule() to re-schedule processes, which may switch process if the sw_flag is set, meaning that there are processes in the readyQueue with higher priority. Then the current running process restores [usp, upc, cpsr] from its PROC structure and returns to Umode by

ldmfd sp!, {r0–r12, pc}^

Upon return to Umode, r0 contains the return value of the syscall.

IRQ Entry: irq_handler is the entry point of IRQ interrupts. Unlike syscalls, which can only originate from Umode, IRQ interrupts may occur in either Umode or Kmode. EOS is a uniprocessor OS. The EOS kernel is non-preemptive, which means it does not switch process while in kernel mode. However, it may switch process if the interrupted process was running in Umode. This is necessary in order to support process scheduling by time-slice and dynamic priority. Task switching in ARM IRQ mode poses a unique problem, which we shall elaborate shortly. Upon entry to irq_handler, it first saves the context of the interrupted process in the IRQ mode stack. Then it checks whether the interrupt occurred in Umode. If so, it also saves the Umode [usp, upc, cpsr] into the PROC structure. Then it calls irq_chandler() to process the interrupt. The timer interrupt handler may set the switch process flag, sw_flag, if the time-slice of the current running process has expired. Similarly, a device interrupt handler may also set the sw_flag if it wakes up or unblocks processes with higher priority. At the end of interrupt processing, if the interrupt occurred in Kmode or the sw_flag is off, there should be no task switch, so the process returns normally to the originally point of interruption. However, if the interrupt occurred in Umode and the sw_flag is set, the kernel switches process to run the process with the highest priority.

8.5.7.5 IRQ and Process Preemption

Unlike syscalls, which always use the process kstack in SVC mode, task switch in IRQ mode is complicated by the fact that, while interrupt processing uses the IRQ stack, task switch must be done in SVC mode, which uses the process kstack. In this case, we must perform the following operations manually.

(1). Transfer the INTERRUPT stack frame from IRQ stack to process (SVC) kstack;
(2). Flatten out the IRQ stack to prevent it from overflowing;
(3). Set SVC mode stack pointer to the INTERRUPT stack frame in the process kstack.
(4). Switch to SVC mode and call tswitch() to give up CPU, which pushes a RESUME stack frame onto the process kstack pointed by the saved PROC.ksp.
(5). When the process regains CPU, it resumes in SVC mode by the RESUME stack frame and returns to where it called tswitch() earlier.
(6). Restore Umode [usp, cpsr] from saved [usp, ucpsr] in PROC structure
(7). Return to Umode by the INTERRUPT stack frame in kstack.

Task switch in IRQ mode is implemented in the code segment irq_tswitch(), which can be best explained by the following diagrams.

(1) Copy INTERRUPT frame from IRQ stack to SVC stack, set SVC_sp,

```
      IRQ_stack                          SVC_stack
 --------------------        copy   --------------------
   |ulr|ur12 - ur0|          ===>   |ulr|ur12 - ur0|
 --------------------               --------------|-----
   |-- INT frame -|                           SVC_sp
```

(2). Call tswitch() to give up CPU, which pushes a RESUME frame onto the SVC stack and sets the saved PROC.ksp pointing to the resume stack frame.

```
              |-- INT frame -|-RESUME frame-|
              ------------------------------
SVC_stack =|ulr|ur12 - ur0|klr|kr12 - kr0|
              ----------------------------|---------
                                     SVC_sp = PROC.ksp
```

(3). When the process is scheduled to run, it resumes in SVC mode and returns to

```
     here: // return from tswitch()
    restore Umode.[usp,cpsr] from PROC.[usp,ucpsr]
     ldmfd sp, {r0-r12, pc}^  // return to Umode
// ---------------- ts.s.4 --------------------------
tswitch:                 // tswitch() in Kmode
   mrs r0, cpsr          // disable interrupts
   orr r0, r0, #0xC0     // I and F bits=1: mask out IRQ, FIQ
   mrs cpsr, r0          // I and F interrupts are disabled
   stmfd sp!, {r0-r12, lr} // save context in kstack
   ldr r0, =running      // r0=&running; access running->PROC
   ldr r1, [r0, #0]      // r1->running PROC
   str sp, [r1, #4]      // running->ksp = sp
   bl  scheduler         // call scheduler() to pick next running
   ldr r0, =running      // resume CURRENT running
   ldr r1, [r0, #0]      // r1->runningPROC
   ldr sp, [r1, #4]      // sp = running->ksp
   mrs r0, cpsr          // disable interrupts
```

```
    bic r0, r0, #0xC0   // enable I and F interrupts
    mrs cpsr, r0
    ldmfd sp!, {r0-r12, pc}
irq_tswitch:          // irq_tswitch: task switch in IRQ mode
    mov r0, sp        // r0 = IRQ mode current sp
    bl copyistack     // transfer INT frame from IRQ stack to SVC stack
    mrs r7, spsr      // r7 = IRQ mode spsr, which must be Umode cpsr
// flatten out irq stack
    ldr sp, =irq_stack_top
// change to SVC mode
    mrs r0, cpsr
    bic r1, r0, #0x1F  // r1 = r0 = cspr's lowest 5 bits cleared to 0
    orr r1, r1, #0x13  // OR in 0x13=10011 = SVC mode
    msr cpsr, r1       // write to cspr, so in SVC mode now
    ldr r5, =running   // r5 = &running
    ldr r6, [r5, #0]   // r6 -> PROC of running
// svc stack already has an irq frame, set SVC sp to kstack[-14]
    ldr sp, [r6, #4]   // SVC mode sp= &running->kstack[SSIZE-14]
    bl tswitch         // switch task in SVC mode
    ldr r5, =running   // r5=&running
    ldr r6, [r5, #0]   // r6 -> PROC of running
    ldr r7, [r6, #16]  // r7 = saved Umode cpsr
// restore spsr to saved Umode cpsr
    msr spsr, r7
// go to SYS mode to access user mode sp
    mrs r7, cpsr       // r7 = SVC mode cpsr
    mov r8, r7         // save a copy of cpsr in r8
    orr r7, r7, #0x1F  // r7 = SYS mode
    msr cpsr, r7       // change cpsr to SYS mode
// now in SYS mode; restore Umode usp
    ldr sp, [r6, #8]   // restore usp
    ldr lr, [r6, #12]  // restore upc; REALLY need this?
// back to SVC mode
    msr cpsr, r8       // go back to IRQ mode
    ldmfd sp!, {r0-r12, pc}^ // return via INT frame in SVC stack
switchPgdir: // switch pgdir to new PROC's pgdir; passed in r0
// r0 contains new PROC's pgdir address
   mcr p15, 0, r0, c2, c0, 0    // set TTBase
   mov r1, #0
   mcr p15, 0, r1, c8, c7, 0    // flush TLB
   mcr p15, 0, r1, c7, c10, 0   // flush cache
   mrc p15, 0, r2, c2, c0, 0
// set domain: all 01=client(check permission)
   mov r0, #0x5                 //01|01 for CLIENT|client
   mcr p15, 0, r0, c3, c0, 0
   mov pc, lr                   // return
```

ts.s.4: The fourth part of the assembly code implements task switching. It consists of three functions. tswitch() it for task switch in Kmode, irq_tswitch() is for task switch in IRQ mode and switchPgdir() is for switching process pgdir during task switch. Since all these functions are already explained before, we shall not repeat them here.

```
//-------------ts.s.5 --------------------------
// IRQ interrupt mask/unmask functions
int_on:               // int int_on(int cpsr)
   msr cpsr, r0
   mov pc,lr
```

```
int_off:              // int cpsr = int_off();
  mrs r4, cpsr
  mov r0, r4
  orr r4, r4, #0x80   // set bit means MASK off IRQ interrupt
  msr cpsr, r4
  mov pc,lr
unlock:               // enable IRQ directly
  mrs r4, cpsr
  bic r4, r4, #0x80   // clear bit means UNMASK IRQ interrupt
  msr cpsr, r4
  mov pc,lr
lock:                 // disable IRQ directly
  mrs r4, cpsr
  orr r4, r4, #0x80   // set bit means MASK off IRQ interrupt
  msr cpsr, r4
  mov pc,lr
get_cpsr:
  mrs r0, cpsr
  mov pc, lr
get_spsr:
  mrs r0, spsr
 mov pc, lr
setulr: // setulr(oldPC): set Umode lr=oldPC for signal catcher()
  mrs r7, cpsr    // to SYS mode
  mov r8, r7      // save cpsr in r8
  orr r7, #0x1F  //
  msr cpsr, r7
// in SYS mode now
  mov lr, r0      // set Umode lr to oldPC
  msr cpsr, r8    // back to original mode
  mov pc, lr      // return
vectors_start:
  LDR PC, reset_handler_addr
  LDR PC, undef_handler_addr
  LDR PC, svc_handler_addr
  LDR PC, prefetch_abort_handler_addr
  LDR PC, data_abort_handler_addr
  B .
  LDR PC, irq_handler_addr
  LDR PC, fiq_handler_addr
reset_handler_addr:          .word reset_handler
undef_handler_addr:          .word undef_abort_handler
svc_handler_addr:            .word svc_entry
prefetch_abort_handler_addr: .word prefetch_abort_handler
data_abort_handler_addr:     .word data_abort_handler
irq_handler_addr:            .word irq_handler
fiq_handler_addr:            .word fiq_handler
vectors_end:
// end of ts.s file
```

The last part of the assembly code implements various utility functions, such as lock/unlock, int_off/int_on, and getting CPU status register, etc. Note the difference between lock/unlock and int_off/int_on. Whereas lock/unlock disable/enable IRQ interrupts unconditionally, int_off disables IRQ interrupts but returns the original CPSR, which is restored in int_on.

These are necessary in device interrupt handlers, which run with interrupts disabled but may issue V operation on semaphores to unblock processes.

8.5.8 Kernel Files of EOS

Part 2. t.c file: The t.c file contains the main() function, which is called from reset_handler when the system starts.

8.5.8.1 The main() Function

The main() function consists of the following steps

(1). Initialize the LCD display driver to make printf() work.
(2). Initialize block device I/O buffers for file I/O on SDC.
(3). Configure VIC, SIC interrupt controllers and devices for vectored interrupts.
(4). Initialize device drivers and start timer.
(5). Call kernel_init() to initialize kernel data structures. Create and run the initial process P0. Construct pgdirs and pgtables for processes. Construct free page frame list for dynamic paging. Switch pgdir to use two-level dynamic paging.
(6). Call fs_init() to initialize file system and mount the root file system.
(7). Create the INIT process P1 and load /bin/init as its Umode image.
(8). P0 switches task to run the INIT process P1.

P1 forks login processes on console and serial terminals for users to login. Then it waits for any ZOMBIE children, which include the login processes as well as any orphans, e.g. in multi-stage pipes. When the login processes start up, the system is ready for use.

```
/*********************** t.c file ****************************/
#include "../type.h"
int main()
{
   fbuf_init();           //initialize LCD frame buffer: driver/vid.c
   printf("Welcome to WANIX in Arm\n");
   binit();               // I/O buffers: fs/buffer.c
   vectorInt_init();      // Vectored Interrupts: driver/int.c
   irq_init();            // Configure VIC,SIC,deviceIRQs: driver/int.c
   kbd_init();            // Initialize KBB driver: driver/kbd.c
   uart_init();           // initialize UARTs:      driver/uart.c
   timer_init();          // initialize timer:      driver/timer.c
   timer_start(0);        // start timer0           driver/timer.c
   sdc_init();            // initialize SDC driver: driver/sdc.c
   kernel_init();         // initialize kernel structs:kernel/kernel.c
   fs_init();             // initialize FS and mount root file system
   kfork("/bin/init"); // create INIT proc P1: kernel/fork.c
   printf("P0 switch to P1\n");
   while(1){              // P0 code
     while(!readyQueue); // loop if no runnable procs
     tswitch();           // P0 switch task if readyQueue non-empty
   }
}
```

8.5.8.2 Kernel Initialization

kernel_init() function: The kernel_init() function consists of the following steps.

(1). Initialize kernel data structures. These include free PROC lists, a readyQueue for process scheduling and a FIFO sleepList containing SLEEP processes.
(2). Create and run the initial process P0, which runs in Kmode with the lowest priority 0. P0 is also the idle process, which runs if there are no other runnable processes, i.e. when all other processes are sleeping or blocked. When P0 resumes, it executes a busy waiting loop until the readyQueue is non-empty. Then it switches process to run a ready process with the highest priority. Instead of a busy waiting loop, P0 may put the CPU in a power-saving WFI state with interrupts enabled. After processing an interrupt, it tries to run a ready process again, etc.
(3). Construct a Kmode pgdir at 32 KB and 258 level-2 page table at 5 MB. Construct level-1 pgdirs for (64) processes in the area of 6 MB and their associated level-2 page tables in 7 MB. Details of the pgdirs and page tables will be explained in the next section on memory management.
(4). Switch pgdir to the new level-1 pgdir at 32 KB to use 2-level paging.
(5). Construct a pfreeList containing free page frames from 8 MB to 256 MB, and implement palloc()/pdealloc() functions to support dynamic paging.
(6). Initialize pipes and message buffers in kernel.
(7). Return to main(), which calls fs_init() to initialize the file system and mount the root file system. Then it creates and run the INIT process P1.

```
/********************** kernel.c file **********************/
#include "../type.h"
PROC proc[NPROC+NTHREAD];
PRES pres[NPROC];
PROC *freelist, *tfreeList, *readyQueue, *sleepList, *running;;
int sw_flag;
int procsize = sizeof(PROC);
OFT  oft[NOFT];
PIPE pipe[NPIPE];

int kernel_init()
{
  int i, j;
  PROC *p; char *cp;
  printf("kernel_init()\n");
  for (i=0; i<NPROC; i++){ // initialize PROCs in freeList
    p = &proc[i];
    p->pid = i;
    p->status = FREE;
    p->priority = 0;
    p->ppid = 0;
    p->res = &pres[i];    // res point to pres[i]
    p->next = p + 1;
    // proc[i]'s umode pgdir and pagetable are at 6MB + pid*16KB
    p->res->pgdir = (int *)(0x600000 + (p->pid-1)*0x4000);
  }
  proc[NPROC-1].next = 0;
  freeList = &proc[0];
  // similar code to initialize tfreeList for NTHREAD procs
  readyQueue = 0;
  sleepList = 0;
  // create P0 as the initial running process;
  p = running = get_proc(&freeList);
```

```
    p->status = READY;
    p->res->uid = p->res->gid = 0;
    p->res->signal = 0;
    p->res->name[0] = 0;
    p->time = 10000;         // arbitrary since P0 has no time limit
    p->res->pgdir = (int *)0x8000;  // P0's pgdir at 32KB
    for (i=0; i<NFD; i++)   // clear file descriptor array
        p->res->fd[i] = 0;
    for (i=0; i<NSIG; i++) // clear signals
        p->res->sig[i] = 0;
    build_ptable();         // in mem.c file
    printf("switch pgdir to use 2-level paging : ");
    switchPgdir(0x8000);

    // build pfreelist: free page frames begin from 8MB end = 256MB
    pfreeList = free_page_list((int *)0x00800000, (int *)0x10000000);
    pipe_init();        // initialize pipes in kernel
    mbuf_init();        // initialize message buffers in kernel
}
```

8.5.8.3 Process Scheduling Functions

```
int scheduler()
{
   PROC *old = running;
   if (running->pid == 0 && running->status == BLOCK){// P0 only
      unlock();
      while(!readyQueue);
      return;
   }
   if (running->status==READY)
      enqueue(&readyQueue, running);
   running = dequeue(&readyQueue);
   if (running != old){
      switchPgdir((int)running->res->pgdir);
   }
   running->time = 10; // time slice = 10 ticks;
   sw_flag = 0;         // turn off switch task flag
}
int schedule(PROC *p)
{
   if (p->status ==READY)
      enqueue(&readyQueue, p);
   if (p->priority > running->priority)
      sw_flag = 1;
}
int reschedule()
{
  if (sw_flag)
    tswitch();
}
```

The remaining functions in t.c include scheduler(), schedule() and reschedule(), which are parts of the process scheduler in the EOS kernel. In the scheduler() function, the first few lines of code apply only to the initial process P0. When the system starts up, P0 executes mount_root() to mount the root file system. It uses an I/O buffer to read the SDC, which causes P0 to block on the I/O buffer until the read operation completes. Since there is no other process yet, P0 can not switch process when it becomes blocked. So it busily waits until the SDC interrupt handler executes V to unblock it. Alternatively, we may modify the SDC driver to use polling during system startup, and switch to interrupt-driven mode after P0 has created P1. The disadvantage is that it would make the SDC driver less efficient since it has to check a flag on every read operation.

8.5.9 Process Management Functions

8.5.9.1 fork-exec

EOS supports dynamic process creation by fork, which creates a child process with an identical Umode image as the parent. It allows process to change images by exec. In addition, it also supports threads within the same process. These are implemented in the following files.

fork.c file: this file contains fork1(), kfork(), fork() and vfork(). **fork1()** is the common code of all other fork functions. It creates a new proc with a pgdir and pgtables. **kfork()** is used only by P0 to create the INIT proc P1. It loads the Umode image file (/bin/init) of P1 and initializes P1's kstack to make it ready to run in Umode. **fork()** creates a child process with an identical Umode image as the parent. **vfork()** is the same as fork() but without copying images.

exec.c file: this file contains **kexec(),** which allows a process to change Umode image to a different executable file and pass command-line parameters to the new image.

threads.c file: this file implements threads in a process and threads synchronization by mutexes.

8.5.9.2 exit-wait

The EOS kernel uses sleep/wakeup for process synchronization in process management and also in pipes. Process management is implemented in the wait.c file, which contains the following functions.

ksleep(): process goes to sleep on an event. Sleeping PROCs are maintained in a FIFO sleepList for waking up in order
kwakeup(): wakeup all PROCs that are sleeping on an event
kexit(): process termination in kernel
kwait(): wait for a ZOMBIE child process, return its pid and exit status

8.5.10 Pipes

The EOS kernel supports pipes between related processes. A pipe is a structure consisting of the following fields.

```
typedef struct pipe{
  char  *buf; // data buffer: dynamically allocated page frame;
  int   head, tail;         // buffer index
  int   data, room;         // counters for synchronization
  int   nreader, nwriter;   // number of READER,WRITER on pipe
  int   busy;               // status of pipe
}PIPE;
```

The syscall int r = pipe(int pd[]);

creates a pipe in kernel and returns two file descriptors in pd[2], where pd[0] is for reading from the pipe and pd[1] is for writing to the pipe. The pipe's data buffer is a dynamically allocated 4 KB page, which will be released when the pipe is deallocated. After creating a pipe, the process typically forks a child process to share the pipe, i.e. both the parent and the child have the same pipe descriptors pd[0] and pd[1]. However, on the same pipe each process must be either a READER or a WRITER, but not both. So, one of the processes is chosen as the pipe WRITER and the other one as the pipe READER. The pipe WRITER must closes its pd[0], redirects its stdout (fd = 1) to pd[1], so that its stdout is connected to the write end of the pipe. The pipe READER must closes its pd[1] and redirects its stdin (fd = 0) to pd[0], so that its stdin is connected to the read end of the pipe. After these, the two processes are connected by the pipe. READER and WRITER processes on the same pipe are synchronized by sleep/wakeup. Pipe read/write functions are implemented in the pipe.c file. Closing pipe descriptor functions are implemented in the open_close.c file of the file system. A pipe is deallocated when all the file descriptors on the pipe are closed. For more information on the implementation of pipes, the reader may consult (Chap. 6.14, Wang 2015) or the pipe.c file for details.

8.5.11 Message Passing

In addition to pipes, the EOS kernel supports inter-process communication by message passing. The message passing mechanism consists of the following components.

(1). A set of NPROC message buffers (MBUFs) in kernel space.
(2). Each process has a message queue in PROC.res.mqueue, which contains messages sent to but not yet received the process. Messages in the message queue are ordered by priority.
(3). send(char *msg, int pid): send a message to a target process by pid.
(4). recv(char *msg): receive a message form proc's message queue.

In EOS, message passing is synchronous. A sending process waits if there are no free message buffers. A receiving process waits if there are no messages in its message queue. Process synchronization in send/recv is by semaphores. The following lists the mes.c file.

```
/************** mes.c file: Message Passing ************/
#include "../type.h"
/******** message buffer type in type.h ********
typedef struct mbuf{
  struct mbuf *next;   // next mbuf pointer
  int sender;          // sender pid
  int priority;        // message priority
  char text[128];      // message contents
} MBUF;
**************************************************/
MBUF mbuf[NMBUF], *freeMbuflist;   // free mbufs; NMBUF=NPROC
SEMAPHORE mlock; // semaphore for exclusive access to mbuf[ ]
int mbuf_init()
{
  int i; MBUF *mp;
  printf("mbuf_init\n");
  for (i=0; i<NMBUF; i++){ // initialize mbufs
      mp = &mbuf[i];
      mp->next = mp+1;
      mp->priority = 1;    // for enqueue()/dequeue()
  }
  freeMbuflist = &mbuf[0];
  mbuf[NMBUF-1].next = 0;
  mlock.value = 1; mlock.queue = 0;
}
MBUF *get_mbuf()              // allocate a mbuf
{
  MBUF *mp;
  P(&mlock);
  mp = freeMbuflist;
  if (mp)
     freeMbuflist = mp->next;
  V(&mlock);
  return mp;
}
```

```
int put_mbuf(MBUF *mp)      // release a mbuf
{
  mp->text[0] = 0;
  P(&mlock);
    mp->next = freeMbuflist;
    freeMbuflist = mp;
  V(&mlock);
}
int ksend(char *msg, int pid)    // send message to pid
{
  MBUF *mp; PROC *p;
  // validate receiver pid
  if ( pid <= 0 || pid >= NPROC){
     printf("sendMsg : invalid target pid %d\n", pid);
     return -1;
  }
  p = &proc[pid];
  if (p->status == FREE || p->status == ZOMBIE){
     printf("invalid target proc %d\n", pid);
     return -1;
  }
  mp = get_mbuf();
  if (mp==0){
     printf("no more mbuf\n");
     return -1;
  }
  mp->sender = running->pid;
  strcpy(mp->text, msg);  // copy text from Umode to mbuf
  // deliver mp to receiver's message queue
  P(&p->res->mlock);
    enqueue(&p->res->mqueue, mp);
  V(&p->res->mlock);
  V(&p->res->message);     // notify receiver
  return 1;
}
int krecv(char *msg)       // receive message from OWN mqueue
{
  MBUF *mp;
  P(&running->res->message);  // wait for message
  P(&running->res->mlock);
    mp = (MBUF *)dequeue(&running->res->mqueue);
  V(&running->res->mlock);
  if (mp){                       // only if it has message
     strcpy(msg, mp->text);      // copy message contents to Umode
     put_mbuf(mp);               // release mbuf
     return 1;
  }
  return -1; // if proc was killed by signal => no message
}
```

8.5.12 Demonstration of Message Passing

In the USER directory, the programs, send.c and recv.c, are used to demonstrate the message passing capability of EOS. The reader may test send/recv messages as follows.

(1). login to the console. Enter the command line recv &. The sh process forks a child to run the recv command but does not wait for the recv process to terminate, so that the user may continue to enter commands. Since there are no messages yet, the recv process will be blocked on its message queue in kernel.
(2). Run the send command. Enter the receiving proc's pid and a text string, which will be sent to the recv process, allowing it to continue. Alternatively, the reader may also login from a different terminal to run the send command.

8.6 Memory Management in EOS

8.6.1 Memory Map of EOS

The following shows the memory map of the EOS system.

```
------------------ Memory Map of EOS --------------------
    0-2MB  :  EOS Kernel
  2MB-4MB  :  LCD display frame buffer
  4MB-5MB  :  Data area of 256 I/O buffers
  5MB-6MB  :  Kmode level-2 page tables; 258 (1KB) pgtables
  6MB-7MB  :  pgdirs for (64) processes, each pgdir=16KB
  7MB-8MB  :  unused; for expansion, e.g. to 128 PROC pgdirs
 8MB-256MB: free page frames for dynamic paging
 256-258MB: 2MB I/O space
---------------------------------------------------------
```

The EOS kernel code and data structures occupy the lowest 2 MB of physical memory. The memory area from 2 to 8 MB are used by the EOS kernel as LCD display buffer, I/O buffers, level-1 and level-2 page tables of processes, etc., The memory area from 8 to 256 MB is free. Free page frames from 8 to 256 MB are maintained in a pfreeList for allocation/dealloction of page frames dynamically.

8.6.2 Virtual Address Spaces

EOS uses the KML virtual address space mapping scheme, in which the kernel space is mapped to low Virtual Address (VA), and User mode space is mapped to high VA. When the system starts, the Memory Management Unit (MMU) is off, so that every address is a real or physical address. Since the EOS kernel is compile-linked with real addresses, it can execute the kernel's C code directly. First, it sets up an initial one-level page table at 16 KB to create an identity mapping of VA to PA and enables the MMU for VA to PA translation.

8.6.3 Kernel Mode Pgdir and Page Tables

In reset_handler, after initializing the stack pointers of the various privileged modes for exception processing, it constructs a new pgdir at 32 KB and the associated level-2 page tables in 5 MB. The low 258 entries of the new pgdir point to their level-2 page tables at 5 MB+i*1 KB (0<=i<258). Each page table contains 256 entries, each pointing to a 4 KB page frame in memory. All other entries of the pgdir are 0's. In the new pgdir, entries 2048–4095 are for User mode VA space. Since the high 2048 entries are all 0's, the pgdir is good only for the 258 MB kernel VA space. It will be the pgdir of the initial process P0, which runs only in Kmode. It is also the prototype of all other pgdirs since their Kmode entries are all identical. Then it switches to the new pgdir to use 2-level paging in Kmode.

8.6.4 Process User Mode Page Tables

Each process has a pgdir at 6 MB+pid*16 KB. The low 258 entries of all pgdirs are identical since their Kmode VA spaces are the same. The number of pgdir entries for Umode VA depends on the Umode image size, which in turn depends on the executable image file size. For simplicity, we set the Umode image size, USZIE, to 4 MB, which is big enough for all the Umode programs used for testing and demonstration. The Umode pgdir and page tables of a process are set up only when the process is created. When creating a new process in fork1(), we compute the number of Umode page tables needed as npgdir = USIZE/1 MB. The Umode pgdir entries point to npgdir dynamically allocated page frames. Each page table uses only the low 1 KB space of the (4 KB) page frame. The attributes of Umode pgdir entries are set to 0x31 for domain 1. In the Domain Access Control register, the access bits of both domains 0 and 1 are set to b01 for client mode, which checks the Access Permission (AP) bits of the page table entries. Each Umode page table contains pointers to 256 dynamically allocated page frames. The attributes of the page table entries are set to 0xFFE for AP = 11 for all the (1 KB) subpages within each page to allow R|W access in User mode.

8.6.5 Switch Pgdir During Process Switch

During process switch, we switch pgdir from the current process to that of the next process and flush the TLB and I and D buffer caches. This is implemented by the switchPgdir() function in ts.s.

8.6.6 Dynamic Paging

In the mem.c file, the functions free_page_list(), palloc() and pdealloc() implement dynamic paging. When the system starts, we build a pfreeList, which threads all the free page frames from 8 to 256 MB in a link list. During system operation, palloc() allocates a free page frame from pfreeList, and pdealloc() releases a page frame back to pfreeList for reuse. The following shows the mem.c file.

```
/***************** mem.c file: Dynamic Paging ***************/
int *pfreeList, *last;
int mkPtable()  // called from ts.s, create initial ptable at 16KB
{
  int i;
  int *ut = (int *)0x4000; // at 16KB
  u32 entry = 0 | 0x41E;  // AP=01(Kmode R|W; Umode NO) domaian=0
  for (i=0; i<4096; i++)   // clear 4096 entries to 0
    ut[i] = 0;
  for (i=0; i<258; i++){  // fill in low 258 entries ID map to PA
    ut[i] = entry;
    entry += 0x100000;
  }
}
int *palloc()                 // allocate a page frame
{
  int *p = pfreeList;
  if (p)
     pfreeList = (int *)*p;
  return p;
}
```

```
void pdealloc(int *p)        // deallocate a page frame
{
  *last = (int)(*p);
  *p = 0;
  last = p;
}
// build pfreeList of free page frames
int *free_page_list(int *startva, int *endva)
{
  int *p;
  printf("build pfreeList: start=%x end=%x : ", startva, endva);
  pfreeList = startva;
  p = startva;
  while(p < (int *)(endva-1024)){
    *p = (int)(p + 1024);
     p += 1024;
  }
  last = p;
  *p = 0;
  return startva;
}
int build_ptable()
{
   int *mtable = (int *)0x8000; // new pgdir at 32KB
   int i, j, *pgdir, paddr;
   printf("build Kmode pgdir at 32KB\n");
   for (i=0; i<4096; i++){      //zero out mtable[ ]
     mtable[i] = 0;
   }
   printf("build Kmode pgtables in 5MB\n");
   for (i=0; i<258; i++){      // point to 258 pgtables in 5MB
     pgtable = (int *)(0x500000 + i*1024);
     mtable[i] = (int)pgtable | 0x11; // 1KB entry in 5MB
     paddr = i*0x100000 | 0x55E;       // AP=01010101 CB=11 type=10
     for (j=0; j<256; j++){
         pgtable[j] = paddr + j*4096; // inc by 4KB
     }
   }
  printf("build 64 proc pgdirs at 6MB\n");
  for (i=0; i<64; i++){
     pgdir = (int *)(0x600000 + i*0x4000); // 16KB each
     for (j=0; j<4096; j++){ // zero out pgdir[ ]
         pgdir[j] = 0;
     }
     for (j=0; j<258; j++){  // copy low 258 entries from mtable[]
        pgdir[j] = mtable[j];
     }
  }
}
```

8.7 Exception and Signal Processing

During system operation, the ARM processor recognizes six types of exceptions, which are FIQ, IRQ, SWI, data_abort, prefecth_abort and undefined exceptions. Among these, FIQ and IRQ are for interrupts and SWI is for system calls. So the only true exceptions are data_abort, prefeth_abort and undefined exceptions, which occur under the following circumstances.

A **data_abort** event occurs when the memory controller or MMU indicates that an invalid memory address has been accessed. Example: attempt to access invalid VA.

A **prefetch_abort** event occurs when an attempt to load an instruction results in a memory fault. Example: if 0x1000 is outside of the VA range, then BL 0x1000 would cause a prefetch abort at the next instruction address 0x1004.

An **undefined** (instruction) event occurs when a fetched and decoded instruction is not in the ARM instruction set and none of the coprocessors claims the instruction.

In all Unix-like systems, exceptions are converted to signals, which are handled as follows.

8.7.1 Signal Processing in Unix/Linux

(1). Signals in Process PROC: Each PROC has a 32-bit vector, which records signals sent to a process. In the bit vector, each bit (except bit 0) represents a signal number. A signal n is present if bit n of the bit vector is 1. In addition, it also has a MASK bit-vector for masking out the corresponding signals. A set of syscalls, such as sigmask, sigsetmask, siggetmask, sigblock, etc. can be used to set, clear and examine the MASK bit-vector. A pending signal becomes effective only if it is not masked out. This allows a process to defer processing masked out signals, similar to CPU masking out certain interrupts.
(2). Signal Handlers: Each process PROC has a signal handler array, int sig[32]. Each entry of the sig[32] array specifies how to handle a corresponding signal, where 0 means DEFault, 1 means IGNore, other nonzero value means by a preinstalled signal catcher (handler) function in Umode.
(3). Trap Errors and signals: When a process encounters an exception, it traps to the exception handler in the OS kernel. The trap handler converts the exception cause to a signal number and delivers the signal to the current running process. If the exception occurs in kernel mode, which must be due to hardware error or, most likely, bugs in the kernel code, there is nothing the process can do. So it simply prints a PANIC error message and stops. Hopefully the problem can be traced and fixed in the next kernel release. If the exception occurs in User mode, the process handles the signal by the signal handler function in its sig[] array. For most signals, the default action of a process is to terminate, with an optional memory dump for debugging. A process may replace the default action with IGNore(1) or a signal catcher, allowing it to either ignore the signal or handle it in User mode.
(4). Change Signal Handlers: A process may use the syscall

int r = signal(int signal_number, void *handler);

to change the handler function of a selected signal number except SIGKILL(9) and SIGSTOP(19). Signal 9 is reserved as the last resort to kill a run-away process, and signal 19 allows a process to stop a child process during debugging. The installed handler, if not 0 or 1, must be the entry address of a function in User space of the form

void catcher(int signal_number){............}

(5). Signal Processing: A process checks and handles signals whenever it is in Kmode. For each outstanding signal number n, the process first clears the signal. It takes the default action if sig[n] = 0, which normally causes the process to terminate. It ignores the signal if sig[n] = 1. If the process has a pre-installed catcher function for the signal, it fetches the catcher's address and resets the installed catcher to DEFault (0). Then it manipulates the return path in such a way that it returns to execute the catcher function in Umode, passing as parameter the signal number. When the catcher function finishes, it returns to the original point of interruption, i.e. to the place from where it lastly entered Kmode. Thus, the process takes a detour to execute the catcher function first. Then it resumes normal execution.
(6). Reset user installed signal catchers: User installed catcher functions are intended to deal with trap errors in user program code. Since the catcher function is also executed in Umode, it may commit the same kind of trap error again. If so, the process would end up in an infinite loop, jumping between Umode and Kmode forever. To prevent this, the process typically resets the handler to DEFault (0) before executing the catcher function. This implies that a user installed catcher function is valid for only one occurrence of the signal. To catch another occurrence of the same signal, the Umode program

must install the catcher again. However, the treatment of user installed signal catchers is not uniform as it varies across different versions of Unix. For instance, in BSD the signal handler is not reset but the same signal is blocked while executing the signal catcher. Interested readers may consult the man pages of signal and sigaction of Linux for more details.
(7). Inter-Process Signals: In addition to handling exceptions, signals may also be used for inter-process communication. A process may use the syscall

$$\textbf{int r} = \textbf{kill(pid, signal\_number)};$$

to send a signal to another process identified by pid, causing the latter to execute a pre-installed catcher function in Umode. A common usage of the kill operation is to request the targeted process to terminate, thus the (somewhat misleading) term kill. In general, only related processes, e.g. those with the same uid, may send signals to each other. However, a superuser process (uid=0) may send signals to any process. The kill syscall may use an invalid pid, to mean different ways of delivering the signal. For example, pid = 0 sends the signal to all processes in the same process group, pid = -1 for all processes with pid > 1, etc. The reader may consult Linux man pages on signal/kill for more details.
(8). Signal and Wakeup/Unblock: kill only sends a signal to a target process. The signal does not take effect until the target process runs. When sending signals to a target process, it may be necessary to wakeup/unblock the target process if the latter is in a SLEEP or BLOCKed state. For example, when a process waits for terminal inputs, which may not come for a long time, it is considered as interruptible, meaning that it can be woken up or unblocked by arriving signals. On the other hand, if a process is blocked for SDC I/O, which will come very soon, it is non-interruptible, which should not be unblocked by signals.

8.8 Signal Processing in EOS

8.8.1 Signals in PROC Resource

In EOS, each PROC has a pointer to a resource structure, which contains the following fields for signals and signal handling.

int signal; //31 signals; bit 0 is not used.
int sig[32]; //signal handlers : 0 = default, 1 = ignore, else a catcher in Umode.

For the sake of simplicity, EOS dose not support signal masking. If desired, the reader may add signal masking to the EOS kernel.

8.8.2 Signal Origins in EOS

(1). Hardware: EOS supports the Control-C key from terminals, which is converted to the interrupt signal SIGINT(2) delivered to all processes on the terminal, and the interval timer, which is converted to the alarm signal SIGALRM(14) delivered to the process.
(2). Traps: EOS supports data_abort, prefetch and undefined instruction exceptions.
(3). From Other Process: EOS supports the kill(pid, signal) syscall, but it does not enforce permission checking. Therefore, a process may kill any process. If the target process is in the SLEEP state, kill() wakes it up. It the target process is BLOCKed for inputs in either the KBD of a UART driver, it is unblocked also.

8.8.3 Deliver Signal to Process

The kill syscall delivers a signal to a target process. The algorithm of the kill syscall is

```
/************** Algorithm of kill syscall  ***************/
int kkill(int pid, int sig_number)
```

```
{
  (1). validate signal number and pid;
  (2). check permission to kill; // not enforced, may kill any pid
  (3). set proc.signal.[bit_sig_number] to 1;
  (4). if proc is SLEEP, wakeup pid;
  (5). if proc is BLOCKed for terminal inputs, unblock proc;
}
```

8.8.4 Change Signal Handler in Kernel

The signal() syscall changes the handler function of a specified signal. The algorithm of the signal syscall is

```
/*********** Algorithm of signal syscall ***************/
int ksignal(int sig_number, int *catcher)
{
  (1). validate sig number, e.g. cannot change signal number 9;
  (2). int oldsig = running->sig[sig_number];
  (3). running->sig[sig_number] = catcher;
  (4). return oldsig;
}
```

8.8.5 Signal Handling in EOS Kernel

A CPU usually checks for pending interrupts at the end of executing an instruction. Likewise, it suffices to let a process check for pending signals at the end of Kmode execution, i.e. when it is about to return to Umode. However, if a process enters Kmode via a syscall, it should check and handle signals first. This is because if a process already has a pending signal, which may cause it to die, executing the syscall would be a waste of time. On the other hand, if a process enters Kmode due to an interrupt, it must handle the interrupt first. The algorithm of checking for pending signals is

```
/************ Algorithm of Check Signals ***********/
int check_sig()
{
  int i;
  for (i=1; i<NSIG; i++){
     if (running->signal & (1 << i)){
        running->signal &= ~(1 << i);
        return i;
     }
  }
  return 0;
}
```

A process handles outstanding signals by the code segment

```
if (running->signal)
   psig();
```

The algorithm of psig() is

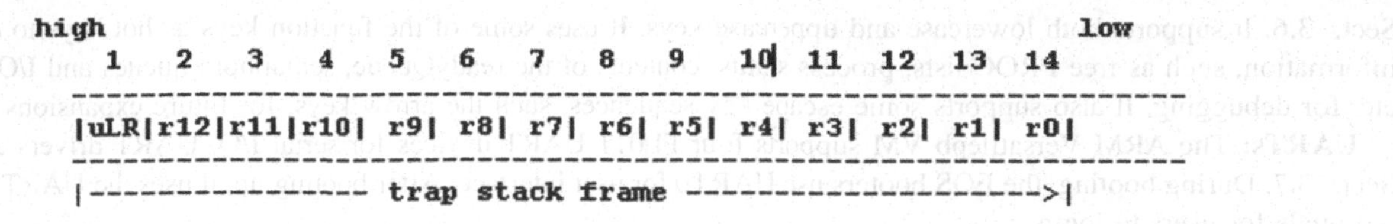

Fig. 8.1 Process trap stack frame

```
/************** Algorithm psig() *****************/
int psig(int sig)
{
   int n;
   while(n=check_sig()){ // for each pending signal do
(1).  clear running PROC.signal[bit_n]; // clear the signal bit
(2).  if (running->sig[n] == 1)         // IGNore the signal
         continue;
(3).  if (running->sig[n] == 0)         // DEFault : die with sign#
         kexit(n<<8);     // high byte of exitCode=signal number
(4).  // execute signal handler in Umode
      fix up running PROC's "interrupt stack frame" for it to return
      to execute catcher(n) in Umode;
   }
}
```

8.8.6 Dispatch Signal Catcher for Execution in User Mode

In the algorithm of psig(), only step (4) is interesting and challenging. Therefore, we shall explain it in more detail. The goal of step (4) is to let the process return to Umode to execute a catcher(int sig) function. When the catcher() function finishes, it should return to the point where the process lastly entered Kmode. The following diagrams show how to accomplish these. When a process traps to kernel from Umode, its privileged mode stack top contains a **"trap stack frame"** consisting of 14 entries, as shown in Fig. 8.1.

In order for the process return to execute catcher(int sig) with the signal number as a parameter, we modify the trap stack frame as follows.

(1). Replace the **uLR** (at index 1) with the entry address of catcher();
(2). Replace **r0** (at index 14) with sig number, so that upon entry to catcher(), **r0 = sig**;
(3). Set **User mode lr (r14)** to **uLR**, so that when catcher() finishes it would return by **uLR** to the original point of interruption.

8.9 Device Drivers

The EOS kernel supports the following I/O devices.

LCD display: The ARM Versatilepb VM uses the ARM PL110 Color LCD controller [ARM primeCell Color LCD Controller PL110] as the primary display device. The LCD driver used in EOS is the same driver developed in Sect. 2.8.4. It can display both text and images.

Keyboard: The ARM Versatilepb VM includes an ARM PL050 Mouse-Keyboard Interface (MKI) which provides support for a mouse and a PS/2 compatible keyboard [ARM PL050 MKI]. EOS uses a command-line user interface. The mouse device is not used. The keyboard driver of EOS is an improved version of the simple keyboard driver developed in

Sect. 3.6. It supports both lowercase and uppercase keys. It uses some of the function keys as hot keys to display kernel information, such as free PROC lists, process status, contents of the readyQueue, semaphore queues and I/O buffer usage, etc. for debugging. It also supports some escape key sequences, such the arrow keys, for future expansions of EOS.

UARTs: The ARM Versatilepb VM supports four PL011 UART devices for serial I/O. UART drivers are covered in Sect. 3.7. During booting, the EOS booters use UART0 for user interface. After booting up, it uses the UART ports as serial terminals for users to login.

Timer: The ARM Versatilepb VM contains two SB804 dual timer modules (ARM 926EJ-S 2016). Each timer module contains two timers, which are driven by the same clock. Timer drivers are covered in Sect. 3.5. Among the four timers, EOS uses only timer0 to provide timer service functions.

SDC: The SDC driver supports read/write of multi-sectors of file block size.

With the exception of the LCD display, all device drivers are interrupt-driven and use semaphores for synchronization between interrupt handlers and processes.

8.10 Process Scheduling in EOS

Process scheduling in EOS is by time-slice and dynamic process priority. When a process is scheduled to run, it is given a time-slice of 5–10 timer ticks. While a process runs in Umode, the timer interrupt handler decrements its time-slice by 1 at each timer tick. When the process time-slice expires, it sets a switch process flag and calls resechdule() to switch process when it exits Kmode. In order to keep the system simple, EOS uses a simplified priority scheme. Instead of re-computing process priorities dynamically, it uses only two distinct priority values. When a process is in Umode, it runs with the user-level priority of 128. When it enters Kmode, it continues to run with the same priority. If a process becomes blocked (due to I/O or file system operation), it is given a fixed Kmode priority of 256 when it is unblocked to run again. When a process exits Kmode, it drops back to the user level priority of 128.

Exercise 2: Modify the EOS kernel to implement dynamic process priority by CPU usage time.

8.11 Timer Service in EOS

The EOS kernel provides the following timer service to processes.

(1). pause(t): the process goes to sleep for t seconds.
(2). itimer(t): sets an interval timer of t seconds. Send a SIGALRM (14) signal to the process when the interval timer expires.

To simplify the discussion, we shall assume that each process has only one outstanding timer request and the time unit is in seconds in real time, i.e. the virtual timer of a process continues to run whether the process is executing or not.

(1). Timer Request Queue: Timer service provides each process with a virtual or logical timer by a single physical timer. This is achieved by maintaining a timer queue to keep track of process timer requests. A timer queue element (TQE) is a structure

```
typedef struct tq{
        struct tq *next;     // next element pointer
        int       time;      // requested time
        struct PROC *proc;   // pointer to PROC
        int     (*action)(); // 0|1|handler function pointer
}TQE;
TQE *tq, tqe[NPROC];         // tq = timer queue pointer
```

In the TQE, action is a function pointer, where 0 means WAKEUP, 1 means NOTIFY, other value = entry address of a handler function to execute. Initially, the timer queue is empty. As processes invoke timer service, their requests are added to the timer queue. Figure 8.2 shows an example of the timer queue.

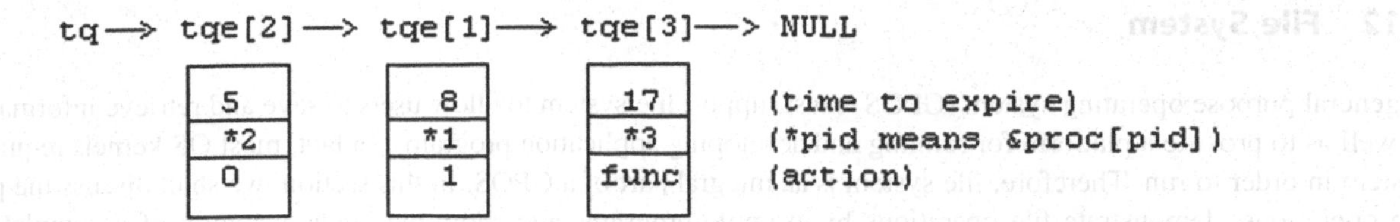

Fig. 8.2 Timer request queue

At each second, the interrupt handler decrements the time field of each TQE by 1. When a TQE's time reaches 0, the interrupt handler deletes the TQE from the timer queue and invokes the action function of the TQE. For example, after 5 s, it deletes tqe[2] from the timer queue and wakes up P2. In the above timer queue, the time field of each TQE contains the exact remaining time. The disadvantage of this scheme is that the interrupt handler must decrement the time field of each and every TQE. In general, an interrupt handler should complete the interrupt processing as quickly as possible. This is especially important for the timer interrupt handler. Otherwise, it may loss ticks or even never finish. We can speed up the timer interrupt handler by modifying the timer queue as shown in Fig. 8.3.

In the modified timer queue, the time field of each TQE is relative to the cumulative time of all the preceding TQEs. At each second, the timer interrupt handler only needs to decrement the time of the first TQE and process any TQE whose time has expired. With this setup, insertion and deletion of a TQE must be done carefully. For example, if a process P4 makes an itimer(10) request, its TQE should be inserted after TQ[1] with a time = 2, which changes the time of TQ[3] to 7. Similarly, when P1 calls itimer(0) to cancel its timer request, its TQE[1] will be deleted from the timer queue, which changes the time of TQE[3] to 12, etc. The reader is encouraged to figure out the general algorithms for insertion and deletion of TQEs.

(2). Timer Queue as a Critical Region: The timer queue data structure is shared by processes and the timer interrupt handler. Accesses to the timer queue must be synchronized to ensure its integrity. EOS is a uniprocessor OS, which allows only one process to execute in kernel at a time. In the EOS kernel a process can not be interfered by another process, so there is no need for process locks. However, while a process executes, interrupts may occur. If a timer interrupt occurs while a process is in the middle of modifying the timer queue, the process would be diverted to execute the interrupt handler, which also tries to modify the timer queue, resulting in a race condition. To prevent interference from interrupt handler, the process must mask out interrupts when accessing the timer queue. The following shows the algorithm of itimer().

```
/*************** Algorithm of itimer() ********************/
int itimer(t)
{
  (1). Fill in TQE[pid] information, e.g. proc pointer, action.
  (2). lock();        // mask out interrupts
  (3).   traverse timer queue to compute the position to insert TQE;
  (4).   insert the TQE and update the time of next TQE;
  (5). unlock();      // unmask interrupts
}
```

```
tq —> TQE[2] —> TQE[1] —> TQE[3] —> NULL

  5        3        9      (relative time)
  *2       *1       *3     (pointer to proc[pid])
  0        1        func   (action)
```

Fig. 8.3 Improved timer request queue

8.12 File System

A general purpose operating system (GPOS) must support file system to allow users to save and retrieve information as files, as well as to provide a platform for running and developing application programs. In fact, most OS kernels require a root file system in order to run. Therefore, file system is an integral part of a GPOS. In this section, we shall discuss the principles of file operations, demonstrate file operations by example programs and show the implementation of a complete EXT2 file system in EOS.

8.12.1 File Operation Levels

File operations consist of five levels, from low to high, as shown in the following hierarchy.

(1). Hardware Level:

File operations at hardware level include

fdisk : divide a mass storage device, e.g. a SDC, into partitions.
mkfs : format a partition to make it ready for file system.
fsck : check and repair file system.
defragmentation : compact files in a file system.

Most of these are system-oriented utility programs. An average user may never need them, but they are indispensable tools for creating and maintaining file systems.

(2). File System Functions in OS Kernel:

Every general purpose OS kernel provides support for basic file operations. The following lists some of these functions in a Unix-like system kernel, where the prefix k denotes kernel functions, which rely on device drivers for I/O on real devices.

```
kmount(), kumount()              (mount/umount file systems)
kmkdir(), krmdir()               (make/remove directory)
kchdir(), kgetcwd()              (change directory, get CWD pathname)
klink(),  kunlink()              (hard link/unlink files)
kchmod(), kchown(), ktouch()     (change r|w|x permissions,owner,time)
kcreat(), kopen()                (create/open file for R,W,RW,APPEND)
kread(),  kwrite()               (read/write opened files)
klseek(); kclose()               (lseek/close opened file descriptors)
ksymlink(), kreadlink()          (create/read symbolic link files)
kstat(),  kfstat(), klstat()     (get file status/information)
kopendir(), kreaddir()           (open/read directories)
```

(3). System Calls:

User mode programs use system calls to access kernel functions. As an example, the following program reads the second 1024 bytes of a file.

```
/******* Example program: read first 1KB of a file *******/
#include <stdio.h>
#include <stdlib.h>
```

```
#include <fcntl.h>
int main(int argc, char *argv[ ])    // run as a.out filename
{
    int fd, n;
    char buf[1024];
    if ((fd = open(argv[1], O_RDONLY)) < 0) // if open fails
        exit(1);
    lseek(fd, (long)1024, SEEK_SET);      // lseek to byte 1024
    n = read(fd, buf, 1024);              // read 1024 bytes of file
    close(fd);
}
```

In the above example program, the functions open(), read(), lseek() and close() are C library functions. Each library function issues a system call, which causes the process to enter kernel mode to execute a corresponding kernel function, e.g. open() goes to kopen(), read() goes to kread(), etc. When the process finishes executing the kernel function, it returns to user mode with the desired results. Switch between user mode and kernel mode requires a lot of actions (and time). Data transfer between kernel and user spaces is therefore quite expensive. Although it is permissible to issue a read(fd, buf, 1) system call to read only one byte of data, it is not very wise to do so since that one byte of data would come with a terrific cost. Every time we have to enter kernel mode, we should do as much as we can to make the journey worthwhile. In the case of read/write files, the best way is to match what the kernel does. The kernel reads/writes files by block size, which ranges from 1 to 8 KB. For instance, in Linux, the default block size is 4 KB for hard disks and 1 KB for floppy disks. Therefore, each read/write system call should also try to transfer one block of data at a time.

(4). Library I/O Functions:

System calls allow user mode programs to read/write chunks of data, which are just a sequence of bytes. They do not know, nor care, about the meaning of the data. A user mode program often needs to read/write individual chars, lines or data structure records, etc. With only system calls, a user mode program must do these operations from/to a buffer area by itself. Most users would consider this too inconvenient. The C library provides a set of standard I/O functions for convenience, as well as for run-time efficiency. Library I/O functions include:

```
FILE mode I/O: fopen(),fread();  fwrite(),fseek(),fclose(),fflush()
char mode I/O: getc(), getchar() ugetc(); putc(),putchar()
line mode I/O: gets(), fgets();  puts(), fputs()
formatted I/O: scanf(),fscanf(),sscanf(); printf(),fprintf(),sprintf()
```

With the exceptions of sscanf()/sprintf(), which read/write memory locations, all other library I/O functions are built on top of system calls, i.e. they ultimately issue system calls for actual data transfer through the system kernel.

(5). User Commands:

Instead of writing programs, users may use Unix/Linux commands to do file operations. Examples of user commands are

mkdir, **rmdir**, **cd**, **pwd**, **ls**, **link**, **unlink**, **rm**, **cat**, **cp**, **mv**, **chmod**, etc.

Each user command is in fact an executable program (except cd), which typically calls library I/O functions, which in turn issue system calls to invoke the corresponding kernel functions. The processing sequence of a user command is either

```
   Command => Library I/O function => System call => Kernel Function
OR Command ======================== > System call => Kernel Function
```

Most contemporary OS supports Graphic User Interface (GUI), which allows the user to do file operations through the GUI. For instance, clicking a mouse button on a program name icon may invoke the program execution. Similarly, clicking a pointing device on a filename icon followed by choosing Copy of a pull-down menu copies the file contents to a global buffer, which can be transferred to a target file by Paste, etc. Many users are so accustomed to using the GUI that they often ignore and do not understand what is really going on underneath the GUI interface. This is fine for most naïve computer users, but not for computer science and computer engineering students.

(6). Sh Scripts:

Although much more convenient than system calls, commands must be entered manually, or by repeatedly dragging and clicking a pointing device as in the case of using GUI, which is tedious and time-consuming. Sh scripts are programs written in the sh programming language, which can be executed by the command interpreter sh. The sh language include all valid Unix/Linux commands. It also supports variables and control statements, such as if, do, for, while, case, etc. In practice, sh scripts are used extensively in systems programming on all Unix-like systems. In addition to sh, many other script languages, such as Perl and Tcl, are also in wide use.

8.12.2 File I/O Operations

Figure 8.4 shows the diagram of file I/O operations. In Fig. 8.4, the upper part above the double line represents kernel space and the lower part represents user space of a process. The diagram shows the sequence of actions when a process read/write a file stream. Control flows are identified by the labels (1) to (10), which are explained below.

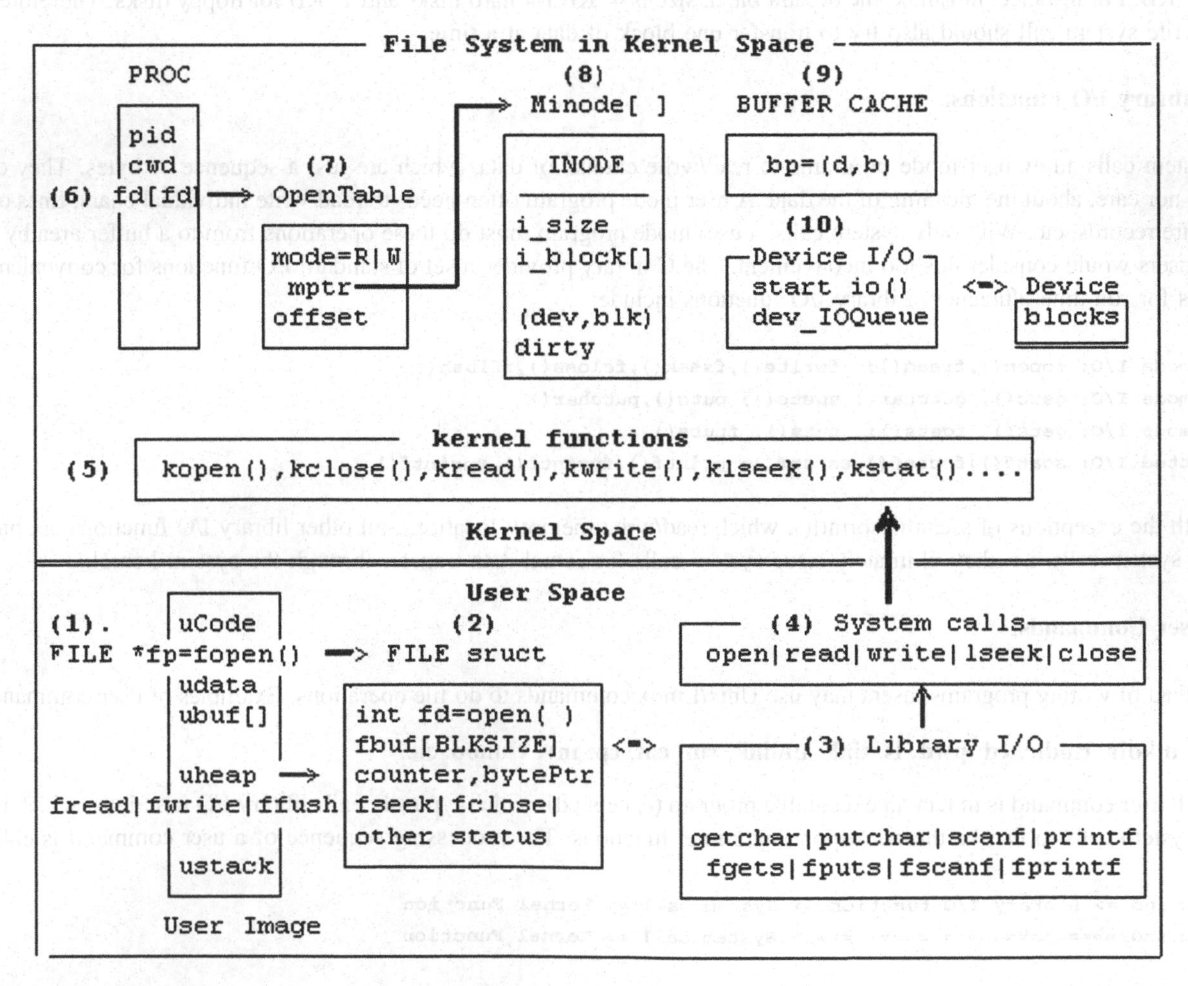

Fig. 8.4 File operation diagram

---------- User Mode Operations ----------

(1). A process in User mode executes

```
FILE *fp = fopen("file", "r"); or FILE *fp = fopen("file", "w");
```

which opens a **file stream** for READ or WRITE.

(2). fopen() creates a FILE structure in user (heap) space containing a file descriptor, fd, and a fbuf[BLKSIZE]. It issues fd = open("file", flags = READ or WRITE) syscall to kopen() in kernel, which constructs an OpenTable to represent an instance of the opened file. The OpenTable's mptr points to the file's INODE in memory. For non-special files, the INODE's i_block array points to data blocks on the storage device. On success, fp points to the FILE structure, in which fd is the file descriptor returned by the open() syscall.
(3). fread(ubuf, size, nitem, fp): READ nitem of size each to ubuf by
. copy data from FILE structure's fbuf to ubuf, if enough, return;
. if fbuf has no more data, then execute (4a).
(4a). issue read(fd, fbuf, BLKSIZE) syscall to read a file block from kernel to fbuf, then copy data to ubuf until enough or file has no more data.
(4b). fwrite(ubuf, size, nitem, fp): copy data from ubuf to fbuf;
. if (fbuf has room): copy data to fbuf, return;
. if (fbuf is full): issue write(fd, fbuf, BLKSIZE) syscall to write a block to kernel,
then write to fbuf again.

Thus, fread()/fwrite() issue read()/write() syscalls to kernel, but they do so only when necessary and they transfer chunks of data from/to kernel in BLKSIZE for better efficiency. Similarly, other Library I/O Functions, such as fgetc/fputc, fgets/fputs, fscanf/fprintf, etc. also operate on fbuf in the FILE structure, which is in user space.

```
=================== Kernel Mode Operations ===================
```

(5). File system functions in kernel:
Assume read(fd, fbuf[], BLKSIZE) syscall of non-special file.
(6). In a read() syscall, fd is an opened file descriptor, which is an index in the running PROC's fd array, which points to an OpenTable representing the opened file.
(7). The OpenTable contains the files's open mode, a pointer to the file's INODE in memory and the current byte offset into the file for read/write. From the OpenTable's offset, the kernel kread() function
. Compute logical block number, lbk;
. Convert logical block to physical block, blk, via INODE.i_block array.
(8). Minode contains the in-memory INODE of the file. The INODE.i_block array contains pointers to physical disk blocks. A file system may use the physical block numbers to read/write data from/to the disk blocks directly, but these would incur too much physical disk I/O.

Processing of the write(fd, fbuf[], BLKSIZE) syscall is similar, except that it may allocate new disk blocks and grow the file size as new data are written to the file.

(9). In order to improve disk I/O efficiency, the OS kernel usually uses a set of I/O buffers as a cache between kernel memory and I/O devices to reduce the number of physical I/O. We shall discuss I/O buffering in later sections.

We illustrate the relationship between the various levels of file operations by the following examples.

Example 1. The example program C8.1 shows how to make a directory by the syscall

```
        int mkdir(char *dirname, int mode)
/**** Program C8.1: mkdir.c: run as a.out dirname ****/
```

```
#include <stdio.h>
#include <stdlib.h>
#include <errno.h>
int main(int argc, char *argv[])
{    int r;
     if (argc < 2){
        printf("Usage: a.out dirname); exit(1);
     }
     if ((r = mkdir(argv[1], 0755)) < 0)
        perror("mkdir"); // print error message
     }
}
```

In order to make a directory, the program must issue a mkdir() system call since only the kernel knows how to make directories. The system call is routed to kmkdir() in kernel, which tries to create a new directory with the specified name and mode. The system call returns r = 0 if the operation succeeds or -1 if it fails. If the system call fails, the error number is in a (extern) global variable errno. The program may call the library function perror("mkdir"), which prints the program name followed by a string describing the cause of the error, e.g. mkdir: File exists, etc.

Exercise 1: Modify the program C8.1 to make many directories by a single command, e.g. mkdir dir1 dir2 .... dirn //with command-line parameters

Exercise 2: The Linux system call **int r = rmdir(char *pathname)** removes a directory, which must be empty. Write a C program rmdir.c, which removes a directory.

Example 2: This example develops a ls program which mimics the ls –l command of Unix/Linux: In Unix/Linxu, the stat system calls

```
int  stat(const char *file_name, struct stat *buf);
int lstat(const char *file_name, struct stat *buf);
```

return information of the specified file. The difference between stat() and lstat() is that the former follows symbolic links but the latter does not. The returned information is in a stat structure (defined in stat.h), which is

```
struct stat {
  dev_t      st_dev;      /* device */
  ino_t      st_ino;      /* inode number */
  mode_t     st_mode;     /* file type and permissions */
  nlink_t    st_nlink;    /* number of hard links */
  uid_t      st_uid;      /* user ID of file owner */
  gid_t      st_gid;      /* group ID of owner */
  dev_t      st_rdev;     /* device type (if inode device) */
  off_t      st_size;     /* total size, in bytes */
  blksize_t  st_blksize;  /* blocksize for filesystem I/O */
  blkcnt_t   st_blocks;   /* number of 512-byte sectors */
  time_t     st_atime;    /* time of last access */
  time_t     st_mtime;    /* time of last modification */
  time_t     st_ctime;    /* time of last status change */
};
```

In the stat structure, st_dev identifies the device (number) on which the file resides and st_ino is its inode number on that device. The field st_mode is a 2-byte integer, which specifies the file type, special usage and permission bits for protection. Specifically, the bits of st_mode are

```
        4   3   3   3   3
       |----|---|---|---|---|
       |tttt|fff|rwx|rwx|rwx|
```

The highest 4 bits of st_mode determine the file type. For examples, b1000 = REGular file, b0100 = DIRectory, b1100 = symbolic link file, etc. The file type can be tested by the pre-defined macros S_ISREG, S_ISDIR, S_ISLNK, etc. For example,

```
if (S_ISREG(st_mode))  // test for REGular file
if (S_ISDIR(st_mode))  // test for DIR file
if (S_ISLNK(st_mode))  // test for symbolic link file
```

The low 9 bits of st_mode define the permission bits as r (readable) w (writeable) x (executable) for owner of the file, same group as the owner and others. For directory files, the x bit means whether cd into the directory is allowed or not allowed. The field st_nlink is the number of hard links to the file, st_size is the file size in bytes, st_atime, st_mtime and st_ctime are time fields. In Unix-like systems, time is the elapsed time in seconds since 00:00:00 of January 1, 1970. The time fields can be converted to strings in calendar form by the the library function char *ctime(time_t *time).

Based on the information returned by the stat system call, we can write a ls.c program, which behaves the same as the ls –l command of Unix/Linux. The program is denoted by C8.2, which is shown below.

```
/** Program C8.2: ls.c: run as a.out [filename] **/
#include <stdio.h>
#include <stdlib.h>
#include <string.h>
#include <sys/stat.h>
#include <time.h>
#include <sys/types.h>
#include <dirent.h>
#include <errno.h>

char *t1 = "xwrxwrxwr-------";
char *t2 = "----------------";
struct stat mystat, *sp;
int ls_file(char *fname)      // list a single file
{
  struct stat fstat, *sp = &fstat;
  int r, i;
  char sbuf[4096];
  r = lstat(fname, sp);       // lstat the file
  if (S_ISDIR(sp->st_mode))
     printf("%c",'d');        // print file type as d
  if (S_ISREG(sp->st_mode))
     printf("%c",'-');        // print file type as -
  if (S_ISLNK(sp->st_mode))
     printf("%c",'l');        // print file type as l
  for (i=8; i>=0; i--){
    if (sp->st_mode & (1<<i))
       printf("%c", t1[i]); // print permission bit as r w x
    else
 printf("%c", t2[i]); // print permission bit as -
  }
  printf("%4d ", sp->st_nlink);  // link count
  printf("%4d ", sp->st_uid     // uid
  printf("%8d ", sp->st_size);  // file size
  strcpy(ftime, ctime(&sp->st_ctime));
```

```
    ftime[strlen(ftime)-1] = 0;        // kill \n at end
    printf("%s ",ftime);               // time in calendar form
    printf("%s", basename(fname));     // file basename
    if (S_ISLNK(sp->st_mode)){         // if symbolic link
       r = readlink(fname, sbuf, 4096);
       printf(" -> %s", sbuf);         // -> linked pathname
    }
    printf("\n");
}
int ls_dir(char *dname)            // list a DIR
{
    char name[256];                // EXT2 filename: 1-255 chars
    DIR *dp;
    struct dirent *ep;
    // open DIR to read names
    dp = opendir(dname);           // opendir() syscall
    while (ep = readdir(dp)){      // readdir() syscall
       strcpy(name, ep->d_name);
       if (!strcmp(name, ".") || !strcmp(name, ".."))
          continue;                // skip over . and .. entries
       strcpy(name, dname);
       strcat(name, "/");
       strcat(name, ep->d_name);
       ls_file(name);              // call list_file()
    }
}
int main(int argc, char *argv[])
{
    struct stat mystat, *sp;
    int r;
    char *s;
    char filename[1024], cwd[1024];
    s = argv[1];                  // ls [filename]
    if (argc == 1)                // no parameter: ls CWD
       s = "./";
    sp = &mystat;
    if ((r = stat(s, sp)) < 0){ // stat() syscall
       perror("ls"); exit(1);
    }
    strcpy(filename, s);
    if (s[0] != '/'){             // filename is relative to CWD
       getcwd(cwd, 1024);         // get CWD path
       strcpy(filename, cwd);
       strcat(filename, "/");
       strcat(filename,s);        // construct $CWD/filename
    }
    if (S_ISDIR(sp->st_mode))
       ls_dir(filename);        // list DIR
    else
       ls_file(filename);       // list single file
}
```

The reader may compile and run the program C8.2 under Linux. It should list either a single file or a DIR in the same format as does the Linux ls –l command.

Example 4: File copy programs: This example shows the implementations of two file copy programs; one uses system calls and the other one uses library I/O functions. Both programs are run as **a.out src dest,** which copies src to dest. We list the programs side by side in order to to show their similarities and differences.

```
----------- cp.sysall.c -----------|------ cp.libio.c -----------
#include <stdio.h>                  | #include <stdio.h>
#include <stdlib.h>                 | #include <stdlib.h>
#include <fcntl.h>                  |
main(int argc, char *argv[ ])       | main(int argc, char *argv[ ])
{                                   | {
 int fd, gd;                        |  FILE *fp, *gp;
 int n;                             |  int n;
 char buf[4096];                    |  char buf[4096];
 if (argc < 3) exit(1);             |  if (argc < 3) exit(1);
 fd = open(argv[1], O_RDONLY);      |  fp = fopen(argv[1], "r");
 gd = open(argv[2],O_WRONLY|O_CREAT);| gp = fopen(argc[1], "w+");
 if (fd < 0 || gd < 0) exit(2);     |  if (fp==0 || gp==0) exit(2);
 while(n=read(fd, buf, 4096)){      |  while(n=fread(buf,1,4096,fp)){
    write(gd, buf, n);              |     fwrite(buf, 1, n, gp);
 }                                  |  }
 close(fd); close(gd);              |  fclose(fp); fclose(gp);
}                                   | }
------------------------------------------------------------------
```

The program cp.syscall.c shown on the left-hand side uses system calls. First, it issues open() system calls to open the src file for READ and the dest file foe WRITE, which creates the dest file if it does not exist. The open() syscalls return two (integer) file descriptors, fd and gd. Then it uses a loop to copy data from fd to gd. In the loop, it issues read() syscall on fd to read up to 4 KB data from the kernel to a local buffer. Then it issues write() syscall on gd to write the data from the local buffer to kernel. The loop ends when read() returns 0, indicating that the source file has no more data to read.

The program cp.libio.c shown on the right-hand side uses library I/O functions. First, it calls fopen() to create two file streams, fp and gp, which are pointers to FILE structures. fopen() creates a FILE structure (defined in stdio.h) in the program's heap area. Each FILE structure contains a local buffer, char fbuf[BLKSIZE], of file block size and a file descriptor field. Then it issues an open() system call to get a file descriptor and record the file descriptor in the FILE structure. Then it returns a file stream (pointer) pointing at the FILE structure. When the program calls fread(), it tries to read data from fbuf in the FILE structure. If fbuf is empty, fread() issues a read() system call to read a BLKSIZE of data from kernel to fbuf. Then it transfers data from fbuf to the program's local buffer. When the program calls fwrite(), it writes data from the program's local buffer to fbuf in the FILE structure. If fbuf is full, fwrite() issues a write() system call to write a BLKSIZE of data from fbuf to kernel. Thus, library I/O functions are built on top of system calls, but they issue system calls only when needed and they transfer data from/to kernel in file block size for better efficiency. Based on these discussions, the reader should be able to deduce which program is more efficient if the objective is only to transfer data. However, if a user mode program intends to access single chars in files or read/write lines, etc. using library I/O functions would be the better choice.

Exercise 3: Assume that we rewrite the data transfer loops in the programs of Example 3 as follows, both transfer one byte at a time.

```
       cp.syscall.c             |      cp.syscall.c
------------------------------------------------------------------
while((n=read(fd, buf, 1)){     | while((n=fgetc(fp))!= EOF){
   write(gd, buf, n);           |    fputc(n, gp);
}                               | }
------------------------------------------------------------------
```

Which program would be more efficient? Explain your reasons.

Exercise 4: When copying files, the source file must be a regular file and we should never copy a file to itself. Modify the programs in Example 3 to handle these cases.

8.12.3 EXT2 File System in EOS

For many years, Linux used EXT2 (Card et al. 1995; EXT2 2001) as the default file system. EXT3 (ETX3 2015) is an extension of EXT2. The main addition to EXT3 is a journal file, which records changes made to the file system in a journal log. The log allows for quicker recovery from errors in case of a file system crash. An EXT3 file system with no error is identical to an EXT2 file system. The newest extension of EXT3 is EXT4 (Cao et al. 2007). The major change in EXT4 is in the allocation of disk blocks. In EXT4, block numbers are 48 bits. Instead of discrete disk blocks, EXT4 allocates contiguous ranges of disk blocks, called extents. EOS is a small system intended mainly for teaching and learning the internals of embedded operating systems. Large file storage capacity is not the design goal. Principles of file system design and implementation, with an emphasis on simplicity and compatibility with Linux, are the major focal points. For these reasons, we choose ETX2 as the file system. Support for other file systems, e.g. FAT and NTFS, are not implemented in the EOS kernel. If needed, they can be implemented as user-level utility programs. This section describes the implementation of an EXT2 file system in the EOS kernel. Implementation of EXT2 file system is discussed in detail in (Wang 2015). For the sake of completeness and reader convenience, we include similar information here.

8.12.3.1 File System Organization

Figure 8.5 shows the internal organization of an EXT2 file system. The organization diagram is explained by the labels (1) to (5).

(1) is the PROC structure of the running process. Each PROC has a cwd field, which points to the in-memory INODE of the PROC's Current Working Directory (CWD). It also contains an array of file descriptors, fd[], which point to opened file instances.
(2) is the root directory pointer of the file system. It points to the in-memory root INODE. When the system starts, one of the devices is chosen as the root device, which must be a valid EXT2 file system. The root INODE (inode #2) of the root device is loaded into memory as the root directory (/) of the file system. This operation is known as "mount root file system"
(3) is an openTable entry. When a process opens a file, an entry of the PROC's fd array points to an openTable, which points to the in-memory INODE of the opened file.
(4) is an in-memory INODE. Whenever a file is needed, its INODE is loaded into a minode slot for reference. Since INODEs are unique, only one copy of each INODE can be in memory at any time. In the minode, (dev, ino) identify where the INODE came from, for writing the INODE back to disk if modified. The refCount field records the number of processes that are using the minode. The dirty field indicates whether the INODE has been modified. The mounted flag indicates whether the INODE has been mounted on and, if so, the mntabPtr points to the mount table entry of the mounted file system. The lock field is to ensure that an in-memory INODE can only be accessed by one process at a time, e.g. when modifying the INODE or during a read/write operation.
(5) is a table of mounted file systems. For each mounted file system, an entry in the mount table is used to record the mounted file system information. In the in-memory INODE of the mount point, the mounted flag is turned on and the mntabPtr points to the mount table entry. In the mount table entry, mntPointPtr points back to the in-memory INODE of the mount point. As will be shown later, these doubly-linked pointers allow us to cross mount points when traversing the file system tree. In addition, a mount table entry may also contain other information of the mounted file system, such as the device name, superblock, group descriptor and bitmaps, etc. for quick reference. Naturally, if any such information has been changed while in memory, they must be written back to the storage device when the mounted file system is umounted.

8.12.3.2 Source Files in the EOS/FS Directory

In the EOS kernel source tree, the FS directory contains files which implement an EXT2 file system. The files are organized as follows.

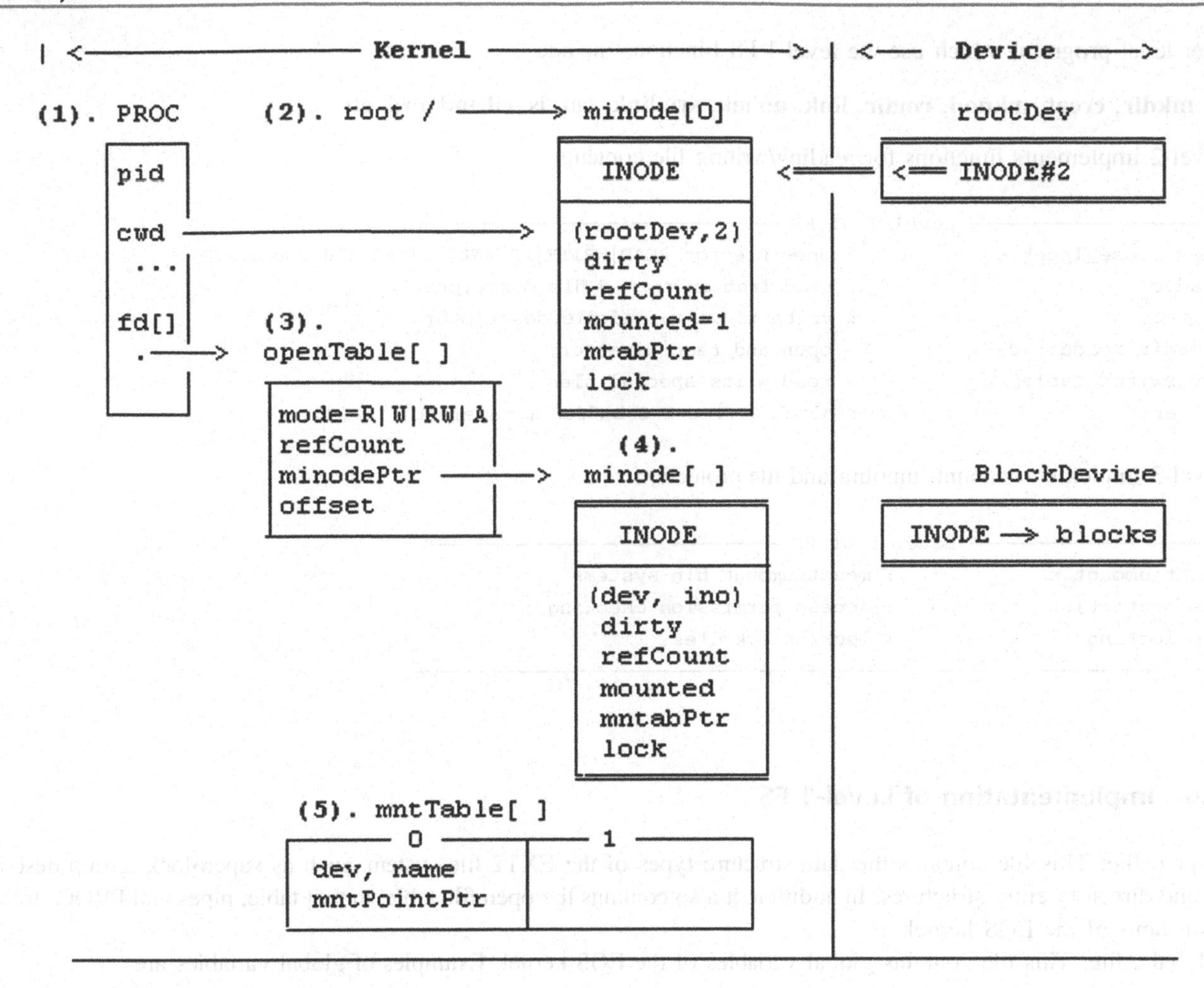

Fig. 8.5 EXT2 file system data structures

Common files of FS	
type.h	: EXT2 data structure types
global.c:	global variables of FS
util.c	: common utility functions: getino(), iget(), iput(), search(), etc.
allocate_deallocate.c	inodes/blocks management functions

Implementation of the file system is divided into three levels. Each level deals with a distinct part of the file system. This makes the implementation process modular and easier to understand. Level-1 implements the basic file system tree. It contains the following files, which implement the indicated functions.

Level-1 of FS	
mkdir_creat.c	: make directory, create regular and special file
cd_pwd.c	: change directory, get CWD path
rmdir.c	: remove directory
link_unlink.c	: hard link and unlink files
symlink_readlink.c	: symbolic link files
stat.c	: return file information
misc1.c	: access, chmod, chown, touch, etc.

User level programs which use the level-1 FS functions include

mkdir, **creat**, **mknod**, **rmdir**, **link**, **unlink**, **symlink**, **rm**, **ls**, **cd and pwd**, **etc**.

Level-2 implements functions for reading/writing file contents.

```
------------------------------ Level-2 of FS ------------------------------
open_close_lseek.c      : open file for READ|WRITE|APPEND, close file and lseek
read.c                  : read from an opened file descriptor
write.c                 : write to an opened file descriptor
opendir_readdir.c       : open and read directory
dev_switch_table        : read/write special files
buffer.c                : block device I/O buffer management
```

Level-3 implements mount, umount and file protection.

```
------------------------------ Level-3 of FS ------------------------------
mount_umount.c          : mount/umount file systems
file protection         : access permission checking
file-locking            : lock/unlock files
----------------------------------------------------------------------------
```

8.12.4 Implementation of Level-1 FS

(1). type.h file: This file contains the data structure types of the EXT2 file system, such as superblock, group descriptor, inode and directory entry structures. In addition, it also contains the open file table, mount table, pipes and PROC structures and constants of the EOS kernel.
(2). global.c file: This file contains global variables of the EOS kernel. Examples of global variables are

```
MINODE minode[NMINODES];     // in memroy INODEs
MOUNT  mounttab[NMOUNT];     // mount table
OFT    oft[NOFT];            // Opened file instance
```

(3). util.c file: This file contains utility functions of the file system. The most important utility functions are getino(), iget() and iput(), which are explained in more detail.
(3).1. u32 getino(int *dev, char *pathname): getino() returns the inode number of a pathname. While traversing a pathname the device number may change if the pathname crosses mounting point(s). The parameter dev is used to record the final device number. Thus, getino() essentially returns the (dev, ino) of a pathname. The function uses tokenize() to break up pathname into component strings. Then it calls search() to search for the component strings in successive directory minodes. Search() returns the inode number of the component string if it exists, or 0 if not.
(3).2. MINODE *iget(in dev, u32 ino): This function returns a pointer to the in-memory INODE of (dev, ino). The returned minode is unique, i.e. only one copy of the INODE exists in kernel memory. In addition, the minode is locked (by the minode's locking semaphore) for exclusive use until it is either released or unlocked.
(3).3. iput(MINODE *mip): This function releases and unlocks a minode pointed by mip. If the process is the last one to use the minode (refCount = 0), the INODE is written back to disk if it is dirty (modified).
(3).4. Minodes Locking: Every minode has a lock field, which ensures that a minode can only be accessed by one process at a time, especially when modifying the INODE. Unix uses a busy flag and sleep/wakeup to synchronize processes accessing the same minode. In EOS, each minode has a lock semaphore with an initial value 1. A process is allowed to access a minode only if it holds the semaphore lock. The reason for minodes locking is as follows.

Assume that a process Pi needs the inode of (dev, ino), which is not in memory. Pi must load the inode into a minode entry. The minode must be marked as (dev, ino) to prevent other processes from loading the same inode again. While loading the inode from disk, Pi may wait for I/O completion, which switches to another process Pj. If Pj needs exactly the same inode, it would find the needed minode already exists. Without the minode lock, Pj would proceed to use the minode before it is even loaded in yet. With the lock, Pj must wait until the minode is loaded, used and then released by Pi. In addition, when a process read/write an opened file, it must lock the file's minode to ensure that each read/write operation is atomic.

(4). allocate_deallocate.c file: This file contains utility functions for allocating and deallocating minodes, inodes, disk blocks and open file table entries. It is noted that both inode and disk block numbers count from 1. Therefore, in the bitmaps bit i represents inode/block number i+1.
(5). mount_root.c file: This file contains the mount_root() function, which is called during system initialization to mount the root file system. It reads the superblock of the root device to verify the device is a valid EXT2 file system. It loads the root INODE (ino=2) into a minode and sets the root pointer to the root minode. Then it unlocks the root minode to allow all processes to access the root minode. A mount table entry is allocated to record the mounted root file system. Some key parameters on the root device, such as the starting blocks of the bitmaps and inodes table, are also recorded in the mount table for quick reference.
(6). mkdir_creat.c file: This file contains mkdir and creat functions for making directories and creating files, respectively. mkdir and creat are very similar, so they share some common code. Before discussing the algorithms of mkdir and creat, we first show how to insert/delete a DIR entry into/from a parent directory. Each data block of a directory contains DIR entries of the form

```
|ino rlen nlen name|ino rlen nlen name| ...
```

where name is a sequence of nlen chars without a terminating NULL byte. Since each DIR entry begins with a u32 inode number, the rec_len of each DIR entry is always a multiple of 4 (for memory alignment). The last entry in a data block spans the remaining block, i.e. its rec_len is from where the entry begins to the end of block. In mkdir and creat, we assume the following.

(a). A DIR file has at most 12 direct blocks. This assumption is reasonable because, with 4 KB block size and an average file name of 16 chars, a DIR can contain more than 3000 entries. We may assume that no user would put that many entries in a single directory.
(b). Once allocated, a DIR's data block is kept for reuse even if it becomes empty.

With these assumptions, the insertion and deletion algorithms are as follows.

```
/************* Algorithm of Insert_dir_entry ******************/
(1). need_len = 4*((8+name_len+3)/4); // new entry need length
(2). for each existing data block do {
        if (block has only one entry with inode number==0)
           enter new entry as first entry in block;
        else{
(3).       go to last entry in block;
           ideal_len = 4*((8+last_entry's name_len+3)/4);
           remain = last entry's rec_len - ideal_len;
           if (remain >= need_len){
              trim last entry's rec_len to ideal_len;
              enter new entry as last entry with rec_len = remain;
           }
(4).       else{
              allocate a new data block;
              enter new entry as first entry in the data block;
```

```
            increase DIR's size by BLKSIZE;
        }
      }
      write block to disk;
    }
(5). mark DIR's minode modified for write back;
```

```
/************* Algorithm of Delete_dir_entry (name) *************/
(1). search DIR's data block(s) for entry by name;
(2). if (entry is the only entry in block)
          clear entry's inode number to 0;
     else{
(3).      if (entry is last entry in block)
              add entry's rec_len to predecessor entry's rec_len;
(4).      else{ // entry in middle of block
              add entry's rec_len to last entry's rec_len;
              move all trailing entries left to overlay deleted entry;
          }
     }
(5). write block back to disk;
```

Note that in the Delete_dir_entry algorithm, an empty block is not deallocated but kept for reuse. This implies that a DIR's size will never decrease. Alternative schemes are listed in the Problem section as programming exercises.

8.12.4.1 mkdir-creat-mknod

mkdir creates an empty directory with a data block containing the default . and .. entries. The algorithm of mkdir is

```
/********* Algorithm of mkdir *********/
int mkdir(char *pathname)
{
 1. if (pathname is absolute) dev = root->dev;
    else                      dev = PROC's cwd->dev
 2. divide pathname into dirname and basename;
 3. // dirname must exist and is a DIR:
    pino = getino(&dev, dirname);
    pmip = iget(dev, pino);
    check pmip->INODE is a DIR
 4. // basename must not exist in parent DIR:
       search(pmip, basename) must return 0;
 5. call kmkdir(pmip, basename) to create a DIR;
    kmkdir() consists of 4 major steps:
    5-1. allocate an INODE and a disk block:
           ino = ialloc(dev); blk = balloc(dev);
           mip = iget(dev,ino);  // load INODE into an minode
    5-2. initialize mip->INODE as a DIR INODE;
           mip->INODE.i_block[0] = blk; other i_block[ ] are 0;
           mark minode modified (dirty);
           iput(mip);  // write INODE back to disk
    5-3. make data block 0 of INODE to contain . and .. entries;
         write to disk block blk.
    5-4. enter_child(pmip, ino, basename); which enters
         (ino, basename) as a DIR entry to the parent INODE;
 6. increment parent INODE's links_count by 1 and mark pmip dirty;
    iput(pmip);
}
```

Creat creates an empty regular file. The algorithm of creat is similar to mkdir. The algorithm of creat is as follows.

```
/******************* Algorithm of creat ()  *******************/
creat(char * pathname)
{
  This is similar to mkdir() except
  (1). the INODE.i_mode field is set to REG file type, permission
       bits set to 0644 for rw-r--r--, and
  (2). no data block is allocated, so the file size is 0.
  (3). Do not increment parent INODE's links_count
}
```

It is noted that the above creat algorithm differs from that in Unix/Linux. The new file's permissions are set to 0644 by default and it does not open the file for WRITE mode and return a file descriptor. In practice, creat is rarely used as a stand-alone syscall. It is used internally by the kopen() function, which may create a file, open it for WRITE and return a file descriptor. The open operation will be described later.

Mknod creates a special file which represents either a char or block device with a device number = (major, minor). The algorithm of mknod is

```
/*********** Algorithm of mknod ***********/
mknod(char *name, int type, int device_number)
{
  This is similar to creat() except
  (1). the default parent directory is /dev;
  (2). INODE.i_mode is set to CHAR or BLK file type;
  (3). INODE.I_block[0] contains device_number=(major, minor);
}
```

8.12.4.2 chdir-getcwd-stat

Each process has a Current Working Directory (CWD), which points to the CWD minode of the process in memory. chdir (pathname) changes the CWD of a process to pathname. getcwd() returns the absolute pathname of CWD. stat() returns the status information of a file in a STAT structure. The algorithm of chdir() is

```
/********** Algorithm of chdir ************/
int chdir(char *pathname)
{
  (1). get INODE of pathname into a minode;
  (2). verify it is a DIR;
  (3). change running process CWD to minode of pathname;
  (4). iput(old CWD); return 0 for OK;
}
```

getcwd() is implemented by recursion. Starting from CWD, get the parent INODE into memory. Search the parent INODE's data block for the name of the current directory and save the name string. Repeat the operation for the parent INODE until the root directory is reached. Construct an absolute pathname of CWD on return. Then copy the absolute pathname to user space. stat(pathname, STAT *st) returns the information of a file in a STAT structure. The algorithm of stat is

```
/********* Algorithm of stat  *********/
int stat(char *pathname, STAT *st) // st points to STAT struct
{
  (1). get INODE of pathname into a minode;
  (2). copy (dev, ino) to (st_dev, st_ino) of STAT struct in Umode
```

```
    (3). copy other fields of INODE to STAT structure in Umode;
    (4). iput(minode); retrun 0 for OK;
}
```

8.12.4.3 rmdir

As in Unix/Linux, in order to rm a DIR, the directory must be empty, for the following reasons. First, removing a non-empty directory implies removing all the files and subdirectories in the directory. Although it is possible to implement a rrmdir() operation, which recursively removes an entire directory tree, the basic operation is still to remove one directory at a time. Second, a non-empty directory may contain files that are actively in use, e.g. opened for read/write, etc. Removing such a directory is clearly unacceptable. Although it is possible to check whether there are any active files in a directory, it would incur too much overhead in the kernel. The simplest way out is to require that a directory to be removed must be empty. The algorithm of rmdir() is

```
/******** Algorithm of rmdir ********/
rmdir(char *pathname)
{
  1. get in-memory INODE of pathname:
     ino = getino(&de, pathanme);
     mip = iget(dev,ino);
  2. verify INODE is a DIR (by INODE.i_mode field);
     minode is not BUSY (refCount = 1);
     DIR is empty (traverse data blocks for number of entries = 2);
  3. /* get parent's ino and inode */
     pino = findino(); //get pino from .. entry in INODE.i_block[0]
     pmip = iget(mip->dev, pino);
  4. /* remove name from parent directory */
     findname(pmip, ino, name); //find name from parent DIR
     rm_child(pmip, name);
  5. /* deallocate its data blocks and inode */
     truncat(mip);   // deallocate INODE's data blocks
  6. deallocate INODE
     idalloc(mip->dev, mip->ino); iput(mip);
  7. dec parent links_count by 1;
     mark parent dirty; iput(pmip);
  8. return 0 for SUCCESS.
}
```

8.12.4.4 link-unlink

The link_unlikc.c file implements link and unlink. link(old_file, new_file) creates a hard link from new_file to old_file. Hard links can only be to regular files, not DIRs, because linking to DIRs may create loops in the file system name space. Hard link files share the same inode. Therefore, they must be on the same device. The algorithm of link is

```
/********* Algorithm of link ********/
link(old_file, new_file)
{
  1. // verify old_file exists and is not DIR;
     oino = getino(&odev, old_file);
     omip = iget(odev, oino);
     check file type (cannot be DIR).
  2. // new_file must not exist yet:
     nion = get(&ndev, new_file) must return 0;
     ndev of dirname(newfile) must be same as odev
```

```
    3. // creat entry in new_parent DIR with same ino
       pmip -> minode of dirname(new_file);
       enter_name(pmip, omip->ino, basename(new_file));
    4. omip->INODE.i_links_count++;
       omip->dirty = 1;
       iput(omip);
       iput(pmip);
}
```

unlink decrements the file's links_count by 1 and deletes the file name from its parent DIR. When a file's links_count reaches 0, the file is truly removed by deallocating its data blocks and inode. The algorithm of unlink() is

```
/*********** Algorithm of unlink *********/
unlink(char *filename)
{
  1. get filenmae's minode:
     ino = getino(&dev, filename);
     mip = iget(dev, ino);
     check it's a REG or SLINK file
  2. // remove basename from parent DIR
     rm_child(pmip, mip->ino, basename);
     pmip->dirty = 1;
     iput(pmip);
  3. // decrement INODE's link_count
     mip->INODE.i_links_count--;
     if (mip->INODE.i_links_count > 0){
        mip->dirty = 1; iput(mip);
     }
  4. if (!SLINK file)  // assume:SLINK file has no data block
        truncate(mip); // deallocate all data blocks
     deallocate INODE;
     iput(mip);
}
```

8.12.4.5 symlink-readlink

symlink(old_file, new_file) creates a symbolic link from new_file to old_file. Unlike hard links, symlink can link to anything, including DIRs or files not on the same device. The algorithm of symlink is

```
Algorithm of symlink(old_file, new_file)
{
  1. check: old_file must exist and new_file not yet exist;
  2. create new_file; change new_file to SLINK type;
  3. // assume length of old_file name <= 60 chars
     store old_file name in newfile's INODE.i_block[ ] area.
     mark new_file's minode dirty;
     iput(new_file's minode);
  4. mark new_file parent minode dirty;
     iput(new_file's parent minode);
}
```

readlink(file, buffer) reads the target file name of a SLINK file and returns the length of the target file name. The algorithm of readlink() is

```
Algorithm of readlink (file, buffer)
{
  1. get file's INODE into memory; verify it's a SLINK file
  2. copy target filename in INODE.i_block into a buffer;
  3. return strlen((char *)mip->INODE.i_block);
}
```

8.12.4.6 Other Level-1 Functions

Other level-1 functions include stat, access, chmod, chown, touch, etc. The operations of all such functions are of the same pattern:

```
(1). get the in-memory INODE of a file by
          ino = getinod(&dev, pathname);
          mip = iget(dev,ino);
(2). get information from the INODE or modify the INODE;
(3). if INODE is modified, mark it DIRTY for write back;
(4). iput(mip);
```

8.12.5 Implementation of Level-2 FS

Level-2 of FS implements read/write operations of file contents. It consists of the following functions: open, close, lseek, read, write, opendir and readdir.

8.12.5.1 open-close-lseek

The file open_close_lseek.c implements open(), close() and lseek(). The system call

int open(char *filename, int flags);

opens a file for read or write, where flags = 0|1|2|3|4 for R|W|RW|APPEND, respectively. Alternatively, flags can also be specified as one of the symbolic constants O_RDONLY, O_WRONLY, O_RDWR, which may be bitwise or-ed with file creation flags O_CREAT, O_APPEND, O_TRUNC. These symbolic constants are defined in type.h. On success, open() returns a file descriptor for subsequent read()/write() system calls. The algorithm of open() is

```
/*************** Algorithm of open() ************/
int open(file, flags)
{
 1. get file's minode:
    ino = getino(&dev, file);
    if (ino==0 && O_CREAT){
       creat(file); ino = getino(&dev, file);
    }
    mip = iget(dev, ino);
 2. check file INODE's access permission;
    for non-special file, check for incompatible open modes;
 3. allocate an openTable entry;
    initialize openTable entries;
    set byteOffset = 0 for R|W|RW; set to file size for APPEND mode;
 4. Search for a FREE fd[ ] entry with the lowest index fd in PROC;
    let fd[fd]point to the openTable entry;
 5. unlock minode;
    return fd as the file descriptor;
}
```

Figure 8.6 show the data structure created by open(). In the figure, (1) is the PROC structure of the process that calls open(). The returned file descriptor, fd, is the index of the fd[] array in the PROC structure. The contents of fd[fd] points to a OFT, which points to the minode of the file. The OFT's refCount represents the number of processes which share the same instance of an opened file. When a process first opens a file, it sets the refCount in the OFT to 1. When a process forks, it copies all opened file descriptors to the child process, so that the child process share all the opened file descriptors with the parent, which increments the refCount of every shared OFT by 1. When a process closes a file descriptor, it decrements the OFT's refCount by 1 and clears its fd[] entry to 0. When an OFT's refCount reaches 0, it calls iput() to dispose of the the minode and deallocates the OFT. The OFT's offset is a conceptual pointer to the current byte position in the file for read/write. It is initialized to 0 for R|W|RW mode or to file size for APPEND mode.

In EOS, lseek(fd, position) sets the offset in the OFT of an opened file descriptor to the byte position relative to the beginning of the file. Once set, the next read/write begins from the current offset position. The algorithm of lseek () is trivial. For files opened for READ, it only checks the position value to ensure it's within the bounds of [0, file_size]. If fd is a regular file opened for WRITE, lseek allows the byte offset to go beyond the current file size but it does not allocate any disk block for the file. Disk blocks will be allocated when data are actually written to the file. The algorithm of closing a file descriptor is

```
/*************  Algorithm of close()  *****************/
int close(int fd)
{
   (1). check fd is a valid opened file descriptor;
   (2). if (PROC's fd[fd] != 0){
   (3).      if (openTable's mode == READ/WRITE PIPE)
                return close_pipe(fd); // close pipe descriptor;
   (4).      if (--refCount == 0){ // if last process using this OFT
                lock(minodeptr);
                iput(minode);         // release minode
             }
        }
   (5). clear fd[fd] = 0;             // clear fd[fd] to 0
   (6). return SUCCESS;
}
```

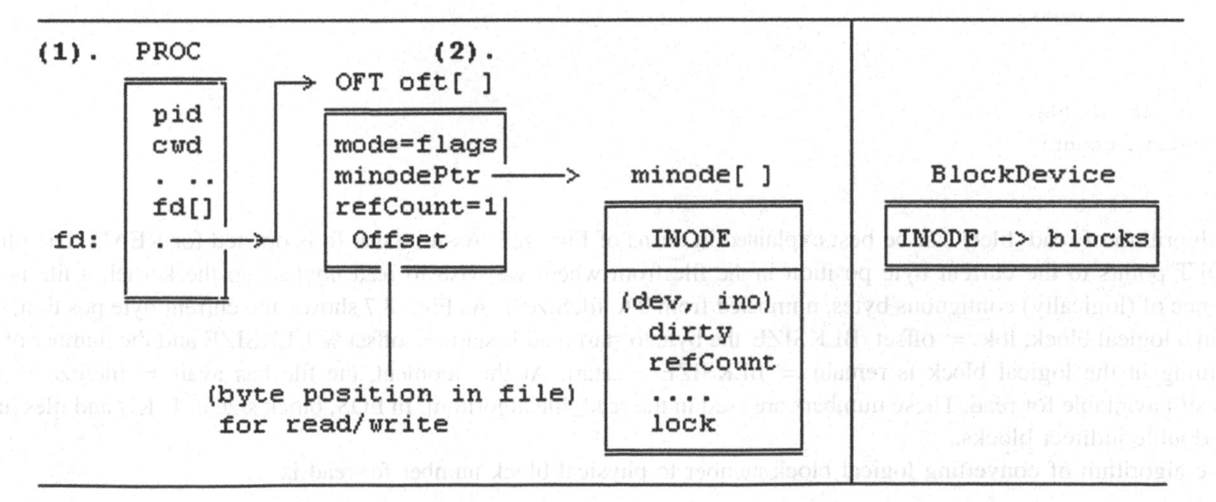

Fig. 8.6 Data structures of open()

8.12.5.2 Read Regular Files

The system call **int read(int fd, char buf[], int nbytes)** reads nbytes from an opened file descriptor into a buffer area in user space. read() invokes kread() in kernel, which implements the read system call. The algorithm of kread() is

```
/**************** Algorithm of kread() in kernel ****************/
int kread(int fd, char buf[ ], int nbytes, int space) //space=K|U
{
 (1). validate fd; ensure oft is opened for READ or RW;
 (2). if (oft.mode = READ_PIPE)
        return read_pipe(fd, buf, nbytes);
 (3). if (minode.INODE is a special file)
        return read_special(device,buf,nbytes);
 (4). (regular file):
        return read_file(fd, buf, nbytes, space);
}
```

```
/*************** Algorithm of read regular files ****************/
int read_file(int fd, char *buf, int nbytes, int space)
{
(1). lock minode;
(2). count = 0; avil = fileSize - offset;
(3). while (nbytes){
       compute logical block: lbk   = offset / BLKSIZE;
       start byte in block:    start = offset % BLKSIZE;
(4).   convert logical block number, lbk, to physical block number,
       blk, through INODE.i_block[ ] array;
(5).   read_block(dev, blk, kbuf); // read blk into kbuf[BLKSIZE];
       char *cp = kbuf + start;
       remain = BLKSIZE - start;
(6)    while (remain){// copy bytes from kbuf[ ] to buf[ ]
          (space)? put_ubyte(*cp++, *buf++) : *buf++ = *cp++;
          offset++; count++;           // inc offset, count;
          remain--; avil--; nbytes--;  // dec remain, avil, nbytes;
          if (nbytes==0 || avail==0)
               break;
       }
    }
(7). unlock minode;
(8). return count;
}
```

The algorithm of read_file() can be best explained in terms of Fig. 8.7. Assume that fd is opened for READ. The offset in the OFT points to the current byte position in the file from where we wish to read nbytes. To the kernel, a file is just a sequence of (logically) contiguous bytes, numbered from 0 to fileSize-1. As Fig. 8.7 shows, the current byte position, offset, falls in a logical block, lbk = offset /BLKSIZE, the byte to start read is start = offset % BLKSIZE and the number of bytes remaining in the logical block is remain = BLKSIZE − start. At this moment, the file has avail = fileSize − offset bytes still available for read. These numbers are used in the read_file algorithm. In EOS, block size is 4 KB and files have at most double indirect blocks.

The algorithm of converting logical block number to physical block number for read is

```
/* Algorithm of Converting Logical Block to Physical Block */
u32 map(INODE, lbk){                    // convert lbk to blk
    if (lbk < 12)                       // direct blocks
        blk = INODE.i_block[lbk];
    else if (12 <= lbk < 12+256){ // indirect blocks
```

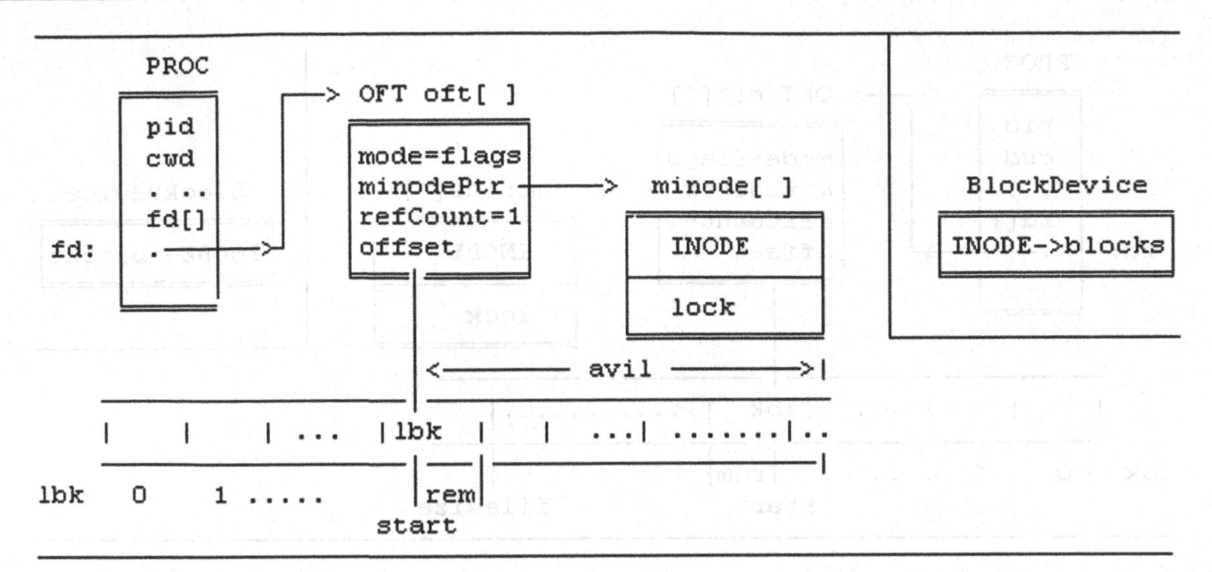

Fig. 8.7 Data structures for Read_file()

```
        read INODE.i_block[12] into u32 ibuf[256];
        blk = ibuf[lbk-12];
    }
    else{                          // doube indirect blocks
        read INODE.i_block[13] into u32 dbuf[256];
        lbk -= (12+256);
        dblk = dbuf[lbk / 256];
        read dblk into dbuf[ ];
        blk  = dbuf[lbk % 256];
    }
    return blk;
}
```

8.12.5.3 Write Regular Files

The system call **int write(int fd, char ubuf[], int nbytes)** writes nbytes from ubuf in user space to an opened file descriptor and returns the actual number of bytes written. write() invokes kwrite() in kernel, which implements the write system call. The algorithm of kwrite() is

```
/*************** Algorithm of kwrite in kernel ************/
int kwrite(int fd, char *ubuf, int nbytes)
{
 (1). validate fd; ensure OFT is opened for write;
 (2). if (oft.mode = WRITE_PIPE)
        return write_pipe(fd, buf, nbytes);
 (3). if (minode.INODE is a special file)
        return write_special(device,buf.nbytes);
 (4). return write_file(fd, ubuf, nbytes);
}
```

The algorithm of write_file() can be best explained in terms of Fig. 8.8.

In Fig. 8.8, the offset in the OFT is the current byte position in the file for write. As in read_file(), it first computes the logical block number, lbk, the start byte position and the number of bytes remaining in the logical block. It converts the

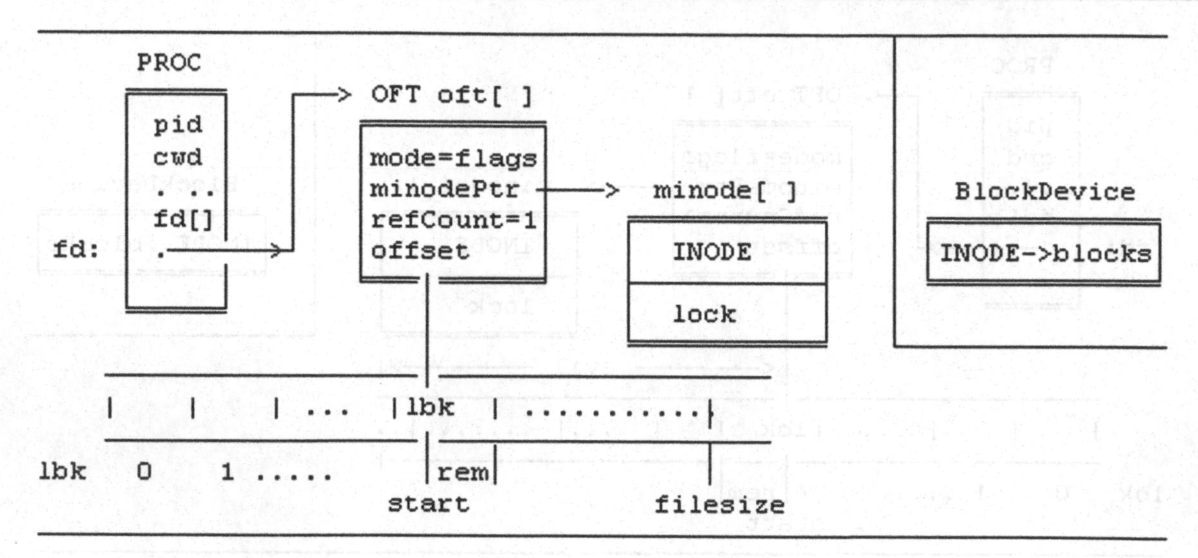

Fig. 8.8 Data structures for write_file()

logical block to physical block through the file's INODE.i_block array. Then it reads the physical block into a buffer, writes data to it and writes the buffer back to disk. The following shows the write_file() algorithm.

```
/************** Algorithm of write regular file ***************/
int write_file(int fd, char *ubuf, int nbytes)
{
(1). lock minode;
(2). count = 0;          // number of bytes written
(3). while (nbytes){
        compute logical block: lbk = oftp->offset / BLOCK_SIZE;
        compute start byte:  start = oftp->offset % BLOCK_SIZE;
(4).    convert lbk to physical block number, blk;
(5).    read_block(dev, blk, kbuf); //read blk into kbuf[BLKSIZE];
        char *cp = kbuf + start; remain = BLKSIZE - start;
(6)     while (remain){  // copy bytes from kbuf[ ] to ubuf[ ]
           put_ubyte(*cp++, *ubuf++);
           offset++;  count++;       // inc offset, count;
           remain --; nbytes--;      // dec remain, nbytes;
           if (offset > fileSize) fileSize++; // inc file size
           if (nbytes <= 0) break;
        }
(7).    wrtie_block(dev, blk, kbuf);
     }
(8). set minode dirty = 1; // mark minode dirty for iput()
     unlock(minode);
     return count;
}
```

The algorithm of converting logical block to physical block for write is similar to that of read, except for the following difference. During write, the intended data block may not yet exist. If a direct block does not exist, it must be allocated and recorded in the INODE. If the indirect block does not exist, it must be allocated and initialized to 0. If an indirect data block does not exist, it must be allocated and recorded in the indirect block, etc. The reader may consult the write.c file for details.

8.12.5.4 Read-Write Special Files

In kread() and kwrite(), read/write pipes and special files are treated differently. Read/write pipes are implemented in the pipe mechanism in the EOS kernel. Here we only consider read/write special files. Each special file has a file name in the /dev directory. The file type in a special file's inode is marked as special, e.g. 0060000 = block device, 0020000 = char device, etc. Since a special file does not have any disk block, i_block[0] of its INODE stores the device's (major, minor) number, where major = device type and minor = unit number of that device type. For example, /dev/sdc0 = (3,0) represents the entire SDC, /dev/sdc1 = (3,1) represents the first partition of a SDC, /dev/tty0 = (4,0) and /dev/ttyS1 = (5,1), etc. The major device number is an index in a device switch table, dev_sw[], which contains pointers to device driver functions, as in

```
struct dev_sw {
    int (*dev_read)();
    int (*dev_write)();
} dev_sw[];
```

Assume that int nocall(){} is an empty function, and

```
sdc_read(), sdc_write(),            // SDC read/write
console_read(), console_write(),  // console read/write
serial_read(), serial_write(),    // serail port read/write
```

are device driver functions. The device switch table is set up to contain the driver function pointers.

```
struct dev_sw dev_sw[ ] =
{ //   read                write
  //--------             --------
     nocall,             nocall,            // 0=/dev/null
     nocall,             nocall,            // 1=kernel memory
     nocall,             nocall,            // 2=FD (no FD in EOS)
     sdc_read,           sdc_write,         // 3=SDC
     console_read,       console_write,     // 4=console
     serial_read,        serial_write       // 5=serial ports
};
```

Then read/write a special file becomes

(1). get special file's (major, minor) number from INODE.i_block[0];
(2). return (*dev_sw[major].dev_read) (minor, parameters); //READ
OR return (*dev_sw[major].dev_write)(minor, parameters); //WRITE

(2). invokes the corresponding device driver function, passing as parameters the minor device number and other parameters as needed. The device switch table is a standard technique used in all Unix-like systems. It not only makes the I/O subsystem structure clear but also greatly reduces the read/write code size.

8.12.5.5 Opendir-Readdir

Unix considers everything as a file. Therefore, we should be able to open a DIR for read just like a regular file. From a technical point of view, there is no need for a separate set of opendir() and readdir() functions. However, different Unix-like systems may have different file systems. It may be difficult for users to interpret the contents of a DIR file. For this reason, POSIX specifies opendir and readdir operations, which are independent of file systems. Support for opendir is trivial; it's the same open system call, but readdir() has the form

```
struct dirent *ep = readdir(DIR *dp);
```

which returns a pointer to a dirent structure on each call. This can be implemented in user space as a library I/O function. Since EOS does not yet support user-level I/O streams by library functions, we shall implement opendir() and readir() as system calls.

```
int opendir(pathaname)
{ return open(pathname, O_RDONLY|O_DIR); }
```

where O_DIR is a bit pattern for opening the file as a DIR. In the open file table, the mode field contains the O_DIR bit, which is used to routes readdir syscalls to the kreaddir() function.

```
int kreaddir(int fd, struct udir *dp) // struct udir{DIR; name[256]};
{
   // same as kread() in kernel except:
   use the current byte offset in OFT to read the next DIR record;
   copy the DIR record into *udir in User space;
   advance offset by DIR entry's rec_len;
}
```

User mode programs must use the readdir(fd, struct udir *dir) system call instead of the readdir(DIR *dp) call.

8.12.6 Implementation of Level-3 FS

Level-3 of FS implements mount and umount of file systems and file protection.

8.12.6.1 mount-umount

The mount command, mount filesys mount_point, mounts a file system to a mount_point directory. It allows the file system to include other file systems as parts of an existing file system. The data structures used in mount are the MOUNT table and the in-memory minode of the mount_point directory. The algorithm of mount is

```
/************** Algorithm of mount **************/
mount()  // Usage: mount [filesys mount_point]
{
1. If no parameter, display current mounted file systems;
2. Check whether filesys is already mounted:
   The MOUNT table entries contain mounted file system (device) names
   and their mounting points. Reject if the device is already mounted.
   If not, allocate a free MOUNT table entry.
3. filesys is a special file with a device number dev=(major,minor).
   Read filesys' superblock to verify it is an EXT2 FS.
4. find the ino, and then the minode of mount_point:
        call ino = get_ino(&dev, pathname);  to get ino:
        call mip = iget(dev, ino); to load its inode into memory;
5. Check mount_point is a DIR and not busy, e.g. not someone's CWD.
6. Record dev and filesys name in the MOUNT table entry;
   also, store its ninodes, nblocks, etc. for quick reference.
7. Mark mount_point's minode as mounted on (mounted flag=1) and let
   it point at the MOUNT table entry, which points back to the
   mount_point minode.
}
```

The operation Umount filesys detaches a mounted file system from its mounting point, where filesys may be either a special file name or a mounting point directory name. The algorithm of umount is

```
/****************** Algorithm of umount *******************/
umount(char *filesys)
{
 1. Search the MOUNT table to check filesys is indeed mounted.
 2. Check (by checking all active minode[].dev) whether any file is
    active in the mounted filesys; If so, reject;
 3. Find the mount_point's in-memory inode, which should be in memory
    while it's mounted on. Reset the minode's mounted flag to 0; then
    iput() the minode.
}
```

8.12.6.2 Implications of mount

While it is easy to implement mount and umount, there are implications. With mount, we must modify the get_ino(&dev, pathname) function to support crossing mount points. Assume that a file system, newfs, has been mounted on the directory /a/b/c/. When traversing a pathname, mount point crossing may occur in both directions.

(1). Downward traversal: When traversing the pathname /a/b/c/x, once we reach the minode of /a/b/c, we should see that the minode has been mounted on (mounted flag = 1). Instead of searching for x in the INODE of /a/b/c, we must
. Follow the minode's mountTable pointer to locate the mount table entry.
. From the newfs's dev number, get its root (ino = 2) INODE into memory.
. Then continue search for x under the root INODE of newfs.

(2). Upward traversal: Assume that we are at the directory /a/b/c and traversing upward, e.g. cd ../../, which will cross the mount point /a/b/c. When we reach the root INODE of the mounted file system, we should see that it is a root directory (ino = 2) but its dev number differs from that of the real root, so it is not the real root yet. Using its dev number, we can locate its mount table entry, which points to the mounted minode of /a/b/c/. Then, we switch to the minode of /a/b/c and continue the upward traversal. Thus, crossing mount point is like a monkey or squirrel hoping from one tree to another and then back.

8.12.6.3 File Protection

In Unix, file protection is by permission checking. Each file's INODE has an i_mode field, in which the low 9 bits are for file permissions. The 9 permission bits are

```
owner  group  other
-----  -----  -----
r w x  r w x  r w x
------ ------ -----
```

where the first 3 bits apply to the owner of the file, the second 3 bits apply to users in the same group as the owner and the last 3 bits apply to all others. For directories, the x bit indicates whether a process is allowed to go into the directory. Each process has a uid and a gid. When a process tries to access a file, the file system checks the process uid and gid against the file's permission bits to determine whether it is allowed to access the file with the intended mode of operation. If the process does not have the right permission, it is not allowed to access the file. For the sake of simplicity, EOS ignores gid. It uses only the process uid to check for file access permission.

8.12.6.4 Real and Effective uid

In Unix, a process has a real uid and an effective uid. The file system checks the access rights of a process by its effective uid. Under normal conditions, the effective uid and real uid are identical. When a process executes a setuid program, which has the setuid bit (bit 11) in the file's i_mode field turned on, the process' effective uid becomes the uid of the program. While executing a setuid program, the process effectively becomes the owner of the program. For example, when a process executes the mail program, which is a setuid program owned by the superuser, it can write to a mail file of another user. When a process finishes executing a setuid program, it reverts back to the real uid. For simplicity reasons, EOS does not yet support effective uid. Permission checking is based on real uid.

8.12.6.5 File Locking

File locking is a mechanism which allows a process to set locks on a file, or parts of a file to prevent race conditions when updating files. File locks can be either shared, which allows concurrent reads, or exclusive, which enforces exclusive write. File locks can also be mandatory or advisory. For example, Linux supports both shared and exclusive files locks but file locking is only advisory. In Linux, file locks can be set by the fcntl() system call and manipulated by the flock() system call. In EOS, file locking is enforced only in the open() syscall of non-special files. When a process tries to open a non-special file, the intended mode of operation is checked for compatibility. The only compatible modes are READs. If a file is already opened for updating mode, i.e. W|RW|APPEND, it cannot be opened again. This does not apply to special files, e.g. terminals. A process may open its terminal multiple times even if the modes are incompatible. This is because access to special files is ultimately controlled by device drivers.

File operations from User mode are all based on system calls. As of now, EOS does not yet support library file I/O functions on file streams.

8.13 Block Device I/O Buffering

The EOS file system uses I/O buffering for block device (SDC) to improve the efficiency of file I/O operations. I/O buffering is implemented in the buffer.c file. When EOS starts, it calls binit() to initialize 256 I/O buffers. Each I/O buffer consists of a header for buffer management and a 4 KB data area for a block of SDC data. The 1 MB data area of the I/O buffers is allocated in 4 MB–5 MB of the EOS system memory map. Each buffer (header) has a lock semaphore for exclusive access to the buffer. The buffer management algorithm is as follows.

8.14 I/O Buffer Management Algorithm

(1). bfreelist = a list of free buffers. Initially all buffers are in the bfreelist.
(2). dev_tab = a device table for the SDC partition. It contains the device id number, the start sector number and size in number of sectors. It also contains two buffer linked lists. The dev_list contains all I/O buffers that are assigned to the device, each identified by the buffer's (dev, blk) numbers. The device I/O queue contains buffers for pending I/O.
(3). When a process needs to read a SDC block data, it calls

```
struct buffer *bread(dev, blk)
{
  struct buffer *bp = getblk(dev, blk); // get a bp =(dev,blk)
  if (bp data invalid){
     mark bp for READ
     start_io(bp);           // start I/O on buffer
     P(&bp->iodone);         // wait for I/O completion
  }
  return bp;
}
```

(4). After reading data from a buffer, the process releases the buffer by brelse(bp). A released buffer remains in the device list for possible reuse. It is also in the bfreelist if it is not in use.
(5). When a process writes data to a SDC block, it calls

```
int bwrite(dev, blk)
{
   struct buffer *bp;
   if (write new block or a complete block)
      bp = getblk(dev, blk); // get a buffer for (dev,blk)
   else                      //write to existing block
      bp = bread(dev,blk);   // get a buffer with valid data
```

```
    write data to bp;
    mark bp data valid and dirty (for delayed write-back)
    brelse(bp);                     // release bp
  }
```

(6). Dirty buffer contain valid data, which can be read/write by any process. A dirty buffer is written back to SDC only when it is to be reassigned to a different block, at which time, it is written out by

```
  awrite(struct buffer *bp)  // for ASYNC write
  {
    mark bp ASYNC write;
    start_io(bp);  // do not wait for completion
  }
```

When an ASYNC write operation completes, the SDC interrupt handler turns off the buffer's ASYNC and dirty flags and releases the buffer.

```
int start_io(struct buf *bp) // start I/O on bp
{
  int ps = int_off();
  enter bp into dev_tab.IOqueue;
  if (bp is first in dev_tab.IOqueue){
     if(bp is for READ)
       get_block(bp->blk, bp->buf);
     else  // WRITE
       put_block(bp->blk, bp->buf);
  }
  int_on(ps);
}
```

```
(7). SDC Interrupt Handler:
  {
     bp = dequeue(dev_tab.IOqueue);
     if (bp==READ){
         mark bp data valid;
         V(&bp->iodone);  // unblock process waiting on bp
     else{
         turn off bp ASYNC flag
         brelse(bp);
     }
     bp = dev_tab.IOqueue;
     if (bp){  // I/O queue non-empty
        if (bp==READ)
          get_block(bp->blk, bp->buf);
        else
          put_block(bp->blk, bp->buf);
        }
  }
```

(8). getblk() and brelse() form the core of buffer management. The following lists the algorithms of getblk() and brelse(), which use semaphores for synchronization.

```
SEMAPHORE freebuf = NBUF; // counting semaphore
Each buffer has SEMAPHOREs lock = 1; io_done = 0;

 struct buf *getblk(int dev, int blk)
{
  struct buf *bp;
  while(1){
    P(&freebuf);              // get a free buf
    bp = search_dev(dev,blk);
    if (bp){                  // buf in cache
       hits++;                // buffer hits number
       if (bp->busy){         // if buf busy
         V(&freebuf);         // bp not in freelist, give up the free buf
         P(&bp->lock);        // wait for bp
         return bp;
       }
       // bp in cache and not busy
       bp->busy = 1;          // mark bp busy
       out_freelist(bp);
       P(&bp->lock);          // lock bp
       return bp;
    }
    // buf not in cache; already has a free buf in hand
    lock();
      bp = freelist;
      freelist = freelist->next_free;
    unlock();
    P(&bp->lock);             // lock the buffer
    if (bp->dirty){           // delayed write buf, can't use it
       awrite(bp);
       continue;              // continue while(1) loop
    }
    // bp is a new buffer; reassign it to (dev,blk)
    if (bp->dev != dev){
      if (bp->dev >= 0)
         out_devlist(bp);
      bp->dev = dev;
      enter_devlist(bp);
    }
    bp->dev = dev; bp->blk = blk;
    bp->valid = 0; bp->async = 0; bp->dirty = 0;
    return bp;
  }
}
int brelse(struct buf *bp)
{
  if (bp->lock.value < 0){ // bp has waiter
      V(&bp->lock);
      return;
  }
  if (freebuf.value < 0 && bp->dirty){
      awrite(bp);
```

```
        return;
    }
    enter_freelist(bp);        // enter b pint bfreeList
    bp->busy = 0;              // bp non longer busy
    V(&bp->lock);              // unlock bp
    V(&freebuf);               // V(freebuf)
}
```

Since both processes and the SDC interrupt handler access and manipulate the free buffer list and device I/O queue, interrupts are disabled when processes operate on these data structures to prevent any race conditions.

8.14.1 Performance of the I/O Buffer Cache

With I/O buffering, the hit ratio of the I/O buffer cache is about 40% when the system starts. During system operation, the hit ratio is constantly above 60%. This attests to the effectiveness of the I/O buffering scheme.

8.15 User Interface

All user commands are ELF executable files in the /bin directory (on the root device). From the EOS system point of view, the most important user mode programs are init, login and sh, which are necessary to start up the EOS system. In the following, we shall explain the roles and algorithms of these programs.

8.15.1 The INIT Program

When EOS starts, the initial process P0 is handcrafted. P0 creates a child P1 by loading the /bin/init file as its Umode image. When P1 runs, it executes the init program in user mode. Henceforth, P1 plays the same role as the INIT process in Unix/Linux. A simple init program, which forks only one login process on the system console, is shown below. The reader may modify it to fork several login processes, each on a different terminal.

```
/********************** init.c file *****************/
#include "ucode.c"
int console;
int parent()        // P1's code
{
  int pid, status;
  while(1){
    printf("INIT : wait for ZOMBIE child\n");
    pid = wait(&status);
    if (pid==console){   // if console login process died
       printf("INIT: forks a new console login\n");
       console = fork(); // fork another one
       if (console)
           continue;
       else
          exec("login /dev/tty0"); // new console login process
    }
    printf("INIT: I just buried an orphan child proc %d\n", pid);
  }
}
main()
{
```

```
    int in, out;    // file descriptors for terminal I/O
    in  = open("/dev/tty0", O_RDONLY); // file descriptor 0
    out = open("/dev/tty0", O_WRONLY); // for display to console
    printf("INIT : fork a login proc on console\n");
    console = fork();
    if (console)  // parent
       parent();
    else           // child: exec to login on tty0
       exec("login /dev/tty0");
}
```

8.15.2 The Login Program

All login processes executes the same login program, each on a different terminal, for users to login. The algorithm of the login program is

```
/****************** Algorithm of login *******************/
// login.c : Upon entry, argv[0]=login, argv[1]=/dev/ttyX
#include "ucode.c"
int in, out, err;   char name[128],password[128]
main(int argc, char *argv[])
{
  (1). close file descriptors 0,1 inherited from INIT.
  (2). open argv[1] 3 times as in(0), out(1), err(2).
  (3). settty(argv[1]); // set tty name string in PROC.tty
  (4). open /etc/passwd file for READ;
       while(1){
  (5).     printf("login:");     gets(name);
           printf("password:"); gets(password);
           for each line in /etc/passwd file do{
               tokenize user account line;
  (6).         if (user has a valid account){
  (7).             change uid, gid to user's uid, gid; // chuid()
                   change cwd to user's home DIR       // chdir()
                   close opened /etc/passwd file       // close()
  (8).             exec to program in user account     // exec()
               }
           }
           printf("login failed, try again\n");
       }
}
```

8.15.3 The sh Program

After login, the user process typically executes the command interpreter sh, which gets command lines from the user and executes the commands. For each command line, if the command is non-trivial, i.e. not cd or exit, sh forks a child process to execute the command line and waits for the child to terminate. For simple commands, the first token of a command line is an executable file in the /bin directory. A command line may contain I/O redirection symbols. If so, the child sh handles I/O redirections first. Then it uses exec to change image to execute the command file. When the child process terminates, it wakes up the parent sh, which prompts for another command line. If a command line contains a pipe symbol, such as cmd1 | cmd2, the child sh handles the pipe by the following do_pipe algorithm.

```
/***************** do_pipe Algorithm ***************/
int pid, pd[2];
pipe(pd);  // create a pipe: pd[0]=READ, pd[1]=WRITE
pid = fork();          // fork a child to share the pipe
if (pid){              // parent: as pipe READER
   close(pd[1]);       // close pipe WRITE end
   dup2(pd[0], 0);     // redirect stdin to pipe READ end
   exec(cmd2);
}
else{                  // child : as pipe WRITER
   close(pd[0]);       // close pipe READ end
   dup2(pd[1], 1);     // redirect stdout to pipe WRITE end
   exec(cmd1);
}
```

Multiple pipes are handled recursively, from right to left.

8.16 Demonstration of EOS

8.16.1 EOS Startup

Figure 8.9 shows the startup screen of the EOS system. After booting up, it first initializes the LCD display, configures vectored interrupts and initializes device drivers. Then it initializes the EOS kernel to run the initial process P0. P0 builds page directories, page tables and switches page directory to use dynamic 2-level paging. P0 creates the INIT process P1 with /bin/init as Umode image. Then it switches process to run P1 in User mode. P1 forks a login process P2 on the console and another login process P3 on a serial terminal. When creating a new process, it shows the dynamically allocated page frames of the process image. When a process terminates, it releases the allocated page frames for reuse. Then P1 executes in a loop, waiting for any ZOMBIE child.

Each login process opens its own terminal special file as stdin (0), stdout (1), stderr (2) for terminal I/O. Then each login process displays a login: prompt on its terminal and waits for a user to login. When a user tries to login, the login process validates the user by checking the user account in the /etc/passwd file. After a user login, the login process becomes the user process by acquiring its uid and changing directory to the user's home directory. Then the user process changes image to execute the command interpreter sh, which prompts for user commands and execute the commands.

8.16.2 Command Processing in EOS

Figure 8.10 shows the processing sequence of the command line "cat f1 | grep line" by the EOS sh process (P2).

For any non-trivial command line, sh forks a child process to execute the command and waits for the child to terminate. Since the command line has a pipe symbol, the child sh (P8) creates a pipe and forks a child (P9) to share the pipe. Then the child sh (P8) reads from the pipe and executes the command grep. The child sh (P9) writes to the pipe and executes command cat. P8 and P9 are connected by a pipe and run concurrently. When the pipe reader (P8) terminates, it sends the child P9 as an orphan to the INIT process P1, which wakes up to free the orphan process P9.

8.16.3 Signal and Exception Handling in EOS

In EOS, exceptions are handled by the unified framework of signal processing. We demonstrate exception and signal processing in EOS by the following examples.

```
QEMU
                Welcome to EOS in ARM                          00:02:20
LCD init(): binit vectorInterrupt_init()
kbd_init(): uart[0-4] init(): timer_init(): timer_start: sdc_init(): done
kernel_init()
building pgdirs at 5MB: building P0's pgdirs at 32KB
build 512 Kmode pgtables at 5MB: build 64 proc pgdir at 6MB
switch pgdir to use 2-level paging : switched pgdir OK
build pfreeList: start=0x800000  end=0x10000000  : 63487  4KB entries
mounting root : SUPER magic=0xEF53  blksize=4096  bmap=9  imap=10  iblock=11
/dev/sda1  mounted on / OK
fork1: pgtables= 0x801000 0x902000 0xA03000 0xB04000
kfork:load /bin/init proc 0  kforked a child 1
readyQueue = [1 256 ]->NULL
P0 switch to P1
proc 1  in Umode argc=2  init start
fork1: pgtables= 0xC06000 0xD07000 0xE08000 0xF09000
FORK: proc 1  forked a child 2
readyQueue = [2 256 ]->[0 0 ]->NULL
fork1: pgtables= 0x100B000 0x110C000 0x120D000 0x130E000
FORK: proc 1  forked a child 3
readyQueue = [2 256 ]->[3 256 ]->[0 0 ]->NULL
KCLOGIN : open /dev/tty0 as stdin=0 , stdout=1 , stderr=2
login:root
password:12345
KCLOGIN : Welcome! root
KCLOGIN : cd to HOME=/  change uid to 0
sh 2 #
```

Fig. 8.9 Startup screen of EOS

```
sh 2 # cat f1 | grep line
parent sh 2 : fork a child
fork1: pgtables= 0x1C1A000 0x1D1B000 0x1E1C000 0x1F1D000
FORK: proc 2  forked a child 6
readyQueue = [6 256 ]->[0 0 ]->NULL
parent sh 2 : wait for child 6  to die
child sh 6  running : cmd=cat f1 | grep line
fork1: pgtables= 0x201F000 0x2120000 0x2221000 0x2322000
FORK: proc 6  forked a child 7
readyQueue = [7 256 ]->[0 0 ]->NULL
proc 6  exec grep line
       KC's grep in action
       KC's cool cat MEOW!
proc 7  in kexit, exitValue=0
with a few lines for testing
proc 6  in kexit, exitValue=0
proc 2  found a ZOMBIE child 6  deallocate 6  kstack at 0x1C19000
sh 2 : child 6  exit status = 0x0
sh 2 # proc 1  found a ZOMBIE child 7  deallocate 7  kstack at 0x201E000
KCINIT: I just buried an orphan child task 7
```

Fig. 8.10 Command processing by the EOS sh

8.16.3.1 Interval Timer and Alarm Signal Catcher

In the USER directory, the itimer.c program demonstrates interval timer, alarm signal and alarm signal catcher.

```
/******************* itimer.c file ***********************/
void catcher(int sig)
{ printf("proc %d in catcher: sig=%d\n", getpid(), sig); }

main(int argc, char *argv[])
{
  int t = 1;
  if (argc>1) t = atoi(argv[1]);   // timer interval
  printf("install catcher? [y|n]");
  if (getc()=='y')
     signal(14, catcher);    // install catcher() for SIGALRM(14)
  itimer(t);                 // set interval timer in kernel
  printf("proc %d looping until SIGALRM\n", getpid());
  while(1);                  // looping until killed by a signal
}
```

In the itimer.c program, it first lets the user to either install or not to install a catcher for the SIGALRM(14) signal. Then, it sets an interval timer of t seconds and executes a while(1) loop. When the interval timer expires, the timer interrupt handler sends a SIGALRM(14) signal to the process. If the user did not install a signal 14 catcher, the process will die by the signal. Otherwise, it will execute the catcher once and continue to loop. In the latter case, the process can be killed by other means, e.g. by the Control_C key or by a kill pid command from another process. The reader may modify the catcher() function to install the catcher again. Recompile and run the system to observe the effect.

8.16.3.2 Exceptions Handling in EOS

In addition to timer signals, we also demonstrate unified exception and signal processing by the following user mode programs. Each program can be run as a user command.

Data.c: This program demonstrates data_abort exception handling. In the data_abort handler, we first read and display the MMU's fault status and address registers to show the MMU status and invalid VA that caused the exception. If the exception occurred in Kmode, it must be due to bugs in the kernel code. In this case, there is nothing the kernel can do. So it prints a PANIC message and stops. If the exception occurred in Umode, the kernel converts it to a signal, SIGSEG(11) for segmentation fault, and sends the signal to the process. If the user did not install a catcher for signal 11, the process will die by the signal. If the user has installed a signal 11 catcher, the process will execute the catcher function in Umode when it gets a number 11 signal. The catcher function uses a long jump to bypass the faulty code, allowing the process to terminate normally.

Prefetch.c: This program demonstrates prefetch_abort exception handling. The C code attempts to execute the in-line assembly code asm(“bl 0x1000”); which would cause a prefetch_abort at the next PC address 0x1004 because it is outside of the Umode VA space. In this case, the process gets a SIGSEG(11) signal also, which is handled the same way as a data_abort exception.

Undef.c: This program demonstrates undefined exception handling. The C code attempts to execute

asm(“mcr p14,0,r1,c8,c7,0”)

which would cause an undef_abort because the coprocessor p14 does not exist. In this case, the process gets an illegal instruction signal SIGILL(4). If the user has not installed a signal 4 catcher, the process will die by the signal.

Divide by zero: Most ARM processors do not have divide instructions. Integer divisions in ARM are implemented by idiv and udiv functions in the aeabi library, which check for divide by zero errors. When a divide by zero is detected, it branches to the __aeabi_idiv0 function. The user may use the link register to identify the offending instruction and take remedial actions. In the Versatilepb VM, divide by zero simply returns the largest integer value. Although it is not possible to generate divide by zero exceptions on the ARM Versatilepb VM, the exception handling scheme of EOS should be applicable to other ARM processors.

8.17 Summary

This chapter presents a fully functional general purpose embedded OS, denoted by EOS. The following is a brief summary of the organization and capabilities of the EOS system.

1. System Images: Bootable kernel image and User mode executables are generated from a source tree by ARM toolchain (of Ubuntu 15.0 Linux) and reside in an EXT2 file system on a SDC partition. The SDC contains stage-1 and stage-2 booters for booting up the kernel image from the SDC partition. After booting up, the kernel mounts the SDC partition as the root file system.
2. Processes: The system supports NPROC=64 processes and NTHRED=128 threads per process, both can be increased if needed. Each process (except the idle process P0) runs in either Kernel mode or User mode. Memory management of process images is by 2-level dynamic paging. Process scheduling is by dynamic priority and time-slice. It supports inter-process communication by pipes and messages passing. The EOS kernel supports fork, exec, vfork, threads, exit and wait for process management.
3. It contains device drivers for the most commonly used I/O devices, e.g. LCD display, timer, keyboard, UART and SDC. It implements a fully Linux compatible EXT2 file system with I/O buffering of SDC read/write to improve efficiency and performance.
4. It supports multi-user logins to the console and UART terminals. The User interface sh supports executions of simple commands with I/O re-directions, as well as multiple commands connected by pipes.
5. It provides timer service functions, and it unifies exceptions handling with signal processing, allowing users to install signal catchers to handle exceptions in User mode.
6. The system runs on a variety of ARM virtual machines under QEMU, mainly for convenience. It should also run on real ARM based system boards that support suitable I/O devices. Porting EOS to some popular ARM based systems, e.g. Raspberry PI-2 is currently underway. The plan is to make it available for readers to download as soon as it is ready.

PROBLEMS

1. In EOS, the initial process P0's kpgdir is at 32KB. Each process has its own pgdir in PROC.res. With a 4MB Umode image size, entries 2048-2051 of each PROC's pgdir define the Umode page tables of the process. When switch task, we use

switchPgdir((int)running–>res–>pgdir);

to switch to the pgdir of the next running process. Modify the scheduler() function (in kernel.c file) as follows.,

```
int *kpgdir = (int *)0x8000; // Kmode pgdir at 32KB
if (running != old){         // truly switch process
   for (i=0; i<npgdir; i++)  // copy Umode pgtables to kpgdir
       kpgdir[2048+i] = running->res->pgdir[2048+i];
   switchPgdir((int)kpgdir); // use the same kpgdir at 32KB
}
```

(1). Verify that the modified scheduler() still works, and explain WHY?
(2). Extend this scheme to use only one kpgdir for ALL processes.
(3). Discuss the advantages and disadvantages of using one pgdir per process vs. one pgdir for all processes.
2. The send/recv operations in EOS use the synchronous protocol, which is blocking and may lead to deadlock due to, e.g. no free message buffers. Redesign the send/recv operations to prevent deadlocks.
3. Modify the Delete_dir_entry algorithm as follows. If the deleted entry is the only entry in a data block, deallocate the data block and compact the DIR INODE's data block array. Modify the Insert_dir_entry algorithm accordingly and implement the new algorithms in the EOS file system.
4. In the read_file algorithm of Sect. 8.12.5.2, data can be read from a file to either user space or kernel space.
(1). Justify why it is necessary to read data to kernel space.

(2). In the inner loop of the read_file algorithm, data are transferred one byte at a time for clarity. Optimize the inner loop by transferring chunks of data at a time (HINT: minimum of data remaining in the block and available data in the file).
5. Modify the write-file algorithm in Sect. 8.12.5.3 to allow
(1). Write data to kernel space, and
(2). Optimize data transfer by copying chunks of data.
6. Assume: dir1 and dir2 are directories. cpd2d dir1 dir2 recursively copies dir1 into dir2.
(1). Write C code for the cpd2d program.
(2). What if dir1 contains dir2, e.g. cpd2d /a/b /a/b/c/d?
(3). How to determine whether dir1 contains dir2?
7. Currently, EOS does not yet support file streams. Implement library I/O functions to support file streams in user space.

References

Android: https://en.wikipedia.org/wiki/Android_operating_system, 2016.
ARM Versatilepb: ARM 926EJ-S, 2016: Versatile Application Baseboard for ARM926EJ-S User Guide, Arm information Center, 2016.
Cao, M., Bhattacharya, S, Tso, T., "Ext4: The Next Generation of Ext2/3 File system", IBM Linux Technology Center, 2007.
Card, R., Theodore Ts'o,T., Stephen Tweedie,S., "Design and Implementation of the Second Extended Filesystem", web.mit.edu/tytso/www/linux/ext2intro.html, 1995.
EXT2: www.nongnu.org/ext2-doc/ext2.html, 2001.
EXT3: jamesthornton.com/hotlist/linux-filesystems/ext3-journal, 2015.
FreeBSD: FreeBSD/ARM Project, https://www.freebsd.org/platforms/arm.html, 2016.
Raspberry_Pi: https://www.raspberrypi.org/products/raspberry-pi-2-model-b, 2016.
Sevy, J., "Porting NetBSD to a new ARM SoC", http://www.netbsd.org/docs/kernel/porting_netbsd_arm_soc.html, 2016.
Wang, K.C., "Design and Implementation of the MTX Operating System", Springer International Publishing AG, 2015.

Multiprocessing in Embedded Systems

9

9.1 Multiprocessing

A multiprocessor system consists of a multiple number of processors, including multi-core processors, which share main memory and I/O devices. If the shared main memory is the only memory in the system, it is called a Uniform Memory Access (UMA) system. If, in addition to the shared memory, each processor also has private local memory, it is called a Non-uniform Memory Access (NUMA) system. If the roles of the processors are not the same, e.g. only some of the processors may execute kernel code while others may not, it is called an Asymmetric MP (ASMP) system. If all the processors are functionally identical, it is called a Symmetric MP (SMP) system. With the current multicore processor technology, SMP has become virtually synonymous with MP. In this chapter, we shall discuss SMP operating systems on ARM based multiprocessor systems.

9.2 SMP System Requirements

A SMP system requires much more than just a multiple numbers of processors or processor cores. In order to support SMP, the system architecture must have additional capabilities. SMP is not new. In the PC world, it started in the early 90's by Intel on their x86 based multicore processors. Intel's Multiprocessor Specification (Intel 1997) defines SMP-compliant systems as PC/AT compatible systems with the following capabilities.

(1). Cache Coherence: In a SMP-compliant system, many CPUs or cores share memory. In order to speed up memory access, the system typically employs several levels of cache memory, e.g. L1 cache inside each CPU and L2 cache in between the CPUs and main memory, etc. The memory subsystem must implement a cache coherence protocol to ensure consistency of the cache memories.
(2). Support interrupts routing and inter-processor interrupts. In a SMP-compliant system, interrupts from I/O devices can be routed to different processors to balance the interrupt processing load. Processors can interrupt each other by Inter-Processor Interrupts (IPIs) for communication and synchronization. In a SMP-compliant system, these are provided by a set of Advanced Programmable Interrupt Controllers (APICs). A SMP-compliant system usually has a system-wide IOAPIC and a set of local APICs of the individual processors. Together, the APICs implement an inter-processor communication protocol, which supports interrupts routing and IPIs.
(3). An extended BIOS, which detects the system configuration and builds SMP data structures for the operating system to use.
(4). When a SMP-compliant system starts, one of the processors is designated as the Boot Processor (BSP), which executes the boot code to start up the system. All other processors are called Application Processors (APs), which are held in the idle state initially but can receive IPIs from the BSP to start up. After booting up, all processors are functionally identical.

In comparison, SMP on ARM based systems is relatively new and still evolving. It may be enlightening to compare ARM's approach to SMP with that of Intel.

K.C. Wang, *Embedded and Real-Time Operating Systems*,
DOI 10.1007/978-3-319-51517-5_9

(1). Cache Coherence: all ARM MPcore based systems include a Snoop Control Unit (SCU), which implements a cache memory coherence protocol to ensure the consistency of the cache memories. Due to the internal pipelines of the ARM CPUs, ARM introduced several kinds of barriers to synchronize both memory access and instruction executions.
(2). Interrupts Routing: Similar to the APICs of Intel, all ARM MPcore systems use a Generic Interrupt Controller (GIC) (ARM GIC 2013) for interrupts routing. The GIC consists of two parts; an interrupt distributor and an interface to CPUs. The GIC's interrupt distributor receives external interrupt requests and sends them to the CPUs. Each CPU has a CPU interface, which can be programmed to either allow or disallow interrupts to be sent to the CPU. Each interrupt has an interrupt ID number, where lower numbers have higher priorities. Interrupts routing may be further controlled by the CPU's interrupt priority mask register. For I/O devices, the interrupt ID numbers correspond roughly to their traditional vector numbers.
(3). Inter-processor Interrupts: The sixteen special GIC interrupts of ID numbers 0 to 15 are reserved for Software Generated Interrupts (SGIs), which correspond to the IPIs in the Intel SMP architecture. In an ARM MPcore based system, a CPU may issue SGIs to wakeup other CPUs from the WFI state, causing them to take actions, e.g. to execute an interrupt handler, as a means of inter-processor communication. In the ARM Cortex-A9 MPcore, the SGI register (GICD_SGIR) is at the offset 0x1F00 from the Peripheral Base Address. The SGI register contents are

```
---------------- SGI Register Contents -------------------
 bits 25-24 : targetListfilter: 00 = to a specified CPU
                                01 = to all other CPUs
                                10 = to the requesting CPU
  bits 23-16  : CPUtargetList; each bit for a CPU interface
  bit  15     : for CPUs with security extension only
  bits 3-0    : Interrupt ID (0-15)
  ---------------------------------------------------------
```

To issue a SGI, simply write an appropriate value to the SGI register. For example, the following code segment sends a SGI of interrupt ID to a specific target CPU.

```
int send_sgi(int intID, int targetCPU, int filter)
{
    int *sgi_reg = (int *)(CGI_BASE + 0x1F00);
    *sgi_reg = (filter<<24)|((1<<targetCPU)<<16)|(intID);
}
send_sgi(0x00, CPUID, 0x00); // intID=0x00, CPUID = 0 to 3, filter=0x00
```

To send a SGI to all other CPUs, change the filter value to 0x01. In this case, the target CPU list should be set to 0x0F (for 4 CPUs). Upon receiving a SGI interrupt, the target CPU can either continue from the WFI state or execute an interrupt handler for the interrupt ID number to perform a prescribed task.

A special usage of SGI is during SMP system booting. When an ARM SMP system starts, the initial booter usually chooses CPU0 as the boot processor, which executes the startup code to initialize the system. Meantime, all other secondary CPUs are held in a WFI loop, waiting for a SGI from CPU0 to start up. After initializing the system, CPU0 writes a starting address for the secondary CPUs to a communication area, which can be either a fixed memory location or a system–wide register accessible by all the CPUs. Then it broadcasts a SGI to the secondary CPUs. Upon waking up from the SGI interrupt, each secondary CPU checks the contents of the communication area. If the contents are zero, it repeats the WFI loop. Otherwise, it executes from the starting address deposited by CPU0. The communication area and the starting address for the secondary CPUs are chosen by the initial booter and the kernel startup code. For example, when booting SMP Linux kernels on ARM using Uboot, CPU0 writes the starting address for the secondary CPUs to the SYS_FLAGSSET register (Booting ARM Linux SMP on MPcore 2010). In the emulated real-view-pbx-a9 VM under QEMU, it also uses the SYS_FLAGSSET regiater as the communication area between CPU0 and the secondary CPUs.

9.3 ARM MPcore Processors

The ARM Cortex-A is a group of 32-bit and 64-bit processor cores that implement the ARMv7 architecture. The group comprises 32-bit ARM MPcore processors labeled Cortex-A5, A8, A9, A12, as well as some 64-bit models, such as the Cortex-A15 and A17. These are not simple microcontrollers but multicore (MPcore) processors intended for general MP applications. In the following, we shall confine our discussions to the 32-bit Cortex-A9 MPcore processor, for the following reasons.

(1). The ARM Cortex-A9 MPcore processor is well documented in [ARM Cortex-A9 MPcore Technical Reference Manual]. A series of ARM RealView baseboards have been implemented based on the Cortex-A9 MPcore, which are also well documented in [ARM RealView Platform Baseboard for Cortex-A9].
(2). The Cortex-A9 processors support CPU synchronization instructions and memory management, which are essential to multitasking OS.
(3). The ARM Realview baseboards support standard ARM peripheral devices, such as UARTs, LCD, keyboard and Multi-media interface for SDC card.
(4). It is fairly easy to talk about the general principles of SMP. However, without programming practice, such knowledge would be superficial at best. Since most readers may not have access to a real ARM MPcore based hardware system, this presents a challenge in learning both the theoretical and practical aspects of SMP. To deal with this problem, we again turn to virtual machines. As of now, QEMU supports several versions of ARM MPcore VMs based on the ARM Cortex-A9 MPcore. For example, it supports the ARM realview-pb-a8, realview-pbx-a9 and vexpress-a9 boards with up to 4 cores. In addition, it also supports the vexpress-a15, which is based on the ARM Cortex-A15 MPcore processor with up to 8 cores. The QEMU emulated ARM MPcore VMs provide us with a convenient environment to develop and run SMP OS systems.

Although we have tested all the programming examples on several versions of ARM MPcore virtual machines, in the following we shall focus primarily on the Realview-pbx-a9 virtual machine under QEMU.

9.4 ARM Cortex-A9 MPcore Processor

The main features of the ARM Cortex-A9 MPcore processors in support of SMP can be summarized as follows.

9.4.1 Processor Cores

1 to 4 cores. The SMP status of each core is controlled by bit 6 of the Auxiliary Control Register (ACTL) of the coprocessor CP15. Each CPU core can be programmed to either participate in SMP or not. Specifically, each CPU core may use the following code segments to either join or disjoin from SMP operation.

```
join_smp:                          // CPU join SMP operation
    MRC p15, 0, r0, c1,c0, 1       // read ACTLR
    ORR r0, r0, #0x040             // set bit 6
    MCR p15, r0, c1, c0, 1         // write to ACTLR
    BX  lr

disjoin_smp:                       // CPU disjoin SMP operation
    MRC 0, r0, c1, c0, 1           // read ACTLR
    BIC r0, #0x040                 // clear bit 6
    MCR p15, 0, r0, c0, 1          // write to ACTLR
    BX  lr
```

If some of the cores do not participate in SMP, they can be designated to run dedicated tasks, or even separate operating systems, outside of the SMP environment, resulting in an **Asymmetric MP (ASMP)** system. Since a MP system should try to utilize the full capabilities of all the cores, we shall assume a default SMP environment, in which all the cores participate in SMP operations.

9.4.2 Snoop Control Unit (SCU)

The SCU ensures L1 cache coherence among the CPUs.

9.4.3 Generic Interrupt Controller (GIC)

The GIC's Interrupt Distributor can be programmed to route interrupts to specific CPUs. Each CPU has its own CPU Interface, which can be programmed to either allow or disallow interrupts with different priorities. On the ARM Realview-PBX-A9 board, the peripheral base address is at 0x1F000000. It can also be read into a general register Rn from the p15 coprocessor by

```
MRC p15, 4, Rn, c15, c0, 0
```

From the peripheral base address, the CPU Interface is at the offset address 0x100, and the Interrupt Distributor is at the offset address 0x1000. Other registers are 32-bit offsets from these base addresses. The following describes the GIC registers.

Interrupt Distributor Registers: These are 32-bit registers for SCU control. The offsets and their functions of these registers are

0x000: distributor control; bit-0 = enable/disable
0x100: 3 (32-bit) set enable registers; each bit enables a corresponding interrupt ID
0x180: 3 (32-bit) clear enable registers; each bit disables a corresponding interrupt ID.
0x800: CPU targets registers; send Interrupt ID to target CPU
0xC00: Configuration register; determine 1-to-N or N-to-N interrupt processing model
0xF00: Soft Generated Interrupts (SGI) register; send SGI to target CPUs
CPU Interface Registers: These are 32-bit CPU interface registers in the SCU. Their offset addresses and functions are
0x00: control register; bit-0 = enable/disable
0x04: priority mask; an interrupt is sent to this CPU only if its priority is higher than the mask. Lower value means higher priority.
0x08: binary point: specify whether pre-emptive interrupts are allowed
0x0C: Interrupt ACK register: contain interrupt ID number
0x10: End of Interrupt register: write interrupt ID to this register to signal EOI.

9.5 GIC Programming Example

Since the GIC is an essential part of ARM MPcore systems, it is necessary and important to understand its functions and usage. Before discussing ARM SMP, we first illustrate GIC programming by an example.

9.5.1 Configure the GIC to Route Interrupts

As usual, the example program C9.1 consists of a ts.s file in assembly and a t.c file in C. To begin with, we shall only use a single CPU to support three kinds of interrupts: input interrupts from two UARTs, the keyboard and a timer. The ARM Realview-pbx-a9 board supports the same kind of I/O devices as in the ARM Versatilepb board (ARM926EJ-S 2010; ARM Timers 2004), but their base addresses and IRQ numbers are different. The following lists the base addresses and GIC interrupt numbers of the intended I/O devices of the ARM Realview-pbx-a9 board.

```
I/O Device  Base Address  GIC IRQ number
--------    ------------  --------------
 Timer0     0x10011000       36
 UART0:     0x10009000       44
 UART1:     0x1000A000       45
 Keyboard   0x10006000       52
 LCD        0x10120000        -
----------------------------------------
```

When the ARM realview-pbx-a9 VM starts, CPU0 executes the reset handler in SVC mode. It sets the SVC and IRQ mode stack pointers and copies the vector table to address 0. Then it enables IRQ interrupts and calls main() in C. Since the object here is to show GIC configuration and programming, all the device drivers are simplified versions of the drivers developed in previous chapters for the ARM Versatilepb board. Each simplified driver only includes an init() function, which initializes the device to generate interrupts, and an interrupt handler, which responds to interrupts. In order to let the reader test run the program directly, we show the complete program code of C9.1.

(1). ts.s file of C9.1

```
/************* ts.s file of C9.1 ***************/
.text
.code 32
.global reset_handler, vectors_start, vectors_end
.global enable_scu, get_cpu_id
reset_handler:
  LDR sp, =svc_stack_top  // set SVC stack
// go in IRQ mode to set IRQ stack
  MSR cpsr, #0x92
  LDR sp, =irq_stack_top  // set IRQ stack
// back to SVC mode
  MSR cpsr, #0x93
  BL  copy_vectors          // copy vectors to address 0
  MSR cpsr, #0x13           // enable IRQ interrupts
  BL main                   // CPU0 call main() in C
  B .
irq_handler:
  sub lr, lr, #4
  stmfd sp!, {r0-r3, r12, lr}
  bl  irq_chandler          // call irq_chandler() in C
  ldmfd sp!, {r0-r3, r12, pc}^
// unused dummy exception handlers
undef_handler:
swi_handler:
prefetch_abort_handler:
data_abort_handler:
fiq_handler:
  B .
// SMP utility functions:
enable_scu:                     // void enable_scu(void)
  MRC p15, 4, r0, c15, c0, 0  // Read periph base address
  LDR r1, [r0, #0x0]          // Read the SCU Control Register
  ORR r1, r1, #0x1            // Set bit 0 (The Enable bit)
  STR r1, [r0, #0x0]          // Write back modifed value
  BX  lr
```

```
// int get_cpu_id(): return ID (0 to 3) of the executing CPU
get_cpu_id:
  MRC p15, 0, r0, c0, c0, 5   // Read CPU ID register
  AND r0, r0, #0x03           // Mask in low 2 bits = CPU ID
  BX  lr
vectors_start:
  LDR PC, reset_handler_addr
  LDR PC, undef_handler_addr
  LDR PC, swi_handler_addr
  LDR PC, prefetch_abort_handler_addr
  LDR PC, data_abort_handler_addr
  B .
  LDR PC, irq_handler_addr
  LDR PC, fiq_handler_addr
reset_handler_addr:          .word reset_handler
undef_handler_addr:          .word undef_handler
swi_handler_addr:            .word swi_handler
prefetch_abort_handler_addr: .word prefetch_abort_handler
data_abort_handler_addr:     .word data_abort_handler
irq_handler_addr:            .word irq_handler
fiq_handler_addr:            .word fiq_handler
vectors_end:
```

(2). uart.c file of C9.1

```
/************** uart.c file of C9.1 *************/
#define UART0_BASE  0x10009000
typedef struct uart{
  u32 DR;          // data register
  u32 DSR;
  u32 pad1[4];     // 8+16=24 bytes to FR register
  u32 FR;          // flag register at 0x18
  u32 pad2[7];
  u32 imsc;        // imsc register at offset 0x38
}UART;
UART *upp[4];      // 4 UART pointers to UART structures
int uputc(UART *up, char c)
{
  int i = up->FR;
  while((up->FR & 0x20));
  (up->DR) = (int)c;
}
void uart_handler(int ID)
{
  UART *up;
  char c;
  int cpuid = get_cpu_id();
  color = (ID==0)? YELLOW : PURPLE;
  up = upp[ID];
  c = up->DR;
  uputc(up, c);
  printf("UART%d interrupt on CPU%d c=%c\n", ID, cpuid, c);
```

```
  if (c=='\r')
     uputc(up, '\n');
  color=RED;
}
int uart_init()
{
  int i;
  for (i=0; i<4; i++){        // uart0 to uart2 are adjacent
    upp[i] = (UART *)(UART0_BASE + i*0x1000);
    upp[i]->imsc |= (1<<4); // enable UART RXIM interrupt
  }
}
```

(3). kbd.c file of C9.1

```
/************** kbd.c file of C9.1 ************/
#include "keymap"
extern int kputc(char);  // in vid.c of LCD driver

typedef struct kbd{        // base = 0x10006000
  u32 control; // 7- 6-     5(0=AT)  4=RxIntEn 3=TxIntEn  2   1   0
  u32 status;  // 7- 6=TxE 5=TxBusy 4=RXFull  3=RxBusy   2   1   0
  u32 data;
  u32 clock;
  u32 intstatus;
  // other fields;
}KBD;
KBD *kbd;

void kbd_handler()
{
  unsigned char scode, c;
  int cpuid = get_cpu_id();
  color = RED;
  scode = kbd->data;
  if (scode & 0x80)          // ignore key release
     goto out;
  c = unsh[scode];
  printf("kbd interrupt on CPU%d: c=%x %c\n", cpuid, c, c);
 out:
  kbd->status = 0xFF;
}
int kbd_init()
{
  kbd = (KBD *)0x10006000; // base address
  kbd->control = 0x14;      // 0001 0100
  kbd->clock = 8;
}
```

(4). timer.c file of C9.1

```
/************ timer.c file of C9.1 ***********/
#define CTL_ENABLE          ( 0x00000080 )
#define CTL_MODE            ( 0x00000040 )
#define CTL_INTR            ( 0x00000020 )
```

```
#define CTL_PRESCALE_1      ( 0x00000008 )
#define CTL_PRESCALE_2      ( 0x00000004 )
#define CTL_CTRLEN          ( 0x00000002 )
#define CTL_ONESHOT         ( 0x00000001 )
#define DIVISOR 64

typedef struct timer{
  u32 LOAD;    // Load Register, TimerXLoad                     0x00
  u32 VALUE;   // Current Value Register, TimerXValue           0x04
  u32 CONTROL;// Control Register, TimerXControl                0x08
  u32 INTCLR;  // Interrupt Clear Register, TimerXIntClr        0x0C
  u32 RIS;     // Raw Interrupt Status Register, TimerXRIS      0x10
  u32 MIS;     // Masked Interrupt Status Register,TimerXMIS 0x14
  u32 BGLOAD;  // Background Load Register, TimerXBGLoad        0x18
  u32 *base;
}TIMER;
TIMER *tp[4];  // 4 timers; 2 timers per unit; at 0x00 and 0x20
int kprintf(char *fmt, ...);
extern int row, col;
int kpchar(char, int, int);
int unkpchar(char, int, int);
char clock[16];
char *blanks = "  :  :  ";
int hh, mm, ss;
u32 tick = 0;
void timer0_handler()
{
   int i;
   tick++;
   if (tick >= DIVISOR){
      tick = 0; ss++;
      if (ss==60){
         ss = 0; mm++;
         if (mm==60){
           mm = 0; hh++;
         }
      }
   }
   if (tick==0){  // every second: display a wall clock
      color = GREEN;
      for (i=0; i<8; i++){
          unkpchar(clock[i], 0, 60+i);
      }
      clock[7]='0'+(ss%10); clock[6]='0'+(ss/10);
      clock[4]='0'+(mm%10); clock[3]='0'+(mm/10);
      clock[1]='0'+(hh%10); clock[0]='0'+(hh/10);
      for (i=0; i<8; i++){
         kpchar(clock[i], 0, 60+i);
      }
   }
   timer_clearInterrupt(0); // clear timer interrupt
}
```

```
void timer_init()
{
  int i;
  printf("timer_init()\n");
  // set timer base address of versatilepb-A9 board
  tp[0] = (TIMER *)0x10011000;
  tp[1] = (TIMER *)0x10012000;
  tp[2] = (TIMER *)0x10018000;
  tp[3] = (TIMER *)0x10019000;
// set control counter regs to defaults
  for (i=0; i<4; i++){
    tp[i]->LOAD = 0x0;    // reset
    tp[i]->VALUE= 0xFFFFFFFF;
    tp[i]->RIS  = 0x0;
    tp[i]->MIS  = 0x0;
    tp[i]->LOAD    = 0x100;
    // 0x62=|011- 0010=|NOTEn|Pe|IntE|-|scal=00|1=32-bit|0=wrap|
    tp[i]->CONTROL = 0x62;
    tp[i]->BGLOAD  = 0xF0000/DIVISOR;
  }
  strcpy(clock, "00:00:00");
  hh = mm = ss = 0;
}
void timer_start(int n) // timer_start(0), 1, etc.
{
  TIMER *tpr;
  printf("timer_start\n");
  tpr = tp[n];
  tpr->CONTROL |= 0x80;  // set enable bit 7
}
int timer_clearInterrupt(int n) // timer_start(0), 1, etc.
{
  TIMER *tpr = tp[n];
  tpr->INTCLR = 0xFFFFFFFF;
}
```

(5). vid.c file of C9.1: same as in previous chapters except LCD base=0x10120000

(6). t.c file of C9.1

```
/************* t.c file of C9.1: ConfigGIC *************/
#define GIC_BASE 0x1F000000

#include "uart.c"
#include "kbd.c"
#include "timer.c"
#include "vid.c"

int copy_vectors(){ // same as before }

int config_int(int intID, int targetCPU)
{
  int reg_offset, index, address;
  char priority = 0x80;
  // set intID BIT in int ID register
  reg_offset = (intID>>3) & 0xFFFFFFFC;
  index = intID & 0x1F;
```

```
  address   =  (GIC_BASE + 0x1100) + reg_offset;
  *(int *)address |= (1 << index);
  // set intID BYTE in processor targets register
  reg_offset = (intID & 0xFFFFFFFC);
  index = intID & 0x3;
  address   = (GIC_BASE + 0x1400) + reg_offset + index;
  // set priority BYTE in priority register
  *(char *)address = (char)priority;
  address   = (GIC_BASE + 0x1800) + reg_offset + index;
  *(char *)address = (char)(1 << targetCPU);
}

int config_gic()
{
  printf("config interrupts 36, 44, 45, 52\n");
  config_int(36, 0);  // Timer0
  config_int(44, 0);  // UART0
  config_int(45, 0);  // UART1
  config_int(52, 0);  // KBD
  // set int priority mask register
  *(int *)(GIC_BASE + 0x104) = 0xFF;
  // set CPU interface control register:  enable interrupts routing
  *(int *)(GIC_BASE + 0x100) = 1;
  // set distributor control register: send pending interrupts to CPUs
  *(int *)(GIC_BASE + 0x1000) = 1;
}

int irq_chandler()
{
   // read ICCIAR of CPU interface in the GIC
   int intID = *(int *)(GIC_BASE + 0x10C);
   switch(intID){
     case 36: timer0_handler(); break; // timer interrupt
     case 44: uart_handler(0);  break; // UART0 interrupt
     case 45: uart_handler(1);  break; // UART1 interrupt
     case 52: kbd_handler();    break; // KBD   interrupt
   }
   *(int *)(GIC_BASE + 0x110) = intID; // write EOI
}

int main()
{
   fbuf_init();     // initialize LCD driver
   printf("***** Config ARM GIC Example ******\n");
   enable_scu();    // enable SCU
   config_gic();    // config GID
   kbd_init();      // initialize KKB driver
   uart_init();     // initialize UART driver
   timer_init();    // initialize timer driver
   timer_start(0); // start timer
   printf("enter a key from KBD or UARTs :\n");
   while(1);        // looping but can respond to interrupts
}
```

(7). Compile-link ts.s and t.c to a t.bin: same as in previous Chapters
(8). Run t.bin on the realview-pbx-a9 VM with 2 UART ports:

```
qemu-system-arm -M realview-pbx-a9 -kernel t.bin \

              -serial mon:stdio -serial /dev/pts/1
```

9.5.2 Explanations of the GIC Configuration Code

In config_gic(), the lines which set the CPU Interface and Distributor enable registers (both at bit0) are obvious. The CPU's interrupt mask register is set to the lowest priority 0xFF, so that the CPU will receive interrupts of priority values < 0xFF. In the Interrupt Distributor (at GIC base + 0x1000), the various registers are at the offsets

```
0x100: Interrupt Set-enable registers:   each bit  = enable an intID
0x400: Interrupt Priority registers:     high 4 bits of each byte=priority of an intID
0x800: Target CPU registers:             each byte= target CPU of an intID
```

There are 3 Interrupt Set-enable registers, denoted by Set-enable0 to Set-enable2, which are at the offsets 0x100 to 0x108. These registers may be regarded as a linear list of bits, in which each bit enables an interrupt ID, as the following diagram shows

```
BITS: |0 1   . . .   31|32 33 . . . 63|64 65 . . .  95|
      ----------------------------------------------------
REGs: |   Set-enable0   |  Set-enable1  |  Set-enable2   |
      ----------------------------------------------------
```

Given an Interrupt ID number, intID, we must determine the bit position to set in the interrupt Set-enable registers. The computation is based on divide and modulo operations, which is called the Mailman's algorithm in (Wang 2015). First, we compute the register and bit offsets of intID as

```
reg_offset        = 4*(intID  /  32);  // same as (intID >> 5) << 2
index             = intID % 32;        // same as intID & 0x1F
```

Then the following lines of code

```
address    =   (GIC_BASE + 0x1100) + reg_offset;
*(int *)address  |= (1 << index);
```

set the enable bit of interrupt ID to 1.

Likewise, we can set the priority and target CPU byte of an interrupt ID by the same algorithm. There are 24 Target CPU registers. Each register holds the CPU data of 4 interrupt IDs, i.e. each byte specifies the target CPUs of an interrupt ID. Similarly, there are 24 Interrupt Priority registers. Each register holds the priority levels of 4 interrupt IDs, i.e. each byte holds the priority of an interrupt ID. The (linear) layout of these registers is similar to the above diagram except that each byte specifies either a list of target CPUs or an interrupt priority. Given an interrupt ID, we compute the register and byte offsets as

```
reg_offset        = intID  /  4;          // same as intID >> 2
index             = intID % 4;            // same as intID & 0x3
```

Then the following lines of code

```
bc
QEMU
********** Config ARM GIC Example ************          00:00:59
config interrupts 36, 44, 45, 52
timer_init()
timer_start
enter a key from KBD or UARTs :
kbd interrupt on CPU0 : c=0x61  a
UART0  interrupt on CPU0  c=b
UART0  interrupt on CPU0  c=c
```

Fig. 9.1 Demonstration of GIC programming

```
address   = (GIC_BASE + 0x1800) + reg_offset + index;
*(char *)address = (char)(1 << targetCPU);
```

set the CPU byte in the target CPU register. Exactly the same algorithm can be used to set the interrupt priority mask byte in the priority mask register.

Since we have not enabled the other CPUs yet, all the interrupts are routed to CPU0 for the time being. We shall show how to route interrupts to different CPUs later. In general, each interrupt ID should be routed to a unique CPU. An interrupt ID may also be routed to multiple CPUs. In that case, the user must ensure that interrupt processing uses the 1-to-N model, in which only one CPU actually handles the interrupt.

9.5.3 Interrupt Priority and Interrupt Mask

Each interrupt ID can be set to a priority value from 0 to 15, where low value means high priority. In order to send an interrupt to a CPU, the interrupt priority must be higher (lower in value) than the CPU's priority mask value. For simplicity, in the example all the interrupts are set to the same priority 8. The CPU's mask register is set to 0xF, so that it will accept any interrupt of priority > 0xF. The reader may try the following experiment. If we set the CPU's priority mask register to a value <= 8, then no interrupts will occur since no interrupts will be sent to the CPU.

9.5.4 Demonstration of GIC Programming

Figure 9.1 shows the sample outputs of running the example program C9.1. For timer interrupts, the timer interrupt handler displays a wall clock. For keyboard and UARTs interrupts, it shows the interrupts and the input keys. For each UART interrupt, it also echoes the input key to the UART port, which is shown at the top of Figure 9.1.

9.6 Startup Sequence of ARM MPcores

The startup sequence of Intel x86 based SMP systems is well defined (Intel 1997). All Intel x86 based SMP systems contain a standard BIOS in ROM, which runs when the system is powered on or following a reset. When an Intel x86 based SMP system starts, BIOS configures the system for SMP operation first. It designates one of the CPUs, usually CPU0, as the Boot Processor (BSP), which executes the boot code to bring up the system. All other CPUs are called Application Processors (APs), which are held in an inactive state, waiting for an Inter-Processor Interrupt (IPI) from the BSP to start up. For more information on the startup sequence of Intel x86 based SMP systems, the reader may consult (Intel 1997; Wang 2015).

In contrast, the startup sequence of ARM SMP systems is somewhat murky and in many cases ad-hoc. The main reason for this lack of standard is because ARM based systems do not have a standard BIOS. Most ARM systems have an onboard booter implemented in firmware. The startup sequence of ARM SMP systems relies heavily on the onboard booter, which varies widely, depending on the specific board or vendor. After surveying the literatures, we may classify the startup sequence of ARM SMP systems into three categories.

9.6.1 Raw Startup Sequence

When an ARM based SMP system starts, all CPUs execute from the vector address 0 (assuming no vector relocation during booting). Each CPU may get its CPUID number from the coprocessor p15. Depending on the CPUID number, only one of the CPUs, typically CPU0, is chosen to be the BSP, which initializes itself and executes the system initialization code. All other secondary CPUs (APs) put themselves into a busy-waiting loop or a power-saving WFI state, waiting for a SGI (Software Generated Interrupt) from the BSP to truly start up. After initializing the system, the BSP writes a starting address for the APs to a communication area, which may be a fixed location in memory or a system-wide register accessible to all CPUs. Most ARM SMP systems use the system-wide SYS_FLAGSSET register as the communication area. Then the BSP activates the APs by sending them SGIs. As in Intel SMP systems, ARM SGIs can be sent to each individual AP (CPU filterList=00) or to all APs by broadcasting (CPU targetList=0xF, filterList=01). Upon waking up from the WFI state, each AP examines the contents of the communication area. If the contents are zero, it repeats the WFI loop. Otherwise, it starts to execute from the starting address deposited by the BSP.

9.6.2 Booter Assisted Startup Sequence

When an ARM SMP system starts, the onboard booter chooses CPU0 as the BSP and puts other Secondary CPUs on hold until they are activated by the BSP. In many cases the onboard booter is too primitive to boot up a real OS. So it loads a stage-2 booter from a storage device and turning control over to the stage-2 booter. The stage-2 booter is designed to boot up a specific SMP kernel. It activates the APs only after the BSP has initialized the kernel. In doing so, it may deposit the startup information of the APs in a different communication area. This scheme is used by most ARM SMP Linux using Das Uboot as the stage-2 booter (Booting ARM Linux SMP on MPcore 2010).

9.6.3 SMP Booting on Virtual Machines

Many ARM virtual machines (VMs) support ARM MPcores for SMP. To be more specific, we shall consider the ARM realview-pbx-a9 VM, which is based on the ARM Cortex-A9 MPcore with up to 4 cores. When a QEMU emulated VM starts, QEMU loads an executable image, by the –kernel IMAGE option, to 0x10000 and starts to execute the IMAGE. The loaded image may be an OS kernel or a stage-2 booter. Thus, for emulated VMs, QEMU acts like an onboard or stage-1 booter. It puts the APs in the WFI state, waiting for SGI from the BSP. In this case, the CPUs use the SYS_FLAGSSET register as the communication area, which is at the memory mapped address 0x10000030 on the realview-pbx-a9 board. To activate the APs, the BSP writes a starting address to the SYS_FLAGSSET register, followed by broadcasting a SGI to all the APs.

9.7 ARM SMP Startup Examples

Although it is fairly easy to start up a SMP system, the operations of SMP systems are much more complex than that of uniprocessor systems. In order to let the reader have a better understanding of SMP operations, we first illustrate the startup sequence of ARM SMP systems by a series of examples. To be more specific, we shall use the emulated realview-pbx-a9 VM under QEMU as the implementation platform.

9.7.1 ARM SMP Startup Example 1

In the first example, denoted by C9.2, we show the minimum amount of code that is needed to start up an ARM based SMP system. For simplicity, we shall only support input interrupts from a UART and interrupts of a timer. In order to display outputs, we also implement a uprintf() function for formatted printing to the UART port.

(1). ts.s file:

```
/****************** ts.s file of C9.2 **********************\
.text
.code 32
.global reset_handler, vectors_start, vectors_end
.global enable_SCU, get_cpuid
reset_handler:              // ALL CPUs start execution here
// get CPU ID and keep it in R11
  MRC p15, 0, r11, c0, c0, 5   // read CPU ID register into R11
  AND r11, r11, #0x03          // mask in only CPUID
// set SVC stack
  LDR r0, =svc_stack           // r0->svc_stack (16KB area in t.ld)
  mov r1, r11                  // r1 = cpuid
  add r1, r1, #1               // cpuid++
  lsl r2, r1, #12              // (cpuid+1)* 4096
  add r0, r0, r2
  mov sp, r0                   // SVC sp=svc_stack[cpuid] high end

// go in IRQ mode with interrupts OFF
  MSR cpsr, #0x92
// set IRQ stack
  LDR r0, =irq_stack           // r0->irq_stack (16KB area in t.ld)
  mov r1, r11
  add r1, r1, #1
  lsl r2, r1, #12              // (cpuid+1) * 4096
  add r0, r0, r2
  mov sp, r0                   // IRQ sp=irq_stack[cpuid] high end

// go back to SVC mode with IRQ ON
  MSR cpsr, #0x13
  cmp r11, #0
  bne APs                      // only CPU0 copy vectors, call main()
  BL copy_vectors              // copy vectors to address 0
  BL main                      // CPU0 call main() in C
  B .
APs:                           // each AP call APstart() in C
  adr r0, APaddr
  ldr pc, [r0]
APaddr: .word APstart
irq_handler:
  sub lr, lr, #4
  stmfd sp!, {r0-r3, r12, lr}
  bl irq_chandler              // call irq_chandler() in C
  ldmfd sp!, {r0-r3, r12, pc}^

vectors_start:
  LDR PC, reset_handler_addr
  LDR PC, undef_handler_addr
  LDR PC, swi_handler_addr
  LDR PC, prefetch_abort_handler_addr
```

```
  LDR PC, data_abort_handler_addr
  B .
  LDR PC, irq_handler_addr
  LDR PC, fiq_handler_addr
reset_handler_addr:           .word reset_handler
undef_handler_addr:           .word undef_handler
swi_handler_addr:             .word swi_handler
prefetch_abort_handler_addr:  .word prefetch_abort_handler
data_abort_handler_addr:      .word data_abort_handler
irq_handler_addr:             .word irq_handler
fiq_handler_addr:             .word fiq_handler
vectors_end:
// unused dummy exception handlers
undef_handler:
swi_handler:
prefetch_abort_handler:
data_abort_handler:
fiq_handler:  B .

enable_scu:                     // enable the SCU
  MRC p15, 4, r0, c15, c0, 0   // Read peripheral base address
  LDR r1, [r0]                  // read SCU Control Register
  ORR r1, r1, #0x1              // set bit0 (Enable bit) to 1
  STR r1, [r0]                  // write back modified value
  BX  lr
get_cpuid:
  MRC p15, 0, r0, c0, c0, 5    // read CPU ID register
  AND r0, r0, #0x03             // mask in the CPUID field
  MOV pc, lr
// ---------------- end of ts.s file -------------------
```

Explanations of the ts.s file: The system is compile-linked to a t.bin executable image with starting address 0x10000. In the linker script file, t.ld, it specifies svc_stack as the beginning address of a 16 KB area, which will be the SVC mode stacks of the 4 CPUs. Similarly, the CPUs will use the 16KB areas at irq_stack as their IRQ mode stacks. Since we do not intend to handle any other kinds of exceptions, the ABT and UND mode stacks, as well as exception handlers are omitted. The image is run under QEMU as

qemu-system-arm **–m realview-pbx-a9 –smp 4** –m 512M –kernel t.bin –serial mon:stdio

For clarity, the machine type **(–m realview-pbx-a9)** and number of CPUs **(-smp 4)** are shown in bold face. When the system starts, the executable image is loaded to 0x10000 by QEMU and runs from there. When CPU0 starts to execute reset_handler, QEMU has started up the other CPUs (CPU1 to CPU3) also but they are held in the WFI state. So, only CPU0 executes the reset_handler at first. It sets up the SVC and IRQ mode stacks, copies the vector table to address 0 and then calls main() in C. CPU0 first initializes the system. Then it writes 0x10000 as the start address of the APs to the SYS_FLAGSSET register (at 0x10000030), and issues a SGI to all the APs, causing them to execute from the same reset_handler code at 0x10000. However, based on their CPU ID number, each AP only sets up its own SVC and IRQ mode stacks and calls APstart() in C, bypassing the system initialization code, such as copying vectors, that are already done by CPU0.

(2). t.c file: in main(), CPU0 enables the SCU and configures the GIC to route interrupts. For simplicity, the system only supports UART0 (0x10009000) and timer0 (0x10011000). In config_gic(), timer interrupts are routed to CPU0 and UART interrupts are routed to CPU1, which are quite arbitrary. If desired, the reader may verify that they can be routed to any CPU. Then CPU0 initializes the device drivers and starts the timer. Then it writes the AP start address to 0x10000030 and issues SGI to activate the APs. After that, all CPUs execute in WFI loops, but CPU0 and CPU1 can respond to and handle timer and UART interrupts.

```
/******************** t.c file of C9.2 ***************************/
#include "type.h"
#define GIC_BASE 0x1F000000
int *apAddr = (int *)0x10000030; // SYS_FLAGSSET register

#include "uart.c"
#include "timer.c"

int copy_vectors(){  // same as before }
int APstart()         // AP startup code
{
   int cpuid = get_cpuid();
   uprintf("CPU%d start: ", cpuid);
   uprintf("CPU%d enter WFI state\n", cpuid);
   while(1){
      asm("WFI");
   }
}
int config_int(int intID, int targetCPU)
{
  int reg_offset, index, address;
  char priority = 0x80;
  reg_offset = (intID>>3) & 0xFFFFFFFC;
  index = intID & 0x1F;
  address   =  (GIC_BASE + 0x1100) + reg_offset;
  *(int *)address = (1 << index);
  // set interrupt ID priority
  reg_offset = (intID & 0xFFFFFFFC);
  index = intID & 0x3;
  address   = (GIC_BASE + 0x1400) + reg_offset + index;
  *(char *)address = (char)priority;
  // set target CPUs
  address   = (GIC_BASE + 0x1800) + reg_offset + index;
  *(char *)address = (char)(1 << targetCPU);
}
int config_gic()
{
  // set int priority mask register
  *(int *)(GIC_BASE + 0x104) = 0xFF;
  // Enable CPU interface control register to signal interrupts
  *(int *)(GIC_BASE + 0x100) = 1;
  // Enable distributor control register to send interrupts to CPUs
  *(int *)(GIC_BASE + 0x1000) = 1;
  config_int(36, 0);  // timer ID=36 to CPU0
  config_int(44, 1);  // UART0 ID=44 to CPU1
}
int irq_chandler()
{
   // read ICCIAR of CPU interface in the GIC
   int intID = *(int *)(GIC_BASE + 0x10C);
   if (intID == 36)
      timer_handler();  // timer0 interrupt
   if (intID == 44)
      uart_handler(0);  // UART0 interrupt
   *(int *)(GIC_BASE + 0x110) = intID; // issue EOI
}
```

```
int main()
{
   enable_scu();                      // enable SCU
   uart_init();                       // initialize UARTs
   uprintf("CPU0 starts\n");
   timer_init();                      // initialize timer
   timer_start(0);                    // start timer
   config_gic();                      // configure GIC
   // send SGI to wakeup APs
   send_sgi(0x00, 0x0F, 0x01);   // intID=0,CPUs=0xF,filter=0x01
   apAddr  = (int *)0x10000030;  // SYS_FLAGSSET register
   *apAddr = (int)0x10000;       // all APs execute from 0x10000
   uprintf("CPU0 enter WFI loop: enter key from UART\n");
   while(1)
     asm("WFI");
}
```

(3). timer.c file: this file implements the timer0 driver. On each second, it displays a line to the UART0 port.

```
//**************** timer.c file of C9.2 ******************
#define DIVISOR 64
typedef struct timer{
  u32 LOAD;      // Load Register
  u32 VALUE;     // Current Value Register
  u32 CONTROL;   // Control Register
  u32 INTCLR;    // Interrupt Clear Register
  u32 RIS;       // Raw Interrupt Status Register
  u32 MIS;       // Masked Interrupt Status Register
  u32 BGLOAD;    // Background Load Register
}TIMER;
TIMER *tp;
extern UART *up;
u32 tick = 0, ss = 0;
int timer_handler()
{
   int cpuid = get_cpuid();
   tick++;
   if (tick >= DIVISOR){
      tick = 0; ss++;
      if (ss==60)
         ss = 0;
   }
   if (tick==0){  // every second: display a line
      uprintf("TIMER interrupt on CPU%d : time = %d\r", cpuid, ss);
   }
   timer_clearInterrupt(0); // clear timer interrupt
}
int timer_init()
{
  tp = (TIMER *)0x10011000; // set timer base address
  // set control and counter registers
  tp->LOAD = 0x0;
  tp->VALUE= 0xFFFFFFFF;
  tp->RIS  = 0x0;
  tp->MIS  = 0x0;
```

```
  tp->LOAD = 0x100;
  tp->CONTROL = 0x62; // |En|Per|Int|-|Sca|00|32B|Wrap|=01100010
  tp->BGLOAD  = 0xF0000/DIVISOR;
}
int timer_start()
{
  TIMER *tpr = tp;
  tpr->CONTROL |= 0x80;       // set enable bit 7
}
int timer_clearInterrupt()
{
  TIMER *tpr = tp;
  tpr->INTCLR = 0xFFFFFFFF;   // write to INTCLR register
}
```

(4). uart.c file: this is the UART driver. The driver is ready to 4 UARTs but it only uses UART0. For the sake of brevity, the uprintf() code is not shown.

```
/************ uart.c file of C9.2 ***********/
#define UART0_BASE 0x10009000
typedef struct uart{
 u32 DR; // data reg
 u32 DSR;
 u32 pad1[4]; // 8+16=24 bytes to FR register
 u32 FR; // flag reg at 0x18
 u32 pad2[7];
 u32 IMSC; // at offset 0x38
}UART;
UART *upp[4]; // 4 UART pointers
UART *up; // active UART pointer
int uprintf(char *fmt, ...){ // same as before }
int uart_handler(int ID)
{
 char c;
 int cpuid = get_cpuid();
 up = upp[ID];
 c = up->DR;
 uprintf("UART%d interrupt on CPU%d : c=%c\n", ID, cpuid, c);
}
int uart_init()
{
 int i;
 for (i=0; i<4; i++){ // UARTs base addresses
     upp[i] = (UART *)(0x10009000 + i*0x1000);
     upp[i]->IMSC |= (1<<4); // enable UART RXIM interrupt
 }
}
```

9.7.2 Demonstration of SMP Startup Example 1

Figure 9.2 shows the outputs of the SMP startup Example program C9.2. The system has 4 CPUs, which are identified by CPU0 to CPU3. In config_gic(), timer interrupts (36) are routed to CPU0 and UART interrupts (44) are routed to CPU1. At

```
CPU0 starts
CPU0 enter WFI loop: enter key from UART
CPU1  start: CPU1  enter WFI state
CPU2  start: CPU2  enter WFI state
CPU3  start: CPU3  enter WFI state
UART0  interrupt on CPU1  : c=t = 10
UART0  interrupt on CPU1  : c=e = 13
UART0  interrupt on CPU1  : c=s
UART0  interrupt on CPU1  : c=t = 14
TIMER interrupt on CPU0  : time = 19
```

Fig. 9.2 ARM SMP startup example 1

each second, the timer display a line showing the elapsed time in seconds. All the CPUs execute in a WFI loop, but CPU0 and CPU1 can respond to and handle interrupts. As an exercise, the reader may modify the config_gic() code to route interrupts to different CPUs to observe the effects.

9.7.3 ARM SMP Startup Example 2

In the second ARM SMP startup example program, denoted by C9.3, we add LCD and keyboard drivers to the program for better user interface and display. These are the same LCD and keyboard drivers developed in previous chapters for the ARM Versatilepb board, except for the following minor differences. On the ARM realview-pbx-a9 board, the LCD display timing parameters are the same as before but its base address is at 0x10120000. For the keyboard, the base address (at 0x1000600) is the same as before but it uses the GIC interrupt number 52 since the realview-pbx-a9 board does not have VIC and SIC interrupt controllers. When the APs start up, we let each AP print a few lines to the LCD display. For the sake of brevity, we only show the modified t.c file.

```
/********* t.c file of C9.3: SMP Startup Example 2 ********/
#include "type.h"
extern int uprintf(char *fmt, ...);
extern int printf(char *fmt, ...);
#define GIC_BASE 0x1F000000
int *apAddr = (int *)0x10000030;

#include "uart.c"
#include "timer.c"
#include "kbd.c"             // KBD dirver
#include "vid.c"             // LCD driver

int copy_vectors(){// same as before }
int sen_sgi(int filter,int targetCPU, int intID){//same as before}

int APstart()
{
   int i, cpuid = get_cpuid();
   printf("CPU%d start\n", cpuid);
   for (i=0; i<2; i++){
      printf("CPU%d before WFI state i=%d\n",cpuid, i);
   }
   printf("CPU%d enter WFI state\n", cpuid);
   while(1){
     asm("WFI");
   }
}
```

```
int config_gic()
{
  // set int priority mask register
  *(int *)(GIC_BASE + 0x104) = 0xFF;
  // set CPU interface control register: enable signaling interrupts
  *(int *)(GIC_BASE + 0x100) = 1;
  // distributo control register to send pending interrupts to CPUs
  *(int *)(GIC_BASE + 0x1000) = 1;
  config_int(36, 0);      // timer interrupts t0 CPU0
  config_int(44, 1);      // UART0 interrupts to CPU1
  config_int(52, 2);      // KBD   interrupts to CPU2
}

int config_int(int intID, int targetCPU){// SAME AS in Example 1}
int irq_chandler()
{
   // read ICCIAR of CPU interface in the GIC
   int intID = *(int *)(GIC_BASE + 0x10C);
   if (intID == 36)
      timer_handler();
   if (intID == 44)
      uart_handler(0);
   if (intID == 52)         // KBD interrupt handler
      kbd_handler();
   *(int *)(GIC_BASE + 0x110) = intID; // issue EOI
}

int main()
{
   enable_scu();
   fbuf_init();                 // initialize LCD display
   printf("********* ARM SMP Startup Example 2 ************\n");
   kbd_init();
   uart_init();
   printf("CPU0 starts\n");
   timer_init();
   timer_start(0);
   config_gic();
   send_sgi(0x00, 0x0F, 0x01); // intID=0,CPUs=0xF,filter=b01
   apAddr  = (int *)0x10000030;
   *apAddr = (int)0x10000;
   printf("CPU0 enter while(1)loop. Enter key from KBD:\n");
   uprintf("Enter key from UART terminal:\n");
   while(1)
      asm("WFI");
}
```

9.7.4 Demonstration of ARM SMP Startup Example 2

Figure 9.3 shows the outputs of running the example program C9.3. As the figure shows, the outputs of the APs are interleaved. For instance, before CPU2 finishes printing, CPU3 starts to print also, resulting in mixed outputs. If we let the APs print more lines, some of the lines may become garbled. This is because the CPUs may write to the same locations in the LCD display memory. This is a typical phenomenon of different CPUs executing in parallel. The CPUs may access and

```
QEMU
************ ARM SMP Startup Example 2 ************          00:01:07
CPU0 starts
timer_init()
timer_start
CPU0 enter while(1) loop. Enter key from KBD:
CPU1  start
CPU1  before WFI state i=0
CPU1  before WFI state i=1
CPU1  enter WFI state
CPU2  start
CPU2 CPU3  start
CPU3  before WFI state i=0
CPU3  before W_before WFI state i=0
CPU2  before WFI state i=1
CPU2  enter WFI state
FI state i=1
CPU3  enter WFI state
KBD interrupt on CPU2 : c=0x74  t
```

Fig. 9.3 ARM SMP startup example 2

modify shared memory locations in any order. When many CPUs try to modify the same memory location, if the outcome depends on the execution order, it is called a **race condition**. In a SMP system, race conditions must not exist because they may corrupt shared data objects, resulting in inconsistent results and causing the system to crash. The above examples are intended to show that while it is very easy to start up multiple CPUs in a SMP system, we must control their executions in order to ensure the integrity of shared data objects. This leads us to re-examine the problem of process synchronization, which is essential to every SMP system.

9.8 Critical Regions in SMP

In a SMP system, each CPU executes a process, which may access the same data objects in shared memory. A **Critical Region (CR)** (Silberschatz et al. 2009; Stallings 2011) is a sequence of executions on shared data objects which can only be performed by one process at a time. Critical regions realize the principle of process mutual exclusion, which is the basis of process synchronization. So the basic problem is how to implement critical regions in SMP systems.

9.8.1 Implementation of Critical Regions in SMP

Assume that x is an addressable memory location, e.g. a byte or word. In every computer system, read(x) and write(x) are atomic operations. No matter how many CPUs try to read or write the same x, even at the same time, the memory controller allows only one CPU to access x at a time. Once a CPU starts to read or write x, it completes the operation before any other CPU is allowed to access the same x. However, the individual atomicity of read(x) and write(x) does not guarantee that x can be updated correctly. This is because updating x requires read(x) and then write(x) in two steps. Between the two steps other CPUs may cut in, which either read the yet to be updated old value of x or write values to x that will be over-written. To remedy this problem, CPUs designed for multiprocessing usually support a Test-and-Set (TS) or equivalent instruction, which works as follows. Assume again that x is an addressable memory location and x=0 initially. The TS(x) instruction performs the following sequence of actions on x as a single indivisible (atomic) operation.

```
TS(x) = {read x from memory; test x for 0 or 1; write 1 to x}
```

No matter how many CPUs try to do TS(x), even at the same time, only one CPU can read x as 0, all others will read x as 1. In the Intel x86 CPU, the equivalent instruction is XCHG, which exchanges a CPU register with a memory location in a single indivisible operation. In earlier versions of ARM CPUs (before ARMv6), the equivalent instruction is SWAP, which swaps a CPU register with a memory location in a single indivisible operation. With TS or equivalent instructions, we can implement a CR associated with x as follows.

```
         Byte x = 0;
(1). int SR = int_off();
(2). while(TS(x));
     ------------------
(3). | Critical Region |
     ------------------
(4). x = 0;
(5). int_on(SR);
```

On every CPU, process switching is usually triggered by interrupts. Step (1) disables CPU interrupts to prevent process switching. This ensures that each process keeps running on a CPU without being switched out. At Step (2), the process loops until TS(x) gets a 0 value. Among the CPUs trying to execute TS(x), only one can get 0 to enter the CR at Step (3). All other CPUs will get the value 1 and continue to execute the while loop. Therefore, only one process can be inside the CR at any time. When the process finishes the CR, it clears x to 0, allowing another process to pass through Step (2) to enter the CR. At Step (5), the process restores the original CPU status register, which may enable interrupts to allow process switching on the CPU, but the process has already exited the CR.

9.8.2 Shortcomings of XCHG/SWAP Operations

Although TS-like instructions can be, and have been, used to implement CRs in SMP systems (Wang 2015), they also have several shortcomings.

(1). When a CPU starts to execute a TS-like instruction, the entire memory bus may be locked until the instruction completes. Meantime, no other CPU can access the memory, which reduces concurrency.
(2). When a CPU starts to execute a TS-like instruction, it must complete the instruction before it can take any interrupt, which increases interrupt processing latency.
(3). Assume that a CPU has set a memory location to 1 and is executing inside a CR. Before the CPU resets the memory location to 0, all other CPUs attempting to set it to 1 must continually execute the TS-like instruction. This increases power consumption, which is undesirable in embedded and mobile systems.

9.8.3 ARM Synchronization Instructions for SMP

To remedy the shortcomings of TS-like instructions, ARM introduced several new instructions in their MPcore processors for process synchronization in SMP.

9.8.3.1 ARM LDREX/STREX Instructions

LDREX: The LDREX instruction loads a word from memory, which is marked for exclusive access. It also initializes the states of the associated hardware units, called the **exclusive monitors**, to keep track of updating operations on such memory locations.

STREX: The STREX instruction attempts to store a word into a memory location that was marked as exclusive access. If the exclusive monitors permit the store, it updates the memory location and returns 0, indicating that the operation succeeded. If the exclusive monitors do not permit the store, it does not update the memory location and returns 1, indicating that the operation failed. In the latter case, the CPU may try strex again later or take alternative actions.

LDREX-STREX effectively split the classical TS-like operation into two separate steps. With the help of the exclusive monitors, they support atomic memory update in ARM MPcore processors. Regardless how many CPUs try to update the same memory location, only one CPU can succeed. Since a CPU executes LDREX-STREX in two steps, it does not have to continually execute the instruction sequence until it succeeds. After LDREX, if a CPU finds the memory location is already 1, rather than repeatedly trying again it may take alternative actions. A better way is to put the CPU in a power saving mode until it is signaled up, at which time it may try to set the memory location again. For this purpose, ARM introduced the WFI, WFE and SEV instructions, which can be used as follows.

9.8.3.2 ARM WFI, WFE, SEV Instructions

WFI (Wait-for-Interrupts): The CPU goes to power saving mode, awaiting any interrupt to wake it up.

WFE (Wait-for-Event): The CPU goes to power saving mode, awaiting any event, which includes interrupts and event caused by another CPU, to wake it up.

SEV (Send-Event); Send an event to wake up other CPUs in WFE mode.

The above instructions would work if the CPU and the (optimizing) compiler do not alter the instruction execution order of programs. However, the ARM-toolchain generated code and the ARM CPU itself may re-order instruction executions, which may result in out-of-order memory accesses different from the instruction sequence in a program. In a SMP system, out-of-order memory accesses may produce inconsistent results. In order to ensure a consistent view of memory contents from different execution entities, ARM introduced memory barriers.

9.8.3.3 ARM Memory Barriers

A memory barrier is a type of instruction that causes a CPU to enforce an ordering constraint on memory operations issued before and after the barrier instruction. This typically means that operations issued priori to the barrier are guaranteed to be performed before operations after the barrier. ARM memory barriers include the following.

DMB (Data Memory Barrier): DMB acts as a memory barrier. It ensures that all explicit data memory transfers before the DMB are completed before any subsequent explicit data memory transactions after the DMB starts. This ensures correct ordering between two memory accesses.

DSB (Data Synchronization Barrier): DSB acts as a special kind of memory barrier. The DSB instruction ensures all explicit data transfers before the DSB are complete before any instruction after the DSB is executed. The instruction completes when all explicit memory accesses before this instruction complete, and all Cache, Branch predictor and TLB maintenance operations before this instruction complete.

ISB (Instruction Synchronization Barrier): ISB flushes the pipeline in the processor, so that all instructions following the ISB are fetched from cache or memory, after the instruction has been completed. It ensures that the effects of context altering operations, such as changing the ASID, or completed TLB maintenance operations, or branch predictor maintenance operations, as well as all changes to the CP15 registers, executed before the ISB instruction are visible to the instructions fetched after the ISB. In addition, the ISB instruction ensures that any branches that appear in program order after the ISB instruction are always written into the branch prediction logic with the context that is visible after the ISB instruction. This ensures correct execution of the instruction stream.

9.9 Synchronization Primitives in SMP

Synchronization primitives are software tools for process synchronization. There are many different kinds of synchronization primitives, ranging from simple spinlocks to very sophisticated high-level synchronizing constructs, such as Condition Variables and Monitors, etc. (Wang 2015). In the following, we shall only discuss synchronization primitives that are most commonly used in SMP operating systems.

9.9.1 Spinlocks

The simplest kind of synchronization primitives is the spinlock, which is a conceptual lock used to protect Critical Regions (CR) of short durations. To access a CR, a process must acquire the spinlock associated with the CR first. It repeatedly tries to acquire the spinlock until it succeeds, hence the name spinlock. After finishes with the CR, the process releases the spinlock, allowing another process to acquire the spinlock to enter the CR. Spinlock operations consist of two main functions.

```
slock(int *spin)   : acquire a spinlock pointed by spin
sunlock(int *spin): release a spinlock pointed by spin
```

They are called as slock(&spin) and sunlock(&spin), respectively, where spin denotes a spinlock initialized to the UNLOCKed state. The following code segments show the implementation of spinlock functions in ARM assembly.

```
UNLOCKED = 0
LOCKED = 1
int spin = 0;             // spinlock, 0=UNLOCKED, 1=LOCKED
slock:                    // slock(int *spin): acquire spinlock
  ldrex r1, [r0]          // read spinlock value
  cmp   r1, #0x0          // compare with 0
  WFENE                   // not 0 means already locked: do WFE
  bne   slock             // try again after woken up by event
  mov   r1, #1            // set r1=1
  strex r2, r1, [r0]      // try to store 1 to [r0]; r2=return value
  cmp   r2, #0x0          // check return value in r2
  bne   slock             // not 0 means failed; retry to lock again
  DMB                     // memory barrier BEFORE accessing CR
  bx    lr                // return only if has acquired the spinlock
sunlock:                  // sunlock(int *spin)
  mov   r1, #0x0          // set r1=0
  DMB                     // memory barrier BEFORE releasing CR
  str   r1, [r0]          // store 0 to [r0]
  DSB                     // ensure update has completed before SEV
  SEV                     // signal event to wakeup CPUs in WFE mode
  bx    lr                // return
```

9.9.2 Spinlock Example

In the SMP startup examples programs, CPU0 is the BSP, which initializes the system first. Then it wakes up other APs by sending them SGI. Upon waking up, all the APs execute the same APstart() code. As Fig. 9.3 shows, their outputs on the LCD screen are most likely inter-mixed. For instance, CPU0's prompt for inputs appears even before all the APs are woken up. In practice, the BSP should wait until all the APs are ready before continuing. To achieve this, we may initialize a (global) variable ncpu=1 and have each AP increment ncpu by 1 when it is up and ready. After sending SGI to wakeup the APs, the BSP executes a busy-wait loop while(ncpu < 4); before displaying the prompt for inputs. However, without proper synchronization, this scheme may not work because the APs may update ncpu in arbitrary order due to race conditions. The final value of ncpu may not be 4, leaving the BSP stuck in the while loop. All these problems can be eliminated by adding a spinlock to ensures that the APs execute APstart() one at a time. To do this, we define the global variables, spin as a spinlock, ncpu as a counter, and modify the APstart() and t.c code as follows.

```
int spin = 0;             // spinlock
volatile int ncpu = 1; // number of CPUs ready

int APstart()             // code for APs to execute
{
   int cpuid = get_cpuid();
   slock(&spin);
    printf("CPU%d in APstart\n", cpuid);
```

```
    ncpu++;
    printf("CPU%d enter WFI loop ncpu=%d\n", cpuid, ncpu);
   sunlock(&spin);
   while(1)
     asm("WFI");
}
main()              // executed by CPU0
{
   // same code as before
   while(ncpu < 4);  // wait until all APs are ready
   printf("enter a key from KBD or UARTs : ");
   while(1) asm("WFI");
}
```

9.9.3 Demonstration of SMP Startup with Spinlock

Then we recompile the modified program, denoted by C9.4, and run it again. Figure 9.4 shows the sample outputs of the modified program C9.4. As the figure shows, all the outputs are now in CPU order.

9.9.4 Mutex in SMP

A mutex is a software synchronization tool used for locking/unlocking critical regions of long durations. A mutex contains a lock field, which indicates whether the mutex is in locked or unlocked state, and an owner field, which identifies the execution entity, e.g. a process or a CPU ID, which currently holds the mutex lock. An unlocked mutex has no owner. If an execution entity has successfully locked a mutex, it becomes the owner. Only the owner can unlock a locked mutex. The simplest form of mutex is just a (global) integer variable visible to all the execution entities. Assume that the execution entity IDs are all >= 0. A mutex with the invalid ID=-1 represents the unlocked state, and a non-negative ID represents the locked state as well as the owner of the mutex. To begin with, we shall assume such a simple mutex form with CPUs as the execution entities.

```
typedef struct mutex{
    int lock;
}MUTEX; MUTEX m;          // m is a mutex
```

The following code segments in ARM assembly show the implementations of mutex operations, which consist of three functions, mutex_init, mutex_lock and mutex_unlock.

```
QEMU
************ ARM SMP Startup Example 2 ************          00:00:29
CPU0 starts
timer_init()
timer_start
CPU1  in APstart: CPU1  enter WFI loop ncpu=2
CPU2  in APstart: CPU2  enter WFI loop ncpu=3
CPU3  in APstart: CPU3  enter WFI loop ncpu=4
CPU0 enter while(1) loop. Enter key from KBD:
KBD interrupt on CPU2 : c=0x74  t
KBD interrupt on CPU2 : c=0x65  e
KBD interrupt on CPU2 : c=0x73  s
KBD interrupt on CPU2 : c=0x74  t
```

Fig. 9.4 SMP Startup Program with spinlock

```
    UNLOCKED=0xFFFFFFFF                     // same as -1
mutex_init:                                 // init_mutex(MUTEX *m)
    MOV     r1, #UNLOCKED                   // Mark as unlocked
    STR     r1, [r0]
    BX      lr

mutex_lock:                                 // int mutex_lock(MUTEX *m)
    LDREX   r1, [r0]                        // Read lock field
    CMP     r1, #UNLOCKED                   // Compare with UNLOCKED
    WFENE                                   // If already locked, WFE
    BNE     mutex_lock                      // On waking up: re-try again
  // Attempt to lock mutex with CPU ID
    MRC     p15, 0, r1, c0, c0, 5   // Read CPU ID register
    AND     r1, r1, #0x03           // Mask in CPU ID field.
    STREX   r2, r1, [r0] // Attempt to write CPU ID to m->lock
    CMP     r2, #0x0     // Check return value: 0=OK, 1=failed
    BNE     mutex_lock   // If store failed, retry again
    DMB
    BX      lr           // return
mutex_unlock:                               // int mutex_unlock(MUTEX *m)
    MRC     p15, 0, r1, c0, c0, 5   // Read CPU ID register to r1
    AND     r1, r1, #0x03           // Mask in CPU ID field r1
    LDR     r2, [r0]                // Read lock field of mutex
    CMP     r1, r2                  // Compare CPU ID with mutex owner
    MOVNE   r0, #0x1                // If not owner: return 1 for fail
    BXNE    lr
    DMB             // Ensure all access to shared resource have completed
    MOV     r1, #unlocked           // Write "unlocked" into lock field
    STR     r1, [r0]
    DSB             // Ensure update complete before waking up other CPUs
    SEV             // Send event to other CPUs that are waiting in WFE
    MOV     r0, #0x0                // Return 0 for success
    BX      lr
```

As can be seen, the implementations of mutex operations are similar to that of spinlocks. Both rely on atomic update and memory barrier operations. The only difference is that a locked mutex has an owner, which can only be unlocked by the current owner. In contrast, spinlocks do not have any owner, so that any execution entity can unlock a spinlock even if it does not possess the spinlock, which may result in misuse of spinlocks.

9.9.5 Implementation of Mutex Using Spinlock

Alternatively, a mutex may be regarded as a structure consisting of the following fields.

```
typedef struct mutex{
        int lock;    // spinlock to access this mutex
        int status;  // 0=UNLOCKED, 1=LOCKED
        int owner;   // owner ID; -1=NOOWNER
}MUTEX; MUTEX m;
```

The lock field is a spinlock, which ensures that any operation on a mutex can only be performed inside the CR of the mutex's spinlock, the status field represents the current state of the mutex, e.g. 0=UNLOCKED, 1=LOCKED, and the owner field identifies the current execution entity that holds the mutex. With a spinlock field, mutex operations can be implemented based on slock()/sunlock() on the spinlock. The following code segments show the implementation of mutex operations in the high-level language C.

```
#define NO_OWNWR -1
int mutex_init(MUTEX *m)
{ m->lock = 0; m->status = UNLOCKED; m->owner = NO_OWNER; }

int mutex_lock(MUTEX *m)
{
   while(1){
     slock(&m->lock);      // acquire spinlock
     if (m->status == UNLOCKED)
        break;
     sunlock(&m->lock);    // release spinlock
     asm("WFE");           // wait for event, then retry
   }
   // holding the spinlock now, update mutex in the CR
   m->status = LOCKED;
   m->owner = get_cpuid();
   sunlock(&m->lock);        // release spinlock
}
int mutex_unlock(MUTEX *m)
{
   slock(&m->lock);          // acquire spinlock
   if (m->owner != get_cpuid()){
      sunlock(&m->lock);     // if not owner: release spinlock
      return -1;             // return -1 for FAIL
   }
   m->status = UNLOCKED;     // mark mutex as UNLOCKED
   m->owner = NO_OWNER;      // and no owner
   sunlock(&m->lock);        // release spinlock
   return 0;                 // return 0 for SUCCESS
}
```

The above implementation of mutex operations in C may be slightly less efficient than that in assembly, but it has two advantages. First, it is easier to understand and therefore less likely to contain errors since the readability of C code is always better than assembly code. Second and more importantly, it shows the hierarchical relation between different kinds of synchronization primitives. Once we have spinlocks to protect CRs of short durations, we may use them as a basis to implement mutexes, which are essentially CRs of longer durations. Likewise, we may also use spinlocks to implement other more powerful synchronization tools. In the hierarchical approach, we don't have to repeat the low level ldrex-strex and memory barrier sequence in the new synchronization tools.

9.9.6 Demonstration of SMP Startup with Mutex Lock

In the example program C9.5, we shall use a mutex to synchronize the executions of the APs. The effect is the same as that of using spinlocks, both enforcing each AP executes APstart() one at a time.

```
/**************** t.c file of C9.5 ******************/
#include "mutex.c"
MUTEX m;
volatile int ncpu = 1;
int APstart()
```

```
{
   int i;
   int cpuid = get_cpuid();
   mutex_lock(&m);
    printf("CPU%d start\n", cpuid);
    for (i=0; i<4; i++){ // each AP prints lines inside the CR
       printf("CPU%d before WFI state i=%d\n", cpuid, i);
    }
    printf("CPU%d enter WFI state\n", cpuid, cpuid);
    ncpu++;               // each AP increment ncpu by 1 inside CR
   mutex_unlock(&m);
   while(1)
      asm("WFI");
}
int main()
{
   // same code as in Example 2
   send_sgi(0x0, 0x0F, 0x01); // intID=0,targetList=0xF,filter=0x01
   apAddr  = (int *)0x10000030;
   *apAddr = (int)0x10000;
   while(ncpu < 4);                // wait until allAPs are ready
   printf("CPU0 enter while(1) loop. Enter key from KBD:\n");
   uprintf("Enter key from UART terminal:\n");
   while(1);
}
```

Figure 9.5 shows the outputs of the ARM SMP Startup Program using mutex for CPU synchronization. As can be seen from the figure, while holding the mutex lock, each CPU can print, or do anything for that matter, without interference from other CPUs.

```
QEMU
************ ARM SMP Startup Example 2 ************          00:00:18
CPU0 starts
timer_init()
timer_start
CPU1  start
CPU1  before WFI state i=0
CPU1  before WFI state i=1
CPU1  before WFI state i=2
CPU1  before WFI state i=3
CPU1  enter WFI state
CPU2  start
CPU2  before WFI state i=0
CPU2  before WFI state i=1
CPU2  before WFI state i=2
CPU2  before WFI state i=3
CPU2  enter WFI state
CPU3  start
CPU3  before WFI state i=0
CPU3  before WFI state i=1
CPU3  before WFI state i=2
CPU3  before WFI state i=3
CPU3  enter WFI state
CPU0 enter while(1) loop. Enter key from KBD:
KBD interrupt on CPU2 : c=0x74  t
```

Fig. 9.5 ARM SMP startup example using Mutex

9.10 Global and Local Timers

The Realview-pbx board supports several different kinds of timers. These include

```
Global timer        : a 64-bit global timer common to all CPUs
Local timers        : each CPU has a 32-bit local timer
Peripheral timers : 4 peripheral timers
```

The global timer is a 64-bit count-up timer, which can be used as a single timing source for all the CPUs in an ARM MPcore system. Each CPU has a private 64-bit comparator register. When the global timer count reaches the local comparator value, it generates an interrupt with ID 27. So far, we have only used the peripheral timer. In the following, we shall use the local timers of the CPUs for SMP operations. The next example program, C9.6, demonstrates local timers of the CPUs. The local timer of each CPU is at Peripheral Base Address + 0x600. It interrupts with ID 29. Each CPU must configure the GIC to route interrupt 29 to itself. For each CPU, we shall display a wall clock based on its local timer at the right upper corner of the LCD screen. The following show the code segments that implement the local timers.

```
/*************** ptimer.c file of C9.6 *****************/
#define DIVISOR 64
#define TIME 0x6000000/DIVISOR
int plock = 0;
typedef struct ptimer{
  u32 load;     // Load Register              0x00
  u32 count;    // Current count Register    0x04
  u32 control;  // Control Register           0x08
  u32 intclr;   // Interrupt Clear Register 0x0C
}PTIMER;
PTIMER *ptp;
// private timer at BASE+0x600 interruptID=29
int printf(char *fmt, ...);
char clock[4][16];
char *blanks = "  :  :  ";
//              01234567
struct tt{
  int hh, mm, ss, tick;
}tt[4];

int ptimer_handler()  // local timer interrupt handler
{
   int i, id;
   struct tt *tp;
   int cpuid = id = get_cpuid();
   tp = &tt[cpuid];
   slock(&plock);
   tp->tick++;
   if (tp->tick >= DIVISOR){
      tp->tick = 0; tp->ss++;
      if (tp->ss==60){
         tp->ss = 0;    tp->mm++;
         if (tp->mm==60){
             tp->mm = 0; tp->hh++;
         }
      }
   }
```

```
    if (tp->tick==0){  // every second: display a wall clock
       color = GREEN+cpuid;
       for (i=0; i<8; i++){
           unkpchar(clock[cpuid][i], 0+cpuid, 70+i);
       }
       clock[id][0]='0'+(tp->hh/10); clock[id][1]='0'+(tp->hh%10);
       clock[id][3]='0'+(tp->mm/10); clock[id][4]='0'+(tp->mm%10);
       clock[id][6]='0'+(tp->ss/10); clock[id][7]='0'+(tp->ss%10);
       for (i=0; i<8; i++){
          kpchar(clock[id][i], 0+cpuid, 70+i);
       }
    }
    sunlock(&plock);
    ptimer_clearInterrupt(); // clear timer interrupt
}
int ptimer_init()
{
    int i;
    printf("ptimer_init() ");
    // set timer base address
    ptp = (PTIMER *)0x1F000600;
    // set control counter regs to defaults
    ptp->load = TIME;
    ptp->control = 0x06; // IAE bits = 110, not eanbled yet
    for (i=0; i<4; i++){
       tt[i].tick = tt[i].hh = tt[i].mm = tt[i].ss = 0;
       strcpy(clock[i], "00:00:00");
    }
}
int ptimer_start()          // start local timer
{
   PTIMER *tpr;
   printf("ptimer_start\n");
   tpr = ptp;
   tpr->control |= 0x01;    // set enable bit 0
}
int ptimer_stop()           // stop local timer
{
   PTIMER *tptr = ptp;
   tptr->control &= 0xFE;   // clear enable bit 0
}
int ptimer_clearInterrupt() // clear interrupt
{
   PTIMER *tpr = ptp;
   ptp->intclr = 0x01;
}

int APstart()               // AP startup code
{
    int cpuid = get_cpuid();
    mutex_lock(&m);
    printf("CPU%d start: ", cpuid);
    config_int(29, cpuid);  // need this per CPU
    ptimer_init();
    ptimer_start();
```

```
QEMU
*************** ARM SMP Startup Example 4 *************
CPU0 starts
ptimer_init() ptimer_start
CPU1  locked mutex OK
CPU1  start: ptimer_init() ptimer_start
CPU1  enter WFI state
CPU1  unlocked mutex OK
CPU2  locked mutex OK
CPU2  start: ptimer_init() ptimer_start
CPU2  enter WFI state
CPU2  unlocked mutex OK
CPU3  locked mutex OK
CPU3  start: ptimer_init() ptimer_start
CPU3  enter WFI state
CPU3  unlocked mutex OK
CPU0 enter while(1) loop. Enter key from KBD:
KBD interrupt on CPU3 : c=0x74  t
KBD interrupt on CPU3 : c=0x65  e
KBD interrupt on CPU3 : c=0x73  s
KBD interrupt on CPU3 : c=0x74  t
                                                   00:00:11
                                                   00:00:11
                                                   00:00:11
                                                   00:00:11
```

Fig. 9.6 Demonstration of local timers in SMP

```
    ncpu++;
    printf("CPU%d enter WFI state\n", cpuid, cpuid);
    mutex_unlock(&m);
    while(1){
       asm("WFI");
    }
}
int irq_chandler()
{
   int int_ID = *(int *)(GIC_BASE + 0x10C);
   if (int_ID == 29){
      ptimer_handler();
   }
   *(int *)(GIC_BASE + 0x110) = int_ID;
}
```

9.10.1 Demonstration of Local Timers in SMP

Figure 9.6 shows the outputs of running the example program C9.6, which shows the ARM SMP startup sequence with mutex for CPU synchronization and local timers.

9.11 Semaphores in SMP

A (counting) semaphore is a structure

```
typedef struct sem{
        int lock;       // spinlock
        int value;      // value
```

```
  struct proc *queue;    // FIFO PROC queue
}SEMAPHORE;
SEMAPHORE s = VALUE;     // s.lock=0; s.value=VALUE; s.queue=0;
```

In the semaphore structure, the lock field is a spinlock, which ensures that any operation on a semaphore must be performed inside the CR of the semaphore's spinlock, the value field is the initial value of the semaphore, which represents the number of available resources protected by the semaphore, and queue is a FIFO queue of blocked processes waiting for available resources. The most often used semaphore operations are P and V, which are defined as follows.

```
// running is a pointer to the current executing process
--------------------------------------------------------------------
P(SEMAPHORE *s)                  | V(Semaphore *s)
{                                | {
  int sr = int_off();            |   int sr = int_off();
  slock(&s->lock);               |   slock(&s->lock);
  if (--s->value < 0)            |   if (++s->value <= 0)
     BLOCK(s);                   |      SIGNAL(s);
  else                           |
     sunlock(&s->lock);          |   sunlock(&s->lock);
  int_on(sr);                    |   int_on(sr);
}                                | }
--------------------------------------------------------------------
int BLOCK(SEMAPHORE *s)          | int SIGNAL(SEMAPHORE *s)
{                                | {
   running->status = BLOCK;      |   PROC *p = dequeue(&s->queue);
   enqueue(&s->queue, running);  |   p->status = READY;
   sunlcok(s->lock);             |   enqueue(&readyQueue, p);
   tswitch(); // switch process  |
}                                | }
--------------------------------------------------------------------
```

In a SMP system, process switching is usually triggered by interrupts. In both P and V functions, the process first disables interrupts to prevent process switching on the CPU. Then it acquires the semaphore's spinlock, which prevents other CPUs from entering the same CR. Therefore, regardless of how many CPUs, only one process can execute any function operating on the same semaphore. In P(s), the process decrements the semaphore value by 1. If the value is non-negative, the process releases the spinlock, enables interrupts and returns. In this case, the process completes the P operation without being blocked. If the value becomes negative, the process blocks itself in the (FIFO) semaphore queue, releases the spinlock and switches process. In V(s), the process disables interrupts, acquires the semaphores's spinlock and increments the semaphore value by 1. If the value is non-positive, which means there are blocked processes in the semaphore queue, it unblocks a process from the semaphore queue and makes the process ready to run again. When a blocked process resumes running in P (s), it enables interrupts and returns. In this case, the process completes the P operation after being blocked. It is noted again that the semaphores defined in this book are counting semaphores, which are more general than traditional binary semaphores. As a counting semaphore, the value may go negative. At any time, the following invariants hold.

```
   if (s.value >= 0) : value   = number of available resources
   else              : |value| = number of blocked processes in the semaphore queue
```

9.11.1 Applications of Semaphores in SMP

Unlike mutexes, which can only be used as locks, semaphores are more flexible since they can be used as locks as well as for process cooperation. For a list of semaphore applications, the reader may consult Chap. 6 of (Wang 2015). In this book, we shall use semaphores as synchronizing tools in resource management, device drivers and the file system.

9.12 Conditional Locking

Spinlocks, mutexes and semaphores all use the locking protocol, which blocks a process until it succeeds. Any locking protocol may lead to deadlocks (Silberschatz et al. 2009; Stallings 2011; Wang 2015). Deadlock is a condition in which a set of processes mutually wait for one another so that no process can continue. A simple way to prevent deadlock is to ensure that the order of acquiring different locks is always unidirectional, so that crossed or circular locking can never occur. However, this may not be possible in all concurrent programs. A practical approach to deadlock prevention is to use conditional locking and back-off. In this scheme, if a process, which already holds some locks, attempts to acquire another lock, it tries to acquire the next lock conditionally. If the locking attempt succeeds, it proceeds as usual. If the locking attempt fails, it takes remedial actions, which typically involves releasing some of the locks it already holds and retries the algorithm. In order to implement this scheme, we introduce the following conditional locking operations.

9.12.1 Conditional Spinlock

When a process attempts to acquire a regular spinlock, it does not return until it has acquired the spinlock. This may lead to deadlocks due to crossed locking attempts by different processes. Assume that a process Pi has acquired a spinlock spin1, and it tries to acquire another spinlock spin2. Another process Pj has acquired spin1 but tries to acquire the spinlock spin2. Then Pi and Pj would mutually wait for each other forever, so they end up in a deadlock due to crossed-locking attempts. To prevent such deadlocks, one of the processes may use conditional spinlock, which is defined as follows.

```
// int cslock(int *spin): conditional acquiring a spinlock
// return 0 if locking fails; return 1 if locking succeeds
cslock:
  ldrex r1, [r0]       // read spinlock value
  cmp   r1, #UNLOCKED  // compare with UNLOCKED(0)
  beq   trylock
  mov   r0, #0         // return 0 if fails
  bx    lr
trylock:               // try to lock
  mov   r1, #1         // set r1=1
  strex r2, r1, [r0]   // try to store 1 to [r0]; r2=return value
  cmp   r2, #0x0       // check strex return value in r2
  bne   cslock         // strex failed; retry to lock again
  DMB                  // memory barrier BEFORE accessing CR
  Mov   r0, #1
  bx    lr             // return 1 for SUCCESS
```

The conditional cslock() returns 0 (for FAIL) if the spinlock is already locked. It returns 1 (for SUCCESS) if it has acquired the spinlock. In the former case, the process may release some locks it already holds and retries the algorithm again, thus preventing any possible deadlock. Similarly, we can also implement other kinds of conditional locking mechanisms.

9.12.2 Conditional Mutex

```
int mutex_clock(MUTEX *m)
{
    slock(&m->lock);      // acquire spinlock
    if (m->status == LOCKED || m->owner != running->pid){
      sunlock(&m->lock);   // release spinlock
     return 0;             // return 0 for fail
    }
```

```
// holding the spinlock now, update mutex in the CR
m->status = LOCKED;
m->owner = get_cpuid();
sunlock(&m->lock);          // release spinlock
return 1;                   // return 1 for SUCCESS
```

9.12.3 Conditional Semaphore Operations

```
-----------------------------------------------------------------
int CP(SEMAPHORE *s)             |   int CV(Semaphore *s)
{                                |   {
  int sr = int_off();            |     int sr = int_off();
  slock(&s->lock);               |     slock(&s->lock);
  if (s->value <= 0){            |     if (s->value >= 0){
    sunlock(&s->lock)            |       sunlock(&s->lock);
    return 0;                    |       return 0;
  }                              |     }
  if (--s->value < 0)            |     if (++s->value <= 0)
    BLOCK(s);                    |       SIGNAL(s);
  int_on(sr);                    |     int_on(sr);
  return 1;                      |     return 1;
}                                |   }
-----------------------------------------------------------------
```

The conditional CP operates exactly like P and returns 1 for SUCCESS if the semaphore value is > 0. Otherwise, it returns 0 for FAIL without changing the semaphore value. The conditional CV operates exactly like V and returns 1 if the semaphore value is < 0. Otherwise, it returns 0 without changing the semaphore value or unblocking any process from the semaphore queue. We shall demonstrate the use of these conditional locking operations later in the SMP OS kernel.

9.13 Memory Management in SMP

The Memory Management Units (MMUs) in Cortex-A9 (ARM Cortex-A9 MPcore Technical Reference Manual r4p1 2012) and Cortex-A11 (ARM11 MPcore Processor, r2p0 2008) have the following additional capabilities in support of SMP.

(1). Level-1 Descriptors: The level-1 page table descriptor has the additional bits: NS(19), NG(17), S(16), APX(15), TEX (14-12) and P(9). The other bits in AP(11-10), DOMain(8-5), CB(3-2), ID(1-0) are unchanged. Among the added bits, the S bit determines if the memory region is shared or not. The TEX (Type Extension) bits are used to classify the memory region as **shared, non-shared** or **device** type. Shared memory regions are strongly ordered global memory areas accessible by all CPUs. Non-shared memory regions are private to specific CPUs. Device memory regions are memory-mapped peripherals accessible by all CPUs. For example, when using paging, the memory region attributes in the Level-1 page table entries should be set to

```
Shared       0x14C06
Non-shared   0x00C1E
Device       0x00C06
```

rather than the values of 0x412 or 0x41E. Otherwise, the MMU would forbid memory access unless the domain access AP bits are set to 0x3 for manager mode, which does not enforce permission checking at all. When using 2-level paging, the page size can be 4KB, 64KB, 1MB or super pages of 16MB. The accessibility bits of domains and 4KB small pages remain unchanged.

(2). More TLB Features: In Cortex-A9 to Cortex-A11, the TLB consists of micro TLBs for both instructions and data, and also a unified main TLB. These essentially divide the TLB into two-level caches, allowing quicker resolution of TLB entries. In addition to TLB entry lock-downs, main TLB entries can be associated with particular processes or applications by using the **Address Space Identifiers (ASIDs),** which enable such TLB entries to remain resident during context switches, avoiding the need for reloading them again.

(3). Program flow prediction: The ARM CPU usually predicts branch instructions. In Cortex-A9, program flow prediction can be disabled and enabled explicitly. These features, together with Data Synchronization Barriers (DSB), can be used to ensure consistent maintenance of TLB entries.

9.13.1 Memory Management Models in SMP

When configuring the ARM MMU for SMP, the Virtual Address (VA) spaces of the CPUs can be configured as uniform or non-uniform VA spaces. In the uniform VA space model, VA to PA mapping is identical for all the CPUs. In the non-uniform VA space model, each CPU may map the same VA range to different PA areas for private usage. We demonstrate these memory mapping models by the following examples.

9.13.2 Uniform VA Spaces

In this example, denoted by C9.7, we assume 4 CPUs, CPU0 to CPU3. When the system starts, CPU0 begins to execute reset_handler, which is loaded to 0x10000 by QEMU. It initializes the SVC and IRQ mode stacks, copies the vector table to address 0 and creates a Level-1 page table using 1MB sections to identity map 512MB VA to PA. All memory regions are marked as SHARED (0x14C06) except the 1MB I/O space at 256MB, which is marked as DEVICE (0x00C06). All the memory regions are in domain 0, whose AP bits are set to Client mode (b01) to enforce permission checking. Then it enables the MMU for VA to PA translation and calls main() in C. After initializing the system, CPU0 issues SGI to wakeup the secondary CPUs.

All the secondary CPUs or APs begin to execute the same reset_handler code at 0x10000. Each AP sets its own SVC and IRQ mode stacks, but does not copy the vector table and create page table again. Each AP uses the same page table created by CPU0 at 0x4000 to turn on the MMU. Thus, all the CPUs share the same VA space since their page tables are identical. Then each AP calls APstart() in C.

```
/******************** ts.s file of C9.7 **********************/
.text
.code 32
.global reset_handler, vectors_start, vectors_end
.global apStart, slock, sunlock
// ALL CPUs execute reset_handler at 0x10000
reset_handler:
// get CPU ID and keep it in R11
  MRC p15, 0, r11, c0, c0, 5   // Read CPU ID register
  AND r11, r11, #0x03          // Mask off, leaving the CPU ID field
// set CPU SVC stack
  LDR r0, =svc_stack        // r0->16KB svc_stack in t.ld
  mov r1, r11               // r1 = cpuid
  add r1, r1, #1            // cpuid++
  lsl r2, r1, #12           // (cpuid+1) * 4096
  add r0, r0, r2
  mov sp, r0                // SVC sp=svc_stack[cpuid] high end
// go in IRQ mode
  MSR cpsr, #0x12
// set CPU IRQ stack
```

```
  LDR r0, =irq_stack          // r0->16KB irq_stack in t.ld
  mov r1, r11
  add r1, r1, #1
  lsl r2, r1, #12             // (cpuid+1) * 4096
  add r0, r0, r2
  mov sp, r0                  // IRQ sp=irq_stack[cpuid] high end
// back to SVC mode
  MSR cpsr, #0x13
  cmp r11, #0
  bne skip                    // APs skip over
  BL copy_vectors             // only CPU0 copy vectors to address 0
  BL mkPtable                 // only CPU0 create pgdir and pgtable in C
skip:
  ldr r0, Mtable              // all CPUs enable MMU
  mcr p15, 0, r0, c2, c0, 0  // set TTBase
  mcr p15, 0, r0, c8, c7, 0  // flush TLB
// set domain0 AP to 01=client(check permission)
  mov r0,  #0x1                // AP=b01 for CLIENT
  mcr p15, 0, r0, c3, c0, 0
// enable MMU
  mrc p15, 0, r0, c1, c0, 0
  orr r0, r0, #0x00000001      // set bit0
  mcr p15, 0, r0, c1, c0, 0  // write to c1
  nop
  nop
  nop
  mrc p15, 0, r2, c2, c0
  mov r2, r2
  cmp r11, #0
  bne APs
  BL main                     // CPU0 call main() in C
  B .
APs:                          // each AP call APstart() in C
  adr r0, APaddr
  ldr pc, [r0]
APaddr:  .word  APstart
Mtable:  .word  0x4000

/******************** t.c file of C9.7 *******************/
int *apAddr = (int *)0x10000030; // SYS_FLAGSSET register
int aplock = 0;                  // spinlock for APs
volatile int ncpu = 1;           // number of CPUs ready
#include "uart.c"                // UARTs driver
#include "kbd.c"                 // KBD   driver
#include "ptimer.c"              // local timer driver
#include "vid.c"                 // LCD   driver
int copy_vectors()        { // SAME AS BEFORE }
int config_gic(int cpuid){ // SAME as before }
int irq_chandler()        ( // SAME as before )
int send_sgi()            { // SAME as before }
// Template descriptor of various memory regions
#define SHARED    0x14C06
#define NONSHARED 0x00C1E
#define DEVICE    0x00C06
int mkPtable()                // create pgdir by CPU0
```

```
{
  int i;
  u32 *ut = (u32 *)0x4000; // Mtable at 16KB
  for (i=0; i<4096; i++)    // clear pgdir entries to 0
      ut[i] = 0;
  u32 entry = SHARED;       // start with default=SHARED
  for (i=0; i<512; i++){    // 512 entries: ID map 512MB VA to PA
    ut[i] = entry;
    entry += 0x100000;
  }
  // mark the 1MB I/O space at 256MB as DEVICE
  ut[256] = (256*0x100000) | DEVICE;
}
int APstart()                 // APs startup code
{
   slock(&aplock);
   int cpuid = get_cpuid();
   printf("CPU%d in APstart()\n", cpuid);
   config_int(29, cpuid);  // configure per CPU local timer
   ptimer_init(); ptimer_start();
   ncpu++;
   sunlock(&aplock);
   printf("CPU%d enter WFI loop ncpu=%d\n", cpuid, ncpu);
   while(1){
     asm("WFI");
   }
}
int main()
{
   enable_scu();
   fbuf_init();
   printf("********** ARM SMP Startup Example 5 ***********\n");
   kbd_init();
   uart_init();
   ptimer_init(); ptimer_start();
   config_gic();
   *apAddr = (int)0x10000;      // APs begin to execute at 0x10000
   send_sgi(0x01, 0x0F, 0x00); // wakeup APs
   printf("CPU0: wait for APs ready\n");
   while(ncpu < 4);
   printf("CPU0: continue ncpu=%d\n", ncpu);
   printf("enter a key from KBD or UARTs : ");
   while(1) asm("WFI");
```

9.13.2.1 Demonstration of Uniform VA Space Mapping

Figure 9.7 shows the outputs of the SMP system with uniform VA space mapping.

9.13.3 Non-uniform VA Spaces

In the non-uniform VA space model, the CPUs may map the same VA range to different PA ranges to create private memory regions containing CPU specific information. In this case, each CPU has its own page table, which can be created either by CPU0 or by each CPU itself during system startup. We illustrate this technique by the following example.

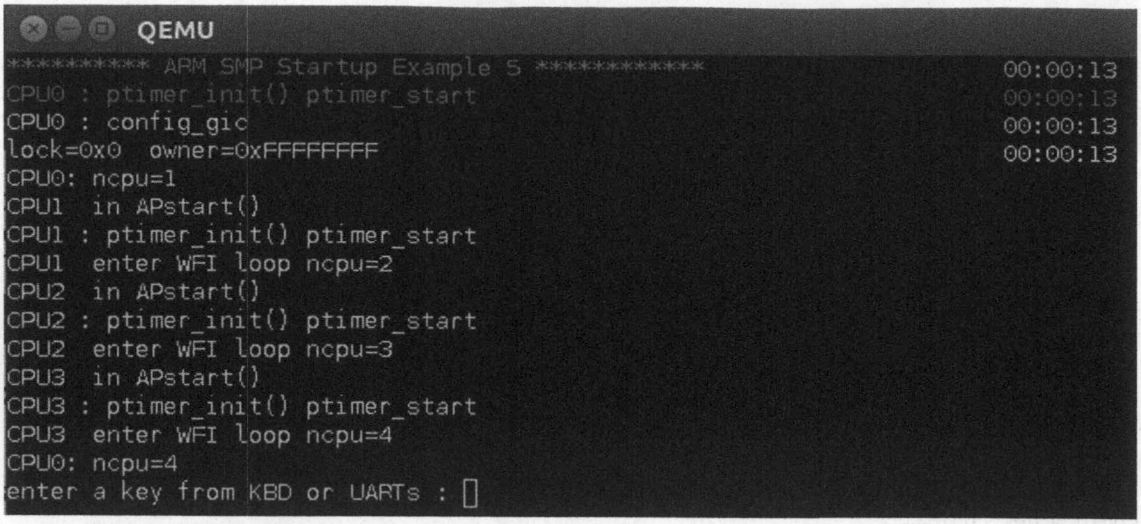

```
QEMU
********** ARM SMP Startup Example 5 ************          00:00:13
CPU0 : ptimer_init() ptimer_start                           00:00:13
CPU0 : config_gic                                           00:00:13
lock=0x0  owner=0xFFFFFFFF                                  00:00:13
CPU0: ncpu=1
CPU1  in APstart()
CPU1 : ptimer_init() ptimer_start
CPU1  enter WFI loop ncpu=2
CPU2  in APstart()
CPU2 : ptimer_init() ptimer_start
CPU2  enter WFI loop ncpu=3
CPU3  in APstart()
CPU3 : ptimer_init() ptimer_start
CPU3  enter WFI loop ncpu=4
CPU0: ncpu=4
enter a key from KBD or UARTs :
```

Fig. 9.7 Demonstration of uniform VA space mapping

In this example, C9.8, we assume that the system executes in the lowest 1MB of physical memory. When the system starts, we shall let CPU0 create 4 level-1 page tables, one for each CPU. In the page tables, we map the same VA=1MB to different PAs based on the CPU ID. Since the system is loaded by QEMU at 0x10000 (64KB), there is not enough space to create four (16KB) page tables at 0x4000 (16KB). We shall use the 64KB space below 1MB (at 0xF0000) as the Level-1 page tables (pgdirs) of the CPUs. When the system starts, CPU0 creates four pgdirs, one per CPU, at 0xF0000 + cpuid*16KB to ID map 512MB VA to PA. However, for each CPU, the VA space at 1MB is mapped to a different 1MB area in physical memory. Specifically, the VA=1MB of cpuid is mapped to PA=(cpuid+1)*1MB. Before enabling the MMU, CPU0 can access all the PAs. It writes different strings to the PAs at 1M to 4MB, each for a different CPU. Then CPU0 initializes the system and issues SGI to wakeup the APs.

When the APs start to execute reset_handler, they only initialize the SVC and IRQ mode stacks. Each AP gets its cpuid and uses the pgdir at 0xF0000+cpuid*16K to enable the MMU. In APstart(), each AP prints the string at the same VA=1MB. The assembly code of the system is the same as before, except for the trivial modifications mentioned above. Therefore, we only show the modified C code.

```
/******************* t.c file of C9.8 *******************/
int *apAddr = (int *)0x10000030;
int aplock = 0;
volatile int ncpu = 1;
#include "uart.c"
#include "kbd.c"
#include "ptimer.c"
#include "vid.c"
int copy_vectors() { // SAME as before }
int config_gic()   { // SAME as before }
int irq_chandler() { // SAME as before }
int seng_sgi()     { // SAME as before }

#define SHARED      0x00014C06
#define NONSHARED   0x00000C1E
#define DEVICE      0x00000C06
int mkPtable()      // CPU0 creates 4 pgdirs
{
  int i, j, *ut;
  for (i=0; i<4; i++){          // create 4 pgdirs
     ut = (u32 *)(0xF0000 + i*0x4000); // 16KB apart
```

```
      u32 entry = SHARED;
      for (j=0; j<4096; j++)       // clear pgdir entries
          ut[j] = 0;
      for (j=0; j<512; j++){       // ID map 512MB VA to PA
          ut[j] = entry;
          entry += 0x100000;
      }
      ut[256] = 256*0x100000 | DEVICE; // mark I/O space as DEVICE
      if (i){                          // for CPU1 to CPU3:
         ut[1] = (i+1)*0x100000 | NONSHARED;
         ut[2] = ut[3] = ut[4] = 0;    // no longer valid entries
      }
   }
   // BEFORE enable MMU: write different strings to PA=1MB to 4MB
   char *cp = (char *)0x100000;        // PA=1MB
   strcpy(cp, "initial string of CPU0");
   cp += 0x100000;                     // PA=2MB
   strcpy(cp, "INITIAL string of CPU1");
   cp += 0x100000;                     // PA=3MB
   strcpy(cp, "initial STRING of CPU2");
   cp += 0x100000;                     // PA=4MB
   strcpy(cp, "INITIAL STRING of CPU3");
}
int APstart()
{
    char *cp;
    slock(&aplock);
    int cpuid = get_cpuid();
    cp = (char *)0x100000;    // SAME VA=1MB
    printf("CPU%d in APstart(): string at VA 1MB = %s\n", cpuid, cp);
    config_int(29, cpuid);      // configure per CPU local timer
    ptimer_init(); ptimer_start();
    ncpu++;
    sunlock(&aplock);
    printf("CPU%d enter WFI loop ncpu=%d\n", cpuid, ncpu);
    while(1){
       asm("WFI");
    }
}
int main()
{
    enable_scu();
    fbuf_init();
    printf("********** ARM SMP Startup Example 6 ************\n");
    kbd_init();
    uart_init();
    ptimer_init();  ptimer_start();
    config_gic();
    apAddr  = (int *)0x10000030;
    *apAddr = (int)0x10000;
    send_sgi(0x01, 0x0F, 0x01);
    printf("CPU0: ncpu=%d\n", ncpu);
    while(ncpu < 4);
    printf("CPU0: ncpu=%d\n", ncpu);
    char *cp = (char *)0x100000;       // at VA=1MB
```

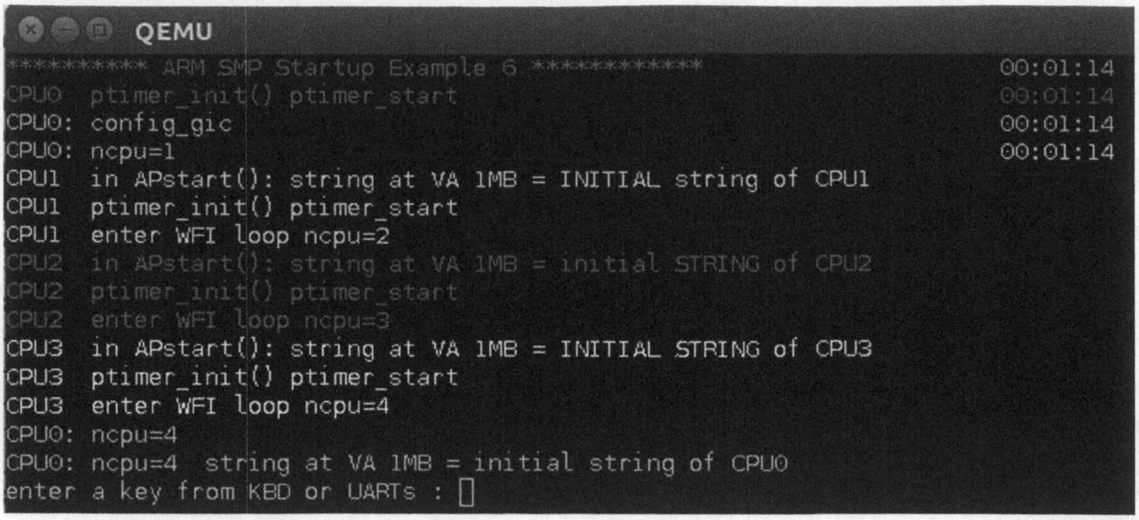

```
QEMU
********** ARM SMP Startup Example 6 ************            00:01:14
CPU0  ptimer_init() ptimer_start                              00:01:14
CPU0: config_gic                                              00:01:14
CPU0: ncpu=1                                                  00:01:14
CPU1  in APstart(): string at VA 1MB = INITIAL string of CPU1
CPU1  ptimer_init() ptimer_start
CPU1  enter WFI loop ncpu=2
CPU2  in APstart(): string at VA 1MB = initial STRING of CPU2
CPU2  ptimer_init() ptimer_start
CPU2  enter WFI loop ncpu=3
CPU3  in APstart(): string at VA 1MB = INITIAL STRING of CPU3
CPU3  ptimer_init() ptimer_start
CPU3  enter WFI loop ncpu=4
CPU0: ncpu=4
CPU0: ncpu=4  string at VA 1MB = initial string of CPU0
enter a key from KBD or UARTs :
```

Fig. 9.8 Demonstration of non-uniform VA space mapping

```
    printf("CPU0: ncpu=%d string at VA 1MB = %s\n", ncpu, cp);
    printf("enter a key from KBD or UARTs : ");
    while(1)
        asm("WFI");
}
```

9.13.3.1 Demonstration of Non-uniform VA Space Mapping

Figure 9.8 shows the outputs of the SMP system with non-uniform VA space mapping. As the figure shows, each AP displays a different string in its private memory region at VA=1MB.

9.13.4 Parallel Computing in Non-uniform VA Spaces

We demonstrate a parallel computing system that uses the non-uniform VA space memory model. In this example system, we let the CPUs perform parallel computations as follows. Before setting up page tables and enabling the MMU, CPU0 deposits a sequence of N integers in the private memory area of each CPU. After enabling the MMU for VA to PA translation, each CPU can access its own private memory area by the same VA. Then each CPU computes the sum of the N integers in its private memory area. When a CPU has computed the (local) partial sum, it adds the partial sum to a (global) total variable, which is initialized to 0. When all the CPUs have updated the total sum, CPU0 prints the final results. The example program is intended to show the following important principles of parallel processing in SMP.

(1). Each CPU may execute a process in its private memory area.
(2). When a CPU intends to modify any shared global variable, it must perform the update in a Critical Region (CR) protected by either a spinlock or a mutex lock.

The following lists the major code segments of such a parallel computing system.

(1). ts.s file: the ts.s file is identical to that of the example program 9.8.
(2) t.c file:

```
/************** t.c file of Program 9.9 ***************/
int aplock = 0;    // spinlock for APs
MUTEX m;           // mutex for locking
volatile int total = 0;
volatile int ncpu = 1;
int *apAddr = (int *)0x10000030;

#include "uart.c"   // UARTs driver
#include "kbd.c"    // KBD driver
#include "ptimer.c" // local timer driver
#include "vid.c"    // LCD driver
#include "mutex.c"  // mutex functions

int copy_vectors(){ // same as before }
int config_gic()  { // same as before }
int irq_chandler(){ // same as before }
int send_sgi()    { // same as befor  }

#define SHARED    0x14C06
#define NONSHARED 0x00C1E
#define DEVICE    0x00C06

int mkPtable()       // CPU0 creates 4 pgdirs
{
  int i, j, *ut;
  for (i=0; i<4; i++){            // create 4 pgdirs
     ut = (u32 *)(0xF0000 + i*0x4000); // 16KB apart
     u32 entry = SHARED;
     for (j=0; j<4096; j++)       // clear pgdir entries
         ut[j] = 0;
     for (j=0; j<512; j++){       // ID map 512MB VA to PA
         ut[j] = entry;
         entry += 0x100000;
     }
     ut[256] = 256*0x100000 | DEVICE; // mark I/O space as DEVICE
     if (i){                              // for CPU1 to CPU3:
        ut[1] = (i+1)*0x100000 | NONSHARED;
        ut[2] = ut[3] = ut[4] = 0;     // no longer valid entries
     }
  }
}

int fill_values()   // CPU0: fill 1MB to 4MB with numbers
{
  int i, j, k = 1;
  int *ip = (int *)0x100000; // 1MB PA
  for (i=1; i<5; i++){        // 4 data areas
    for (j=0; j<8; j++){      // 8 numbers per area
      ip[j] = k++;            // values
    }
    ip += 0x100000/4;         // ip point to next 1MB
  }
}
```

```
// ALL CPUs call compute() to compute local sum and update total
int compute(int cpuid)
{
  int i, *ip, sum = 0;
  printf("CPU%d compute partial sum\n", cpuid);
  ip = (int *)0x100000;      // VA at 1MB
  for (i=0; i<8; i++){
    printf("%d ", ip[i]); // show local data in VA=1MB
    sum += ip[i];          // compute local sum
  }
  printf("CPU%d: sum=%d\n", cpuid, sum);
  mutex_lock(&m);            // lock mutex m
   total += sum;             // update global total
  mutex_unlock(&m);          // unlock mutex m
}

int APstart()
{
   char *cp; int sum;
   slock(&aplock);
   int cpuid = get_cpuid();
   config_int(29, cpuid);  // configure per CPU local timer
   ptimer_init(); ptimer_start();
   compute(cpuid);           // call compute()
   printf("CPU%d enter WFI loop ncpu=%d\n", cpuid, ncpu);
   ncpu++;
   sunlock(&aplock);
   while(1) asm("WFI");
}

int main()
{
   enable_scu();
   fbuf_init();
   printf("********** ARM SMP Startup Example 6 ************\n");
   kbd_init();
   uart_init();
   ptimer_init();
   ptimer_start();
   config_gic();
   mutex_init(&m);                // initialize mutex m
   apAddr  = (int *)0x10000030;
   *apAddr = (int)0x10000;
   send_sgi(0x01, 0x0F, 0x00); // wakeup APs
   printf("CPU0: ncpu=%d\n", ncpu);
   while(ncpu < 4); // wait until all APs have finished computing
   printf("CPU0: ncpu=%d\n", ncpu);
   compute(0);
   printf("total = %d\n", total);
   printf("enter a key from KBD or UARTs : ");
   while(1);
}
```

Figure 9.9 shows the outputs of running the parallel computing system C9.9.

```
QEMU
*********** ARM SMP Startup Example 6 ***********        CPU0: 00:00:15
ptimer_start                                             CPU1: 00:00:15
CPU0: config_gic                                         CPU2: 00:00:15
CPU0: ncpu=1                                             CPU3: 00:00:15
ptimer_start
CPU1  compute partial sum
9  10  11  12  13  14  15  16  CPU1 : sum=100
CPU1  locked mutex OK
CPU1  enter WFI loop ncpu=1
ptimer_start
CPU2  compute partial sum
17  18  19  20  21  22  23  24  CPU2 : sum=164
CPU2  locked mutex OK
CPU2  enter WFI loop ncpu=2
ptimer_start
CPU3  compute partial sum
25  26  27  28  29  30  31  32  CPU3 : sum=228
CPU3  locked mutex OK
CPU3  enter WFI loop ncpu=3
CPU0: ncpu=4
CPU0  compute partial sum
1  2  3  4  5  6  7  8  CPU0 : sum=36
CPU0  locked mutex OK
total = 528
enter a key from KBD or UARTs :
```

Fig. 9.9 Outputs of a parallel computing system

9.14 Multitasking in SMP

In the simple parallel computing example, each CPU executes only a single task, which does not fully utilizes the capabilities of the CPUs. A better way of using the CPUs more effectively is to let each CPU execute multiple processes by multitasking. Each process may execute in two different modes, kernel mode and user mode. While in kernel mode, all processes execute in the same address space of the kernel. While in user mode, each process has a private memory area protected by the MMU. In a single CPU system, only one process can execute in kernel mode at a time. As a result, there is no need to protect kernel data structures from concurrent executions of processes. In SMP, many processes may execute on different CPUs in parallel. All data structures in a SMP kernel must be protected to prevent corruptions from race conditions. In a uniprocessor system, the kernel usually uses a single running pointer to point at the currently executing process. In SMP, a single running pointer is no longer adequate because many processes may execute on different CPUs at the same time. We must devise a way to identify the PROCs that are currently executing on the CPUs. There are two possible approaches to this problem. The first one is to use virtual memory. Define a CPU structure, as in

```
struct cpu{
    struct cpu *cpu;        // pointer to this CPU struct
    PROC       *proc;       // pointer to PROC on this CPU
    int        cpuid;       // CPU ID
    int        *pgdir;      // pgdir of this CPU
    PROC       *readyQueue; // readyQueue of this CPU
    // other fields
}cpu;
```

Each CPU has a CPU structure in a private memory area, which is mapped to the same VA, e.g. at 1MB by all the CPUs. Define the symbols

```
#define cpu (struct cpu *)0x100000
#define running cpu->proc
```

Then we may use the same running symbol to access the PROC currently executing on each CPU. The drawback of this scheme is that it would make dispatching processes to run on different CPUs rather complex. This is because the level-1 page tables (pgdirs) and level-2 page tables (if using 2-level paging) of the CPUs are all different. In order to dispatch a process to a CPU, we must change the process pgdir to that of the target CPU, which may require flushing or invalidating many entries in the CPU's TLB. So this scheme is only suited to each CPU running a fixed set of processes which do not migrate to other CPUs.

The second scheme is use multiple PROC pointers, each pointing to the process that's currently executing on a CPU. To do this, we define the following symbols.

```
#define PROC *run[NCPU]
#define running run[get_cpuid()]
```

Then we may use the same running symbol in all C code to refer to the PROC that's currently executing on a CPU. In this case, all the CPUs may use the uniform VA space memory model. The kernel mode VA spaces of all the processes are identical, so that they can share the same pgdir and page tables in kernel mode. Only their User mode page table entries may be different. In this scheme, dispatching a process to a CPU is much easier. Although we still have to change the process pgdir to suit the target CPU, it only involves a few changes of the User mode entries in the CPU's TLB. Thus, the uniform VA memory model is more suited to running processes on different CPUs by dynamic process scheduling.

9.15 SMP Kernel for Process Management

In Chap. 7, we developed a simple kernel for uniprocessor (UP) systems. The simple kernel supports dynamic processes in both kernel and user modes. As a UP system, it allows only one process to execute in kernel at a time. In this section, we shall show how to extend the UP kernel for SMP. The following is a list of planned gaols and modifications of the resulting SMP system.

(1). The system shall boot up from an EXT2/3 file system in a SDC partition. When running on the ARM realview-pbx-a9 VM under QEMU, the kernel image is loaded to 0x10000 by QEMU and runs from there.
(2). When running on the ARM realview-pbx-a9 VM, the system supports 4 CPUs.
(3). The system uses the uniform VA space memory model with 2-level dynamic paging. All processes share the same VA space in kernel, which is ID mapped to the low 512MB PA. Each CPUi begins by running an initial process, iproc[i], which runs only in kernel mode with the lowest priority 0. The initial process of each CPU is also the idle process of the CPU, which runs whenever there are no other runnable processes.
(4). The system maintains a single process scheduling queue, which is ordered by process priority. All CPUs try to run processes from the same readyQueue. If the readyQueue is empty, each CPU runs an idle process, which puts the CPU in the WFI power-saving state, waiting for interrupts.
(5). All other processes normally run in User mode in the VA range of [2GB to 2GB + image size]. Executable User mode images are in the /bin directory of the file system on the SDC. For simplicity, we assume a fixed run-time image size of 1 to 4 MB. The User mode image of a process consists of dynamically allocated page frames. When a process terminates, it releases the page frames back to the free page list for reuse.
(6). User mode processes use system calls (via SWI) to enter kernel to execute kernel functions. Each system call returns a value back to User mode, except exit() which never returns, and exec(), which returns to a different image if the operation succeeds.
(7). CPU0 uses the console (keyboard + LCD) for I/O. Each AP uses a dedicated UART terminal, e.g. CPUi uses UARTi, for I/O.

Since the main objective here is to demonstrate process management in SMP, the system supports only two User mode images, denoted by u1 and u2. They are used to illustrate the change image operations by exec. If desired, the reader may generate additional User mode images for processes to execute. The following lists the major code segments of the SMP system, denoted by C9.10. For brevity, device drivers and exceptions handlers are not shown.

9.15.1 The ts.s file

```
/*************** ts.s file of C9.10 **************/
        .text
.code 32
.global reset_handler, vectors_start, vectors_end
.global proc, iproc, iprocsize, run
.global tswitch, scheduler, goUmode, switchPgdir
.global int_on, int_off, setUlr
.global apStart, slock, sunlock, lock, unlock
.global get_cpuid, enable_scu, send_sgi

reset_handler: // all CPUs begin to execute reset_handler
// set SVC stack to HIGH END of iproc[CPUID]
  MRC  p15, 0, r11, c0, c0, 5   // read CPU ID register into R11
  AND  r11, r11, #0x03          // mask in CPU ID field; R11=CPUID
  LDR sp, =iproc      // r0 points to iproc's
  LDR r1, =procsize   // r1 -> procsize
  LDR r2, [r1, #0]    // r2 = procsize
  mov r3 ,r2          // r3 = procsize
  mov r1, r11         // get CPU ID
  mul r3, r1          // procsize*CPUID
  add sp, r3, r3      // sp += r3
  add sp, r2, r2      // sp->iproc[cpuid] high end
// set previous mode of SVC mode to USER mode
  MSR spsr, #0x10     // write to previous mode spsr
// go in IRQ mode to set IRQ stack
  MSR cpsr, #0xD      // write to cspr, so in IRQ mode now
  mov r1, r11
  add r1, r1, #1      // CPUID+1
  ldr r0, =irq_stack  // 16 KB svc_stack area in t.ld
  lsl r2, r1, #12     // r2 = (CPUID+1)*4096
  ADD r0, r0, r2      // r0 -> high end irq_stack[cpuid]
  MOV sp, r0
// go in ABT mode to set ABT stack
  MSR cpsr, #0xD7
  mov r1, r11
  add r1, r1, #1      // CPUID+1
  ldr r0, =abt_stack  // 16 KB abt_stack area in t.ld
  lsl r2, r1, #12     // r2 = (CPUID+1)*4096
  ADD r0, r0, r2      // r0 -> high end irq_stack[cpuid]
  MOV sp, r0
// go in UND mode to set UND stack
  MSR cpsr, #0xDB
  mov r1, r11
  add r1, r1, #1      // CPUID+1
  ldr r0, =und_stack  // 16 KB und_stack area in t.ld
  lsl r2, r1, #12     // r2 = (CPUID+1)*4096
  ADD r0, r0, r2      // r0 -> high end irq_stack[cpuid]
  MOV sp, r0
```

```
// go back to SVC mode
  MSR cpsr, #0x13       // write to CPSR
  mov r1, r11           // CPUID
  cmp r1, #0            // if not CPU0, skip
  bne skip
// only CPU0 copy vector table and create initial pgdir
  BL copy_vector_table
  BL mkPtable    // create pgdir and pgtable in C
skip:
  ldr r0, Mtable // all CPUs use the same pgdir to enable MMU
  mcr p15, 0, r0, c2, c0, 0  // set TTBase
  mcr p15, 0, r0, c8, c7, 0  // flush TLB
// set domain 0: 01=client(check permission)
  mov r0,  #0x01             // 01 for CLIENT
  mcr p15, 0, r0, c3, c0, 0
// enable MMU
  mrc p15, 0, r0, c1, c0, 0
  orr r0, r0, #0x00000001    // set bit0
  mcr p15, 0, r0, c1, c0, 0  // write to c1
  nop
  nop
  nop
  mrc p15, 0, r2, c2, c0
  mov r2, r2
// enable IRQ interrupts
  mrs r0, cpsr
  BIC r0, r0, #0x80  // I bit=0 enable IRQ
  MSR cpsr, r0       // write to CPSR
  mov r1, r11
  cmp r1, #0
  bne skip2
  adr r0, mainstart  // CPU0 call main()
  ldr pc, [r0]
  B .
skip2:
  adr r0, APgo        // APs call APstart()
  ldr pc, [r0]

Mtable:     .word 0x4000
mainstart:  .word main
APgo:       .word APstart

irq_handler:                 // IRQ interrupts entry point
  sub lr, lr, #4
  stmfd sp!, {r0-r12, lr}  // save all Umode regs in kstack
  bl irq_chandler          // call irq_chandler() in C in svc.c file
  ldmfd sp!, {r0-r12, pc}^ // pop from kstack but restore Umode SR

data_handler:
  sub lr, lr, #4
  stmfd sp!, {r0-r12, lr}
  bl data_abort_handler
  ldmfd sp!, {r0-r12, pc}^

tswitch:                      // tswitch() in Kmode
// mask out IRQ interrupts
```

```
  MRS r0, cpsr
  ORR r0, r0, #0x80
  MSR cpsr, r0

  stmfd sp!, {r0-r12, lr}
  LDR r4, =run                 // r4=&run
  MRC p15, 0, r5, c0, c0, 5 // read CPUID register to r5
  AND r5, r5, #0x3           // mask in the CPUID
  mov r6, #4                 // r6 = 4
  mul r6, r5                 // r6 = 4*cpuid
  add r4, r6                 // r4 = &run[cpuid]
  ldr r6, [r4, #0]           // r6->running PROC
  str sp, [r6, #4]           // save sp into PROC.ksp

  bl  scheduler              // call scheduler() in C

  LDR r4, =run               // r4=&run
  MRC p15, 0, r5, c0, c0, 5 // read CPUID register to r5
  AND r5, r5, #0x3           // only the CPU ID
  mov r6, #4                 // r6 = 4
  mul r6, r5                 // r6 = 4*cpuid
  add r4, r6                 // r4 = &run[cpuid]
  ldr r6, [r4, #0]           // r6->running PROC
  ldr sp, [r6, #4]           // restore sp from PROC.ksp

// enable IRQ interrupts
  MRS r0, cpsr  BIC r0, r0, #0x80
  MSR cpsr, r0
  Ldmfd sp!, {r0-r12, pc}    // resume next running PROC

klr:  .word 0

svc_entry: // syscall entry point; syscall params in r0-r3
  stmfd sp!, {r0-r12, lr}
  LDR r4, =run                // r4=&run
  MRC p15, 0, r5, c0, c0, 5   // read CPUID register to r5
  AND r5, r5, #0x3            // mask in CPUID
  mov r6, #4                  // r6=4
  mul r6, r5                  // r6 = 4*cpuid
  add r4, r4, r6              // r4 = &run[cpuid]
  ldr r6, [r4, #0]            // r6 -> PROC of running
  mrs r7, spsr                // User mode cpsr
  str r7, [r6, #16]           // save into PROC.spsr
// get usp=r13 from USER mode
  mrs r7, cpsr        // r7 = SVC mode cpsr
  mov r8, r7          // save a copy in r8
  orr r7, r7, #0x1F   // r7 = SYS mode
  msr cpsr, r7        // change cpsr to SYS mode
// now in SYS mode, r13 same as User mode sp r14=user mode lr
  str sp, [r6, #8]    // save usp into PROC.usp at offset 8
  str lr, [r6, #12]   // save Umode PC into PROC.ups at offset 12
// change back to SVC mode
  msr cpsr, r8
// saved lr in kstack return to svc entry, NOT Umode PC at syscall
// replace saved lr in kstak with Umode PC at syscall
  mov r5, sp
  add r5, r5, #52     // offset = 13*4 bytes from sp
```

```
  ldr r7, [r6, #12]  // lr in Umode at syscall
  str r7, [r5]
// enable IRQ interrupts
  MRS r7, cpsr
  BIC r7, r7, #0x80  // I bit=0 enable IRQ
  MSR cpsr, r7

  bl  svc_handler     // call svc_chandler in C

 // replace saved r0 on stack with return value r from svc_handler()
  add sp, sp, #4      // effectively pop saved r0 off stack
  stmfd sp!,{r0}      // push r as the saved r0 to Umode

goUmode:
  LDR r4, =run                 // r4=&run
  MRC p15, 0, r5, c0, c0, 5  // read CPUID register to r5
  AND r5, r5, #0x3             // mask in CPUID
  mov r6, #4                   // r6 = 4
  mul r6, r5                   // r6 = 4*cpuid
  add r4, r4, r6               // r4 = &run[cpuid]
  ldr r6, [r4, #0]             // r6->running PROC
  ldr r7, [r6, #16]            // restore spsr from PROC.spsr
  msr spsr, r7                 // restore spsr
// set cpsr to SYS mode to access user mode sp
  mrs r2, cpsr                 // r2 = SVC mode cpsr
  mov r3, r2                   // save a copy in r3
  orr r2, r2, #0x1F            // r0 = SYS mode
  msr cpsr, r2                 // change cpsr to SYS mode
// now in SYS mode
  ldr sp, [r6, #8]             // restore usp from PROC.usp
// back to SVC mode
  msr cpsr, r3                 // back to SVC mode
  ldmfd sp!, {r0-r12, pc}^     // return to Umode

// utility functions
int_on:                  // int_on(int sr)
  MSR cpsr, r0
  mov pc,lr
int_off:                 // int sr=int_off()
  MRS r1, cpsr
  mov r0, r1             // r0 = r1
  ORR r1, r1, #0x80
  MSR cpsr, r1
  mov pc, lr             // return r0=original cpsr
unlock:                  // mask in IRQ in cpsr
  MRS r0, cpsr
  BIC r0, r0, #0x80
  MSR cpsr, r0
  mov pc,lr
lock:                    // maks out IRQ in cpsr
  MRS r0, cpsr
  ORR r0, r0, #0x80
  MSR cpsr, r0
  mov pc,lr
switchPgdir: // switch pgdir to new PROC's pgdir
 // r0 contains address of PROC's pgdir
```

```
  mcr p15, 0, r0, c2, c0, 0    // set TTBase
  mov r1, #0
  mcr p15, 0, r1, c8, c7, 0    // flush TLB
  mcr p15, 0, r1, c7, c10, 0  // flush TLB
  mrc p15, 0, r2, c2, c0, 0
// set domain 0,1: all 01=client(check permission)
  mov r0, #0xD                 //11|01 for manager|client
  mcr p15, 0, r0, c3, c0, 0
  mov pc, lr                   // return
setUlr:     // for threads in process
// to SYS mode; set ulr to r0
  mrs r7, cpsr        // r7 = SVC mode cpsr
  mov r8, r7          // save a copy in r8
  orr r7, r7, #0x1F   // r7 = SYS mode
  msr cpsr, r7
// now in SYS mode
  mov lr, r0          // set r13 = r0 = VA(4)
// back to SVC mode
  msr cpsr, r8        // back to SVC mode
  mov pc, lr
get_cpuid:
  MRC  p15, 0, r0, c0, c0, 5   // Read CPU ID register
  AND  r0, r0, #0x03           // Mask off, leaving the CPU ID field
  BX   lr
enable_scu:
  MRC  p15, 4, r0, c15, c0, 0  // Read periph base address
  LDR  r1, [r0, #0x0]          // Read the SCU Control Register
  ORR  r1, r1, #0x1            // Set bit 0 (The Enable bit)
  STR  r1, [r0, #0x0]          // Write back modifed value
  BX   lr
send_sgi: // sgi(filter=r0, CPUs=r1, intID=r2)
  AND  r0, r0, #0x0F  // low 4 bits of filter_list
  AND  r1, r1, #0x0F  // low 4 bits of CPUs
  AND  r3, r2, #0x0F  // low 4 bits of intID into r3
  ORR  r3, r3, r0, LSL #24     // fill in the filter field
  ORR  r3, r3, r1, LSL #16     // fill in CPUs field
  // Get address of the GIC
  MRC  p15, 4, r0, c15, c0, 0  // Read peripheral base address
  ADD  r0, r0, #0x1F00         // Add offset of the sgi_trigger reg
  STR  r3, [r0]   // Write to the SGI Register (ICDSGIR)
  BX   lr
slock:                    // int slock(&spin)
  ldrex r1, [r0]
  cmp   r1, #0x0
  WFENE
  bne   slock
  mov   r1, #1
  strex r2, r1, [r0]
  cmp   r2, #0x0
  bne   slock
  DMB                     // barrier
  bx    lr
sunlock:                  // sunlock(&spin)
  mov   r1, #0x0
  DMB
```

```
  str   r1, [r0]
  DSB
  SEV
  bx    lr
vectors_start:
  LDR PC, reset_handler_addr
  LDR PC, undef_handler_addr
  LDR PC, svc_handler_addr
  LDR PC, prefetch_abort_handler_addr
  LDR PC, data_abort_handler_addr
  B .
  LDR PC, irq_handler_addr
  LDR PC, fiq_handler_addr
reset_handler_addr:            .word reset_handler
undef_handler_addr:            .word undef_handler
svc_handler_addr:              .word svc_entry
prefetch_abort_handler_addr:   .word prefetch_abort_handler
data_abort_handler_addr:       .word data_handler
irq_handler_addr:              .word irq_handler
fiq_handler_addr:              .word fiq_handler
vectors_end:
.end
```

Explanations of the ts.s file:

In order to support SMP, the system defines NCPU = 4 globals (in kernel.c file)

```
PROC iporc[NCPU];     // initial PROCs of the CPUs
PROC  *run[NCPU];     // pointers to PROCs executing CPUs
```

In kernel_init(), we initialize run[i]=&iproc[i], so that each run[i] points to the initial PROC structure of CPUi. During system operation, each run[i] points to the process executing on CPUi. Throughout the system's C code, it uses the symbol running

```
#define running run[get_cpuid()]
```

to access the executing PROC on a CPU. The assembly code in ts.s consists of the following parts.

(1). Reset_handler: All CPUs begin by executing the reset_handler in SVC mode with interrupts off and MMU disabled. First, it gets the CPU ID and saves it in R11, so that it does not have to repeatedly get CPU ID again later. It sets the SVC mode sp to the high end of iproc[CPUID], which makes iproc[CPUID]'s kstack as the initial stack of the CPU. It sets the previous mode (SPSR) to User mode, which makes the CPU ready to run images in User mode later. Then it sets the stacks of other privileged modes. Among the CPUs, only CPU0 copies the vectors table to address 0 and creates an initial one-level page table using 1MB sections at 16KB, which ID maps the low 512MB VA to PA. Subsequently, all CPUs use the same page table to configure and enable MMU for VA to PA translation. Then it enables IRQ interrupts and calls functions in the t.c file. CPU0 calls main() and all others APs call APstart() to continue the initialization until they are ready to run tasks.

(2). IRQ and Exception handlers: These are the same as before. In order to keep the system simple, the kernel is non-preemptive, i.e. IRQ interrupts do not switch process in kernel mode. We shall extend the kernel to allow pre-emptive process scheduling later.

(3). tswitch: tswitch is for process switch in kernel (SVC) mode. In tswitch, the calling process first disables interrupts and acquires a spinlock associated with the readyQueue. After saving the process kernel mode context (in PROC.kstack) and the stack pointer, it calls scheduler() to pick the next running process. The spinlock is released by the next running process when it resumes.

(4). svc_entry: this is the system call entry point. Processes in User mode issue system calls (via SWI) to enter kernel to execute kernel functions. Upon entry, it save the User mode context, (upc, ucpsr, usp), into the PROC structure for return to User mode later. Then it enables interrupts and calls svc_chandler() in C to process the system call.
(5). goUmode: When a process finishes a system call, it exits kernel to return to User mode. In goUmode, it restores the saved User mode context from the PROC structure, followed by LDMFD SP!, {R0-R12, SP}^, causing return to User mode. Each system call returns a value back to User mode, except kexit() which never returns. The return value is through R0 in the process kstack.
(6). switchPgdir: When the system starts, it uses an initial one-level page table at 16KB with 1MB sections. In kernel_init(), it builds a 2-level page table (pgdir) at 32KB and the level-2 page tables at 5MB. Then it switches pgdir to use 2-level dynamic paging. Each process has its own pgdir and level-2 page tables. During process switch, the scheduler uses switchPgdir to switch the CPU's pgdir to the next running process.
(7). Utility functions: the remaining assembly code contains utility functions, such as slock/sunlock, enable_scu, get_cpuid, send_sgi, etc. in support of SMP operations.

9.15.2 The t.c file

```
/****************** t.c file of C9.10 ****************/
#include "type.h"
#define NCPU 4
#define running run[get_cpuid()]

#include "uart.c"         // UART driver
#include "kbd.c"          // KBD driver
#include "ptimer.c"       // local timers driver
#include "vid.c"          // LCD driver
#include "except.c"       // exceptions handlers
#include "queue.c"        // queue functions
#include "kernel.c"       // kernel code
#include "wait.c"         // sleep/wakeup, exit, wait
#include "fork.c"         // fork functions
#include "exec.c"         // change image
#include "svc.c"          // system call routing table
#include "loadelf.c"      // ELF image loader
#include "thread.c"       // threads
#include "sdc.c"          // SDC driver
int copy_vector_table(){  // SAME as before }
int config_int(int N, int targetCPU){ // SAME as before }

int aplock = 0;           // spinlock for APs
volatile int ncpu = 1;    // number of CPUs ready
int APstart()
{
   slock(&aplock);         // acquire spinlock
   int cpu = get_cpuid();
   printf("CPU%d in Apstart switchPgdir to 0x8000\n", cpu);
   switchPgdir(0x8000);   // switch to 2-level paging
   config_int(29, cpu);   // configure and start local timer
   ptimer_init();
   ptimer_start();
   printf("CPU%d enter run_task() loop\n", cpu);
   ncpu++;
   sunlock(&aplock);      // releasse spinlock
```

```
    run_task();          // all Aps call run_task()
}
int config_gic()
{
   int cpuid = get_cpuid();
   // set int priority mask register
   *(int *)(GIC_BASE + 0x104) = 0xFFFF;
   // set CPU interface control register: enable signaling interrupts
   *(int *)(GIC_BASE + 0x100) = 1;
   // set distributor control register to route interrupts to CPUs
   *(int *)(GIC_BASE + 0x1000) = 1;
   config_int(29, 0);        // CPU0 local timer at intID=29
   config_int(44, 0);        // UART0
   config_int(45, 1);        // UART1
   config_int(46, 2);        // UART2
   config_int(47, 3);        // UART3
   config_int(49, 0);        // SDC interrupts to CPU0
   config_int(52, 0);        // KBD
}
#define SHARED    0x00014c06
#define NONSHARED 0x00000c1e
#define DEVICE    0x00000c06
int mkPtable()    // create initial one-level pgdir
{
   int i;
   u32 *ut = (u32 *)0x4000; // at 16KB
   u32 entry = 0 | SHARED;  // domain 0
   for (i=0; i<4096; i++)    // clear pgdir entris
     ut[i] = 0;
   for (i=0; i<512; i++){    // ID map 512MB
     ut[i] = entry;
     entry += 0x100000;
   }
   ut[256] = (256*0x100000) | DEVICE; // I/O page at 256MB
}
int irq_chandler()
{
     int intID = *(int *)0x1F00010C; // read intID register
     switch(intID){
        case 29 : ptimer_handler(); break;
        case 44 : uart_handler(0);  break;
        case 45 : uart_handler(1);  break;
        case 46 : uart_handler(2);  break;
        case 47 : uart_handler(3);  break;
        case 49 : sdc_handler();    break;
        case 52 : kbd_handler();    break;
     }
     *(int *)0x1F000110 = intID;     // write EOF
}
int main()
{  int i;
   fbuf_init();   // initialize LCD
   printf("========== Welcome to Wanix in ARM ===========\n");
   enable_scu();           // enable SCU
   sdc_init();             // initialize SDC
```

```
  kbd_init();              // initialize KBD
  uart_init();             // initialize UARTs
  config_gic();            // configure GIC
  ptimer_init();           // configure local timer
  ptimer_start();          // start local timer
  kernel_init();           // initialize kernel
  printf("CPU0 startup APs\n");
  int *APaddr = (int *)0x10000030;
  *APaddr = (int)0x10000;       // AP begin execution address
  send_sgi(0x00, 0x0F, 0x01); // send SGI to wakeup APs
  printf("CPU0 waits for APs ready\n");
  while(ncpu < 4);
  printf("CPU0 continue\n");
  // CPU0: create NCPU PROCs into readyQueue
  for (i=0; i<NCPU; i++)
      kfork("/bin/u1");
  run_task();              // CPU0 call run_task()
}
```

9.15.3 Kernel.c file

```
/****************** kernel.c file of C9.10 ******************/
PROC proc[NPROC+NTHREAD];
PROC *run[NCPU]          // executing PROC pointers
PROC iproc[NCPU];        // initial PROCs of CPUs
PROC *freeList, *tfreeList, *readyQueue, *sleepList;
int  freelock, tfreelock, readylock, sleeplock; // spinlocks
int  procsize = sizeof(PROC);
char *pname[NPROC]={"sun", "mercury", "venus", "earth", "mars",
      "jupiter", "saturn", "uranus", "neptune", "Pluto"};

// pfreeList,free_Page_list(),palloc(),pdealloc(): dynamic paging
int *pfreeList;          // free page frame list
int pfreelock;           // spinlock for pfreeList
int *palloc()            // allocate a free page frame
{
  slock(&pfreelock);
  int *p = pfreeList;
  if (p)
     pfreeList = (int *)(*p);
  sunlock(&pfreelock);
  return p;
}
int pdealloc(int *p)     // deallocate a page frame
{
  slock(&tfrelock);
  int *a = (int *)((int)p & 0xFFFFF000);
  *a = (int)pfreeList;
  pfreeList = a;
  sunlock(&pfreelock);
}
```

```
int *free_page_list(int *startva, int *endva)
{
  int *p;
  printf("build pfreeList: start=%x end=%x : ", startva, endva);
  pfreeList = startva;
  p = startva;
  while(p < (int *)(endva-1024)){
    *p = (int)(p + 1024);
    p += 1024;
    i++;
  }
  *p = 0;
  printf("%d 4KB entries\n", i);
  return startva;
}

u32 *MTABLE = (u32 *)0x4000;
int *kpgdir = (int *)0x8000;
int kernel_init()
{
  int i, j, *ip;
  PROC *p;
  int *MTABLE, *mtable, *ktable, *pgtable;
  int paddr;
  printf("kernel_init(): init procs\n");
  for (i=0; i<NCPU; i++){ // initialize per CPU varaibles
    p = &iproc[i];
    p->pid = 1000 + i;     // special pid of initial PROCs
    p->status = READY;
    p->priority = 0;
    p->ppid = p->pid;
    p->pgdir = (int *)0x8000; // pgdir at 0x8000
    run[i] = p;               // run[i]=&iproc[i]
  }
  for (i=0; i<NPROC; i++){    // initialize procs
    p = &proc[i];
    p->pid = i;
    p->status = FREE;
    p->priority = 0;
    p->ppid = 0;
    strcpy(p->name, pname[i]);
    p->next = p + 1;
    p->pgdir = (int *)(0x600000 + (p->pid-1)*0x4000);  }
  for (i=0; i<NTHREAD; i++){  // threads
    p = &proc[NPROC+i];
    p->pid = NPROC + i;
    p->status = FREE;
    p->priority = 0;
    p->ppid = 0;
    p->next = p + 1;
  proc[NPROC-1].next = 0;
  freeList = &proc[1];  // skip proc[0], so that P1 is proc[1]
  readyQueue = 0;
  sleepList = 0;
  tfreeList = &proc[NPROC];
```

```
  proc[NPROC+NTHREAD-1].next = 0;
  // CPU0: run the initial iproc[0]
  running = run[0];          // CPU0 run iproc[0]
  MTABLE = (int *)0x4000;  // initial pgdir at 16KB
  printf("build pgdir and pgtables at 32KB (ID map 512MB VA to PA)\n");
  mtable = (u32 *)0x8000;  // new pgdir at 32KB
  for (i=0; i<4096; i++){  // zero out 4096 entries
    mtable[i] = 0;
  }
  for (i=0; i<512; i++){   // ASSUME 512MB PA: ID mapped VA to PA
      mtable[i] = (0x500000 + i*1024) | 0x01; // DOMAIN0,Type=01
  }
  printf("build Kmode level-2 pgtables at 5MB\n");
  for (i=0; i<512; i++){
    pgtable = (u32 *)((u32)0x500000 + (u32)i*1024);
    paddr = i*0x100000 | 0x55E;    // all APs=01|01|01|01|CB=11|type=10
    for (j=0; j<256; j++){ // 256 entries, each points to 4KB PA
      pgtable[j] = paddr + j*4096; // inc by 4KB
    }
  }
  printf("build 64 level-1 pgdirs for PROCs at 6MB\n");
  ktable = (u32 *)0x600000; // build 64 proc's pgdir at 6MB
  for (i=0; i<64; i++){      // 512KB area in 6MB
    ktable = (u32 *)(0x600000 + i*0x4000); // each ktable = 16KB
    for (j=0; j<4096; j++){
      ktable[j] = 0;
    }
    // copy low entries of P0's mtable[ ], Kmode spaces are SAME
    for (j=0; j<1024; j++){
      ktable[j] = mtable[j];
    }
    // Umode pgdir entry [2048] will be set when proc is created
    ktable[2048] = 0;
  }
  printf("switch pgdir to use 2-level paging : ");
  switchPgdir((u32)mtable);
  printf("switched pgdir OK\n");
  // build free page frame list: begin at 8MB end = 256MB
  pfreeList = free_page_list((int *)0x00800000, (int *)0x10000000);
}
// all CPUs run task from the same readyQueue
int run_task()
{
  while(1){
    slock(&readylock);
    if (readyqueue == 0){
      sunlock(&readylock);
      asm("WFI");
    }
    else{
      tswitch();
    }
  }
}
```

```
int scheduler()
{
  int cpuid = get_cpuid();
  PROC *old = running;
  // caller already holds the readyQueue spinlock
  if (running->pid < 1000){ // IDLE procs do NOT enter readyQueue
     if (running->status == READY){
        enqueue(&readyQueue, running);
     }
  }
  running = dequeue(&readyQueue);
  if (running == 0){          // if readyQueue empty
     running = &iproc[cpuid];// run the IDLE proc
  }
  if (running != old){
     switchPgdir((u32)running->pgdir);
  }
  sunlock(&readylock);        // next running PROC release spinlock
}
```

9.15.4 Device Driver and Interrupt Handlers in SMP

The sample SMP system supports the following I/O devices.

```
LCD display: vid.c
UARTs:       uart.c
TIMERS:      ptimer.c
Keyboard:    kbd.c
SDC:         sdc.c
```

All device drivers (except LCD display) are interrupt-driven. In an interrupt-driven device driver, processes and interrupt handler share data buffer and control variables, which form a Critical Region (CR). In a uniprocessor kernel, when a process executes a device driver, it can mask out interrupts to prevent interference from the device interrupt handler. In SMP, masking out interrupts is no longer sufficient. This is because while a process executes a device driver on one CPU, another CPU may execute the device interrupt handler at the same time. In order to prevent processes and interrupt handler from interfering with one another, device drivers in SMP must be modified. This can be achieved by requiring processes and interrupt handler to execute inside a common critical region. Since interrupt handlers cannot sleep or block, they must use spinlock or equivalent mechanisms. The process side of a device driver may use blocking mechanism, such as semaphore, to wait but it must release the spinlock before becoming blocked. In the following, we illustrate the principles of SMP driver design by using the SDC driver as a specific example.

9.15.4.1 The SDC Driver in SMP

The SDC driver supports read/write of multi-sector blocks. SDC interrupts are routed to a specific CPU, e.g. CPU0, which handles all SDC interrupts. To serialize the executions of process and interrupt handler, both must execute inside the same critical region of a spinlock (sdclock). When a process executes get_block() or put_block(), it first acquires the spinlock. Before issuing read/write commands to the SDC, if the process is running on CPU0, it must release the spinlock first. Otherwise, the process would lock itself out on the same spinlock which it still holds, resulting in a degenerated form of deadlock. If the process is running on other CPUs, it may release the spinlock after issuing the I/O commands. Then the process waits for data transfer to complete by P(&sdc_sem), which may block the process on the sdc_sem semaphore.

When the interrupt handler executes (on CPU0), it first acquires the spinlock to handle the current interrupt. It releases the spinlock at the end of each interrupt processing. When a SDC block transfer completes, it issues V(&sdc_sem) to unblock the process. The following lists the SDC driver code.

```
/****************** sdc.c file of C9.10 ******************/
#include "sdc.h"  // SDC types and structures
int sdclock = 0;  // SDC driver spinlock
struct semaphore sdc_sem; // semaphore for SDC driver synchronization
int P(struct semaphore *s){ // P operation }
int V(struct semaphore *s){ // V operation }

// shared variables between SDC driver and interrupt handler
volatile char *rxbuf, *txbuf;
volatile int  rxcount, txcount;
int sdc_handler()
{
  u32 status, status_err, *up;
  int i, cpuid=get_cpuid();
  slock(&sdclock);         // acquire spinlock
  // read status register to find out TXempty or RxAvail
  status = *(u32 *)(base + STATUS);
  if (status & (1<<17)){ // RxFull: read 16 u32 at a time;
     up = (u32 *)rxbuf;
     status_err = status & (DCRCFAIL | DTIMEOUT | RXOVERR);
     if (!status_err && rxcount){
        for (i = 0; i < 16; i++)
           *(up + i) = *(u32 *)(base + FIFO);
        up += 16;
        rxcount -= 64;
        rxbuf += 64;
        status = *(u32 *)(base + STATUS); // clear Rx interrupt
     }
     if (rxcount == 0){                   // read block done
         do_command(12, 0, MMC_RSP_R1);   // stop transmission
         V(&sdc_sem);                     // V up process
     }
  }
  else if (status & (1<<18)){ // TXempty: write 16 u32 at a time
    up = (u32 *)txbuf;
    status_err = status & (DCRCFAIL | DTIMEOUT);
    if (!status_err && txcount) {
       for (i = 0; i < 16; i++)
           *(u32 *)(base + FIFO) = *(up + i);
       up += 16;
       txcount -= 64;
       txbuf += 64;              // advance txbuf for next write
       status = *(u32 *)(base + STATUS); // clear Tx interrupt
    }
    if (txcount == 0){                  // write block done
       do_command(12, 0, MMC_RSP_R1); // stop transmission
       V(&sdc_sem);                   // V up process
    }
  }
  //printf("write to SDC status_clear register\n");
  *(u32 *)(base + STATUS_CLEAR) = 0xFFFFFFFF;
  sunlock(&sdclock);      // release spinlock
}
```

```
int delay(){ int i; for (i=0; i<100; i++); }
int do_command(int cmd, int arg, int resp)
{
  *(u32 *)(base + ARGUMENT) = (u32)arg;
  *(u32 *)(base + COMMAND)  = 0x400 | (resp<<6) | cmd;
   delay();
}
int sdc_init()
{
  u32 RCA = (u32)0x45670000; // QEMU's hard-coded RCA
  base    = (u32)0x10005000; // PL180 base address
  printf("sdc_init() ");
  *(u32 *)(base + POWER) = (u32)0xBF; // power on
  *(u32 *)(base + CLOCK) = (u32)0xC6; // default CLK
  // send init command sequence
  do_command(0,  0,   MMC_RSP_NONE);// idle state
  do_command(55, 0,   MMC_RSP_R1);  // ready state
  do_command(41, 1,   MMC_RSP_R3);  // argument must not be zero
  do_command(2,  0,   MMC_RSP_R2);  // ask card CID
  do_command(3,  RCA, MMC_RSP_R1);  // assign RCA
  do_command(7,  RCA, MMC_RSP_R1);  // transfer state: must use RCA
  do_command(16, 512, MMC_RSP_R1);  // set data block length
  // set interrupt MASK0 registers bits = RxFULL(17)|TxEmpty(18)
  *(u32 *)(base + MASK0) = (1<<17)|(1<<18);
  // initialize spinlock and sdc_sem semaphore
  sdclock = 0;
  sdc_sem.lock = 0; sdc_sem.value = 0; sdc_sem.queue = 0;
}
int get_block(int blk, char *buf)
{
  u32 cmd, arg;
  int cpuid=get_cpuid();
  slock(&sdclock);      // process acquire spinlock
  rxbuf = buf; rxcount = FBLK_SIZE;
  *(u32 *)(base + DATATIMER) = 0xFFFF0000;
  // write data_len to datalength reg
  *(u32 *)(base + DATALENGTH) = FBLK_SIZE;
  if (cpuid==0) // CPU0 handles SDC interrupts: must release sdclock
     sunlock(&sdclock);
  cmd = 18;         // CMD18: read multi-sectors
  arg = ((bsector + blk*2)*512);
  do_command(cmd, arg, MMC_RSP_R1);
  // 0x93=|9|0011|=|9|DMA=0,0=BLOCK,1=Host<-Card,1=Enable
  *(u32 *)(base + DATACTRL) = 0x93;
  if (cpuid)   // Other CPUs release spinlock
     sunlock(&sdclock);
  P(&sdc_sem); // wait for read block complete
}
int put_block(int blk, char *buf)
{
  u32 cmd, arg;
  int cpuid = get_cpuid();
  slock(&sdclock); // process acquire spinlock
  txbuf = buf; txcount = FBLK_SIZE;
  *(u32 *)(base + DATATIMER) = 0xFFFF0000;
```

```
    *(u32 *)(base + DATALENGTH) = FBLK_SIZE;
    if (cpuid == 0) // CPU0 handles SDC interrupts, must release sdclock
       sunlock(&sdclock);
    cmd = 25;        // CMD25: write multi-sectors
    arg = (u32)((bsector + blk*2)*512);
    do_command(cmd, arg, MMC_RSP_R1);
    // write 0x91=|9|0001|=|9|DMA=0,BLOCK=0,0=Host->Card, Enable
    *(u32 *)(base + DATACTRL) = 0x91; // Host->card
    if (cpuid)       // other CPUs release spinlock
       sunlock(&sdclock);
    P(&sdc_sem);     // wait for interrupt handler to finish
}
```

9.15.5 Demonstration of Process Management in SMP

Figure 9.10 shows the sample outputs of running the program C9.10, which demonstrates process management is SMP.

When the system starts, CPU0 creates NCPU=4 processes, P1 to P4, all with the same User mode image u1, and enters them into the readyQueue. Then it sends SGI to activate other CPUs (APs). Each AP begins to execute the reset_handler to initialize itself. Then it calls APstart(). Since CPU0 is designated to handle the SDC interrupts, it can load the u1 image file from the SDC before the APs are activated and running. If the SDC interrupts are routed to different CPUs, loading image must be done after the APs are up and running. When the CPUs are ready, they all call run_task(), trying to run tasks from the same readyQueue. If the readyQueue is nonempty, the CPU takes a process from the readyQueue and switches to run that process. If the readyQueue empty, it runs the idle process, which puts the CPU in WFI state, waiting for interrupts. Since each CPU has a local timer, it will get up periodically to handle the timer interrupt and then try run task from the readyQueue again. Alternatively, the CPUs may also use SGIs to communicate with one another. For instance, when a CPU makes a process ready to run, e.g. by fork, wakeup or V operation, it sends a SGI to other CPUs, causing them to run tasks from the readyQueue again. While running in User mode, each process displays a menu, including the CPU ID the process is running on, as in

```
          proc x running on CPUy: enter a command:
  ____________________________________________________________
  | ps  chname switch wait exit fork exec vfork thread sgi  |
  ____________________________________________________________
```

Then it prompts for an input command and executes the command. The actions of the commands are

```
ps :    print status information of all PROCs
chname: change the name string of the running PROC.
switch: enter kernel to switch process
wait:   wait for a ZOMBIE child, return child pid and exit status
exit:   terminate in kernel to become a ZOMBIE, wakeup parent
fork:   fork a child PROC with identical User mode image
exec:   change execution image
vfork:  vfork a child and wait; vforked child exec to run u2 image and terminate
thread: create a thread and wait; thread executes and terminates.
sgi:    send SGI to a target CPU, causing it to execute a SGI handler function
```

The reader may enter commands to test the system. As a natural consequence of SMP, processes may migrate form one CPU to execute on different CPUs.

```
root@D630: ~
proc2 on CPU1 in Umode at VA=0x80000000  PA=0xC05000  parent=1000
-----------------------------------------------------------
ps chname switch wait exit fork exec vfork thread sgi
-----------------------------------------------------------
proc2 on CPU1 : input a command :

root@D630: ~/KCW/n1/fs
proc3 on CPU2 in Umode at VA=0x80000000  PA=0x1009000  parent=1000
-----------------------------------------------------------
ps chname switch wait exit fork exec vfork thread sgi
-----------------------------------------------------------
proc3 on CPU2 : input a command :

root@D630: ~/KCW/xx92/kernel
proc4 on CPU3 in Umode at VA=0x80000000  PA=0x140D000  parent=1000
-----------------------------------------------------------
ps chname switch wait exit fork exec vfork thread sgi
-----------------------------------------------------------
proc4 on CPU3 : input a command :

QEMU
=========== Welcome to Wanix in ARM ============          CPU0: 00:00:43
LCD display initialized : fbuf = 0x220000                 CPU1: 00:00:43
sdc_init() kbd_init() uarts init ptimer_start             CPU2: 00:00:43
kernel_init(): init procs                                 CPU3: 00:00:43
build pgdir and pgtables at 32KB (ID map 512MB VA to PA)
build Kmode level-2 pgtables at 5MB
build 64 level-1 pgdirs for PROCs at 6MB
switch pgdir to use 2-level paging : switched pgdir OK
build pfreeList: start=0x800000  end=0x10000000  : 63487  4KB entries
proc 1000  kforked a child 1  at 0x801000
proc 1000  kforked a child 2  at 0xC05000
proc 1000  kforked a child 3  at 0x1009000
proc 1000  kforked a child 4  at 0x140D000
CPU0 startup APs
CPU0 waits for APs ready
CPU1  in APstart switchPgdir to 0x8000 ... OK ptimer_start
CPU1  enter run_task() loop
CPU2  in APstart switchPgdir to 0x8000 ... OK ptimer_start
CPU2  enter run_task() loop
CPU3  in APstart switchPgdir to 0x8000 ... OK ptimer_start
CPU3  enter run_task() loop
CPU0 continue
CPU mode=0x10  argc=2  argv[0 ] = init argv[1 ] = start
proc1 on CPU0 in Umode at VA=0x80000000  PA=0x801000  parent=1000
-----------------------------------------------------------
ps chname switch wait exit fork exec vfork thread sgi
-----------------------------------------------------------
proc1 on CPU0 : input a command :
```

Fig. 9.10 Demonstration of process management in SMP

9.16 General Purpose SMP Operating System

In this section, we shall present a general purpose SMP operating system, denoted by SMP_EOS, for the ARM architecture.

9.16.1 Organization of SMP_EOS

The SMP_EOS system is essentially the same uniprocessor EOS system of Chap. 8 adapted to SMP.

9.16.1.1 Hardware Platform

SMP_EOS should be able to run on any ARM MPcore based systems that support suitable I/O devices. Since most readers may not have access to a real ARM MPcore based hardware system, we shall use the emulated ARM Realview-pbx-a9 VM under QEMU as the platform for implementation and testing. The emulated Realview-pbx-a9 VM supports the following I/O devices.

(1). SDC: SMP_EOS uses a SDC virtual disk as the primary mass storage device. For simplicity, the SDC has only one partition, which begins from the (fdisk default) sector 2048. After creating the virtual SDC file we set up a loop device for the partition and format it as an EXT2 file system with 4KB block size and one blocks-group. Then we mount the loop device and populate it with DIRs and files, making it ready for use. The resulting file system size is 128MB, which should be big enough for most applications. The single blocks-group assumption on the disk image is only for convenience since it simplifies both the file system traversal and the inodes and disk blocks management algorithms. If desired, the SDC can be created with multiple blocks groups for larger file systems or even with multiple partitions. The following diagram shows the SDC contents.

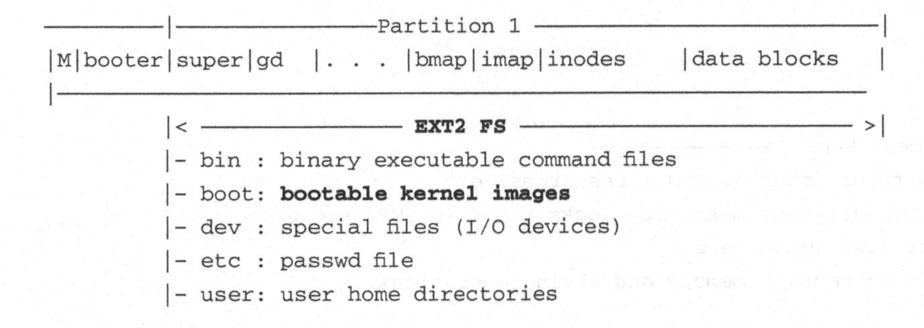

On the SDC, the MBR sector (0) contains the partition table and the beginning part of a booter. The entire booter is installed in sectors 2 to booter_size, assuming that the booter size is no more than 2046 sectors. (The actual booter size is less than 10KB). The booter is designed to boot up a kernel image from an EXT2 file system in a SDC partition. After booting up, the SMP_EOS kernel mounts the SDC partition as the root file system and runs on the SDC partition.

(2). LCD: the LCD is the primary display device. The LCD and the keyboard play the role of the system console.
(3). Keyboard: this is the keyboard device of the Realview-pbx-a9 VM. It is the input device for both the console and UART serial terminals.
(4). UARTs: these are the (4) UARTs of the Realview-pbx-a9 VM, which are used as serial terminals for users to login. Although it is highly unlikely that an embedded system will ever have multiple users, our purpose is to show that SMP_EOS has the capability to support multiple users at the same time.
(5). Timer: The Realview-pbx-a9 VM has four timers in the peripheral device memory area. In addition, each CPU also has a private local timer. In SMP_EOS, each CPU uses its own local timer for timing and process scheduling. Timer service functions to all processes are provided by CPU0.

9.16.2 SMP_EOS Source File Tree

The source files of SMP_EOS are organized as a file tree.

```
SMP_EOS
   |- booter1, booter2: stage1 and stage2 booters
   |- type.h, include.h, mk script
   |- kernel     : kernel source files
   |- driver     : device driver files
   |- fs         : file system files
   |- USER       : commands and user mode programs
```

Booter1 contains the source code of the stage1 booter. When running SMP_EOS on the Realview-pbx-a9 VM, QEMU loads the stage1 booter to 0x10000 and executes it first. Stage1 booter loads the stage2 booter from the beginning part of the SDC to 2MB and executes it next. Booter2 contains the source code of the stage2 booter. It is designed to boot up a kernel image from the /boot directory in an EXT2 partition. Since the SDC has only one partition, which begins from sector 2048, there is no need for the stage2 booter to find out which partition to boot when it starts. So we simply place the stage2 booter in sectors 2 and beyond, which simplifies the loading task of the stage1 booter. The stage2 booter loads the kernel image to 1MB and transfers control to the loaded kernel image, causing the kernel to start up.

```
type.h       : SMP_EOS kernel data structure types, EXT2 file system types, etc.
include.h    : constants and function prototypes.
mk           : sh script for recompile SMP_EOS and install bootable image to SDC partition.
```

9.16.3 SMP_EOS Kernel Files

```
———————— Kernel: Process Management Part ————————
type.h       : kernel data structure types, such as PROC, resources, etc.
ts.s         : reset_handler, tswitch, interrupt mask, spinlocks, cpus in SMP, svc and
               interrupt handler entry/exit code, etc.
eoslib.c     : kernel library functions; memset, memcpy and string operations.
except.c     : exceptions handlers
io.c         : kernel I/O functions
irq.c        : config SCU, GIC and IRQ interrupt hander entry points.
queue.c      : enqueue, deque, list operation functions
mem.c        : memory management functions using dynaminc 2-level paging
wait.c       : ksleep, kwakeup, kwait, kexit functions
loader.c     : ELF executable image loader
fork.c       : kfork, fork, vfork functions
exec.c       : kexec function
threads.c    : threads and mutex functions
pipe.c       : pipe creation and read/write functions
mes.c        : message passing: send/recv functions
signal.c     : signals and signal processing
syscall.c    : simple syscall functions
svc.c        : syscall routing table
t.c          : main entry, initialization, parts of process scheduler
```

```
-------------- Device Drivers ---------------------------------
vid.c         : LCD display driver
ptimer.c      : timer and timer service functions
pv.c          : semaphore operations
kbd.c         : console keyboard driver
uart.c        : UART ports driver
-------------- File system -------------------------------------
fs            : implementation of a simple EXT2 file system with I/O buffering
----------------------------------------------------------------
USER          : source code of user-level programs.
                All user commands are ELF executables in the /bin directory
----------------------------------------------------------------
```

SMP_EOS is implemented mostly in C, with less than 2 percent of assembly code.

9.16.4 Process Management in SMP_EOS

This section describes process management in the SMP_EOS kernel

9.16.4.1 PROC structure

In SMP_EOS, each process or thread is represented by a PROC structure, which consists of three parts.

```
. fields for process management,
. pointer to a per-process resource structure,
. kernel mode stack: a dynamically allocated 4KB page frame.
```

The PROC and resource structures are identical to those in the EOS system of Chap. 8.

9.16.4.2 Running PROC Pointers

In order to reference processes running on different CPUs, the SMP_EOS kernel defines NCPU=4 PROC pointers, each pointing to the PROC that's currently executing on a CPU.

```
#define PROC *run[NCPU]
#define running run[get_cpuid()]
```

In the kernel's C code, it uses the symbol running to access the PROC that's currently executing on a CPU. In the kernel's assembly code, it uses the following code segment to access the PROC executing on a CPU.

```
.global run                    // PROC *run[4] global
 LDR r4, =run                  // r4=&run
 MRC p15, 0, r5, c0, c0, 5 // read CPUID register to r5
 AND r5, r5, #0x3              // mask in CPUID
 mov r6, #4                    // r6 = 4
 mul r6, r5                    // r6 = 4*cpuid
 add r4, r4, r6                // r4 = &run[cpuid]
 ldr r6, [r4, #0]              // r6->running PROC
```

9.16.4.3 Spinlocks and Semaphores

In SMP, processes may run in parallel on different CPUs. All data structures in a SMP kernel must be protected to prevent corruptions from race conditions. Typical tools used to achieve this are spinlocks and semaphores. Spinlocks are suitable for CPUs to wait for critical regions of short durations in which task switching is either unnecessary or not allowed, e.g. in an interrupt handler. To access a critical region, a process must acquire the spinlock associated with the critical region first, as in

```
int spin = 0;              // initial value = 0
slock(&spin);              // acquire spinlock
  // access critical region CR
sunlock(&spin);            // release spinlock
```

For critical regions of longer durations, it is preferable to let the process wait by giving up the CPU. In such cases, mutex and semaphores are more suitable. In SMP, each semaphore must have a spinlock field to ensure that all operations on the semaphore can only be performed in the critical region of the spinlock. Similar modifications are also needed for mutex and mutex operations. In SMP_EOS, we shall use both spinlocks and semaphores to protect kernel data structures.

9.16.4.4 Use sleep/wakeup in SMP

The SMP_EOS kernel uses (modified) sleep/wakeup in process management and pipes. As a synchronization mechanism, sleep/wakeup works well in uniprocessor (UP) systems but is unsuited to SMP. This is because an event is just a value, which does not have an associated memory location to record the occurrence of the event. When an event occurs, wakeup tries to wake up all processes sleeping on the event. If no process is sleeping on the event, wakeup has no effect. This requires a process to sleep first before another process or an interrupt handler tries to wake it up later. This sleep-first and wakeup-later order can always be achieved in UP but not in SMP. In a SMP system, processes may run on different CPUs in parallel. It is impossible to guarantee the process execution order. Therefore, in their original form, sleep/wakeup cannot be used in SMP. In the modified sleep/wakeup, both operations must go through a common critical region protected by a spinlock (Cox et al. 2011; Wang 2015). While holding the spinlock, if a process must go to sleep, it completes the sleeping operation and releases the spinlock in a single indivisible (atomic) operation.

9.16.5 Protect Kernel Data Structures in SMP

In a UP kernel, only one process executes at a time. Therefore, data structures in a UP kernel do not need any protection against concurrent process executions. When adapt a UP kernel for SMP, all kernel data structures must be protected to ensure that processes can only access them one at a time. The required modifications can be classified into two categories.

(1). The first category includes kernel data structures that are used for allocation and deallocation of resources. These include

free PROC list, free page frame list, pipe structures, message buffers, bitmaps for
inodes and disk blocks, in-memory inodes, open file table, mount table, etc.

Each of these data structures can be protected by a spinlock or a locking semaphore. Then modify the allocation/deallocation algorithms as critical regions of the form

```
allocate(resource)
{
   LOCK(resource_lock);
    // allocate resource from the resource data structure;
   UNLOCK(resource_lock);
   retrun allocated resource;
}
deallocate(resource)
{
   LOCK(resource_lock)
    // release resource to the resource data structure;
   UNLOCK(resource_lock);
}
```

where LOCK/UNLOCK denote either slock/sunlock on spinlocks or P/V on locking semaphores. Since P/V require implicit operations on the semaphore's spinlock, it is more efficient to use spinlocks directly. For example, in SMP_EOS we define a spinlock freelock = 0 to protect the free PROC list and modify get_proc()/put_proc() as follows.

```
PROC *get_proc(PROC **list)
{
   slock(&freelock);
    // PROC *p = get a PROC pointer from *list;
   sunlock(&freelosck);
   return p;
}
void put_proc(PROC **list, PROC *p)
{
    slock(sfreelist);
     // enter p into *list;
    sunlock(sfreelist);
}
```

While holding the spinlock, the operations on the free PROC list are exactly the same as that in a UP kernel. Similarly, we can use spinlocks to protect other resources. In short, this category includes all the kernel data structures for which the behavior of a process is to access the data structure without pausing.

(2). The second category includes the cases in which a process must acquire a lock first in order to search for a needed item in a data structure. If the needed item already exists, the process must not create the same item again, but it may have to wait for the item. If so, it must release the lock to allow concurrency. However, this may lead to the following race condition. After releasing the lock but before the process has completed the wait operation, the item may be changed by other processes running on different CPUs, causing it to wait for the wrong item. This kind of race condition cannot occur in UP but is very likely in SMP. There are two possible ways to prevent such race conditions.

(2).1. Set a reservation flag on the item before releasing the lock. Ensure that the reservation flag can only be manipulated inside the critical region of the lock. For example, in the iget() function of the file system, which returns a locked minode, each minode has a refCount field, which represents the number of processes still using the minode. When a process executes iget(), it first acquires a lock and then searches for a needed minode. If the minode already exists, it increases the minode's refCount by 1 to reserve the minode. Then it releases the lock and tries to lock the minode. When a process releases a minode by iput(), it must execute in the same critical region as iget(). After decrementing the minode's refCount by 1, if the refCount is non-zero, meaning that the minode still has users, it does not free the minode. Since both iget() and iput() execute in the same critical region of a lock, race condition cannot occur.

(2).2. Ensure that the process completes the wait operation on the needed item before releasing the lock, which eliminates the time gap. When using spinlocks this is the same technique as requiring both sleep and wakeup to execute in the same critical region of a spinlock. When using semaphore locks we may use the conditional CP operation to test whether a semaphore is already locked. If the semaphore is already locked, we use the PV(s1,s2) operation, which atomically blocks a process on semaphore s1 before releasing semaphore s2, to wait for locked semaphore. As an example, consider the iget()/iput() functions again. Assume that mlock is a lock semaphore for all the minodes in memory and each minode has a lock semaphore minode.sem=1. We only need to modify iget()/iput() slightly as follows.

```
MINODE *iget(int dev, int ino) // return a locked minode=(dev,ino)
{
   P(mlock);                          // acquire minodes lock
   if (needed minode already exists){
      if (!CP(minode.sem)             // if minode is already locked
         PV(minode.sem, mlock);       // atomically (P(s1),V(s2))
      return minode;                  // return locked minode
   }
```

```
    // needed minode not in memory, still holds mlock
    allocate a free minode;
    P(minode.sem);                    // lock minode
    V(mlock);                         // release minodes lock
    load inode from disk into minode;
    return minode;                    // return locked minode
}
void iput(MINDOE minode)
{
    // caller already holds minode.sem lock
    P(mlock);
    // do release minode operations as usual;
    V(minode.sem);
    V(mlock);
}
```

9.16.6 Deadlock Prevention in SMP

A SMP kernel relies on locking mechanisms to protect data structures. In general, any locking mechanism may lead to deadlock (Silberschatz et al. 2009; Wang 2015). There are several schemes to handle deadlocks, which include deadlock prevention, deadlock avoidance and deadlock detection with recovery. In a real operating system, the only practical scheme is deadlock prevention. In this scheme, a SMP system can be designed to prevent deadlocks from occurring. In the SMP_EOS kernel, we use the following strategies to ensure that the system is free of deadlocks.

(1). When using spinlocks or semaphores, ensure that the locking order is always unidirectional, so that circular waiting can never occur.
(2). If cross locking is unavoidable, use conditional locking and back off to prevent any chance of lock.

The effectiveness of (1) is obvious. We illustrate the technique of (2) by an example.

9.16.6.1 Deadlock Prevention in Parallel Algorithms

In an OS kernel, PROC structures are the focal points of process creation and termination. In a UP kernel, it suffices to maintain all free PROC structures in a single free list. In SMP, a single free list may be a serious bottleneck since it greatly reduces concurrency. In order to improve concurrency, we may divide free PROCs into separate free lists, each associated with a CPU. This allows processes executing on different CPUs to proceed in parallel during fork and wait for child termination. If a CPU's free PROC list becomes empty, we let it obtain PROCs from other free lists dynamically. Thus, in a SMP kernel free PROCs may be managed by parallel algorithm as follows.

```
Define: PROC *freelist[NCPU];            // free PROC lists, one per CPU
        int    procspin[NCPU]={0};       // spinlocks for PROC lists
PROC *get_proc()  // allocate a free PROC during fork/vfork;
{
  int cpuid = get_cpuid();              // CPU id
  while(1){
  (1). slock(procspin[cpuid]);          // acquire CPU's spinlock
  (2). if (freelist[cpuid]==0){         // if CPU's free list is empty
  (3).     if (refill(cpuid)==0){       // refill CPU's freelist
               sunlock(procspin[cpuid]);
               continue;                // retry
           }
       }
```

```
  (4). allocate a PROC *p from LOCAL freelist[cpuid];
  (5). sunlock(procspin[cpuid]);      // release CPU's spinlock;
  }
  return p;
}

int refill(int cpuid)                 // refill a CPU's free PROC list
{
  int i, n = 0;
  for (i=0; i<NCPU && i!=cpuid; i++){// try other CPU's free list
     if (!cslock(procspin[i]))        // if conditional lock fails
         continue;                    // try next free PROC list
     if (freelist[i]==0){             // if other free list empty
         sunlock(procspin[i])         // release spinlock
         continue;                    // try next CPU's free list
     }
     remove a PROC from freelist[i];// get a free PROC
     insert into freelist[cpuid];     // add to CPU's list
     n++;
     sunlock(procspin[i]);            // release spinlock
  }
  return n;
}

void put_proc(PROC *p) // release p into CPU's free PROC list
{
  int cpuid = get_cpuid();
  (1). slock(procspin[cpuid];           // acquire CPU's spinlock
  (2). enter p into freelist[cpuid];
  (3). sunlock(procspin[cpuid]);        // release CPU's spinlock
 }
```

In get_proc(), a process first locks the free PROC list of the CPU it is executing on. If the CPU's free PROC list is nonempty, it gets a free PROC, releases the lock and returns. Similarly, in put_proc(), a process only needs to lock the per-CPU free PROC list. Under normal conditions, processes running on different CPUs may proceed in parallel since they do not compete for the same free PROC list. The only possible contention is when a CPU's free PROC list becomes empty. In that case, the process executes the refill() operation, which tries to get a free PROC from every other CPU into the free PROC list of the current CPU. Since the process already holds a CPU's spinlock, trying to acquire the spinlock of another CPU may lead to deadlock. So it uses the conditional locking, cslock(), instead. If the conditional locking fails, the process backs off to prevent any chance of deadlock. If after refill the CPU's free PROC list is still empty, it releases the spinlock and retries the algorithm. This prevents self-deadlock on the same CPU.

9.16.7 Adapt UP Algorithms for SMP

In addition to using locks to protect kernel data structures, many algorithms used in UP kernels must be modified to suit SMP. We illustrate this by examples.

9.16.7.1 Adapt UP Process Scheduling Algorithm for SMP

A UP kernel usually has only a single process scheduling queue. We may adapt the UP process scheduling algorithm for SMP as follows. Define a spinlock, readylock, to protect the scheduling queue. During task switch, a process must acquire the readylock first, which is released by the next running process when it resumes.

9.16.7.2 Adapt UP Pipe Algorithm for SMP

In the EOS of Chap. 8, pipes are implemented by a UP algorithm, which uses the conventional sleep/wakeup for synchronization. We may adapt the UP pipe algorithm for SMP by adding a spinlock to each pipe and requiring pipe readers and writers to execute in the same critical region of the spinlock. While holding the spinlock, if a process has to sleep for data or room in the pipe, it must complete the sleep operation before releasing the spinlock. These can be done by replacing the conventional sleep(event) with the modified sleep(event, spinlock) operation.

9.16.7.3 Adapt UP I/O Buffer Management Algorithm for SMP

The EOS file system uses I/O buffers for block devices. The I/O buffer management algorithm uses semaphores for process synchronization. The algorithm works only in UP because it assumes only one process runs at a time. We may adapt the algorithms for SMP by adding a spinlock and ensuring that both getblk() and brelse() are executed in the same critical region. In getblk(), while holding the spinlock, if a process finds a needed buffer already exists but is busy, it increments the buffer's usercount by 1 to reserve the buffer. Then it releases the spinlock and waits for the buffer by P on the buffer's lock semaphore. When a process (or interrupt handler) releases a buffer by brelse(), it must execute in the same critical region of getblk(). After decrementing the buffer's usercount by 1, if the usercount is non-zero, meaning that the buffer still has users, it does not free the buffer but V the buffer's lock semaphore to unblock a process. Similarly, we may adapt other UP algorithms for SMP.

9.16.8 Device Driver and Interrupt Handlers in SMP

In an interrupt-driven device driver, process and interrupt handler usually share data buffer and control variables, which form a critical region between the process and the interrupt handler. In a UP kernel, when a process executes a device driver, it can mask out interrupts to prevent interference from the interrupt handler. In SMP, masking out interrupts is no longer sufficient. This is because while a process executes a device driver, another CPU may execute the device interrupt handler at the same time. In order to serialize the executions of processes and interrupt handlers, device drivers in SMP must be modified also. Since interrupt handlers cannot sleep or become blocked, they must use spinlock or equivalent mechanisms. In the following, we illustrate the principles of SMP driver design by specific examples in the SMP_EOS kernel.

(1). The LCD display is a memory-mapped device, which does not use interrupts. To ensure processes executes kputc() one at a time, it suffices to protect the driver by a spinlock.
(2). In the timer driver, process and interrupt handler share timer service and process scheduling queues but the process never waits for timer interrupts. In this case, it suffices to protect each timer dependent data structure by a spinlock.
(3). Char device drivers also use I/O buffers for better efficiency. In EOS, all char device drivers use semaphores for synchronization. To adapt the derivers to SMP, each driver uses a spinlock to serialize the executions of process and interrupt handler. While holding the spinlock, if a process has to wait for data or room in the I/O buffer, it uses PU(s, spin), which waits for a semaphore s and releases the spinlock in a single atomic operation.
(4). In the SDC driver, process and interrupt handler shared data structures. For such drivers, it suffices to use a spinlock to serializes the executions of process and interrupt handler.
(5). Process and interrupt handler in a SMP device driver must obey the following timing order.

```
            Process                       Interrupt Handler
 ---------------------------------------------------------
 (a). disable interrupts     |
 (b). acquire spinlock       |  (a). acquire spinlock
 (c). start I/O operation    |  (b). process the interrupt
                             |  (c). start next I/O, if needed
 (d). release spinlock       |  (d). release spinlock
 (e). enable interrupts      |  (e). issue EOI
 ---------------------------------------------------------
```

In both cases, the order of (d) and (e) must be strictly observed in order to prevent a process from locking itself out on the same spinlock.

9.17 SMP_EOS Demonstration System

SMP_EOS is a general purpose SMP operating system designed for the ARM MPcode architecture. This section describes the operations and capabilities of the SMP_EOS system.

9.17.1 Startup Sequence of SMP_EOS

The startup sequence of the SMP_EOS system consists of the following steps.

(1). Booting SMP_EOS kernel: QEMU loads the stage-1 booter to 16KB and executes it first. The stage-1 booter loads stage-2 booter from the SDC to 2MB and executes it next. The stage-2 booter loads the SMP_EOS kernel to 1MB and transfers control to the loaded kernel image. In the following, Steps (2) to (7) are executed only by the boot processor CPU0.
(2). Reset_handler: CPU0 execute reset_handler to set SVC and privileged modes stack pointers, copy vectors to address 0, create the initial one-level page table, enable MMU, call main() in C.
(3). Main(): Enable SCU, configure GIC for IRQ interrupts, initialize devices drivers, call kernel_init() to initialize the kernel.
(4). Kernel_init(): initialize kernel data structures, set run pointers of CPUs to their initial PROCs with special pids 1000 to 1003, run the initial process with pid=1000. Build 2-level page tables and free page list, switch to 2-level page table to use dynamic paging.
(5). Fs_init(): Initialize file system and mount the root file system.
(6). Create INIT Process: Create the INIT process P1 and enter it into readyQueue.
(7). Activate Other CPUs: Send SGI to activate the secondary CPUs (CPU1 to CPU3), then call run_task() to run the INIT process P1.

`Secondary CPUs in SMP`

(8). Each secondary CPU: begin to execute reset_hanlder at 1MB, set SVC and privileged modes stacks, use the one-level page table at 16KB to turn on MMU, then call APstart().
(9). APstart(): each secondary CPU: switch to 2-level page table at 32KB, configure and start local timer, then call run_task (), trying to run tasks from the same readyQueue.

---

(10). INIT process P1 (running on CPU0): fork login processes on console and serial terminals to allow user logins. Then P1 waits for any child to terminate. When the login processes start up, the system is ready for use.
(11). User login: After a user login, the login process becomes the user process, which executes the command interpreter sh.
(12). Sh process: User enters commands for sh to execute. When the user sh process terminates (by user logout or control-D), it wakes up the INIT process P1, which forks another login process on the terminal.

9.17.2 Capabilities of SMP_EOS

The SMP_EOS kernel consists of process management, memory management, device drivers and a complete file system. It supports dynamic process creation and termination. It allows process to change image to execute different files. Each process runs in a private virtual address space in User mode. Memory management is by two-level dynamic paging. Process scheduling is by both time-slice and dynamic process priority. As a SMP system, the SMP_EOS kernel is capable of supporting preemptive process scheduling. For simplicity, preemptive process scheduling is disabled in order to avoid excessive task switches in kernel mode. In the demonstration system, process switch occurs only when processes exit kernel to return to user mode. It supports a complete EXT2 file system that is totally Linux compatible. It uses block device I/O buffering between the file system and SDC driver to improve efficiency and performance. It supports multiple user logins from the console and serial terminals. The command interpreter sh supports executions of single commands with I/O redirections, as well as multiple commands connected by pipes. It provides interval timer service to processes, and it supports inter-process communication by signal, pipes and message passing. It unifies exceptions with signal processing, allowing users to install signal catchers to handle exceptions in User mode.

9.17.3 Demonstration of SMP_EOS

Figure 9.11 shows the sample outputs of running the SMP_EOS system. It uses UART0 to display information during booting, and also as a serial terminal for users to login after the system starts up. After login, the reader may enter commands to test the system. All user commands are in the /bin directory. As the figure shows, processes may run on different CPUs.

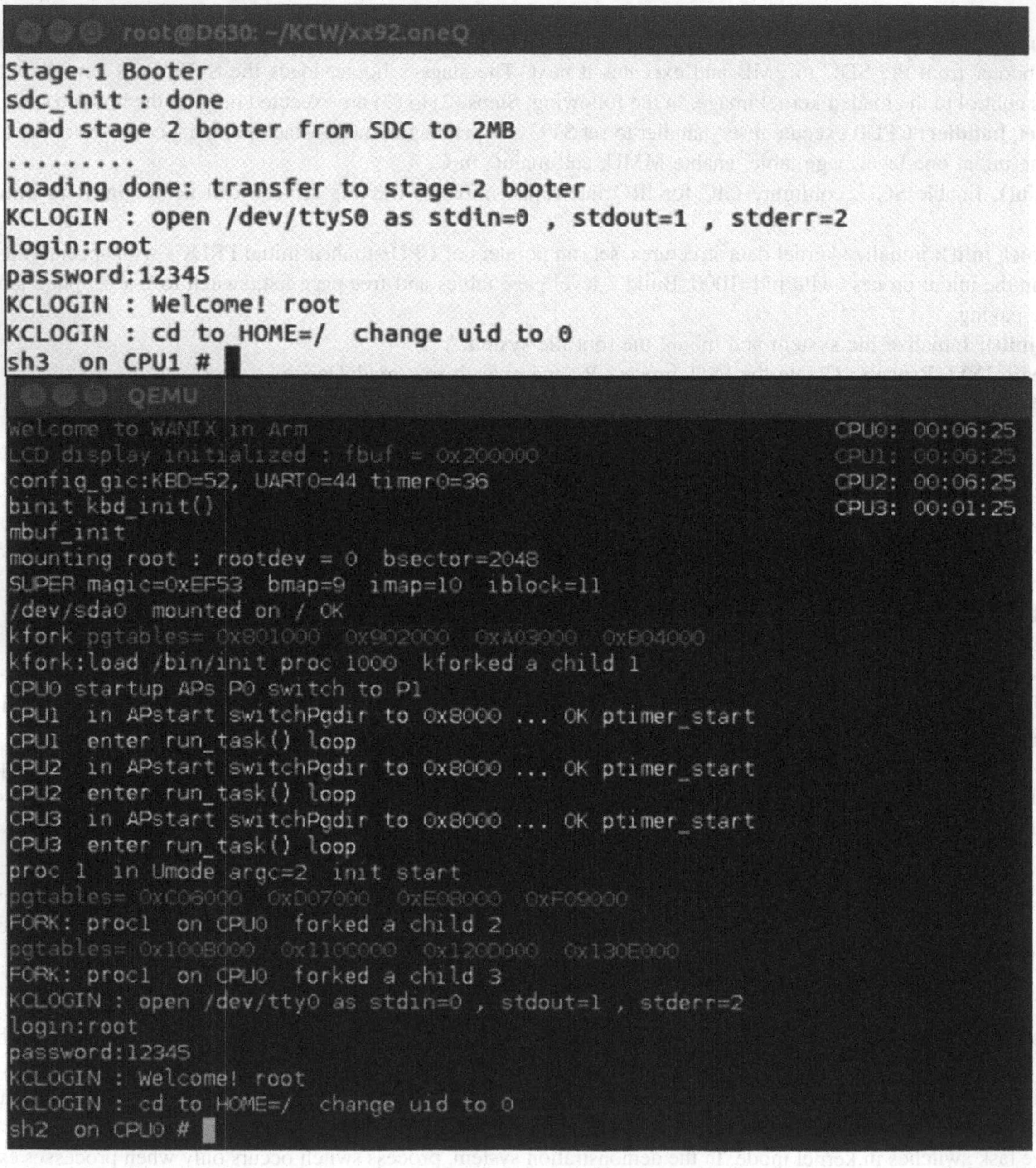

```
root@D630: ~/KCW/xx92.oneQ
Stage-1 Booter
sdc_init : done
load stage 2 booter from SDC to 2MB
.........
loading done: transfer to stage-2 booter
KCLOGIN : open /dev/ttyS0 as stdin=0 , stdout=1 , stderr=2
login:root
password:12345
KCLOGIN : Welcome! root
KCLOGIN : cd to HOME=/  change uid to 0
sh3  on CPU1 #
```

```
QEMU
welcome to WANIX in Arm                                  CPU0: 00:06:25
LCD display initialized : fbuf = 0x200000                CPU1: 00:06:25
config_gic:KBD=52, UART0=44 timer0=36                    CPU2: 00:06:25
binit kbd_init()                                         CPU3: 00:01:25
mbuf_init
mounting root : rootdev = 0  bsector=2048
SUPER magic=0xEF53  bmap=9  imap=10  iblock=11
/dev/sda0  mounted on / OK
kfork pgtables= 0x801000  0x902000  0xA03000  0xB04000
kfork:load /bin/init proc 1000  kforked a child 1
CPU0 startup APs P0 switch to P1
CPU1  in APstart switchPgdir to 0x8000 ... OK ptimer_start
CPU1  enter run_task() loop
CPU2  in APstart switchPgdir to 0x8000 ... OK ptimer_start
CPU2  enter run_task() loop
CPU3  in APstart switchPgdir to 0x8000 ... OK ptimer_start
CPU3  enter run_task() loop
proc 1  in Umode argc=2  init start
pgtables= 0xC06000  0xD07000  0xE08000  0xF09000
FORK: proc1  on CPU0  forked a child 2
pgtables= 0x100B000  0x110C000  0x120D000  0x130E000
FORK: proc1  on CPU0  forked a child 3
KCLOGIN : open /dev/tty0 as stdin=0 , stdout=1 , stderr=2
login:root
password:12345
KCLOGIN : Welcome! root
KCLOGIN : cd to HOME=/  change uid to 0
sh2  on CPU0 #
```

Fig. 9.11 Demonstration of SMP_EOS system

9.18 Summary

This chapter covers multiprocessing in embedded systems. It points out the requirements of Symmetric Multiprocessor (SMP) systems and compares the approaches to SMP of Intel with ARM. It lists ARM MPcore processors and describes the components and functions of ARM MPcore processors in support SMP. All ARM MPcore based systems depends on the Generic Interrupt Controller (GIC) for interrupts routing and inter-processor communication. It shows how to configure the GIC to route interrupts and demonstrates GIC programming by examples. It shows how to start up ARM MPcores and points out the need for synchronization in a SMP environment. It shows how to use the classic test-and-set or equivalent instructions to implement atomic update and critical regions and points out their shortcomings. Then it explains the new features of ARM Mpcores in support of SMP. These include LDRES/STRES instructions and memory barriers. It shows how to use the new features of ARM Mpcores to implement spinlocks, mutexes and semaphores for process synchronization in SMP. It defines conditional locks for deadlock prevention in SMP kernels. It also covers the additional features of the ARM MMU for SMP. It presents a general methodology for adapting uniprocessor OS kernel to SMP. Then it applies the principles to develop a complete SMP OS for embedded systems.

List of Sample Programs

C9.1. GIC Programming Example Program
C9.2. ARM SMP Startup Example 1
C9.3. ARM SMP Startup Example 2
C9.4. SMP Startup with Spinlock
C9.5. SMP Startup with Mutex Lock
C9.6. Local timers in SMP
C9.7. Uniform VA Space Mapping
C9.8. Non-uniform VA Space Mapping
C9.9. A parallel computing system
C9.10. Process Management in SMP
C9.11.Demonstration of SMP_EOS

Problems

1. In the example program C9.1, modify the config_gic() code to route interrupts to different CPUs to observe the effect.
2. Instead of the local timers of the CPUs, use the global timer to provide a single timing source to all CPUs
3. Instead of a single scheduling queue, implement multiple scheduling queues, e.g. one schedule queue per CPU, to speed up task scheduling in a SMP system.
4. The SMP_EOS uses the KML virtual memory mapping scheme, in which the VA space of kernel is mapped to low virtual addresses. Re-implement the system by using the KMH virtual memory mapping scheme, which maps the VA space of kernel to 2GB.

References

ARM 926EJ-S : ARM Versatile Application Baseboard for ARM926EJ-S User guide, ARM Information Center, 2010
ARM11: ARM11 MPcore Processor Technical Reference Manual, r2p0, ARM Information Center, 2008
ARM Cortex-A9 MPCore: Technical Reference Manual Revision: r4p1, ARM information Center, 2012
ARM GIC: ARM Generic Interrupt Controller (PL390) Technical Reference Manual, ARM Information Center, 2013
ARM Linux SMP: Booting ARM Linux SMP on MPCore http://www.linux-arm.org/LinuxBootLoader/SMPBoot, 2010
ARM Timers: ARM Dual-Timer Module (SP804) Technical Reference Manual, Arm Information Center, 2004
Cox, R., Kaashoek, F., Morris, R. "xv6 a simple, Unix-like teaching operating system, xv6-book@pdos.csail.mit.edu, Sept. 2011
Intel: MultiProcessor Specification, v1.4, Intel, 1997
Stallings, W. "Operating Systems: Internals and Design Principles (7th Edition)", Prentice Hall, 2011
Silberschatz, A., P.A. Galvin, P.A., Gagne, G, "Operating system concepts, 8th Edition", John Wiley & Sons, Inc. 2009
Wang, K.C., "Design and Implementation of the MTX Operating Systems, Springer International Publishing AG, 2015

10 Embedded Real-Time Operating Systems

10.1 Concepts of RTOS

A real-time operating system (RTOS) (Dietrich and Walker 2015) is an operating system intended for real-time applications. Real-time applications usually have very stringent timing requirements. First, a RTOS must be able to respond to external events quickly, e.g. within a very short time, known as the **interrupt latency**. Second, it must complete every requested service within a prescribed time limit, known as the **task deadline**. If a real-time system can always meet these critical timing requirements, it is called a **hard real-time** system. If it can only meet the requirements most of the time but not always, it is called a **soft real-time** system. In order to meet the stringent timing requirements, real-time OS are usually designed with the following capabilities.

. **Minimum interrupt latency:** Interrupt latency is the amount of time from the moment an interrupt is received to the moment the CPU starts to execute the interrupt handler. In order to minimize interrupt latency, a RTOS kernel must not mask out interrupts for long periods of time. This usually means the system must support nested interrupts to ensure processing low priority interrupts do not delay the processing of high priority interrupts.

. **Short critical regions:** All OS kernels rely on critical regions to protect shared data objects as well as for process synchronization. In a RTOS kernel, all critical regions must be as short as possible.

. **Preemptive task scheduling:** Preemption means a higher priority task can preempt a lower priority task at any time. In order to meet task deadlines, a RTOS kernel must support preemptive task scheduling. Task switching time must be short also.

. **Advanced task scheduling algorithm:** Preemptive scheduling is a necessary but not sufficient condition for real-time systems. Without preemptive scheduling, high priority tasks may be delayed by low priority tasks for a variable amount of time, making it impossible to meet task deadlines. However, even with preemptive scheduling, there is no guarantee that the tasks will be able to meet their deadlines. The system must use a suitable scheduling algorithm that helps achieve this goal.

10.2 Task Scheduling in RTOS

In a general purpose OS, the task scheduling policy is usually designed to achieve a balanced system performance among the various conflicting goals, such as throughput, resource utilization and quick response to interactive users, etc. In contrast, in an RTOS, the only goals are to ensure fast response and guarantee task deadlines. The scheduling policy must favor those tasks with the most urgent timing constraints. Such constraints can be translated into task priorities. Therefore, the scheduling algorithm of RTOS must be preemptive based on task priorities. There are two main types of task scheduling algorithms for RTOS, known as Rate-Monotonic Scheduling (RMS) and Earliest-Deadline-first (EDF) scheduling, and their variant forms.

K.C. Wang, *Embedded and Real-Time Operating Systems*,
DOI 10.1007/978-3-319-51517-5_10

10.2.1 Rate-Monotonic Scheduling (RMS)

Rate-Monotonic Scheduling (RMS) (Liu and Layland 1973) is a static priority based scheduling algorithm for real time systems. The RMS model assumes the following conditions.

(1). Periodic tasks: tasks are periodic with deadlines exactly equal to periods.
(2). Static priorities: tasks with shorter periods are assigned higher priorities.
(3). Preemption: the system always runs the highest priority task, which immediately preempts other tasks of lower priorities.
(4). Tasks do not share resources that would cause them to block or wait.
(5). Context switch and other task operation time, e.g. release and startup time, are zero.

RMS analyzes such a system model and derives the condition under which the tasks can be scheduled to meet their deadlines. Liu and Layland proved that for a set of *n* periodic tasks with unique periods, a feasible schedule that will always meet task deadlines exists if the CPU utilization factor U is below a specific bound. The schedulability condition of RMS is

$$\mathbf{U = \underset{i=1}{\overset{n}{SUM}}(Ci/Ti) <= n(2 * *(1/n) - 1)}$$

where Ci is the computation time of taski, Ti is the period of taski, and n is the number of tasks to be scheduled. For example, U <= 0.8284 for two tasks. When the number of tasks n becomes large, the value of U tends toward a limit value of

$$\mathbf{\ln(2) = 0.693247\ldots}$$

This implies that RMS can meet all the task deadlines if CPU utilization is less than 69.32%. Furthermore, RMS is optimal for preemptive uniprocessor systems in the sense that, if any static-priority scheduling algorithm can meet all the deadlines, then so can the RMS algorithm. It is worth noting that the schedulability test of RMS is only a sufficient, but not necessary, condition. For instance, for the task set task1 = (C1=2, T1=4), task2 = (C2=4, T2=8), the value of U is 1.0, which is greater than the RMS bound 0.828, but the tasks are schedulable with 100% CPU utilization, as the following timing diagram shows.

```
time: 0  1  2  3  4  5  6  7  8
      -------------------------
      |task1|      |task1|     |
      |     |task2|      |task2|
      -------------------------
```

In general, if the task periods are **harmonic**, meaning that for each task its period is an exact multiple of every task that has a shorter period, the tasks are schedulable with a higher utilization factor than the RMS bound.

10.2.2 Earliest-Deadline-First (EDF) Scheduling

In the Earliest-Deadline-First (EDF) model (Leung et al. 1982), tasks can be periodic or non-periodic. Each task has a deterministic deadline. Tasks are maintained in a priority queue ordered by their closest deadlines, i.e. shorter deadlines tasks have higher priorities. The EDF scheduling algorithm always runs the task with the closest deadline. Like RMS, EDF is also optimal for preemptive uniprocessor systems. When scheduling periodic tasks with deadlines equal to periods, EDF has a CPU utilization bound of 100%. The schedulability condition of EDF is

$$\mathbf{U = \underset{i=1}{\overset{n}{SUM}}(Ci/Ti) <= 1}$$

where Ci and Ti are the worst-case computation-time and inter-arrival period of the tasks. Compared with RMS, EDF can guarantee task deadlines with a higher CPU utilization factor but it also has two drawbacks. First, EDF is more difficult to implement because task priorities are no longer static but dynamic. It must keep track of task deadlines and update the task queue whenever a reschedule event occurs. Second, when the system is overloaded (CPU utilization factor > 1) the tasks that will miss deadlines in RMS are usually the ones with the longer periods (low priorities). In the EDF model, such tasks are largely unpredictable, meaning that any task may miss its deadline. However, there are also comparative analyses of EDF vs. RMS which tends to refute such claims (Buttazzo 2005). Despite this, most real RTOS prefer RMS over EDF, mainly because of the static task priorities and deterministic nature of RMS.

10.2.3 Deadline-Monotonic Scheduling (DMS)

Deadline-Monotonic Scheduling (DMS) (Audsley 1990) is a generalized form of RMS. In RMS, task deadline and period are assumed to be exactly equal. The DMS model relaxes this condition to allow task deadlines to be less than or equal to task periods, i.e.

task computation time <= deadline <= period

In DMS, tasks with shorter deadlines are assigned higher priorities. For this reason, DMS is also known as Inverse-Deadline Scheduling (IDS). Run-time scheduling is the same as in RMS, i.e. by preemptive task priority. When task deadlines and periods are equal, DMS reduces to RMS as a special case. In (Audsley et al. 1993), the RMS condition is relaxed to allow task deadlines to be less than task periods. It also extended the DMS model to include non-periodic tasks and derived schedulability tests for such cases.

Despite these real time system models and analytical results, both RMS and EDF can only be used as general guidelines in the design of real RTOS. The problem becomes more prominent when it comes to the implementation of real RTOS systems, primarily due to the following reasons. A fundamental flaw in both the RMS and EDF models is that tasks can not share resources that would cause them to block or wait, and task switching incurs zero overhead. These conditions are unrealistic since in a real system resource sharing among tasks is inevitable and task switching time can never be zero. Although there are some attempts to extend both the RMS and EDF models to allow resource sharing by including task blocking time in schedualability analyses, the results usually involve many variables that are difficult to quantify, so they just assume the worst case values. Besides, allowing resource sharing also leads to other problems, such as priority and deadline inversions, which must be handled properly in a real RTOS.

10.3 Priority Inversion

As in any OS, a RTOS kernel must use critical regions (CRs) to protect shared resources. Software tools used to enforce critical regions include Event-Control-Bocks (ECBs), mutexes, semaphores and message queues, etc. All these mechanisms are based on the locking protocol, in which a task becomes blocked if it can not acquire a CR. In addition to the usual problems of locking, such deadlock and starvation, allowing preemption in critical regions leads to a unique problem known as priority inversion (Sha et al. 1990), which can be best described as follows.

Let TL, TM, TH denote tasks of low, medium and high priorities, respectively. Assume that TL has acquired a CR and is executing inside the CR. Next, when TH becomes ready to run, it preempts TL. Assume that TH also needs the same CR, which is still held by TL. So TH blocks itself on the CR, waiting for TL to release the CR. Then, TM becomes ready to run, which does NOT need the CR. Since TM has higher priority than TL, it immediately preempts TL. Now, TM is running but its priority is lower TH, resulting in a priority inversion. In the worst case, TH may be delayed for an unknown amount of time since TM may be preempted by other tasks with priorities between TH and TM, etc. The phenomenon is known as **unbounded priority inversion**. Similarly, in the EDF model, task priorities are assigned dynamically by their closest deadlines. If tasks share resources, which may cause higher priority tasks to block, the same priority inversion problem would occur, resulting in a **deadline-inversion**.

It is noted that in an ordinary OS kernel, priority inversion may also occur but its effect is usually unnoticeable and harmless since all it does is to delay some higher priority tasks for awhile, which eventually acquire their needed CRs and continue. In a RTOS, delaying high priority tasks may cause them to miss deadlines, which may trigger system failure alarms. The most famous example of priority inversion is the system reset problem that occurred in the **Mars Pathfinder mission** (Jones 1997; Reeves 1997). The problem was resolved only after the engineers at JPL reproduced the problem on earth, identified the cause and modified the onboard task scheduler to avoid priority inversion.

10.4 Priority Inversion Prevention

There are many ways to prevent priority inversion. The first one is not to let tasks share resources, as required by both the RMS and EDF models, but this is clearly impractical. The second one is to assign all the tasks with the same priority, which is also impractical. So far, the only practical ways of preventing priority inversion are the following schemes.

10.4.1 Priority Ceiling

In the priority ceiling scheme, it assumes that for each CR the ceiling priority of the CR is greater than the highest priority of all the tasks that may compete for the CR. Whenever a task gains control of a CR, its priority is immediately raised to the ceiling priority of the CR, thus preventing preemption from any other tasks with a priority lower than the ceiling priority. This also implies that a task can lock a CR if its priority is higher than the ceiling priorities of all the CRs locked by other tasks. When a task exits a CR, it reverts back to its original priority. Priority ceiling is easy to implement but it may prevent preemption unnecessarily. For instance, while holding a CR, a low priority task will prevent preemption from any other task lower than the ceiling priority even if it does not need the same CR.

10.4.2 Priority Inheritance

In the priority inheritance (Sha et al. 1990) scheme, while a task holds a CR, if another task with higher priority tries to acquire the same CR, it temporarily raises the priority of the task that holds the CR to that of the requesting task. This ensures that the priority of a task executing inside a CR is always equal to the highest priority of the tasks blocked on the CR. When the executing task exits the CR, it reverts back to its original priority. Priority inheritance is more flexible than priority ceiling but it also incurs more overhead to implement. In the priority ceiling scheme, the priority of a task executing inside a CR is static, which does not change until it exits the CR. In the priority inheritance scheme, whenever a task is about to become blocked on a CR, it must check whether its priority is the highest among all the tasks waiting for the CR and, if so, it must pass the priority to the task that holds the CR. Thus, while executing inside a CR, the priority of a task may change dynamically.

10.5 Survey of RTOS

Unlike general purpose operating systems, which require a wide range of capabilities, real-time operating systems are usually designed to provide only limited functionalities intended for specialized environments. As a result, RTOS are usually much simpler than general purpose OS. For instance, in most RTOS all tasks run in the same address space, so they do not have separate kernel and user modes. Furthermore, most ROTS do not support file systems and user interface, etc. Despite their stringent timing requirements, RTOS are actually easier to develop as compared with general purpose OS. This is evidenced by the large number of RTOS posted on the Internet, ranging from open source ROTS intended for hobbyists to proprietary RTOS intended for commercial markets. In this section, we shall present a brief case study of some of the popular RTOS.

10.5.1 FreeRTOS

FreeRTOS (2016) is an open-source real time kernel designed specifically for small embedded systems. FreeRTOS is essentially a bare-bone kernel, which provides basic support for developing real-time applications. Key features of FreeRTOS are

Tasks: The execution units in FreeRTOS are called tasks. Each task is represented by a Task Control Block (TCB). All tasks executes in the same address space of the kernel. The kernel provides support for task creation, suspension, resumption, change priority and deletion. It also supports co-routines, which are executable units that do not need much stack space.

Scheduling: Task scheduling in FreeRTOS is by preemptive priority. For tasks with the same priority, it also supports cooperative and round-robin with time-slice options. Task switch is not allowed during nested interrupts processing. Task scheduler can be disabled to prevent task switch in long critical regions.

Synchronization: Task synchronization in FreeRTOS is based on the queue operation. It uses both binary and counting semaphores for general task synchronization, and it uses recursive mutexes with priority inheritance to protect critical regions. A recursive mutex can be locked/unlocked by the owner recursively for up to 256 levels deep.

Memory Protection: FreeRTOS has no memory protection in general, but it supports memory protection on some specific ARM boards, e.g. FreeRTOS-MPU supports ARM Cortex-M3 Memory Protection Unit (MPU).

Timer Service: The FreeRTOS kernel supports both timer ticks and software timers.
It also supports a tickless mode which suppresses periodic timer tick interrupts to reduce power consumption.

Portability: The FreeRTOS kernel consists of only a few files, most of which are written in C. It has been ported to several different architectures, including ARM, Intel x86 and PowerPC, etc. DNX (DNX 2015) is a RTOS based on freeRTOS. It adds Unix-like API interface, file system support and new device drivers to the basic freeRTOS kernel.

10.5.2 MicroC/OS (μC/OS)

MicroC/OS (uC/OS) (Labrosse 1999) is a preemptive, real-time multitasking kernel for microprocessors, microcontrollers and Digital signal Processors (DSPs). The current version is uC/OS-III, which is maintained and marketed by Micrium (2016).

(1). Tasks: In uC/OS, the execution units are called tasks, which are essentially threads executing in the same address space of the kernel. The uC/OS kernel provides support for task creation, suspension, resumption, deletion and task statistics. uC/OS-II supports up to 256 tasks. In uC/OS-III, the number of tasks is variable, limited only by the amount of available memory.
(2). Scheduling: Task scheduling is by preemptive static priority. Task priorities are assigned by the user, presumably by the RMS algorithm. In μC/OS-II, all tasks have different priorities (by task ID). μC/OS-III allows multiple tasks to run at the same priority, so it also supports round-robin scheduling by time slice.
(3). Memory Management: The uC/OS kernel does not provide any memory protection. It allows the user to define partitions of memory areas consisting of fixed sized memory blocks. Memory allocation is by fixed sized blocks.
(4). Synchronization: The uC/OS kernel relies on disable/enable interrupts to protect short critical regions, and it uses disable/enable task scheduler to protect long critical regions. Other mechanisms used for task synchronization include event control blocks (ECBs), semaphores, mailboxes and message queues. The uC/OS-III kernel uses mutual-exclusive semaphores (mutexes) with priority inheritance to prevent priority inversion.
(5). Interrupts: The uC/OS kernel supports nested interrupts. Interrupt processing is performed inside ISR routines directly. Task scheduling is disabled during nested interrupts processing. Task switch occurs only when all nested interrupts have ended.
(6). Timer Service: μC/OS requires a periodic time source to keep track of time delays and timeouts. The kernel only provides functions for a task to suspend itself for a specified number of timer ticks. A suspended task becomes ready to run again when its delay time expires. Before the requested time expires, a suspend task can also be resumed by another task. It does not provide general timer service functions, such as interval timers with notification and cancellation, etc.
(7). Port: uC/OS is written mainly in ANSI-C. Its syntax, file naming convention and development environment are based on Microsoft IDE on the Intel-x86 architecture. It's reported that uS/OS-III has been ported to several other platforms, such as the ARM architecture.

10.5.3 NuttX

NuttX (Nutt 2016) is a RTOS with an emphasis on compliance with POSIX and ANSI standards. It includes standard POSIX 1003 APIs and also APIs adopted from other common RTOS. Some of the APIs, such as task_creat, waitpid, vfork, execv, etc. are adapted to suit the embedded environment.

(1). Tasks: The NuttX kernel supports task creation, termination, deletion, init, activate and restart, etc. Tasks can be created by task_creat in one step. Alternatively, they can also be created and started up by task_init followed by task_activate in two steps. Unlike most other RTOS, NuttX tries to comply with the POSIX standard closely. For instance, tasks in NuttX obey the parent-child relation. Child task may inherit file streams, e.g. stdin, stdout and stderr, from the parent, and parent may wait for child termination. It allows tasks to change execution images to different files by execv. On some hardware platforms it even supports vfork, which creates a task skeleton without an execution image. A vforked task may use execv to create its own image from an executable file.
(2). Scheduling: Task scheduling is by preemptive task priority. Each task can set its scheduling policy as either FIFO (First-In-First-Out) or RR (Round-Robin). Tasks with the same priority are scheduled by FIFO or Round-robin with a prescribed time slice. In addition, tasks can also change priority and yield CPU to other tasks of the same priority.
(3). Synchronization: The NuttX kernel uses counting semaphore with priority inheritance, which is available as a configurable option. It supports named message queues of POSIX for inter-task communication. Any task may send/receive messages to/from named message queues. Interrupt handlers may also send messages via named message queues. To prevent task blocking on full message queues, messages can be sent with a timeout option. In NuttX, timeout is implemented by POSIX signals as in Unix.
(4). Signals: In addition to semaphores and message queues, NuttX also uses signals for inter-task communication. It allows any task or interrupt handler to post a signal to any task (by task ID). A signal is a (software) interrupt to a task, causing the task to execute a prescribed signal handler function. Unlike Unix, which has predefined signal handlers, there are no predefined actions for signals in NuttX. The default action for all signals is to ignore the signal. It allows the user to install signal handlers to handle the signals.
(5). Clock and Timer Service: The NuttX kernel supports POSIX compatible timer and interval timer service functions. Each task may create a per-task timer based on a clock, which provides clock ticks. A task may set an interval timer request. When the interval timer expires, a timeout signal is delivered to the task, allowing it to handle the signal by a pre-installed signal handler.
(6). File system and Network Interface: NuttX includes an optional file system, which is not needed for NuttX to run. If file system is enabled, NuttX starts with a pseudo root file system in memory. Real file systems can be mounted on the pseudo root file system. File system interface is by a set of standard POSIX APIs, such as open, close, read, write, etc. It provides a limited network capability by a subset of socket interface functions.

10.5.4 VxWorks

VxWorks (2016) is a proprietary RTOS developed for embedded systems by Wind River. Due to its proprietary nature, we can only gather some general information about the system based on published documents and user guides that are available in the public domain. Key features of the system include the following.

(1). Tasks and Scheduling: multitasking kernel with preemptive and round-robin scheduling and fast interrupt response.
(2). Synchronization: Binary and counting semaphores, mutexes with priority inheritance.
(3). Interprocess Communication: Local and distributed message queues.
(4). Development Environment: As a common practice in embedded systems, VxWorks uses a cross-compiling development environment. Application software is developed on a host system, e.g. Linux. The host provides an integrated development environment (IDE) consisting of editor, compiler toolchain, debugger, and emulators. Applications are cross-compiled to run on the target systems, which include ARM, Intel x86 and PowerPC. In addition to the IDE, VxWorks also include board support packages, TCP/IP networking stack, error detection/reporting and symbolic debugging.
(5). File systems: VxWorks support several file systems, which include the Reliability File System (HRFS), FAT file system (DOSFS), Network File system (NFS) and TFFS for flash devices. These are presumably parts of the IDE in support of the development environment. It is unclear whether any real-time applications may include file system support.

10.5.5 QNX

QNX (2015) is a proprietary Unix-like RTOS aimed primarily at the embedded systems market. A unique feature of QNX is that it is a microkernel based system. The QNX kernel contains only CPU scheduling, interprocess communication, interrupt redirection and timers. All other functions are executed as user processes outside of the microkernel. Key features of QNX are:

(1). Interprocess Communication (IPC) The QNX microkernel supports processes. Each process resides in a unique address space. Processes communicate with one another by exchanging messages through the microkernel. The QNX IPC consists of sending a message from one process to another and waiting for a reply. Because of the overhead in message exchange, most microkernel based systems do not perform well. QNX remedies this problem by using a more efficient message passing mechanism. In QNX, msgSend, which allows a process to send a message and wait for a reply, is a single operation. The message is copied by the kernel from the address space of the sending process to that of the receiving process. If the receiving process is waiting for the message, control of the CPU is transferred to the receiving process at the same time, which eliminates the need for explicitly unblocking the receiving process and invoking the scheduler. This tight integration between message passing and CPU scheduling is a key mechanism that makes the QNX microkernel work. All I/O operations, file system operations, and network operations are based on message passing. Message handling is prioritized by thread priority. Since I/O requests are performed using message passing, high priority threads receive I/O service before low priority threads. Later versions of QNX reduce the number of separate processes and integrate the network stack and other function blocks into single applications to improve the system performance.
(2). Threads: In QNX the smallest execution entities are threads. Each process contains a number of threads, which are independent execution units in the same address space of a process. The QNX microkernel supports Pthreads compliant APIs for threads creation, management and synchronization.
(3). Scheduling: Threads scheduling is by preemptive priority. In addition, it also supports adaptive partition scheduling (APS), which guarantees minimum CPU percentages to selected groups of threads, even though others may have higher priority.
(4). Synchronization: In QNX, IPC is for message passing between processes in different address spaces. It is not for threads within the same address space of a process. For threads synchronization, QNX uses POSIX compliant APIs to support mutexes, condition variables, semaphores, barriers and reader-writer locks, etc. When processes are allowed to share memory, most of the mechanisms are also applicable to threads in different processes.
(5). Boot Loader: Another key component of QNX is the boot loader, which can load an image containing not only the kernel but also any desired collection of user programs and shared libraries. It allows user programs, device drivers and supporting libraries to be built into the same boot image.
(6). Platforms: According to the latest QNX documentation, QNX Neutrino supports SMP and MP with processor affinity, which locks each application to a specific CPU. Due to its microkernel architecture, it should also be easier to adapt QNX to distributed environment.

10.5.6 Real-time Linux

Standard Linux is a general purpose operating system, which is not designed for real-time applications. Despite this, standard Linux has excellent average performance and can even provide millisecond level task scheduling precision. However, it is not designed to provide real-time services that require sub-millisecond precision and reliable timing guarantees. The fundamental cause is because the Linux kernel is not preemptive. Traditionally, the Linux kernel allows task preemption only under certain circumstances:

. When the task is running in user-mode
. When the task returns from a system call or interrupt processing back to user mode
. When the task sleeps or blocks in kernel to explicitly yield control to another process

While a task executes in the Linux kernel, if an event occurs that makes a high priority task ready to run, the high priority task can not preempt the running task until the latter explicitly yields control. In the worst case, the latency of switching to the high priority task could be hundreds milliseconds or more. Thus, in a standard Linux kernel high priority tasks can be delayed for an arbitrary amount of time, causing the system unable to respond to events quickly or guarantee task deadlines. There are many attempts to modify the Linux kernel to improve its real-time capabilities. In the following, we shall discuss two different approaches to this problem.

10.5.6.1 Real-time Linux Patches

The Linux 2.6 kernel has a configuration option, CONFIG_PREEMPT_VOLUNTARY, which can be enabled when compiling the kernel image. It introduces checks to the most common causes of long latencies in the kernel code, allowing the kernel to yield control to a higher priority task voluntarily. The advantage of this scheme is that it is very easy to implement and it has only limited impact on the system throughput. The disadvantage is that, although it reduces the occurrences of long latencies, it does not totally eliminate them. To further remedy the problem, the Linux 2.6 kernel provides an additional option, CONFIG_PREEMPT, which causes all kernel code outside of spinlock-protected regions and interrupt handlers to be eligible for preemption by higher priority tasks. With this option, worst case latency drops to around single digit milliseconds (Hagen 2005), although some device drivers may have interrupt handlers that still cause much longer latency. To support real-time tasks that require latencies below single-digit milliseconds, current Linux kernel has yet another option, CONFIG_PREEMPT_RT, known as the RT-Preempt patch, which converts the Linux kernel into a fully preemptible kernel by the following means.

. Make in-kernel locking-primitives (spinlocks) preemptible by a re-implementation with real-time mustexs (rt-mutexes). Rt_mutexes extend the semantics of simple mutexes with priority inheritance, in which a low priority owner of a rt-mutex inherits the highest priority of all the tasks waiting for the rt-mutex. In a chain of requests for rt-mutexes, if the owner of a rt-mutex gets blocked on another rt-mutex, it propagates the boosted priority to the owner of the other rt_mutex. The priority boosting is immediately removed once the rt_mutex has been unlocked. Implementation of rt-mutexes is made more efficient by using the kernel's p-list, which keeps track of the highest priority of blocked tasks. On architectures that support the cmp-xhg (compare & exchange) atomic operation, lock/unlock operations on rt-mutex can be further optimized.
. Convert interrupt handlers into preemptible kernel threads. The RT-Preempt patch treats executions of interrupt handlers as kernel pseudo-threads, which have higher priority than all regular threads but they can be preempted (by other pseudo-threads of higher priority), allowing the Linux kernel to support nested interrupts by interrupt priorities.
. Use a high resolution real-time timer. Convert the old Linux timer API into separate infrastructures for high resolution kernel timers plus a watch-dog like timer for timeouts, allowing high resolution POSIX timers in user space.

The performance of Linux 2.6 kernel with the RT-Preempt patch has been studied in (Hagen 2005). The test results showed a significant reduction of jitters in interrupts processing, resulting in a much more responsive and predictable Linux system.

10.5.6.2 RTLinux

RTLinux (Yodaiken Yodaiken 1999) is a RTOS that runs special real-time tasks and interrupt handlers on the same machine as standard Linux. It treats Linux as a preemptive task with the lowest priority, which runs only if there are no runnable real-time tasks, and it can be preempted whenever a real-time task becomes ready to run. The basic principle of RTLinux is rather simple. It places a layer of emulation software, the RTLinux kernel, between Linux and the interrupt controller hardware. All hardware interrupts are caught by the emulator first. It handles real-time related interrupts directly, and it forwards other non-real time related interrupts to the Linux kernel. In the Linux kernel code (on the Intel x86 architecture) all cli (disable interrupts), sti (enable interrupts) and iret (return from interrupts) instructions are replaced with emulating macros S_CLI, S_STI and S_IRET, respectively. The S_CLI macro clears a global variable SFIF to zero, indicating that the Linux kernel has just executed cli to have interrupts disabled. The S_STI macro simulates a real interrupt by creating a stack frame consisting of saved CPU FLAG register, Linux kernel DS register and a return address but executes the S_IRET macro instead. When a Linux interrupt occurs, the emulator checks the SFIF variable. If it is set, i.e. Linux kernel has interrupts enabled, it invokes the Linux interrupt handler immediately. Otherwise, it sets a bit in the SFIF variable to represent a pending Linux interrupt. When the Linux kernel enables interrupts by sti, the S_IRET macro scans the SFIF variable for pending Linux interrupts (non-zero bits). For each pending Linux interrupt, it invokes the corresponding Linux interrupt handler until all pending interrupts are processed.

RTLinux is structured as a small core component and a set of optional components. The core component permits installation of very low latency interrupt handlers that cannot be delayed or preempted by Linux itself and some low level synchronization and interrupt control routines. Inside RTLinux, real-time tasks are installed as Linux modules which execute in the same address space as the Linux kernel. Communication between Linux process and real-time tasks is by shared memory or dedicated FIFO pipes. Some earlier tests (Yokaiken 1999) showed that the RTLinux core can support real-time tasks with latency in the order of tens of microseconds on older Intel x86 CPUs. However, it was also reported that very frequent real-time interrupts would prevent Linux from running altogether.

10.5.7 Critique of Existing RTOS

In this section, we shall evaluate the various RTOS and formulate a set of general guidelines for the design and implementation of RTOS.

10.5.7.1 RTOS Organizations

Based on the above case studies, we can see that most RTOS are based on the bottom-up approach, in which a RTOS is built with a basic kernel to support tasks, task scheduling, critical regions and task synchronization. Then add additional functionalities to the basic kernel, e.g. execution tracing, events logging, debugging, file system and networking, etc. to improve the capability of the system. Most proprietary RTOS aimed at commercial markets usually also provide an integrated development environment (IDE) to facilitate the development of user applications. The main drawback of the bottom-up approach is the lack of uniformity. Different RTOS may develop and prompt their own proprietary system interfaces, complicating the process of developing user applications. To remedy this problem, many RTOS try to comply with the POSIX 1003.1b real-time extensions. In spite of these efforts, the availability of standard system service functions still varies from system to system.

An alternative approach to RTOS design is the top-down approach, which is aimed at converting an existing operating system, such as Linux, to support real-time operations. The advantage of the approach is obvious. In addition to adding real-time capabilities, it would make all the functionalities of a complete operating system directly accessible. The main drawback of this approach is the large system size, which may be unsuited to small embedded or real-time systems.

10.5.7.2 Tasks in RTOS

In an operating system, processes refer to execution entities with distinct address spaces, and threads are execution units in the same address of a process. In order to provide each process with a unique address space, processes usually executes in two different modes; kernel mode and user mode. While in kernel mode, all processes share the same address space of the kernel. While in user mode, each process executes in a separate address space that is isolated and protected from other processes. This is usually achieved through virtual address mapping by the memory management hardware. In all Unix-like systems, processes are created by the fork-exec paradigm. Fork creates a child process with an identical (user mode) image as the parent. Exec allows a process to change execution image to a different file. In addition, processes also obey the parent-child relation. A parent process may wait for child process termination. When a process terminates, it notifies the parent, which collects the child exit status and eventually releases the child process for reuse. However, the process model is unsuited to simple embedded systems and real-time systems, except in microkernel based systems, such as QNX. In almost all RTOS, tasks are essentially threads since they all execute in the same address space of the system kernel.

10.5.7.3 Memory Management in RTOS

Memory management refers to three distinct aspects: virtual address spaces mapping, dynamic memory allocation during execution and run-time stack overflow checking. In the following, we shall discuss each of the memory management schemes used in most RTOS.

Real Address Space

Tasks in most RTOS execute in the same address space of the kernel. The single address space environment has many advantages. First, it eliminates the overhead associated with virtual address mapping by the memory management hardware. Second, it allows tasks to share memory directly for fast inter-task communication. The main disadvantage of the single address space scheme is the lack of memory protection. Any task may corrupt the shared memory, causing other tasks or the entire system to fail.

Virtual Address Spaces
Most RTOS do not allow tasks to have separate address spaces. So there are no virtual address mapping and memory protection. For security and reliability reasons, memory protection may be necessary. If so, a RTOS should use the simplest memory mapping scheme of the Memory Management Unit (MMU) hardware, for better efficiency. For example, on the Intel architecture, it should use segmentation rather than (2-level) paging. Similarly, on the ARM architecture, it should use one-level paging with large page size rather than two-level paging. Some of the page entries, e.g. shared kernel pages, can be placed in the TLB as lock-down entries to minimize the MMU overhead during task switch.

Dynamic Memory Allocation
Many RTOS allow tasks to allocate memory dynamically during execution. Some RTOS even support the standard malloc()/free() functions of C library for allocate/free memory dynamically, presumably from a heap area of the system. However, a closer examination of real-time system requirements should reveal that allowing dynamic memory allocation at run-time may not be a good idea. Unlike conventional tasks, a key requirement of real-time tasks is that their behavior must be deterministic and predictable. Dynamic memory allocation at run-time would introduce a variable amount of delay to task executions, making them anything but predictable. The only justifiable need for memory allocation is to provide tasks with shared memory for fast inter-task communication. In that case, the needed memory should be allocated either statically or as fixed size blocks rather than in chunks of variable size.

Stack Overflow Checking
All RTOS support task creation. When creating a new task, the user may specify a function to be executed by the task, a task stack size and a pointer to the initial parameters of the function. Most RTOS support stack overflow checking at run-time. This seems to be a good feature but it is really superficial, for the following reasons. First, stack usage of any program should be controlled carefully during program design. Once a program is written, the stack size needed during execution can be estimated by the length of function calls and the amount of local variable spaces in the functions. The actual stack size of a program can be observed through testing. The maximum stack size can be set to the observed size, plus a safety factor, in the final program code. After all, this is the standard practice used by all operating system kernel designers. If every program is developed this way, there should be no reason for any task to run out of stack space during execution. Second, without memory protection hardware, stack overflow checking must be done by software, e.g. check the stack pointer against a preset stack limit value upon entry to every function, but this would introduces additional run-time overhead and delay. Third, even if we include run-time stack overflow checking, it remains unclear what can be done if a task causes a stack overflow. To abort the task may be out of the question. To extend the stack space and allow the task to continue may introduce unacceptable delays. Most RTOS simply leave this question to the user. As usual, the best approach to dealing with stack overflow is to prevent it from occurring during program development through testing.

10.5.7.4 POSIX Compliant APIs

POSIX (2016) specifies a set of standards for Unix-like systems. POSIX.1 specifies the core service functions, such as signals, pipes, file and directory operations. POSIX.1b adds real-time extensions, and POSIX1.c adds threads support. The goal of POSIX standards is to provide a uniform user interface to facilitate the development of portable application programs on Unix-like system. Many real-time systems try to be POSIX compliant. However, strict compliance with the POSIX standards may actually hinder real-time operations. In standard Unix, the execution units are processes. In most real-time systems, tasks are essentially threads since they execute in the same address space of the kernel. It makes sense to provide threads synchronization mechanisms, such as mutex, barrier and condition variables, for task synchronization, but it would be very hard to justify why tasks must use open-close-read-write on file descriptors for I/O. If a real-time task needs I/O, it would be much more efficient to invoke the device drivers directly, rather than going through additional layers of mappings through file descriptors.

10.5.7.5 Synchronization Primitives

All RTOS provide a set of tools for task synchronization. In the following, we discuss the suitability of such tools in RTOS.

. Disable interrupts and task scheduler: Some ROTS allow user programs to disable interrupts when entering critical regions, but this contradicts the short interrupt latency requirement of real-time systems. Many RTOS allow user programs to disable task scheduler to prevent task switch in long critical regions, but this contradicts preemptive task scheduling. In ROTS, user applications should be shielded from, or even not allowed to do, these low level operations.

. Mutex locks: Almost all RTOS use simple mutexes with priority inheritance to ensure exclusive access to critical regions. Some RTOS supports recursive mutexes, which can be locked/unlocked by the owner recursively. When writing concurrent programs using any kind of locking mechanism, a fundamental requirement is that the program must be free of deadlocks. This is usually achieved by ensuring that the locking order is always unidirectional. It is inconceivable that any task will ever need to acquire the same lock it already holds. Therefore, there is no real need for recursive mutexes.

. Binary and counting semaphores: Semaphores are convenient tools for general task synchronization and cooperation. Many RTOS support both binary and counting semaphores, which require different semantics and implementations. Since counting semaphores are more general than binary semaphores, there is no need for two kinds of semaphores. In some RTOS, when a task tries to acquire a semaphore, it may specify a timeout parameter. If the task gets blocked on the semaphore, it will be unblocked when the timeout value expires. The usefulness of this feature is highly questionable due to many problems it may cause. First, this would introduce additional delays to the timer interrupt handler, which must process the remaining time of all blocked tasks and unblock them when their time expires. Other questions include: what should be the timeout value? What should the task do in case of a timeout? A better solution is to use the conditional CP operation on semaphores. It allows the user to specify an alternative action immediately if a task can not acquire a semaphore, rather than waiting for a timeout.

. Event Control Blocks (ECBs): Unlike mutexes and semaphores, event flags allow tasks to wait for a variable number of events, which add flexibility to the system.

. Inter-task communication: Using shared memory, in conjunction with mutexes to provide protection, is both convenient and efficient for fast inter-task communication. Many RTOS provide static or dynamic message queues for tasks to exchange messages. These are less efficient than shared memory and pipes but they allow more flexibility in application programs. Some RTOS, e.g. NuttX, try to use signals, which would be a poor choice since regular signals are unsuited to interprocess communication (Wang 2015), and extended signals are less efficient than messages.

10.5.7.6 Interrupts Handling in RTOS

One of the fundamental requirements of RTOS is minimal interrupt latency, which implies that a RTOS must allow nested interrupts. All ROTS support nested interrupts, but they may handle nested interrupts in different ways. In most RTOS, interrupts processing are performed inside interrupt handlers directly. Task switching is disabled until all nested interrupt processing have completed. In Linux with real-time patches, interrupts are handled by pseudo-threads in kernel. This allows for faster interrupt response since each interrupt handler only needs to activate a pseudo-thread rather than actually processing the interrupt. In this scheme, nested interrupts processing is pushed to the pseudo-threads level. The choice of whether to handle nested interrupts directly or by pseudo-tasks should be based on the interrupt hardware. For Intel x86 CPUs, which do not have a separate interrupt mode stack, interrupts are handled in the context of the interrupted task, which can use the same stack to process nested interrupts. In this case, it would be better to handle nested interrupts directly inside ISRs. For ARM CPUs, which use a separate IRQ mode stack, the interrupted context must be transferred to a different privileged mode stack before allowing another interrupt. In this case, it would be better to handle nested interrupts by pseudo-tasks.

10.5.7.7 Task Deadlines

Although all RTOS are intended for (or claim to be) hard real-time systems, in reality most RTOS only provide a basic framework for developing real-time applications. As such, none of the published RTOS can actually guarantee task deadlines. The analytical results of RMS and EDF only provide general guidelines for RTOS under the simplest and idealistic conditions. They do not account for the overhead of interrupts processing, task blocking time due to resource sharing, task scheduling and switching time, etc. in a real ROTS system. When using a RTOS to develop real-time applications, it is entirely up to the user to determine whether tasks can meet their deadlines. This is no surprise, considering the wide range of possible real-time applications. To help remedy this problem, most RTOS provide run-time tracing facilities, which allow the user to monitor the time during which interrupts are disabled and/or the amount of time tasks spent inside critical regions, etc. Despite these efforts, performance evaluation of real-time systems essentially boils down to individual case analysis.

10.6 Design Principles of RTOS

Based on the above discussions, we propose a set of general guidelines for the design and implementation of the key components of RTOS.

10.6.1 Interrupts Processing

A RTOS must support nested interrupts. Depending on the interrupt hardware, interrupts processing can be performed in interrupt handlers directly or as pseudo-tasks with higher priorities than ordinary tasks.

10.6.2 Task Management

A RTOS should support task creation. Whenever possible, tasks should be static. Dynamic tasks should be regarded as nonessential. For simple RTOS, tasks should execute in the same address space of the kernel. For RTOS with high security and reliability requirements, tasks should run in user mode in separate virtual address spaces, but the system should use the simplest virtual address mapping scheme for better efficiency.

10.6.3 Task Scheduling

Task scheduling must be preemptive based on task priority. Although RMS is static and simpler, task priority should be based on EDF because it's more realistic in terms of meeting task deadlines.

10.6.4 Synchronization Tools

Use simple mutex with priority inheritance to protect critical regions. Use counting semaphore for task cooperation. For added flexibility, use event flags to allow tasks to wait for a variable number of events.

10.6.5 Task Communication

Use shared memory protected by simple mutxes for direct task communication. Use pipes for tasks to share streams of data. Use synchronous message passing for tasks to exchange messages.

10.6.6 Memory Management

Avoid virtual address mapping if possible. Allow tasks to allocate memory in fixed size blocks, but not in chunks of arbitrary sizes. Support stack overflow checking during development but not in the final system.

10.6.7 File System

If a file system is needed, implement the file system as a RAM disk in memory. Load the file system from a SDC to the RAM disk when the system starts and flushes any changes to the file system back to the SDC periodically.

10.6.8 Tracing and Debugging

A RTOS should provide tracing and debugging facilities to allow user to monitor the progress of tasks, at least during development.

In the following sections, we shall show the design and implementation of two RTOS, one for uniprocessor (UP) systems and the other one for multiprocessor (MP) systems.

10.7 Uniprocessor RTOS (UP_RTOS)

UP_RTOS is a real-time operating system for uniprocessor (UP) systems. It is based on the fully preemptive UP kernel developed in Chap. 5 (Sect. 5.14.3) but with the following extensions to support real-time applications.

(1). Preemptive task scheduling by (static) task priority
(2). Support nested interrupts
(3). Mutexes with priority inheritance for critical regions and resource sharing
(4). Semaphores with priority inheritance for process cooperation
(5). Shared memory, pipes and messages for inter-process communication
(6). A logging task, which records task activities to a SDC for tracing and debugging

The following describes the UP_RTOS kernel.

10.7.1 Task Management in UP_RTOS

UP_RTOS supports a variable number of tasks. The maximum number of tasks (NPROC) in the system is a configurable parameter, which can be set to a value suited to the needs. Tasks can be either static or dynamic. When using static tasks, all tasks are created during system initialization and they exist in the system permanently. When using dynamic tasks, tasks can be created on demand and they terminate after completing their work. Each task is created with a static priority. All tasks execute in the same address space of the kernel. Task creation is by the API

kfork(int * f() task_function, int priority);

which creates a task to execute a task_function() with a specified priority. Each task is represented by a PROC structure.

```
#define NPROC 256
typedef struct proc{
  struct proc *next;     // pointer to next PROC
  int     *ksp;          // saved sp when not running
  int     status;        // status
  int     pid;           // task ID
  int     pause;         // pausue time
  int     ready_time;    // task release or ready time
  int     priority;      // effective priority
  int     rpriority;     // real priority
  MSG     *mqueue        // message queue
  SEMAPHORE nmsg;        // number of message in messge queue
  MUTEX   mqlock         // message queue lock
  SEMAPHORE wchild;      // wait for ZOMBIE child
  int     kstack[SSIZE]; // 4KB to 8KB task stack area
```

```
}PROC;
PROC proc[NPROC];          // NPROC PROC structures
PROC *readyQueue;          // ready queue by PROC priority
PROC *running;             // current running PROC pointer
```

10.7.2 Task Synchronization in UP_RTOS

UP_RTOS uses sleep/wakeup for task synchronization only in pipes. It uses mutex and counting semaphore for general task synchronization. Mutexes are used as exclusive locks to protect critical regions. Semaphores are used for task cooperation. Both are implemented with priority inheritance to prevent priority inversion. Each mutex already has a owner field, which identifies the current task that holds the mutex lock. In order to support priority inherence in semaphores, we modify the semaphore structure to contain an owner field also, which identifies the current task that holds the semaphore. For simplicity, priority inheritance is only one-level, i.e. it only applies to each mutex or semaphore but is not transitive in a sequence of mutex or semaphore requests. Extension of priority inheritance to nested mutex or semaphore requests is left as an exercise in the Problem section.

10.7.3 Task Scheduling in UP_RTOS

Task scheduling is by preemptive priority. Task preemption is implemented as follows. First, the kernel uses a global counter to keep track of interrupts nesting levels. When entering an interrupt handler, the counter is incremented by 1. When exiting an interrupt handler, the counter is decremented by 1, etc. Task switch is not allowed if the interrupt nesting level is nonzero. Second, the only operations that can make a task ready to run are task creation, mutex_unlock and V on semaphores. Whenever a ready task is entered into the readyQueue, it invokes a reschedule() function to reschedule tasks. The following shows the algorithm of task preemption.

```
/********* Algorithm of task preemption ********/
int intnest; // interrupt nesting counter, initially 0
int swflag;  // switch task flag
reschedule
{
  if (readyQueue->priority > running->priority){
     if (intsest==0)// if not in interrupt handler
          tswitch();  // preempt running task immediately
       else{
          swflag = 1; // defer preemption until end of interrupts
}
interrupt_handler_exit // switch task if swflag set & end of IRQs
{
  if (!intnest && swflag)
     tswtich() in SVC mode;
  }
}
```

In reschedule(), if the current running task is no longer the highest priority and execution is not inside any interrupt handler, it preempts the current running task immediately. If execution is still inside an interrupt handler, it sets a switch task flag, which defers task switch until all nested interrupts processing have ended.

10.7.4 Task Communication in UP_RTOS

The UP_RTOS kernel provides three kinds of mechanisms for inter-task communication.

10.7.4.1 Shared Memory

When the system starts, it initializes a fixed number (32) of 64KB memory regions, e.g. from 32MB to 34MB, for inter-task communication through shared memory. Each shared memory region is represented by a structure

```
#define NPID NPROC/sizeof(int)
struct shmem{
   int procID[NPID];   // bit vector for NPROC task IDs
   MUTEX mutex;        // mutex lock for shared region
   char *address;      // start address of memory region
}shmem[32];
```

When the system starts, it initializes the shmem structures to contain

```
procID = {0};              // no task using the shmem yet
mutex = UNLOCKED;          // for exclusive access to the shared region
address = pointer to a unique 64KB memory area
```

To use a shared memory, a task must first attach itself to a shmem structure by

```
int shmem_attach(struct shmem *mp);
```

shmem_attach() records the task ID bit in procID (for accessibility checking) and returns the number of tasks currently attached to the shared memory. After attaching to a shared memory, tasks may use

```
shmem_read( struct shmem *mp, char buf[ ], int nbytes);
shmem_write(struct shmem *mp, char buf[ ], int nbytes);
```

to read/write data from/to the shared memory. The read/write functions only guarantee each read/write operation is atomic (by the mutex lock of the shared memory). The data format and meaning of the shared memory contents are entirely up to the user to decide.

10.7.4.2 Pipes for Data Streams

This is the same pipe mechanism of Chap. 5 (Sect. 5.13.2). It allows tasks to use pipes to read/write streams of data.

10.7.4.3 Message Queues

This is the same synchronous message passing mechanism of Chap. 5 (Sect. 5.13.4). It allows tasks to communicate by exchange messages. As in the case of shared memory, the user may design the message format and contents to suit the needs.

10.7.5 Protection of Critical Regions

In UP_RTOS, all critical regions are protected by mutex locks. The mutex locking order is always unidirectional, so deadlocks can never occur in the UP_RTOS kernel.

10.7.6 File System and Logging

Logging typically requires writing information to a log file in a file system, which may incur a variable amount of delay time to tasks. We do not see any justifiable needs for tasks in a real-time system to perform file operations. For simplicity and efficiency, we implement logging by a special logging task, which receives logging requests from other tasks by messages and writes the logging information to a storage device, such as a SDC, directly. When the system starts, it creates a logging task with the second lowest priority 1 (higher than the idle task at priority 0). Other tasks uses

$$\textbf{log(char} * \textbf{log\_information)}$$

to record a line in the log. The logging operation sends a message to the logging task, which formats the log information in the form

timestamp : taskID : line

and writes it to a SDC (1KB) block whenever it runs. It also writes the logged lines to a UART port to let the user see the logging activities on-line. When the system terminates, the log information can be retrieved from the SDC for tracing and analysis.
We show the implementation of the UP_RTOS kernel and demonstrate its capabilities by the following example programs.

C10.1: UP_RTOS with static periodic tasks and round-robin scheduling
C10.2: UP_RTOS with static periodic tasks and preemptive scheduling
C10.3: UP_RTOS with dynamic tasks and preemptive scheduling

10.7.7 UP_RTOS with Static Periodic Tasks and Round-Robin Scheduling

The first UP_RTOS example, denoted by C10.1, demonstrates static periodic tasks with round-robin scheduling. We assume that all tasks are periodic with the same period, hence the same priority in accordance with the RMS scheduling algorithm. As a uniprocessor (UP) system, the UP_RTOS kernel maintains a single readyQueue for task scheduling. In the readyQueue, tasks are ordered by priority. Tasks with the same priority are ordered First-In-first-Out (FIFO). Since all the tasks have the same priority, they are scheduled to run by round-robin. When the system starts, it creates 4 static tasks, all execute the same taskCode() function with the same period. Each task executes an infinite loop, in which a task first blocks itself on a unique semaphore (initialized to 0). A blocked task will be V-ed up by a timer periodically by the task period. When a task is unblocked and becomes ready to run, we get its ready time from a global time and record it in the task PROC structure. When a task runs, it first gets the start time. Then it does some computation, which is simulated by a delay loop. At the end of the execution loop, each task gets the end time and computes its execution and deadline times as follows.

$$\text{Execution time Ci} = \text{end\_time} - \text{start\_time};$$
$$\text{Deadline time Di} = \text{end\_time} - \text{ready\_time};$$

It compares the deadline time with task period to see whether it has met the deadline (equal to task period). Then it repeats the loop again. The following lists the code of the C10.1 program. In order to keep the program simple, the system only supports nested interrupts, preemptive task scheduling, but without priority inheritance which will be implemented later.
(1) . ts.s file of C10.1: The main feature of the ts.s file is that it supports nested IRQ interrupts. Task switch is deferred until nested interrupts processing have ended. Details of task preemption are explained later.

```
/************* ts.s file of C10.1 ***************/
        .text
        .code 32
.global vectors_start, vectors_end
.global proc, procsize
.global tswitch, scheduler, running
.global int_off, int_on, lock, unlock
.global swflag, intnest, int_end
.set vectorAddr, 0x10140030 // VIC vector base address
reset_handler:
// set SVC stack to high end of proc[0].kstack
   ldr r0, =proc
   ldr r1, =procsize
   ldr r2, [r1, #0]
```

```
  add r0, r0, r2
  mov sp, r0
// copy vector table to address 0
  bl copy_vectors
// go in IRQ mode, set IRQ stack
  msr cpsr, #0x92
  ldr sp, =irq_stack_top
// call main() in SVC mode with IRQ on: all tasks run in SVC mode
  msr cpsr, #0x13
  bl main
  b .
irq_handler:              // support nested interrupts in SVC mode
  sub      lr, lr, #4
  stmfd sp!, {r0-r12, lr}
  mrs r0, spsr
  stmfd sp!, {r0}         // push SPSR
  ldr r0, =intnest        // r0->intnest
  ldr r1, [r0]
  add r1, #1
  str r1, [r0]            // intnest++
  mov r1, sp              // get irq sp into r1
  ldr sp, =irq_stack_top // reset IRQ stack poiner to IRQ stack top
// switch to SVC mode     // to allow nested IRQs: clear IRQ source
  MSR cpsr, #0x93         // to SVC mode with interrupts OFF
  sub sp, #60             // dec SVC mode sp by 15 entries
  mov r0, sp              // r0=SVC stack top
// copy IRQ stack to SVC stack
  mov r3, #15         // 15 times
copy_stack:
  ldr r2, [r1], #4    // get an entry from IRQ stack
  str r2, [r0], #4    // write to proc's kstack
  sub r3, #1          // copy 15 entries from IRQ stack to PROC's kstack
  cmp r3, #0
  bne copy_stack
// read vectoraddress register: MUST!!! else no interrupts
  ldr  r1, =vectorAddr
  ldr  r0, [r1]  // read vectorAddr register to ACK interrupt
  stmfd sp!, {r0-r3, lr}
  msr cpsr, #0x13      // still in SVC mode but enable IRQ
  bl irq_chandler      // handle interrupt in SVC mode, IRQ off
  msr cpsr, #0x93
  ldmfd sp!, {r0-r3, lr}
  ldr r0, =intnest     // check interrupt nest level
  ldr r1, [r0]
  sub r1, #1
  str r1, [r0]         // intnest--
  cmp r1, #0           // if intnest != 0 => no_switch
  bne no_switch
// intnest==0: END OF IRQs: if swflag=1: switch task
  ldr r0, =swflag
  ldr r0, [r0]
  cmp r0, #0
  bne do_switch        // if swflag=0: no task switch
no_switch:
  ldmfd sp!, {r0}
```

```
  msr   spsr, r0        // restore SPSR
// irq_chandler() already issued EOI
  ldmfd sp!, {r0-r12, pc}^  // return via SVC stack
do_switch:                  // still in IRQ mode
  bl endIRQ                 // show "at end of IRQs"
  bl tswitch                // call tswitch(): resume to here
// will switch task, so must issue EOI
  ldr  r1, =vectorAddr
  str  r0, [r1]         // issue EOI
  ldmfd sp!, {r0}
  msr   spsr, r0
  ldmfd sp!, {r0-r12, pc}^  // return via SVC stack

tswitch:                    // for task switch in SVC mode
// disable IRQ interrupts
  mrs r0, cpsr
  orr r0, r0, #0x80  // set I bit to MASK out IRQ interrupts
  msr cpsr, r0
  stmfd     sp!, {r0-r12, lr}
  ldr r0, =running  // r0=&running
  ldr r1, [r0, #0]  // r1->runningPROC
  str sp, [r1, #4]  // running->ksp = sp
  bl  scheduler
  ldr r0, =running
  ldr r1, [r0, #0]  // r1->runningPROC
  ldr sp, [r1, #4]
  // enable IRQ interrupts
  mrs r0, cpsr
  bic r0, r0, #0x80   // clear bit means UNMASK IRQ interrupt
  msr cpsr, r0
  ldmfd     sp!, {r0-r12, pc}

// utility functions: int_on/int_off/lock/unlock: NOT shown
vectors_start:
  LDR PC, reset_handler_addr
  LDR PC, undef_handler_addr
  LDR PC, swi_handler_addr
  LDR PC, prefetch_abort_handler_addr
  LDR PC, data_abort_handler_addr
  B .
  LDR PC, irq_handler_addr
  LDR PC, fiq_handler_addr
reset_handler_addr:           .word reset_handler
undef_handler_addr:           .word undef_handler
swi_handler_addr:             .word swi_handler
prefetch_abort_handler_addr:  .word prefetch_abort_handler
data_abort_handler_addr:      .word data_abort_handler
irq_handler_addr:             .word irq_handler
fiq_handler_addr:             .word fiq_handler
vectors_end:
```

Explanations of the ts.s file:

Reset_handler: As usual, reset_handler is the entry point. First, it sets the SVC mode stack pointer to the high end of proc[0] and copies the vector table to address 0. Next, it changes to IRQ mode to set the IRQ mode stack. Then it calls main () in SVC mode. During system operation, all tasks run in SVC mode in the same address of the kernel.

Irq_handler: Task switch is usually triggered by interrupts., which may make a blocked task ready to run. Thus, irq_handler is the most important piece of assembly code relevant to process preemption. Therefore, we only focus on the irq_handler code. As pointer out in Chap. 2, the ARM CPU can not handle nested interrupts in IRQ mode. Nested interrupts processing must be performed in a different privileged mode. In order to support process preemption due to interrupts, we choose to handle IRQ interrupts in SVC mode. The algorithm of irq_handler can be best described by the following algorithm.

/********* **Algorithm of IRQ handler for full task preemption** ********/
(1). Upon entry, adjust return lr; save context, including spsr, in IRQ stack
(2). Increment interrupt nesting counter by 1
(3). Switch to SVC mode with interrupts disabled
(4). Transfer context from IRQ stack to proc's SVC stack; reset IRQ stack
(5). Ack and clear interrupt source (prevent infinite loops from the same interrupt source)
(6). Enable IRQ interrupts; save working registers in SVC stack
(7). Call ISR in SVC mode with IRQ interrupts enabled
(8). (return from ISR): disable IRQ interrupts; restore working registers
(9). Decrement interrupt nesting counter by 1
(10). If still inside interrupt handler (counter nonzero): goto no_switch
(11). (end of IRQs): if swflag is set: goto do_switch
(12). no_switch: return via saved context in SVC stack
(13). do_switch: write EOI to interrupt controller; call tswitch() to switch task
(14). (when switched-out task resumes): return via saved interrupt context in SVC stack

(2). uart.c file: UARTs driver: for outputs only by TX interrupts

(3). vid.c file: LCD driver: SAME as before except frame buffer is at 4MB

(4). timer.c file: Use timer0 to activate tasks periodically

```
/** timer.c file of C10.1 **/
#define TLOAD    0x0
#define TVALUE   0x1
#define TCNTL    0x2
#define TINTCLR 0x3
#define TRIS     0x4
#define TMIS     0x5
#define TBGLOAD 0x6

typedef struct timer{
  u32 *base;              // timer's base address
  int tick, hh, mm, ss; // per timer data area
  char clock[16];
}TIMER;
TIMER timer[4];           // 4 timers, use only timer0

void timer_init()
{
  int i;
  TIMER *tp;
  printf("timer_init(): ");
  gtime = 0;
  for (i=0; i<4; i++){ // 4 timers but use only timer0
    tp = &timer[i];
    if (i==0) tp->base = (u32 *)0x101E2000;
    if (i==1) tp->base = (u32 *)0x101E2020;
    if (i==2) tp->base = (u32 *)0x101E3000;
    if (i==3) tp->base = (u32 *)0x101E3020;
    *(tp->base+TLOAD) = 0x0;    // reset
```

```
      *(tp->base+TVALUE)= 0xFFFFFFFF;
      *(tp->base+TRIS)  = 0x0;
      *(tp->base+TMIS)  = 0x0;
      *(tp->base+TLOAD) = 0x100;
      //0x62=|011-0000=|NOTEn|Pe|IntE|-|scal=00|32-bit|0=wrap|
      *(tp->base+TCNTL) = 0x62;
      *(tp->base+TBGLOAD) = 0xF00; // timer count
      tp->tick = tp->hh = tp->mm = tp->ss = 0;
      strcpy((char *)tp->clock, "00:00:00");
   }
}
void timer_handler(int n)
{
      int i;
      TIMER *t = &timer[n];
      gtime++;                       // increment global time
      t->tick++;                     // for local wall clock
      if (t->tick >= 64){
        t->tick=0; t->ss++;
        if (t->ss == 60){
           t->ss=0; t->mm++;
         if (t->mm==60){
            t->mm=0; t->hh++;
         }
        }
      }
// display a wall clock
      if (t->tick == 0){ // display wall clock
           for (i=0; i<8; i++){
              unkpchar(t->clock[i], 0, 70+i);
           }
           t->clock[7]='0'+(t->ss%10); t->clock[6]='0'+(t->ss/10);
           t->clock[4]='0'+(t->mm%10); t->clock[3]='0'+(t->mm/10);
           t->clock[1]='0'+(t->hh%10); t->clock[0]='0'+(t->hh/10);
           for (i=0; i<8; i++){
               kpchar(t->clock[i], 0, 70+i);
           }
      }
// unblock tasks by period
      if ((gtime % period)==0){            // period = N*T timer ticks
         for (i=1; i<=4; i++){
            V(&ss[i]);                     // activate task i
            proc[i].start_time = gtime;    // task ready to run time
         }
      }
      timer_clearInterrupt(n);
}
void timer_start(int n) // timer_start(0), 1, etc.
{
   TIMER *tp = &timer[n];
   printf("timer_start %d\n", n);
   *(tp->base+TCNTL) |= 0x80;     // set enable bit 7
}
```

```
int timer_clearInterrupt(int n) // timer_start(0), 1, etc.
{
  TIMER *tp = &timer[n];
  *(tp->base+TINTCLR) = 0xFFFFFFFF;
}
```

(5). Kernel files of C10.1:

```
/*********** pv.c file of C10.1 ***********/
extern PROC *running;
extern PROC *readyQueue;
extern int swflag;
extern int intnest;
int P(struct semaphore *s) // no priority inheritance
{
  int SR = int_off();
  s->value--;
  if (s->value < 0){
     running->status = BLOCK;
     enqueue(&s->queue, running);
     int_on(SR);
     tswitch();
  }
  int_on(SR);
}
int V(struct semaphore *s) // no priority inherence
{
  PROC *p; int cpsr;
  int SR = int_off();
  s->value++;
  if (s->value <= 0){
     p = dequeue(&s->queue);
     p->status = READY;
     enqueue(&readyQueue, p);
     printf("V up task%d pri=%d; running pri=%d\n",
            p->pid, p->priority, running->priority);
     reschedule();   // may preempty running task
  }
  int_on(SR);
}

/**************** kernel.c file of C10.1 **************/
#define NPROC 256
PROC proc[NPROC], *running, *freeList, *readyQueue;
int procsize = sizeof(PROC);
int swflag = 0;  // switch task flag
int intnest;     // interrupt nesting level

int kernel_init()
{
  int i, j;
  PROC *p;
  kprintf("kernel_init()\n");
  for (i=0; i<NPROC; i++){
    p = &proc[i];
    p->pid = i;
```

```
    p->status = READY;
    p->run_time = 0;
    p->next = p + 1;
  }
  proc[NPROC-1].next = 0;
  freeList = &proc[0];
  sleepList = 0;
  readyQueue = 0;
  intnest = 0;
  running = getproc(&freeList); // create and run P0
  running->priority = 0;
  printf("running = %d\n", running->pid);
}

int scheduler()
{
  printf("task%d switch task: ", running->pid);
  if (running->status==READY)
     enqueue(&readyQueue, running);
  printQ(readyQueue);
  running = dequeue(&readyQueue);
  printf("next running = task%d pri=%d realpri=%d\n",
        running->pid, running->priority, running->realPriority);
  color = RED+running->pid;
  swflag = 0;
}

int reschedule() // called from inside V()with IRQ disabled
{
  if (readyQueue && readyQueue->priority > running->priority){
     if (intnest==0){
        printf("task%d PREEMPT task%d IMMEDIATELY\n", readyQueue->pid,
               running->pid);
        tswitch();
     }
     else{
        printf("task%d DEFER PREEMPT task%d ", readyQueue->pid,
               running->pid);
        swflag = 1;  // IRQ are disabled, so no need to lock/unlock
     }
  }
}

// kfork() create a new task and enter into readyQueue
PROC *kfork(int func, int priority)
{
  int i;
  PROC *p = getproc(&freeList);
  if (p==0){
     kprintf("kfork failed\n");
     return (PROC *)0;
  }
  p->ppid = running->pid;
  p->parent = running;
  p->status = READY;
  p->realPriority = p->priority = priority;
```

```
  p->run_time = 0;
  p->ready_time = 0;
  // set kstack to resume to execute func()
  for (i=1; i<15; i++)
      p->kstack[SSIZE-i] = 0;
  p->kstack[SSIZE-1] = (int)func;
  p->ksp = &(p->kstack[SSIZE-14]);
  enqueue(&readyQueue, p);
  printf("task%d create a child task%d\n", running->pid, p->pid);
  reschedule();
  return p;
}
/******************* t.c file of C10.1 ****************/
// set T=5 timer ticks, task period = 8*T
#define T 5
#define period 8*T

#include "type.h"
#include "string.c"
#include "queue.c"
#include "pv.c"
#include "uart.c"
#include "vid.c"
#include "exceptions.c"
#include "kernel.c"
#include "timer.c"
// globals
struct semaphore ss[5];    // semaphores for task to block
volatile u32 gtime;        // global time
UART *up0, *up1;
int tcount = 0;            // number of timer interrupts in low IRQ
void copy_vectors(void) { // same as before }

int enterint() // for nested IRQs: clear interrupt source
{
  int status, ustatus, scode;
  status = *((int *)(VIC_BASE_ADDR)); // read status register
  if (status & (1<<4)){ // timer0 at IRQ 4
     tcount++;
  }
  if (status & (1<<12)){ // uart0 at IRQ 12
     ustatus = *(up0->base + UDS);  // read UDS register
  }
  if (status & (1<<13)){ // uart1 at IRQ 13
    ustatus = *(up1->base + UDS);  // read UDS register
  }
}
int endIRQ() { printf("until END of IRQ\n"); }
int int_end(){ printf("task switch at end of IRQ\n"); }

// use vectored interrupts of PL190
void timer0_handler()
{
   timer_handler(0);
}
```

```
void uart0_handler()
{
  uart_handler(&uart[0]);
}
void uart1_handler()
{
  uart_handler(&uart[1]);
}
int vectorInt_init()
{
  printf("vectorInterrupt_init()\n");
  *((int *)(VIC_BASE_ADDR+0x100)) = (int)timer0_handler;
  *((int *)(VIC_BASE_ADDR+0x104)) = (int)uart0_handler;
  *((int *)(VIC_BASE_ADDR+0x108)) = (int)uart1_handler;
  //(2). write to intControlRegs = E=1|IRQ# =  1xxxxx
  *((int *)(VIC_BASE_ADDR+0x200)) = 0x24;  //0100100 at IRQ 4
  *((int *)(VIC_BASE_ADDR+0x204)) = 0x2C;  //0101100 at IRQ 12
  *((int *)(VIC_BASE_ADDR+0x208)) = 0x2D;  //0101101 at IRQ 13
  // write 32-bit 0's to IntSelectReg to generate IRQ interrupts
  *((int *)(VIC_BASE_ADDR+0x0C)) = 0;
}
void irq_chandler()
{
  int (*f)();                              // f is a function pointer
  f =(void *)*((int *)(VIC_BASE_ADDR+0x30)); // get ISR address
  f();                                     // call the ISR function
  *((int *)(VIC_BASE_ADDR+0x30)) = 1; // write to vectorAddr as EOI
}
int delay(int pid)          // delay loop: simulate task compute time
{
  int i, j;
  for (i=0; i<1000; i++){
    // may use pid for different task delay time
    for (j=0; j<1000; j++); // change this line for different dealys
  }
}

/*********** Static Periodic Tasks ***********/
int taskCode()
{
  int pid = running->pid;
  u32 time1, time2, t, ready_time, complete_time;
  while(1){
    lock(); // reset task semaphore and ready_time
     ss[running->pid].value = 0;
     ss[running->pid].queue = 0;
     running->start_time = 0;
    unlock();

    // block on per-proc semaphore until Ved up by timer
    P(&ss[running->pid]);

    ready_time = running->start_time;
     printf("%d ready=%d", pid, ready_time); // to LCD
    uprintf("%d ready=%d", pid, ready_time); // to UART0
    time1 = gtime;
```

```
    printf("start=%d", time1);
   uprintf("start=%d", time1);
     delay(pid);
   time2 = gtime;

   t = time2 - time1;
   complete_time = time2 - ready_time;

    printf("end=%d%d[%d %d]", time2, t, ready_time, complete_time);
   uprintf("end=%d%d[%d %d]", time2, t, ready_time, complete_time);

   if (complete_time > period){ // if task missed deadline
       printf(" PANIC:miss deadline!\n");
      uprintf(" PANIC:miss deadline!\n");
   }
   else{
      printf(" OK\n");
     uprintf(" OK\n");
   }
}
int main()
{
   int i;
   PROC *p;
   color = WHITE;
   fbuf_init();
   uart_init();
   up0 = &uart[0];
   up1 = &uart[1];
   kprintf("Welcome to UP_RTOS in ARM\n");
   /* enable timer0,1, uart0,1 interrupts */
   VIC_INTENABLE = 0;
   VIC_INTENABLE |= (1<<4);    // timer0,1 at bit4
   VIC_INTENABLE |= (1<<12);   // UART0 at bit12
   VIC_INTENABLE |= (1<<13);   // UART2 at bit13
   vectorInt_init();
   timer_init();
   timer_start(0);
   kernel_init();
   printf("P0 kfork tasks\n"); // create 4 tasks
   for (i=1; i<=4; i++){
     ss[i].value = 0;          // initialize task semaphhores
     ss[i].queue = (PROC *)0;
     kfork((int)taskCode, 1); // all SAME priority = 1
   }
   unlock();
   while(1){                   // idle task P0 loop
     if (readyQueue)
        tswitch();
   }
}
```

Deadline Analysis of the C10.1 system: With 4 tasks of equal period, according to the schedulability condition of RMS, all the tasks should be able to meet their deadlines if

$$\mathbf{(C1 + C2 + C3 + C4)/period < 4 * (2 * *(1/4) - 1) = 0.7568} \quad (10.1)$$

```
1  ready= 696 start= 720 end= 724 4 [ 696  28 ] OK
2  ready= 704 start= 729 end= 734 5 [ 704  30 ] OK
3  ready= 712 start= 739 end= 742 3 [ 712  30 ] OK
4  ready= 720 start= 748 end= 753 5 [ 720  33 ] OK
1  ready= 736 start= 759 end= 763 4 [ 736  27 ] OK
2  ready= 744 start= 767 end= 772 5 [ 744  28 ] OK
QEMU
UART init()                                                    00:00:13
readyQueue=[ 1  1 ]->[ 2  1 ]->[ 3  1 ]->[0 0 ]->Nil

task 4  switch task: readyQueue=[ 1  1 ]->[ 2  1 ]->[ 3  1 ]->[0 0 ]->Nil
next running = task 1  pri= 1  realpri= 1
 1  ready= 808 start= 837 V up task 4  pri= 1 ; running pri= 1
readyQueue=[ 2  1 ]->[ 3  1 ]->[ 4  1 ]->[0 0 ]->Nil
end= 840  3 [ 808   32 ] OK
task 1  switch task: readyQueue=[ 2  1 ]->[ 3  1 ]->[ 4  1 ]->[0 0 ]->Nil
task 1  switch task: readyQueue=[ 2  1 ]->[ 3  1 ]->[ 4  1 ]->[0 0 ]->Nil
next running = task 2  pri= 1  realpri= 1
 2  ready= 816 start= 846 V up task 1  pri= 1 ; running pri= 1
readyQueue=[ 3  1 ]->[ 4  1 ]->[ 1  1 ]->[0 0 ]->Nil
end= 850  4 [ 816   34 ] OK
task 2  switch task: readyQueue=[ 3  1 ]->[ 4  1 ]->[ 1  1 ]->[0 0 ]->Nil
next running = task 3  pri= 1  realpri= 1
 V up task 2  pri= 1 ; running pri= 1
readyQueue=[ 4  1 ]->[ 1  1 ]->[ 2  1 ]->[0 0 ]->Nil
3  ready= 832 start= 857 end= 862  5 [ 832   30 ] OK
task 3  switch task: readyQueue=[ 4  1 ]->[ 1  1 ]->[ 2  1 ]->[0 0 ]->Nil
next running = task 4  pri= 1  realpri= 1
 4  ready= 840 start= 867 V up task 3  pri= 1 ; running pri= 1
readyQueue=[ 1  1 ]->[ 2  1 ]->[ 3  1 ]->[0 0 ]->Nil
end= 872  5 [ 840   32 ] OK
task 4  switch task: readyQueue=[ 1  1 ]->[ 2  1 ]->[ 3  1 ]->[0 0 ]->Nil
```

Fig. 10.1 UP_RTOS with round-robin scheduling

If we assume all task compute times Ci=C are the same, then the condition (10.1) becomes

$$\mathbf{C/period < 0.189} \tag{10.2}$$

Figure 10.1 shows the sample outputs of running the example program C10.1. The figure shows that the individual task compute time varies from 3 to 5 timer ticks. The task compute time includes the overhead due to interrupts processing by both timer and I/O interrupts, as well as task scheduling and switching time. In the test program C10.1, we set all the task compute time to C = 5 timer ticks and task period = 8*C, so that C/period = 1/8=0.125, which is within the schedulability bound of RMS. In this case, all the tasks indeed can meet their deadlines, as Figure 10.1 shows.

The condition (10.2) suggests that we should expect all the tasks to meet their deadlines if C/period < 0.189. However, the test results showed otherwise. For instance, if we set the task period to 7*C (C/period = 0.1428), which is still within the RMS bound, some of the tasks will start to miss their deadlines, as shown in Figure 10.2.

In fact, the test results showed that, with period = 6*C (C/period = 0.167), almost all the tasks will miss their deadlines. The example shows that, while it is easy to design and implement a ROTS kernel to support periodic tasks, there is no guarantee that the task can meet their deadlines even if the system follows the principles of RMS strictly.

10.7.8 UP_RTOS with Static Periodic Tasks and Preemptive Scheduling

The second example of UP_RTOS, denoted by C10.2, demonstrates static periodic tasks with preemptive task scheduling. When the system starts, it creates 4 static tasks with different periods. As in RMS, each task is assigned a static priority inversely proportional to its period. Specifically, the task periods and priorities are assigned as follows, where the task periods are in units of T timer ticks.

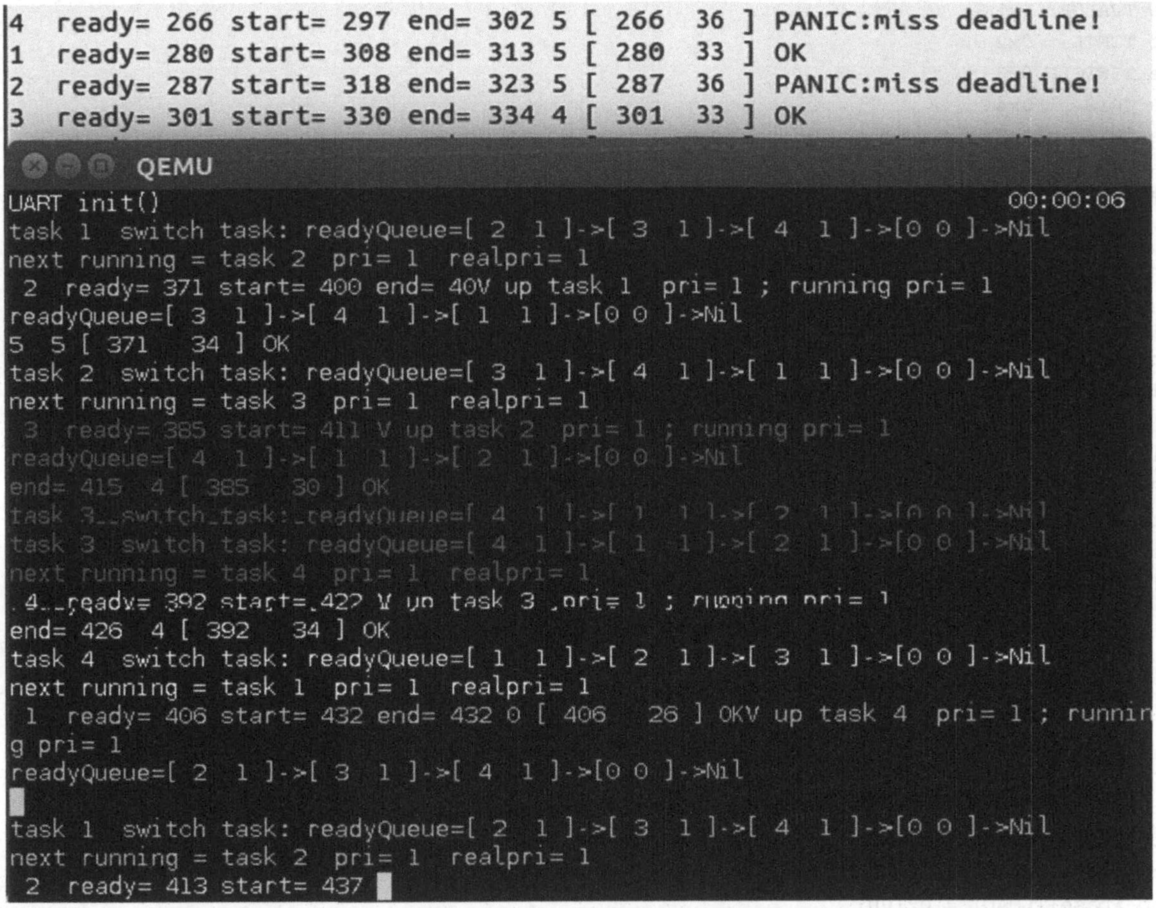

```
4  ready= 266 start= 297 end= 302 5 [ 266  36 ] PANIC:miss deadline!
1  ready= 280 start= 308 end= 313 5 [ 280  33 ] OK
2  ready= 287 start= 318 end= 323 5 [ 287  36 ] PANIC:miss deadline!
3  ready= 301 start= 330 end= 334 4 [ 301  33 ] OK
```

```
QEMU
UART init()                                                      00:00:06
task 1  switch task: readyQueue=[ 2  1 ]->[ 3  1 ]->[ 4  1 ]->[0 0 ]->Nil
next running = task 2  pri= 1  realpri= 1
 2  ready= 371 start= 400 end= 40V up task 1  pri= 1 ; running pri= 1
readyQueue=[ 3  1 ]->[ 4  1 ]->[ 1  1 ]->[0 0 ]->Nil
5  5 [ 371   34 ] OK
task 2  switch task: readyQueue=[ 3  1 ]->[ 4  1 ]->[ 1  1 ]->[0 0 ]->Nil
next running = task 3  pri= 1  realpri= 1
 3  ready= 385 start= 411 V up task 2  pri= 1 ; running pri= 1
readyQueue=[ 4  1 ]->[ 1  1 ]->[ 2  1 ]->[0 0 ]->Nil
end= 415  4 [ 385   30 ] OK
task 3  switch task: readyQueue=[ 4  1 ]->[ 1  1 ]->[ 2  1 ]->[0 0 ]->Nil
task 3  switch task: readyQueue=[ 4  1 ]->[ 1  1 ]->[ 2  1 ]->[0 0 ]->Nil
next running = task 4  pri= 1  realpri= 1
 4  ready= 392 start= 422 V up task 3  pri= 1 ; running pri= 1
end= 426  4 [ 392   34 ] OK
task 4  switch task: readyQueue=[ 1  1 ]->[ 2  1 ]->[ 3  1 ]->[0 0 ]->Nil
next running = task 1  pri= 1  realpri= 1
 1  ready= 406 start= 432 end= 432 0 [ 406   26 ] OKV up task 4  pri= 1 ; runnin
g pri= 1
readyQueue=[ 2  1 ]->[ 3  1 ]->[ 4  1 ]->[0 0 ]->Nil

task 1  switch task: readyQueue=[ 2  1 ]->[ 3  1 ]->[ 4  1 ]->[0 0 ]->Nil
next running = task 2  pri= 1  realpri= 1
 2  ready= 413 start= 437
```

Fig. 10.2 Demonstration of tasks missing deadlines

```
  Task      Period      Priority
-------   ---------   ---------
   P1        5T           1
   P2        4T           2
   P3        3T           3
   P4        2T           4
-------------------------------
```

Each task executes an infinite loop, in which it first blocks itself on a unique semaphore initialized to 0. Each blocked task will be V-ed up by the timer periodically according to the task period. As in the program C10.1, the ready time of a task is the time when the task becomes ready to run, i.e. when it is activated by the timer or when it is entered into the readyQueue. When a task runs, it gets the start time form a global time. Then it does some computation, which is simulated by a delay loop. At the end of the execution loop, each task gets the end time and tests whether it has met the deadline (equal to task period). Then it repeats the loop again. The following shows the code of the C10.2 program. For the sake of brevity, we only show the modified code of C10.2.

(1). ts.s file of C10.2. Same as in C10.1.
(2). uart.c file: UARTs driver: for outputs only by TX interrupts
(3). vid.c file: LCD driver: SAME as before except frame buffer at 4MB
(4). timer.c file: In the timer interrupt handler, tasks are V-ed up by task periods. It also sets the ready_time of each task when the task is entered into the readyQueue.

```
/** timer.c file of C10.2 **/
#define TLOAD   0x0
```

```
#define TVALUE  0x1
#define TCNTL   0x2
#define TINTCLR 0x3
#define TRIS    0x4
#define TMIS    0x5
#define TBGLOAD 0x6
typedef struct timer{
  u32 *base;            // timer's base address
  int tick, hh, mm, ss; // per timer data area
  char clock[16];
}TIMER;
TIMER timer[4];         // 4 timers, use only timer0
void timer_init()
{
  int i;
  TIMER *tp;
  printf("timer_init(): ");
  gtime = 0;
  for (i=0; i<4; i++){ // 4 timers but use only timer0
    tp = &timer[i];
    if (i==0) tp->base = (u32 *)0x101E2000;
    if (i==1) tp->base = (u32 *)0x101E2020;
    if (i==2) tp->base = (u32 *)0x101E3000;
    if (i==3) tp->base = (u32 *)0x101E3020;
    *(tp->base+TLOAD) = 0x0;   // reset
    *(tp->base+TVALUE)= 0xFFFFFFFF;
    *(tp->base+TRIS)  = 0x0;
    *(tp->base+TMIS)  = 0x0;
    *(tp->base+TLOAD) = 0x100;
    //0x62=|011-0000=|NOTEn|Pe|IntE|-|scal=00|32-bit|0=wrap|
    *(tp->base+TCNTL) = 0x62;
    *(tp->base+TBGLOAD) = 0xF00; // timer count
    tp->tick = tp->hh = tp->mm = tp->ss = 0;
    strcpy((char *)tp->clock, "00:00:00");
  }
}
void timer_handler(int n)
{
    int i;
    TIMER *t = &timer[n];
    gtime++;                    // increment global time
    t->tick++;                  // for local wall clock
    if (t->tick >= 64){
      t->tick=0; t->ss++;
      if (t->ss == 60){
         t->ss=0; t->mm++;
        if (t->mm==60){
           t->mm=0; t->hh++;
        }
      }
    }
    if (t->tick == 0){ // display wall clock
         for (i=0; i<8; i++){
            unkpchar(t->clock[i], 0, 70+i);
         }
```

```
        t->clock[7]='0'+(t->ss%10); t->clock[6]='0'+(t->ss/10);
        t->clock[4]='0'+(t->mm%10); t->clock[3]='0'+(t->mm/10);
        t->clock[1]='0'+(t->hh%10); t->clock[0]='0'+(t->hh/10);
        for (i=0; i<8; i++){
            kpchar(t->clock[i], 0, 70+i);
        }
    }
    if ((gtime % (2*T))==0){      // activate P4 every 2T
       V(&ss[4]);
       proc[4].ready_time = gtime;
    }
    if ((gtime % (3*T))==0){      // activate P3 every 3T
       V(&ss[3]);
       proc[3].ready_time = gtime;
    }
    if ((gtime % (4*T))==0){      // activate P2 every 4T
       V(&ss[2]);
       proc[2].ready_time = 0;
    }
    if ((gtime % (5*T))==0){      // activate P1 every 5T
       V(&ss[1]);
       proc[1].ready_time = gtime;
    }
    timer_clearInterrupt(n);
}
void timer_start(int n) // timer_start(0), 1, etc.
{
  TIMER *tp = &timer[n];
  printf("timer_start %d\n", n);
  *(tp->base+TCNTL) |= 0x80;     // set enable bit 7
}
int timer_clearInterrupt(int n) // timer_start(0), 1, etc.
{
  TIMER *tp = &timer[n];
  *(tp->base+TINTCLR) = 0xFFFFFFFF;
}
```

(5). Kernel files of C10.2: Same as in C10.1
(6). t.c file of C10.2

```
/**************** t.c file of C10.2 ****************/
#define T 32
#include "type.h"
#include "string.c"
#include "queue.c"
#include "pv.c"
#include "uart.c"
#include "vid.c"
#include "exceptions.c"
#include "kernel.c"
#include "timer.c"
// globals
```

```
struct semaphore ss[5];     // semaphores for task to block
volatile u32 gtime;         // global time
UART *up0, *up1;
int tcount = 0;             // number of timer interrupts in low IRQ
void copy_vectors(void) { // same as before }

int enterint() // for nested IRQs: clear interrupt source
{
  int status, ustatus, scode;
  status = *((int *)(VIC_BASE_ADDR)); // read status register
  if (status & (1<<4)){ // timer0 at IRQ 4
     tcount++;
  }
  if (status & (1<<12)){ // uart0 at IRQ 12
     ustatus = *(up0->base + UDS);  // read UDS register
  }
  if (status & (1<<13)){ // uart1 at IRQ 13
    ustatus = *(up1->base + UDS);  // read UDS register
  }
}
int endIRQ() { printf("until END of IRQ\n"); }
int int_end(){ printf("task switch at end of IRQ\n"); }

// use vectored interrupts of PL190
void timer0_handler()
{
   timer_handler(0);
}
void uart0_handler()
{
  uart_handler(&uart[0]);
}
void uart1_handler()
{
  uart_handler(&uart[1]);
}
int vectorInt_init()
{
  printf("vectorInterrupt_init()\n");
  *((int *)(VIC_BASE_ADDR+0x100)) = (int)timer0_handler;
  *((int *)(VIC_BASE_ADDR+0x104)) = (int)uart0_handler;
  *((int *)(VIC_BASE_ADDR+0x108)) = (int)uart1_handler;
  //(2). write to intControlRegs = E=1|IRQ# =  1xxxxx
  *((int *)(VIC_BASE_ADDR+0x200)) = 0x24;  //0100100 at IRQ 4
  *((int *)(VIC_BASE_ADDR+0x204)) = 0x2C;  //0101100 at IRQ 12
  *((int *)(VIC_BASE_ADDR+0x208)) = 0x2D;  //0101101 at IRQ 13
  // write 32-bit 0's to IntSelectReg to generate IRQ interrupts
  *((int *)(VIC_BASE_ADDR+0x0C)) = 0;
}
void irq_chandler()
{
  int (*f)();                              // f is a function pointer
  f =(void *)*((int *)(VIC_BASE_ADDR+0x30)); // get ISR address
  f();                                     // call the ISR function
  *((int *)(VIC_BASE_ADDR+0x30)) = 1; // write to vectorAddr as EOI
}
```

```
int delay(int pid) // delay loop: simulate task compute time
{
  int i, j;
  for (i=0; i<1000; i++){
    for (j=0; j<1000; j++); // change this line for different dealys
  }
}
int taskCode()
{
  int pid = running->pid;
  u32 time1, time2, t, period, read_ytime, ctime;
  period = (6 - running->pid)*T; // task period=5,4,3,2
  while(1){
    lock();
     ss[running->pid].value = 0;
     ss[running->pid].queue = 0;
     proc[pid].ready_time = 0;
    unlock();

    P(&ss[running->pid]); // block on semaphore; Ved up by period

    readytime = running->ready_time;
    time1 = gtime;
     printf("P%dready=%dstart=%d", pid, readytime, time1);
    uprintf("P%dready=%dstart=%d", pid, readytime, time1);
       delay(pid); // simulate task compute time
    time2 = gtime;
    t = time2 - time1;
    ctime = time2 - readytime; // task comcpletion time
     printf("end=%d%d[%d%d]", time2, t, period, ctime);
    uprintf("end=%d%d[%d%d]", time2, t, period, ctime);
    if (dtime > period){
        printf(" PANIC! miss deadline!\n"); // to LCD
       uprintf(" PANIC! miss deadline!\n"); // to UART0
    }
    else{
       printf(" OK\n");
      uprintf(" OK\n");
    }
  }
}
int main()
{
  int i;
  PROC *p;
  fbuf_init();
  uart_init();
  up0 = &uart[0];
  up1 = &uart[1];
 kprintf("Welcome to UP_RTOS in ARM\n");
  /* enable timer0,1, uart0,1 SIC interrupts */
  VIC_INTENABLE = 0;
  VIC_INTENABLE |= (1<<4);  // timer0,1 at bit4
  VIC_INTENABLE |= (1<<5);  // timer2,3
  VIC_INTENABLE |= (1<<12); // UART0 at bit12
```

```
    VIC_INTENABLE |= (1<<13); // UART2 at bit13
    VIC_INTENABLE |= (1<<31); // SIC to VIC's IRQ31
    vectorInt_init();
    timer_init();
    timer_start(0);
    kernel_init();
    kprintf("P0 kfork tasks\n");
    for (i=1; i<=4; i++){        // create 4 static tasks
      ss[i].value = 0;
      ss[i].queue = (PROC *)0;
      kfork((int)taskCode, i);  // priority = 1,2,3,4
    }
    printQ(readyQueue);
    unlock();
    while(1){    // P0 loop
      if (readyQueue)
         tswitch();
    }
}
```

Deadline Analysis of the C10.2 system: With 4 tasks and task periods of 2T to 5T, we can derive an upper bound on the task compute time as follows. Assume that Ci is the compute time of taski. According to the schedulability condition of RMS, we expect all the tasks can meet their deadlines if

```
      C1/5T + C2/4T + C3/3T + C4/2T < 4(2**(1/4) - 1)
or    12C1 + 15C2 + 20C3 + 30C4 < 45.41T
```

This condition can be used to estimate the maximum compute times of the tasks without missing deadlines. For instance, if we assume that the compute time of all the tasks are the same, i.e. C=Ci for all i, then the individual task compute time upper bound is

$$C < (45.41/77)T = 0.59T \tag{10.3}$$

However, if we take into account the overhead due to interrupts processing, task blocking time and task switching time, etc. the allowable task compute time must be much smaller. Rather than trying to adjust the compute time of individual tasks, we simply use different values of T in the testing program and observe the run-time task behavior. In the testing program, the unit of task period T is set to 32 timer ticks initially.

Figure 10.3 shows the sample outputs of running the C10.2 program. The figure shows that all 4 tasks are activated to run periodically. Average task compute time is about 5 timer ticks or 0.156T, which is significantly lower than the theoretical bound of 0.59T. In this case, all tasks can indeed meet their deadlines. However, if we shorten T to a smaller value, some of the tasks will start to miss their deadlines. For example, with T=16 timer ticks, C=0.31T, which is still well below the RMS bound of 0.59T, only task4 will run because it is activated so quickly that it's always the ready task with the highest priority. In this case, task1 to taks3 will always miss their deadlines because they can never be scheduled to run. The example shows that, with preemptive task scheduling, it is even harder to determine whether tasks can meet their deadlines based on the theoretical models of RTOS. This is an area of RTOS which definitely needs more study, both analytically and empirically.

10.7.9 UP_RTOS with Dynamic Tasks Sharing Resources

The third example of UP_RTOS, denoted by C10.3, demonstrates dynamic tasks sharing resources. Resources are simulated by mutex locks, which support priority inheritance to prevent priority inversion. In the C10.3 system, tasks are created dynamically, which may terminate after completing their work. Task scheduling is by preemptive (static) task priority. The following lists the implementation code of the C10.3 program.

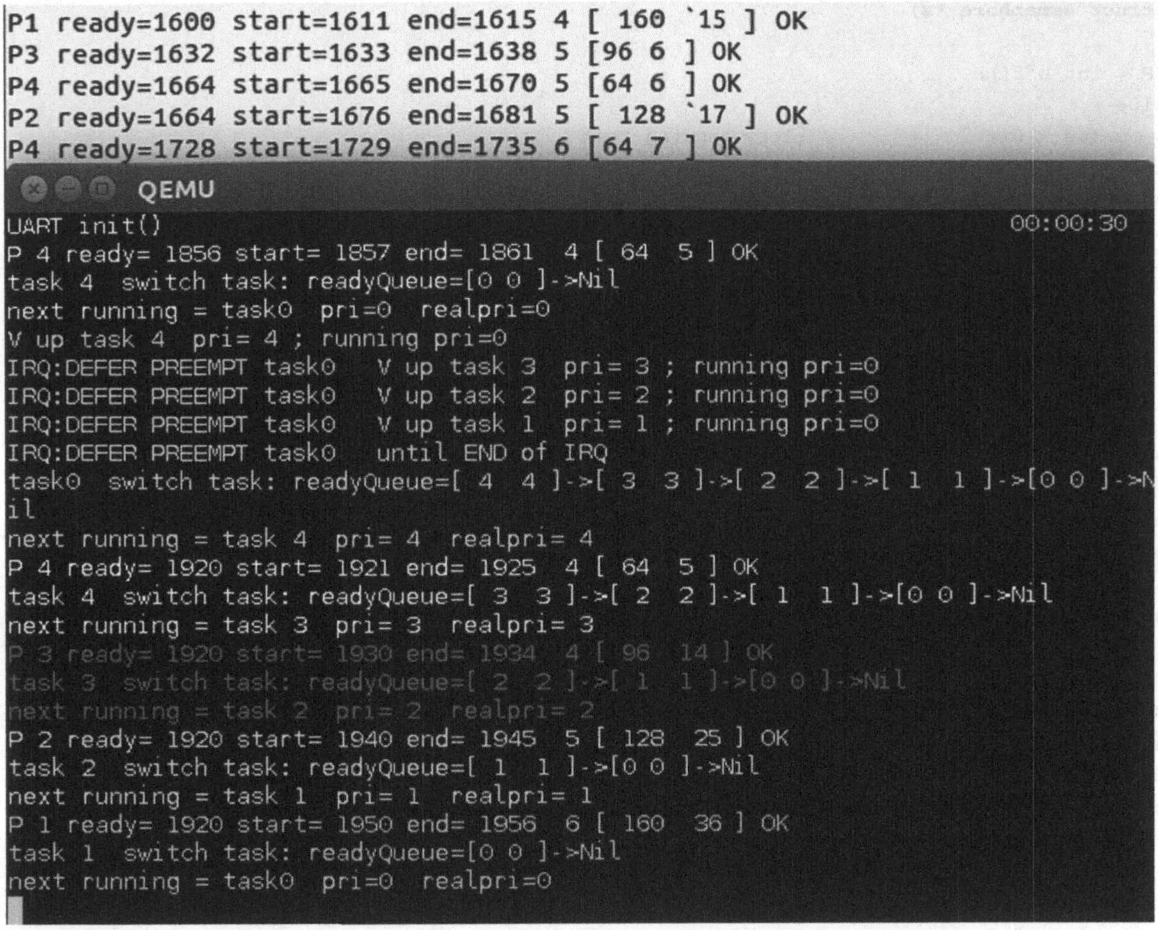

```
P1 ready=1600 start=1611 end=1615 4 [ 160 `15 ] OK
P3 ready=1632 start=1633 end=1638 5 [96 6 ] OK
P4 ready=1664 start=1665 end=1670 5 [64 6 ] OK
P2 ready=1664 start=1676 end=1681 5 [ 128 `17 ] OK
P4 ready=1728 start=1729 end=1735 6 [64 7 ] OK
QEMU
UART init()                                                        00:00:30
P 4 ready= 1856 start= 1857 end= 1861  4 [ 64  5 ] OK
task 4  switch task: readyQueue=[0 0 ]->Nil
next running = task0  pri=0  realpri=0
V up task 4  pri= 4 ; running pri=0
IRQ:DEFER PREEMPT task0   V up task 3  pri= 3 ; running pri=0
IRQ:DEFER PREEMPT task0   V up task 2  pri= 2 ; running pri=0
IRQ:DEFER PREEMPT task0   V up task 1  pri= 1 ; running pri=0
IRQ:DEFER PREEMPT task0   until END of IRQ
task0  switch task: readyQueue=[ 4  4 ]->[ 3  3 ]->[ 2  2 ]->[ 1  1 ]->[0 0 ]->N
il
next running = task 4  pri= 4  realpri= 4
P 4 ready= 1920 start= 1921 end= 1925  4 [ 64  5 ] OK
task 4  switch task: readyQueue=[ 3  3 ]->[ 2  2 ]->[ 1  1 ]->[0 0 ]->Nil
next running = task 3  pri= 3  realpri= 3
P 3 ready= 1920 start= 1930 end= 1934  4 [ 96  14 ] OK
task 3  switch task: readyQueue=[ 2  2 ]->[ 1  1 ]->[0 0 ]->Nil
next running = task 2  pri= 2  realpri= 2
P 2 ready= 1920 start= 1940 end= 1945  5 [ 128  25 ] OK
task 2  switch task: readyQueue=[ 1  1 ]->[0 0 ]->Nil
next running = task 1  pri= 1  realpri= 1
P 1 ready= 1920 start= 1950 end= 1956  6 [ 160  36 ] OK
task 1  switch task: readyQueue=[0 0 ]->Nil
next running = task0  pri=0  realpri=0
```

Fig. 10.3 UP_RTOS with periodic tasks and preemptive scheduling

(1). ts.s file of C10.3: the ts.s file is identical to that of C10.1
(2). Kernel C files of C10.3

For brevity, we only show the modified kernel functions in support of task preemption. These include V operation on semaphores and mutex_unlock, which may make a blocked task ready to run and change the readyQueue. In addition, kfork may create a new task with a higher priority than the current running task. All these functions call reschedule(), which may switch task immediately or defer task switch until the end of IRQ interrupts processing.

```
int reschedule()
{
  int SR = int_off();
  int pid = running->pid;
  if (readyQueue && readyQueue->priority > running->priority){
    if (intnest==0){ // not in IRQ handler: preempt immediately
       printf("%d PREEMPT %d IMMEDIATELY\n", readyQueue->pid, pid);
       tswitch();
    }
    else{              // still in IRQ handler: defer preemption
      printf("%d DEFER PREEMPT %d\n", readyQueue->pid, running->pid);
      swflag = 1;      // set need to switch task flag
    }
  }
  int_on(SR);
}
```

```
int P(struct semaphore *s)
{
  int SR = int_off();
  s->value--;
  if (s->value < 0){              // block running task
     running->status = BLOCK;
     enqueue(&s->queue, running);
     running->status = BLOCK;
     enqueue(&s->queue, running);
     // priority inheritance
     if (running->priority > s->owner->priority){
        s->owner->priority = running->priority; // boost owner priority
        reorder_readyQueue(); // re-order readyQueue
     }
     tswitch();
  }
  s->owner = running;             // as owner of semaphore
  int_on(SR);
}
int V(struct semaphore *s)
{
  PROC *p; int cpsr;
  int SR = int_off();
  s->value++;
  s->owner = 0;                   // clear  s owner field
  if (s->value <= 0){             // s queue has waiters
     p = dequeue(&s->queue);
     p->status = READY;
     s->owner = p;                // new owner
     enqueue(&readyQueue, p);
     running->priority = running->rpriority; // restore owner priority
     printf("timer: V up task%d pri=%d; running pri=%d\n",
             p->pid, p->priority, running->priority);
     resechedule();
  }
  int_on(SR);
}
int mutex_lock(MUTEX *s)
{
  PROC *p;
  int SR = int_off();
  printf("task%d locking mutex %x\n", running->pid, s);
  if (s->lock==0){ // mutex is in unlocked state
     s->lock = 1;
     s->owner = running;
  }
  else{ // mutex is already locked: caller BLOCK on mutex
     running->status = BLOCK;
     enqueue(&s->queue, running);
     // priority inheritance
     if (running->priority > s->owner->priority){
        s->owner->priority = running->priority; // boost owner priority
        reorder_readyQueue(); // re-order readyQueue
     }
     tswitch(); // switch task
  }
```

```
  int_on(SR);
}
int mutex_unlock(MUTEX *s)
{
  PROC *p;
  int SR = int_off();
  printf("task%d unlocking mutex\n", running->pid);
  if (s->lock==0 || s->owner != running){ // unlock error
     int_on(SR); return -1;
  }
  // mutex is locked and running task is owner
  if (s->queue == 0){ // mutex has no waiter
     s->lock = 0;      // clear lock
     s->owner = 0;     // clear owner
  }
  else{ // mutex has waiters: unblock one as new owner
     p = dequeue(&s->queue);
     p->status = READY;
     s->owner = p;
     enqueue(&readyQueue, p);
     running->priority = running->rpriority; // restore owner priority
     resechedule();
   }
   int_on(SR);
   return 1;
}
PROC *kfork(int func, int priority)
{
  // create new task p with priority as before
  p->rpriority = p->priority = priority;
  // with dynamic tasks, readyQueue must be protected by a mutex
  mutex_lock(&readyQueuelock);
   enqueue(&readyQueue, p); // enter new task into readyQueue
  mutex_unlock(&readyQueuelock);
  reschedule();
  return p;
}
```

t.c file of C10.3

```
/**************** t.c file of C10.3 ***************/
#include "type.h"
MUTEX *mp;                    // global mutex
struct semaphore s1;          // global semaphore
#include "queue.c"
#include "mutex.c"            // modified mutex operations
#include "pv.c"               // modified P/V operations
#include "kbd.c"
#include "uart.c"
#include "vid.c"
#include "exceptions.c"
#include "kernel.c"
#include "timer.c"
#include "sdc.c"              // SDC driver
#include "mesg.c"             // message passing
```

```
int klog(char *line){ send(line, 2) } ;// send msg to log task #2
int endIRQ(){ printf("until END of IRQ\n") }
void copy_vectors(void) { // same as before }

// use vectored interrupts of PL190
int timer0_handler(){ timer_handler(0) }
int uart0_handler() { uart_handler(&uart[0]) }
int uart1_handler() { uart_handler(&uart[1]) }
int v31_handler()
{
  int sicstatus = *(int *)(SIC_BASE_ADDR+0);
  if (sicstatus & (1<<3)) { kbd_handler() }
  if (sicstatus & (1<<22)){ sdc_handler() }
}
int vectorInt_init()      // add KBD and SDC interrupts
{
  printf("vectorInt_init() ");
  /***** write to vectorAddr regs ***************************/
  *((int *)(VIC_BASE_ADDR+0x100)) = (int)timer0_handler;
  *((int *)(VIC_BASE_ADDR+0x104)) = (int)uart0_handler;
  *((int *)(VIC_BASE_ADDR+0x108)) = (int)uart1_handler;
  *((int *)(VIC_BASE_ADDR+0x10C)) = (int)kbd_handler;
  *((int *)(VIC_BASE_ADDR+0x10C)) = (int)v31_handler;
  *((int *)(VIC_BASE_ADDR+0x110)) = (int)sdc_handler;
  /***** write to intControlRegs: E=1|IRQ# = 1xxxxx *********/
  *((int *)(VIC_BASE_ADDR+0x200)) = 0x24; //0100100 at IRQ 4
  *((int *)(VIC_BASE_ADDR+0x204)) = 0x2C; //0101100 at IRQ 12
  *((int *)(VIC_BASE_ADDR+0x208)) = 0x2D; //0101101 at IRQ 13
  *((int *)(VIC_BASE_ADDR+0x20C)) = 0x3F; //0111111 at IRQ 31
  *((int *)(VIC_BASE_ADDR+0x210)) = 0x36; //0110110 at IRQ 22
  /***** write 0 to IntSelectReg to generate IRQ interrupts */
  *((int *)(VIC_BASE_ADDR+0x0C)) = 0;
}
int irq_chandler()
{
  int (*f)();                              // f is a function pointer
  f =(void *)*((int *)(VIC_BASE_ADDR+0x30)); // read vectorAddr reg
  f();                                     // call the ISR function
  *((int *)(VIC_BASE_ADDR+0x30)) = 1; // write to vectorAddr as EOI
}

int delay(){ // delay loop to simulate task computation }

int task5()
{
   printf("task%d running: ", running->pid);
   klog("start");
   mutex_lock(mp);
    printf("task%d inside CR\n",  running->pid);
    klog("inside CR");
   mutex_unlock(mp);
   klog("terminate");
   kexit(0);
}
```

```
int task4()
{
    klog("create task5");
    kfork((int)task5, 5);  // create P5 with priority=5
    delay();
    printf("task%d terminte\n", running->pid);
    klog("terminate");
    kexit(0);
}
int task3()                    // Ved up by timer periodically
{
  int pid, status;
  while(1){
    printf("task%d waits for timer event\n", running->pid);
    P(&s1);
    klog("running");
    mutex_lock(mp);
     printf("task%d inside CR\n", running->pid);
     klog("create task4");
     kfork((int)task, 4);  // create P4 while holding mutex lock
     delay();
    mutex_unlock(mp);
    pid = kwait(&status);
  }
}
int task2()  // logging task at priority 1
{
   int blk = 0;
   char line[1024];           // 1KB log line
   printf("task%d: ", running->pid);
   while(1){
     //printf("task%d:recved line=%s\n", running->pid, line);
     r = recv(line);
     put_block(blk, line);    // write log line to SDC
     blk++;
     uprintf("%s\n", line);   // display to UART0 on-line
     printf("log: ");
   }
}
int task1()
{
   int status, pid;
   pid = kfork((int)task2, 2);
   pid = kfork((int)task3, 3);
   while(1)  // task1 waits for any ZOMBIE children
      pid = kwait(&status);
}
int main()
{
   fbuf_init();               // LCD
   uart_init();               // UARTs
   kbd_init();                // KBD
   printf("Welcome to UP_RTOS in ARM\n");
```

```
    /* configure VIC and SIC for interrupts */
    VIC_INTENABLE = 0;
    VIC_INTENABLE |= (1<<4);    // timer0,1 at bit4
    VIC_INTENABLE |= (1<<12);   // UART0 at bit12
    VIC_INTENABLE |= (1<<13);   // UART2 at bit13
    VIC_INTENABLE |= (1<<31);   // SIC to VIC's IRQ31
    UART0_IMSC = 1<<4;          // enable UART0 RXIM interrupt
    SIC_INTENABLE = (1<<3);     // KBD int=bit3 on SIC
    SIC_INTENABLE |= (1<<22);   // SDC int=bit22 on SIC
    SIC_ENSET = (1<<3);         // KBD int=3 on SIC
    SIC_ENSET |= 1<<22;         // SDC int=22 on SIC
    vectorInt_init();           // initialize vectored interrupts
    timer_init();
    timer_start(0);             // timer0
    sdc_init();                 // initialize SDC driver
    msg_init();                 // initialize message buffers
    mp = mutex_create();        // mutex and semaphore
    s1.value = 0; s1.queue = 0;
    kernel_init();              // initialize kernel, run P0
    kfork((int)task1, 1);       // create P1 with priority=1
    while(1){                   // P0 loop
      if (readyQueue)
         tswitch();
    }
}
```

When the C10.3 system starts, it creates and runs the initial task P0, which has the lowest priority 0. P0 creates a new task P1 with a priority=2. Since P1 has a higher priority, it preempts P0 directly, which demonstrates immediate task preemption without any delay. When P1 runs, it creates a child task P2 with a priority=1 as the logging task. In this case, it does not switch process. Then it creates another child task P3 with a priority=3. Since P3 has a higher priority, it immediately preempts P1. When P1 runs again, it executes in a loop, waiting for any ZOMBIE child and frees the child PROC for reuse. When P3 runs, it waits for a timer event, which will be V-ed up periodically by the timer. When P3 runs again, it locks up a mutex and creates P4 with priority=4, which immediately preempts P3. P4 creates a child P5 with priority 5, which immediately preempts P4. When P5 runs, it tries to acquire the same mutex lock, which is still held by P3. So P5 becomes blocked on the mutex but it raises the priority of P3 to 5, thus preventing priority inversion. When P3 unlocks the mutex, it unblocks P5 and restores its own priority back to 3, allowing P4 to run next rather than P3. These demonstrate priority inheritance in UP_RTOS. After unlocking the mutex, P3 has two options. It may repeat the loop or terminate. If P3 repeats the loop, it must wait for its child to terminate and release child PROC structure for reuse. Otherwise, the number of available PROC structures would be one less each time P3 runs. Alternatively, P3 may terminate. In that case, P1 will dispose of all the terminated PROCs and creates a new instance of P3 again. In the demonstration system C10.3, task P2 is the logging task. It runs with the second lowest priority so that it does not impede the executions of other tasks. P2 repeatedly try to receive messages, which are logging requests from other tasks. For each longing request, P2 formats the log request into the form

timestamp : pid:log information

and writes the line to a 1KB block in a SDC by invoking the SDC driver directly. Figure 10.4 shows the sample outputs of running the UP_RTOS system C10.3.

The top part of Fig. 10.4 shows the on-line log information displayed by the logging task to a UART port. The logging information are also saved to a SDC as permanent records. In the demonstration program, the SDC is a virtual SD, which is not an ordinary file. Its contents can be read by the get_block() operation of the SDC driver for inspection.

Schedulability and Deadline Analysis: The schedulability of RTOS comprising tasks sharing resources has been studied by many people. With resource sharing, task priorities are no longer static. So such analyses do not apply to the simple RMS model, but they are applicable to both EDF and DMS models. The following is a (simplified) schedulability analysis based on the work of (Audsley et al. 1993). First, we define the following terms.

```
00:00:05 : 3 : running
00:00:05 : 3 : create task4
00:00:05 : 4 : create task5
00:00:05 : 5 : start
00:00:06 : 5 : inside CR
00:00:06 : 5 : terminate
00:00:07 : 4 : terminate
00:00:07 : 3 : release ZOMBIE child 4
00:00:07 : 1 : release ZOMBIE child 5
```

```
QEMU
welcome to RTOS_UP in ARM
vectorInt_init() timer_init(): timer_start 0                          00:00:09
sdc_init : mesg_init() kernel_init(): task 0 running
fs_init(): mount root: EXT2 MAGIC=0x EF53 bmap= 9 imap= 10 iblk= 11 OK
task0 create a child task 1 : task 1 PREEMPT task0 IMMEDIATELY
task 1 create a child task 2 : task 1 create a child task 3 : task 3 PREEMPT
task 1 IMMEDIATELY
task 3 waits for timer event
task 2 : open: filename=log: found log ino= 76 opened fd = 0x 497E8
V up task 3 pri= 3 ; running pri=0
task 3 DEFER PREEMPT task0 until END of IRQ
V up task 2 pri= 1 ; running pri= 3
task 3 inside CR
task 3 create a child task 4 : task 4 PREEMPT task 3 IMMEDIATELY
task 4 create a child task 5 : task 5 PREEMPT task 4 IMMEDIATELY
task 5 running: TASK 5 BLOCK ON MUTEX
RAISE PROC 3 PRIORITY to 5
3 mutex_unlock: new owner=task 5 restore task 3 priorit to 3
task 5 PREEMPT task 3 IMMEDIATELY
task 5 inside CR
task 4 terminte
task 3 free ZOMBIE child 4
task 3 waits for timer event
task 1 free ZOMBIE child 5
log:log:log:log:log:log:log:log:log:
```

Fig. 10.4 UP_RTOS with dynamic tasks sharing resources

Ci = the worst-case computation time of task i on each release
Ti = the lower bound on the time between successive arrivals of task i. For periodic tasks, Ti is equal to the task period
Di = the deadline requirement of task i, measured relative to a given release time of task i
Bi = the worst case blocking time task i can experience due to priority inheritance. Bi is normally equal to the longest critical section of lower priority tasks accessing semaphores with ceilings higher than or equal to the priority of task i
Ii = the worst case interference a task i can experience from other tasks. Interference on task i is defined as the time higher priority tasks can preempt task i and execute, hence prevent task i from executing
Ri = the worst case response time for a task i measured from the time the task is released. For a schedulable task, Ri < Di or Ri < Ti, if deadline Di is not specified
hp(i) = the set of tasks of higher base priority than the base priority of task i, i.e. tasks which could preempt task i

Based on the terms defined above, the response time of task i can be expressed as

$$\mathbf{Ri = Ci + Bi + Ii} \tag{10.4}$$

The worst case computation time **Ci** can be determined by running each task in an isolated environment. The worst case blocking time **Bi** is equal to the longest critical section of any lower priority task accessing a semaphore or mutex. The worst case interference of task i from task j is [Ri/Tj]Cj, which accounts for the time task i being blocked by task j due to task preemption. The total interference of task i from all other tasks j is given by

$$\mathbf{Ii} = \underset{\mathbf{j\ in\ hp(i)}}{\mathbf{SUM}}\{[\mathbf{Ri/Tj}]\mathbf{Cj}\}; \tag{10.5}$$

Combining equations (10.4) and (10.5) yields the (recurrence) equation

$$\mathbf{Ri(n+1) = Ci + Bi +} \underset{\mathbf{j\ in\ hp(i)}}{\mathbf{SUM}}\{[\mathbf{Ri(n)/Tj}]\mathbf{Cj}\} \tag{10.6}$$

The recurrence equation can be solved by iteration. The iteration starts with Ri(0) = 0 and terminates when Ri(n+1) = Ri (n). The iteration can be terminated earlier if either Ri(n+1) > Di or Ri(n+1) > Ti, indicating that taski will fail the deadline. Furthermore, it has been shown (Joseph and Pandya 1986) that the iteration is guaranteed to converge if the CPU utilization is <100%. After computing the response times of all the tasks, it is a simple matter to test whether they can meet their deadlines. All the tasks can meets their deadlines if

$$\mathbf{Ri < Di\ or\ Ri < Ti(for\ periodic\ tasks)\ for\ all\ i}$$

The analytic results can be used as a guideline when designing a RTOS with resource sharing. To determine whether tasks in a real RTOS can meet their deadlines must still be verified empirically. For the example RTOS system C10.3, we leave the schedulability analysis and verification of task deadlines as an exercise in the Problem section.

10.8 Multiprocessor RTOS (SMP_RTOS)

In Chap. 9, we discussed SMP operations for the ARM MPcore architecture in detail. In this section, we shall show how to design and implement a RTOS for SMP systems. While the design principle is general, which is applicable to any SMP system, we shall use the ARM A9-MPcore architecture as the platform for implementation and testing. Rather than presenting a complete system in one step, we shall develop the SMP_RTOS system in incremental steps. In each step, we shall focus on a problem that is unique to real-time operations in a SMP environment, and demonstrate the solution by a sample system. The steps are:

(1). Develop a SMP kernel for task management with improved concurrency
(2). Adapt UP_RTOS for SMP
(3). Implement preemptive task scheduling by priority in SMP
(4). Implement nested interrupts in SMP
(5). Implement priority inheritance in SMP
(6). Integrate the results into a complete SMP_RTOS

10.8.1 SMP_RTOS Kernel for Task Management

First, we show the design and implementation of an initial SMP_ROTS kernel for task management. The initial SMP_RTOS kernel only supports dynamic task creation, task termination and task synchronization. It should exploit the parallel processing capability of multiple CPUs to improve concurrency. To begin with, we shall assume the following operating environment of tasks.

. Each task is a single execution entity. There are no additional threads within a task.
. All tasks execute in the same address space, so no need for virtual address mapping.
. Tasks do not use signals for inter-task communication and do not need file system support, so they do not have any resources, as in a general purpose OS.

These are reasonable assumptions for real-time applications, which lead to a simplified PROC structure for real-time tasks. The following shows the simplified PROC structure in the SMP_RTOS kernel.

10.8.1.1 PROC Structure in SMP_RTOS

```
#define SSIZE 1024
typedef struct proc{
  struct proc *next;
  int     *ksp;               // saved stack pointer
  int     status;             // FREE|READY|BLOCK|ZOMBIE, etc.
  int     priority;           // effective priority
  int     pid;
  int     ppid;
  struct proc *parent;
  int     exitCode;           // exit code
  int     ready_time;
  struct mbuf *mqueue;        // message queue
  struct mutex mlock;         // message queue mutex lock
  struct semaphore nmsg;      // wait for message semaphore
  int     cpuid;              // which CPU this proc is running on
  int     rpriority;          // for priority inheritance
  int     timeslice;          // for time sliced scheduling
  struct semaphore wchild;    // wait for ZOMBIE children
  int     kstack[SSIZE];
}PROC;
```

In the simplified PROC structure, the fields of virtual address space, user mode context and resources, e.g. file descriptors and signals, are all eliminated. For clarity, the added fields for real-time operations are shown in bold faced lines.

10.8.1.2 Kernel Data Structures: Kernel data structures are defined as follows.

```
#define NCPU     4
#define NPROC 256
```

Each task is represented by a PROC structure. The total number of PROC structures, NPROC, is a configuration parameter, which can be set to suit the actual system needs. Initially, all the PROC structures are in a freeList. When creating a task, we allocate a free PROC from the freeList. When a task terminates, its PROC structure is eventually released back to the freeList for reuse. The freeList is protected by a spinlock during allocation/deallocation of free PROCs.

```
PROC *freeList;     // contain all free PROCs
int freelock = 0; // spinlock for freeList
```

In a UP kernel, it suffices to use a single running pointer pointing at the current executing PROC. In a SMP kernel, we shall identify the tasks executing on different CPUs by the following means.

```
#define running run[get_cpuid()]
PROC *run[NCPU];    // pointer to running PROC on each CPU
```

In a UP kernel, it suffices to use a single queue for task scheduling. In a SMP kernel, a single scheduling queue can be a bottleneck, which severely restricts the concurrency of the system. To improve both concurrency and efficiency, we assume that each CPU has a separate scheduling queue protected by a spinlock.

```
PROC *readyQueue[NCPU];   // per CPU task scheduling queues
int  readylock[NCPU];     // spinlocks for readyQueue[cpuid]
```

Each CPU always tries to run tasks from its own readyQueue[cpuid]. If a CPU's readyQueue is empty, it runs a special idle task, which puts the CPU in a power-saving state, waiting for events or interrupts. When the system starts, each CPU runs a special initial task, which is also the idle task on the CPU.

```
PROC iproc[NCPU];          //  per CPU initial/idle task
```

Tasks normally run in Supervisor (SVC) mode, which use the stack in the PROC structure. A task may enter IRQ mode to handle interrupts, or ABT mode to handle data abort exceptions, etc. Each CPU must have separate stacks for interrupts and exceptions processing. Thus, we define per CPU IRQ and exception mode stacks as

```
int irq_stack[NCPU][1024];  // per CPU IRQ mode stack
int abt_stack[NCPU][1024];  // per CPU ABT mode stack
int und_stack[NCPU][1024];  // per CPU UND mode stack
```

In a ROTS kernel, preemptive task scheduling and nested interrupts are essential. With preemptive task scheduling, each CPU needs a switch task flag. With nested interrupts, each CPU must keep track of the interrupt nesting level. Task switch is not allowed while the CPU is executing inside any interrupt handler. Task switch may occur only if all the nested interrupts processing have ended. Thus, we define

```
volatile int intnest[NCPU]; // per CPU interrupt nesting counter
volatile int swflag[NCPU];  // per CPU switch task flag
```

Since each of these variables is accessed only by a single CPU, either from a task running on a CPU or by an interrupt handler on the same CPU, it suffices to protect them by the usual disable/enable interrupts of the CPU.

10.8.1.3 Protection of Kernel Data Structures

Each kernel data structure is protected by a spinlock. Requests for spinlocks are mostly single level. When tasks must acquire multiple spinlocks, we shall ensure that the locking order is always unidirectional, so that deadlocks can not occur.

10.8.1.4 Synchronization Tools

The SMP_RTOS kernel supports the following tools for task synchronization and inter-task communication.

Spinlocks: spinlocks are used to protect critical regions (CRs) of short durations and also in device interrupt handlers.

Mutexes: mutexes are for protecting CRs of long durations. They are implemented with priority inheritance to prevent priority inversion.

Semaphores: semaphores are used only for process cooperation. They are implemented with priority inheritance also.

Shared Memory: tasks execute in the same address space, so they may use shared memory for inter-task communication. Each shared memory area is protected by a mutex lock to ensure exclusive access.

Messages: tasks may send/recv messages through the per-task message queue.

10.8.1.5 Task Management

tasks can be created dynamically to execute a function with a static priority on a specific CPU. Task creation is by the API

```
int pid = kfork((int)func, int priority, int cpu)
```

To be consistent with the traditional Unix/Linux kernel, tasks obey the usual parent-child relation. When a task has completed its work, it may terminate by the API

```
void kexit(int exitValue)
```

When a task with children terminates, it first releases its own terminated children, if any, and send other children to a special task P1, which behaves like the INIT process in a Unix/Linux kernel. A terminated task becomes a ZOMBIE and notifies its parent, which may be the original parent or the adopted parent P1. A parent task may wait for children termination by the API

```
int kwait(int *status)
```

which returns the terminated child pid, its exit status and releases the ZOMBIE child PROC for reuse. Synchronization between the parent and child is by a semaphore in the parent PROC, which prevents race conditions between the parent and the child tasks even if they run in parallel on different CPUs.

Next, we show the implementation of the task management part of the SMP_RTOS kernel. As usual, the kernel code consists of a ts.s file and a set of files in C. The kernel files are compile-linked to a binary executable t.bin, which is run directly on the emulated realview-pbx-a9 VM under QEMU by a sh script.

```
arm-none-eabi-as -mcpu=cortex-a9 ts.s -o ts.o
arm-none-eabi-gcc -c -mcpu=cortex-a9 t.c -o t.o
arm-none-eabi-ld -T t.ld ts.o t.o -Ttext=0x10000 -o t.elf
arm-none-eabi-objcopy -O binary t.elf t.bin
qemu-system-arm -M realview-pbx-a9 -smp 4 -m 512M -sd ../sdc \
              -kernel t.bin -serial mon:stdio
```

Alternatively, we may also write t.bin as a kernel image to the boot/ directory in a SDC file system and use a separate booter to boot up the kernel image from the SDC.

(1). Startup Sequence: When the realview-pbx-a9 VM starts, CPU0 is the boot processor and all other secondary CPUs (APs) are held in a WFI state, waiting for SGIs from the boot processor to truly start up. The boot processor starts to execute reset_handler in the ts.s file. The actions of CPU0 are

. Set SVC mode stack pointer to the high end of iproc[0], which makes iproc[0].kstack as the initial stack.
. Set IRQ mode stack pointer to the high end of irq_stack[0], which makes irq_stack[0] as its IRQ mode stack. Similarly, it sets up the ABT and UND modes stacks.
. Copy vector table to address 0, then call main() in C to initialize the kernel.

The following shows the reset_handler code of CPU0 and the startup code of the APs.

(2). Reset_handler and apStart Code in Assembly

```
.global reset_handler, apStart           // CPU0, AP code
.global iproc, procsize                  // globals
.global irq_stack, abt_stack, und_stack  // globals
.global tswitch, scheduler
.global int_on, int_off, lock, unlock
.global get_cpsr, get_spsr;
.global slock, sunlock, get_cpuid
.global vectors_start, vectors_end

  reset_handler:

// CPU0: set SVC stack to HIGH END of iproc[0].kstack[]
  ldr r0, =iproc       // r0 points to proc's
  ldr r1, =procsize    // r1 -> procsize
```

```
  ldr r2, [r1, #0]      // r2 = procsize
  add r0, r0, r2
  mov sp, r0            // SVC sp -> high end of iproc[0]
// CPU0: go in IRQ mode to set IRQ stack
  msr cpsr, #0x92
  ldr sp, =irq_stack    // point at int irq_stack[0][1024]
  add sp, sp, #4096     // high end of irq_stack[0]
// CPU0: go in ABT mode to set ABT stack
  msr cpsr, #0x97
  ldr sp, =abt_stack
  add sp, sp, #4096
// CPU0: go in UND mode to set UND stack
  msr cpsr, #0x9B
  ldr sp, =und_stack
  add sp, sp, #4096
// CPU0: go back in SVC mode to set SPSR to SVC mode with IRQ on
  msr cpsr, #0x93
  msr spsr, #0x13
// CPU0: copy vector table to address 0
  bl copy_vectors
  bl  main              // call main() in C
  b   .                 // hang if main() ever returns
apStart:                // AP entry point
  LDR r0, =iproc        // r0 points to iproc
  LDR r1, =procsize     // r1 -> procsize
  LDR r2, [r1, #0]      // r2 = procsize
// get CPU ID
  MRC p15, 0, r1, c0, c0, 5   // read CPU ID register to r1
  AND r1, r1, #0x03           // mask in CPU ID bits only
  mul r3, r2, r1        // procsize*CPUID: r3->high end of iproc[CPUID]
  add r0, r0, r3        // r0 += r3
  mov sp, r0            // sp->iproc[CPUID] high end
//AP: go in IRQ mode to set IRQ stack
  MSR cpsr, #0x92
  ldr r0, =irq_stack    // int irq_stack[4][1024] in C
  lsl r2, r1, #12       // r2 = r1*4096
  ADD r0, r0, r2        // r0 -> high end svcstack[cpuid]
  MOV sp, r0            // IRQ sp of CPU ID -> highend of irq[ID][1024]
//AP: go in ABT mode to set ABT stack
  MSR cpsr, #0x97
  LDR sp, =abt_stack
  lsl r2, r1, #12       // r2 = r1*4096
  ADD r0, r0, r2        // r0 -> high end svcstack[cpuid]
  MOV sp, r0            // IRQ sp of CPU ID -> highend of irq[ID][4096]
//AP:  go in UND mode to set UND stack
  MSR cpsr, #0x9B
  LDR sp, =und_stack
  lsl r2, r1, #12       // r2 = r1*4096
  ADD r0, r0, r2        // r0 -> high end svcstack[cpuid]
  MOV sp, r0            // IRQ sp of CPU ID -> highend of irq[ID][4096]
//AP: go back in SVC mode to set SPSR
  MSR cpsr, #0x93
  MSR spsr, #0x13
```

```
//AP: call APstart() with IRQ on
  MSR cpsr, #0x13
  bl APstart              // call APstart() in C
```

(3). The main() and APstart() functions in C

In main(), CPU0 first initializes the device drivers, configures the GIC for interrupts, initialize the kernel data structures and runs the initial task with pid=1000. Then it issues SGI to activate the APs. After all the APs are ready, it creates a P0 as the logging task and the INIT task P1. Then it execute run_task(), trying to run tasks form the readyQueue of the same CPU. Upon receiving a SGI from CPU0, each AP starts to execute the apStart code in assembly. The actions of each AP are similar to those of CPU0, i.e. it sets SVC mode stack pointer to the high end of iproc[cpuid] and IRQ and other modes stacks by cpuid. Then it calls APstart() in C. In APstart(), each AP configures and starts its local timer, updates the global variable ncpu to synchronize with CPU0. Then it calls run_task() also. At this moment, the readyQueues of the APs are still empty. So each AP runs an idle task with pid=1000+cpuid and enters the WFE state. After creating tasks on the APs, there are two ways to let the APs continue. CPU0 may send a SGI to all the APs, causing them to get up from the WFE state, or each AP will get up when its local timer interrupts. The following shows the code segments of run_task(), APstart() and main().

```
int run_task()  // each CPU tries to run task from its own readyQ
{
   int cpuid = get_cpuid();
   while(1){
      slock(&readylock[cpuid]);
      if (readyQueue[cpuid] == 0){  // check own readyQueue
         sunlock(&readylock[cpuid]);
         asm("WFE");                // wait for events/interrupts
      }
      else{
         sunlock(&readylock[cpuid]);
         ttswitch();                // switch task on CPU
     }
  }
}
int APstart()
{
   int cpuid = get_cpuid();
   slock(&aplock);
   color = YELLOW;
   printf("CPU%d in APstart ... ", cpuid);
   config_int(29, cpuid);          // configure local timer
   ptimer_init(); ptimer_start(); // start local timer
   ncpu++;
   running = &iproc[cpuid];        // run initial iproc[cpuid]
   printf("%d running on CPU%d\n", running->pid, cpuid);
   sunlock(&aplock);
   run_task();
}
int main()
{
   // initialize device drivers
   // configure GIC for interrupts
   kernel_init(); //initialize kernel
   printf("CPU0 startup APs: ");
   int *APaddress = (int *)0x10000030;
```

```
    *APaddress = (int)apStart;
    send_sgi(0x0, 0x0F, 0x01);   // send SGI 0 to all Aps
    printf("CPU0 waits for APs ready\n");
    while(ncpu < 4);
    printf("CPU0 continue\n");
    kfork(int)logtask, 1, 0); // create P0 on CPU0 as logging task
    kfork((int)f1,      1, 0); // create P1 on CPU0 as INIT task
    run_task();
}
```

10.8.1.6 Demonstration of Task Management in SMP_RTOS

We demonstrate task management in the initial SMP_RTOS kernel by a sample program, denoted by C10.4. Task management includes task creation, task termination and wait for child task termination. The program consists of the following components.

(1). ts.s file of C10.4:

```
/*********** ts.s file of C10.4: SMP_RTOS kernel **********/
        .text
        .code 32
.global reset_handler, vectors_start, vectors_end
.global proc, procsize, apStart
.global tswitch, scheduler
.global int_on, int_off, lock, unlock
.global get_fault_status, get_fault_addr, get_cpsr, get_spsr;
.global slock, sunlock, get_cpuid
reset_handler:
   // same as shown above
apStart:                      // AP entry point
   // same as shown above
irq_handler:                  // IRQ interrupts entry point
  sub  lr, lr, #4
  stmfd sp!, {r0-r12, lr}  // save context
  bl  irq_chandler          // call irq_handler()
  ldmfd sp!, {r0-r12, pc}^ // return
data_handler:
  sub lr, lr, #4
  stmfd sp!, {r0-r12, lr}
  bl  data_abort_handler
  ldmfd sp!, {r0-r12, pc}^
tswitch:                      // tswitch() in SVC mode
  msr   cpsr, #0x93         // interrupts off
  stmfd sp!, {r0-r12, lr}
  ldr r4, =run               // r4=&run
  mrc p15, 0, r5, c0, c0, 5 // read CPUID register to r5
  and r5, r5, #0x3           // only the CPU ID
  mov r6, #4                 // r6 = 4
  mul r6, r5                 // r6 = 4*cpuid
  add r4, r6                 // r4 = &run[cpuid]
  ldr r6, [r4, #0]           // r6->running PROC
  str sp, [r6, #4]           // save sp in PROC.ksp
  bl  scheduler              // call scheduler() in C
```

```
  ldr r4, =run                  // r4=&run
  mrc p15, 0, r5, c0, c0, 5 // read CPUID register to r5
  and r5, r5, #0x3              // only the CPU ID
  mov r6, #4                    // r6 = 4
  mul r6, r5                    // r6 = 4*cpuid
  add r4, r6                    // r4 = &run[cpuid]
  ldr r6, [r4, #0]              // r6->running PROC
  ldr sp, [r6, #4]              // restore ksp
  msr cpsr, #0x13               // interrupts on
  ldmfd sp!, {r0-r12, pc}       // resume new running

int_off:                        // int sr=int_off()
  stmfd sp!, {r1}
  mrs r1, cpsr
  mov r0, r1                    // r0 = r1
  orr r1, r1, #0x80
  msr cpsr, r1                  // mask out IRQ
  ldmfd sp!, {r1}
  mov pc, lr                    // return original cpsr
int_on:                         // int_on(SR)
  mrs cpsr, r0
  mov pc, lr
lock:                           // maks out IRQ in cpsr
  mrs r0, cpsr
  orr r0, r0, #0x80
  msr cpsr, r0
  mov pc, lr
unlock:                         // mask in IRQ in cpsr
  mrs r0, cpsr
  bic r0, r0, #0x80
  msr cpsr, r0
  mov pc, lr
  .global send_sgi
// void send_sgi(int ID, int target_list, int filter_list)
send_sgi:                          // send_sgi with CPUTarget list
  and  r3, r0, #0x0F               // Mask off unused bits of ID to r3
  and  r1, r1, #0x0F               // Mask off unused bits of target_filter
  and  r2, r2, #0x0F               // Mask off unused bits of filter_list
  orr  r3, r3, r1, LSL #16         // Combine ID and target_filter
  orr  r3, r3, r2, LSL #24         // and now the filter list
// Get the address of the GIC
  mrc  p15, 4, r0, c15, c0, 0      // Read periph base address
  add  r0, r0, #0x1F00             // Add offset of the sgi_trigger reg
  str  r3, [r0]                    // Write to SGI Register(ICDSGIR)
  bx   lr
slock:                             // int slock(&spin)
  ldrex r1, [r0]
  cmp   r1, #0x0
  WFENE
  bne   slock
  mov   r1, #1
  strex r2, r1, [r0]
  cmp   r2, #0x0
```

```
  bne   slock
  DMB                           // barrier
  bx    lr
sunlock:                        // sunlock(&spin)
  mov   r1, #0x0
  DMB
  str   r1, [r0]
  DSB
  SEV
  bx    lr
get_fault_status:  // read and return MMU reg 5
  mrc p15,0,r0,c5,c0,0    // read DFSR
  mov pc, lr
get_fault_addr:             // read and return MMU reg 6
  mrc p15,0,r0,c6,c0,0    // read DFSR
  mov pc, lr
get_cpsr:
  mrs r0, cpsr
  mov pc, lr
get_spsr:
  mrs r0, spsr
  mov pc, lr
get_cpuid:
  mrc p15, 0, r0, c0, c0, 5    // Read CPU ID register
  and r0, r0, #0x03            // Mask off, leaving the CPU ID field
  bx  lr
  .global  enable_scu
// void enable_scu(void): Enables the SCU
enable_scu:
  mrc p15, 4, r0, c15, c0, 0  // Read periph base address
  ldr r1, [r0, #0x0]          // Read the SCU Control Register
  orr r1, r1, #0x1            // Set bit 0 (The Enable bit)
  str r1, [r0, #0x0]          // Write back modifed value
  bx  lr
vectors_start:
  LDR PC, reset_handler_addr
  LDR PC, undef_handler_addr
  LDR PC, svc_handler_addr
  LDR PC, prefetch_abort_handler_addr
  LDR PC, data_abort_handler_addr
  B .
  LDR PC, irq_handler_addr
  LDR PC, fiq_handler_addr
reset_handler_addr:           .word reset_handler
undef_handler_addr:           .word undef_handler
svc_handler_addr:             .word svc_entry
prefetch_abort_handler_addr: .word prefetch_abort_handler
data_abort_handler_addr:      .word data_handler
irq_handler_addr:             .word irq_handler
fiq_handler_addr:             .word fiq_handler
vectors_end:
// other SMP utility functions: NOT SHOWN
```

(2). Kernel.c file of SMP_RTOS

```
#define NCPU     4
#define NPROC  256
#define SSIZE 1024
#define FREE      0
#define READY     1
#define BLOCK     2
#define ZOMBIE    3
typedef struct proc{
  struct proc *next;
  int     *ksp;              // saved stack pointer
  int     status;            // FREE|READY|BLOCK|ZOMBIE, etc.
  int     priority;          // real priority
  int     pid;
  int     ppid;
  struct proc *parent;
  int     exitCode;          // exit code
  int     ready_time;
  struct mbuf *mqueue;       // message queue
  struct mutex mlock;        // message queue mutex lock
  struct semaphore nmsg;     // wait for message semaphore
  int     cpuid;             // which CPU this proc is on
  int     epriority;         // effective priority for PI
  int     timeslice;         // for time sliced scheduling
  struct semaphore wchild; // wait for ZOMBIE children
  int     kstack[SSIZE];
}PROC;

#define running run[get_cpuid()]
// kernel data structures and spinlocks
PROC iproc[NCPU];              // per CPU initial/idle PROCs
PROC *run[NCPU];               // pointer to running PROC on each CPU
PROC proc[NPROC];              // PROC structures
PROC *freeList;                // freeList
PROC *readyQueue[NCPU];        // per CPU readyQueue
int readylock[NCPU];           // per CPU readyQueue spinlcok
int intnest[NCPU];             // per CPU IRQ nesting level
int swflag[NCPU];              // per CPU switch task flag
int freeListlock = 0;          // freeList spinlock
int procsize = sizeof(PROC); // PROC size
int irq_stack[NCPU][1024];    // per CPU IRQ mode stack
int abt_stack[NCPU][1024];    // per CPU ABT mode stack
int und_stack[NCPU][1024];    // per CPU UND mode stack

int kernel_init()
{
  int i;
  PROC *p; char *cp;
  printf("kernel_init(): init iprocs\n");
  for (i=0; i<NCPU; i++){     // initialize per CPU initial PROCs
    p = &iproc[i];
    p->pid = 1000 + i;        // special pid for per CPU idle PROCs
    p->status = READY;
```

```
    p->ppid = p->pid;
    run[i] = p;                  // run[i] points to idle iPROC[i]
    readyQueue[i] = 0;           // ready queues
    readylock[i] = 0;            // spinlocks
    intnest[i] = 0;              // IRQ nest levels
    swflag[i] = 0;               // switch task flags
  }
  for (i=0; i<NPROC; i++){
    p = &proc[i];
    p->pid = i;
    p->status = FREE;
    p->priority = 0;
    p->ppid = 0;
    p->next = p + 1;
  }
  proc[NPROC-1].next = 0;      // freeList of PROCs
  freeList = &proc[0];
  sleepList = 0;
  freelock = 0;
  running = run[0];            // initial task on CPU0
}

int ttswitch()                 // switch task on CPUID
{
  int cpuid = get_cpuid();
  lock();
  slock(&readylock[cpuid]);
    tswitch();
  sunlock(&readylock[cpuid]);
  unlock();
}

int scheduler()
{
  int pid; PROC *old=running;
  int cpuid = get_cpuid();
  slock(&readylock[cpuid];
  if (running->pid < 1000){  // regular tasks
     if (running->status==READY)
        enqueue(&readyQueue[cpuid], running);
  }
  running = dequeue(&readyQueue[cpuid]);
  if (running == 0)
     running = &iproc[cpuid];// run idle task on CPU
  swflag[cpuid] = 0;
  sunlock(&readylock[cpuid]);
}

int kfork(int func, int priority, int cpu)
{
  int cpuid = get_cpuid();
  PROC *p = getproc(&freeList);  // allocate a free PROC
  if (p==0){ printf("kfork1 failed\n"); return -1 }
  p->ppid = running->pid;
  p->parent = running;
  p->status = READY;
```

```
  p->priority = priority;
  p->cpuid = cpuid
  for (i=1; i<15; i++)              // all 14 entries = 0
      p->kstack[SSIZE-i] = 0;
  p->kstack[SSIZE-1] = (int)func;  // resume to func
  p->ksp = &(p->kstack[SSIZE-14]); // saved ksp
  slock(&readylock[cpu]);
    enqueue(&readyQueue[cpu], p);  // enter p into rQ[cpu]
  sunlock(&readylock[cpu]);
  if (p->priority > running->priority){
    if (cpuid == cpu){              // if on the same CPU
      printf("%d PREEMPT %d on CPU%d\n", p->pid, running->pid, cpuid);
      ttswitch();
    }
  }
  return p->pid;
}

int P(struct semaphore *s)  // P operation in SMP
{
  int SR = int_off();
  slock(&s->lock);
  s->value--;
  if (s->value < 0){
    running->status = BLOCK;
    enqueue(&s->queue, running);
    // priority inheritance only if on the same CPU
    if (running->priority > s->owner->priority){
       if (s->owner->cpu == running->cpu){
          s->owner->priority = running->priority; //raise owner priority
          reorder_readyQueue(); // re-order readyQueue
       }
    }
    sunlock(&s->lock);
    ttswitch();
  }
  else{
    s->owner = running;       // as owner of s
    sunlock(&s->lock);
  }
  int_on(SR);
}

int V(struct semaphore *s)  // V operation in SMP
{
  PROC *p;
  int SR = int_off();
  slock(&s->lock);
  s->value++;
  s->owner = 0;               // clear s owner field
  if (s->value <= 0){         // s queue has waiters
     p = dequeue(&s->queue);
     p->status = READY;
     s->owner = p;            // new owner
```

```
      slock(&readylock[p->cpuid]); // rQ of original CPU
        enqueue(&readyQueue[p->cpuid], p);
      sunlock(&readylock[p->cpuid]);
      running->priority = running->rpriority; // restore owner priority
      resechedule();
   }
   sunlock(&s->lock);
   int_on(SR);
}
int kexit(int value)  // task termination
{
   int i; PROC *p;
   if (running->pid==1){ return -1 } // P1 never dies
   slock(&proclock);                  // acquire proclock
   for (i=2; i<NPROC; i++){
     p = &proc[i];
     if ((p->status != FREE) && (p->ppid == running->pid)){
       if (p->status==ZOMBIE){
          p->status = FREE;
          putproc(&freeList, p); // release any ZOMBIE child
       }
       else{
          printf("send child %d to P1\n", p->pid);
          p->ppid = 1;
          p->parent = &proc[1];
     }
   }
   sunlock(&proclock);                // release proclock
   running->exitCode = value;
   running->status = ZOMBIE;
   V(&running->parent->wchild);
   ttswitch();
}
int kwait(int *status)      // wait for any ZOMBIE child
{
   int i, nChild = 0;
   PROC *p;
   slock(&proclock);          // acquire proclock
   for (i=2; i<NPROC; i++){
      p = &proc[i];
      if (p->status != FREE && p->ppid == running->pid)
         nChild++;
   }
   sunlock(&proclock);
   if (!nChild){               // no child error
      return -1;
   }
   P(&running->wchild);       // wait for a ZOMBIE child
   slock(&proclock);
   for (i=2; i<NPROC; i++){
       p = &proc[i];
       if ((p->status==ZOMBIE) && (p->ppid == running->pid)){
       *status = p->exitCode;
```

```
        p->status = FREE;
        putproc(&freeList, p);
        sunlock(&proclock);
        return p->pid;
    }
}
```

(3). t.c file of SMP_RTOS

```
/************* t.c file of SMP_RTOS ***********/
#include "type.h"
int aplock = 0;
int ncpu = 1;
#include "string.c"
#include "queue.c"
#include "pv.c"
#include "uart.c"
#include "kbd.c"
#include "ptimer.c"
#include "vid.c"
#include "exceptions.c"
#include "kernel.c"
#include "sdc.c"
#include "message.c"

int copy_vectors(){ // same as before }
#define GIC_BASE 0x1F000000
int config_gic()
{
  // set int priority mask register
  *(int *)(GIC_BASE + 0x104) = 0xFFFF;
  // set CPU interface control register: enable signaling interrupts
  *(int *)(GIC_BASE + 0x100) = 1;
  // distributor control register to send pending interrupts to CPUs
  *(int *)(GIC_BASE + 0x1000) = 1;
  config_int(29, 0);     // ptimer at 29 to CPU0
  config_int(44, 1);     // UAR0    to CPU1
  config_int(49, 2);     // SDC     to CPU0
  config_int(52, 3);     // KBD     to CPU1
}
int config_int(int N, int targetCPU)
{
  int reg_offset, index, value, address;
  reg_offset = (N>>3)&0xFFFFFFFC;
  index = N & 0x1F;
  value = 0x1 << index;
  address   =  (GIC_BASE + 0x1100) + reg_offset;
  *(int *)address |= value;
  reg_offset = (N & 0xFFFFFFFC);
  index = N & 0x3;
  address   =  (GIC_BASE + 0x1800) + reg_offset + index;
  *(char *)address = (char)(1 << targetCPU);
  address   =  (GIC_BASE + 0x1400) + reg_offset + index;
  *(char *)address = (char)0x88; // prioirty=8
}
```

```
int irq_chandler()
{
  // read ICCIAR of CPU interface in the GIC
  int intID = *(int *)(GIC_BASE + 0x10C);
  if (intID == 29){   // local timer of CPU
     ptimer_handler();
  }
  if (intID == 44){   // uart0
     uart_handler(0);
  }
  if (intID == 49){   // SDC
     sdc_handler();
  }
  if (intID == 52){   // KBD
     kbd_handler();
  }
  *(int *)(GIC_BASE + 0x110) = intID; // issue EOI
}
int kdelay(int d){ // delay loop }
int taskCode()
{
  int cpuid = get_cpuid();
  int pid = running->pid;
  while(1){
    printf("TASK%d ON CPU%d: ", pid, cpuid);
    kdelay(pid*100000); // simulate task computation
    ttswitch();
  }
}
int INIT_task()  // INIT task P1
{
  int i, pid, status;
  // create 2 tasks on different CPUs
  for (i=0; i<2; i++){
      kfork((int)taskCode, 1, 0);
      kfork((int)taskCode, 1, 1);
      kfork((int)taskcode, 1, 2);
      kfork((int)taskCode, 1, 3);
  }
  while(1){ // task1 waits for ZOMBIE children
     pid = kwait(&status);
  }
}
int logtask()   // logging task P0
{
  int pid;
  int cpuid = get_cpuid();
  char msg[128];
  printf("LOGtask%d start on CPU%d\n", running->pid, cpuid);
  while(1){ // task0 waits for message
    pid = recv(msg);
  }
}
```

```
int run_task()  // each CPU tries to run task from its own readyQ
{
   int cpuid = get_cpuid();
   while(1){
      slock(&readylock[cpuid]);
      if (readyQueue[cpuid] == 0){    // check own readyQueue
         sunlock(&readylock[cpuid]);
         asm("WFE");                  // wait for events/interrupts
      }
      else{
         sunlock(&readylock[cpuid]);
         ttswitch();                  // switch task on CPU
      }
   }
}
int APstart()
{
   int cpuid = get_cpuid();
   slock(&aplock);
   color = YELLOW;
   printf("CPU%d in APstart ... ", cpuid);
   config_int(29, cpuid);   // per CPU local timer
   ptimer_init();
   ptimer_start();
   ncpu++;
   running = &iproc[cpuid]; // run per CPU initial task
   printf("%d running on CPU%d\n", running->pid, cpuid);
   sunlock(&aplock);
   run_task();
}
int main()
{
   int cpuid = get_cpuid();
   enable_scu();
   fbuf_init();
   printf("Welcome to SMP_RTOS in ARM\n");
   sdc_init();
   kbd_init();
   uart_init();
   ptimer_init();
   ptimer_start();
   config_gic();
   printf("CPU0 initialize kernel\n");
   kernel_init();
   printf("CPU0 startup APs: ");
   int *APaddr = (int *)0x10000030;
   *APaddr = (int)apStart;
   send_sgi(0x0, 0x0F, 0x01);
   printf("CPU0 waits for APs ready\n");
   while(ncpu < 4);
   printf("CPU0 continue\n");
   kfork(int)logtask, 1, 0);      // P0 on CPU0 as loggin task
   kfork((int)INIT_task,  1, 0); // P1 on CPU0 as INIT   task
   run_task();
}
```

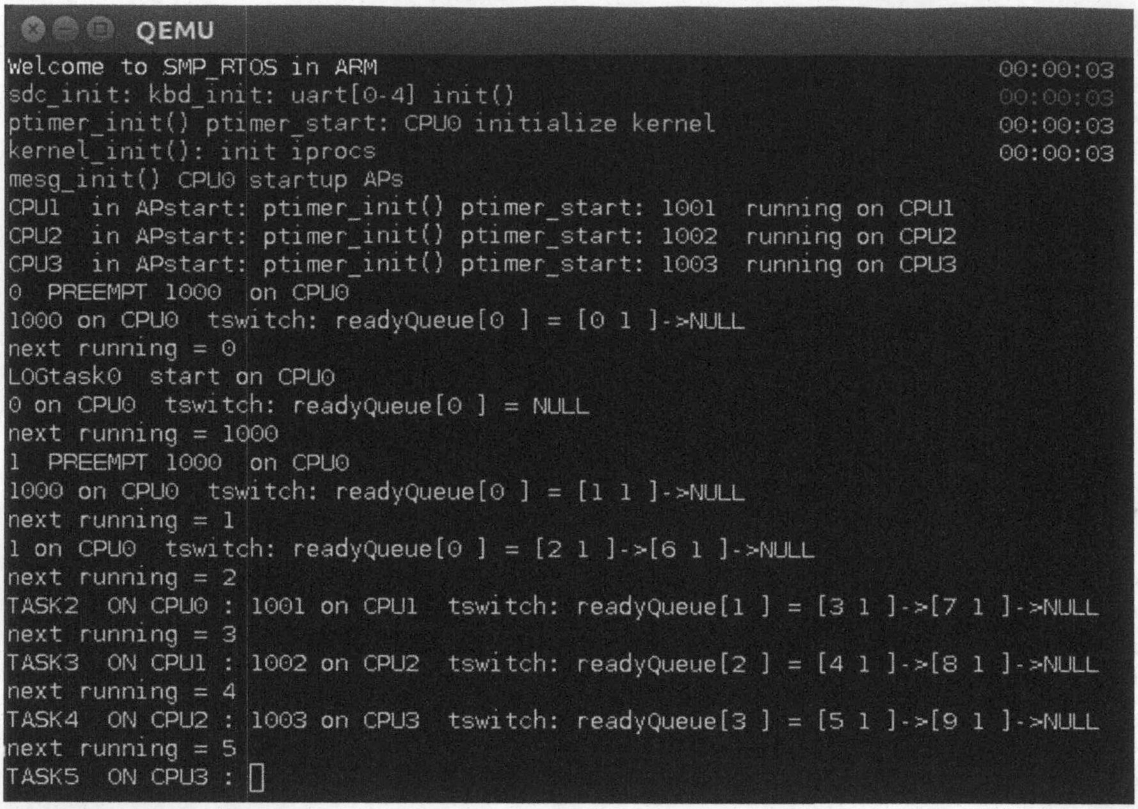

```
QEMU
Welcome to SMP_RTOS in ARM                                            00:00:03
sdc_init: kbd_init: uart[0-4] init()                                  00:00:03
ptimer_init() ptimer_start: CPU0 initialize kernel                    00:00:03
kernel_init(): init iprocs                                            00:00:03
mesg_init() CPU0 startup APs
CPU1  in APstart: ptimer_init() ptimer_start: 1001  running on CPU1
CPU2  in APstart: ptimer_init() ptimer_start: 1002  running on CPU2
CPU3  in APstart: ptimer_init() ptimer_start: 1003  running on CPU3
0  PREEMPT 1000  on CPU0
1000 on CPU0  tswitch: readyQueue[0 ] = [0 1 ]->NULL
next running = 0
LOGtask0  start on CPU0
0 on CPU0  tswitch: readyQueue[0 ] = NULL
next running = 1000
1  PREEMPT 1000  on CPU0
1000 on CPU0  tswitch: readyQueue[0 ] = [1 1 ]->NULL
next running = 1
1 on CPU0  tswitch: readyQueue[0 ] = [2 1 ]->[6 1 ]->NULL
next running = 2
TASK2  ON CPU0 : 1001 on CPU1  tswitch: readyQueue[1 ] = [3 1 ]->[7 1 ]->NULL
next running = 3
TASK3  ON CPU1 : 1002 on CPU2  tswitch: readyQueue[2 ] = [4 1 ]->[8 1 ]->NULL
next running = 4
TASK4  ON CPU2 : 1003 on CPU3  tswitch: readyQueue[3 ] = [5 1 ]->[9 1 ]->NULL
next running = 5
TASK5  ON CPU3 :
```

Fig. 10.5 Task management in SMP_RTOS

The object of the initial SMP_RTOS kernel is to demonstrate that the system can create and run tasks on different CPUs. When the system starts, CPU0 creates and runs the initial task P1000, which calls main(). In main(), P1000 initializes device drivers and the system kernel. Then it activates the secondary CPUs (APs). Each AP initializes itself and runs an initial task P1000+cpuid, which calls run_task(), trying to run tasks from its own readyQueue[cpuid]. After activating the APs, P1000 creates a logging task P0 and an INIT task P1 on CPU0. Then P1000 switches task to run the INIT task P1. P1 creates two children tasks on each CPU and waits for any child task to terminate. Each task executes a loop, in which it delays for some time to simulate task computation and calls ttswitch() to switch tasks. In the sample system, tasks run the loop continually. The reader may test the kwait() operation by modifying the taskCode() function to let tasks terminate. Figure 10.5 shows the sample outputs of running the program C10.4, which demonstrates task management in the SMP_RTOS kernel.

10.8.2 Adapt UP_RTOS to SMP

The simplest way to develop a RTOS for multiprocessors is to adapt the UP_RTOS for MP operations. In this approach, the system comprises multiple CPUs. Each CPU executes a set of tasks that are bound to the CPU and operate in the local environment of the CPU. The main advantage of this approach lies in its simplicity. It requires very little work to convert the UP_ROTS into a MP RTOS. The disadvantage is that the resulting system is only a Asymmetric MP system, not a SMP system. We illustrate such a system, denoted by MP_RTOS, by a sample program C10.5. The program consists of the following components.

(1). Ts.s file: same as in UP_RTOS, except the simplified irq_hanler, which supports task switch but no nested interrupts.

```
irq_handler:                    // IRQ interrupts entry point
 sub   lr, lr, #4
 stmfd sp!, {r0-r12, lr}  // save all Umode regs in kstack
   mrs   r0, spsr
   stmfd sp!, {r0}
   bl  irq_chandler
```

```
// check return value: 0=nornal, 1=swflag was set => swtich task
   cmp r0, #0
   bgt doswitch
   ldmfd sp!, {r0}
   msr   spsr, r0
   ldmfd sp!, {r0-r12, pc}^
doswitch:
   msr cpsr, #0x93
   stmfd sp!, {r0-r3, lr}
   bl ttswitch
   msr cpsr, #0x93
   ldmfd sp!, {r0-r3, lr}
   msr cpsr, #0x92
   ldmfd sp!, {r0}
   msr   spsr, r0
   ldmfd sp!, {r0-r12, pc}^
```

(2). kernel.c file: same as in UP_RTOS for task management, except we define

```
struct semaphore sem[NCPU];
```

and initialize each sem.value = 0. Each semaphore is associated with a CPU, which can only be accessed by tasks executing on the same CPU. If a task becomes blocked on a semaphore, it is unblocked by the CPU's local timer periodically.

```
void ptimer_handler()
{
   int cpuid = id = get_cpuid();
   struct ttt *tp = &tt[cpuid];
   slock(&plock);
   // update tick, ss, mm, hh: same as before
   if (tp->tick==0){  // every second: display a wall clock
      // display wall clock: same as before
      if ((tp->ss % 5)==0)  // every 5 seconds, activate a task
         V(&sem[cpuid]);    // V to unblock a waiting task
   }
   sunlock(&plock);
   ptimer_clearInterrupt(); // clear timer interrupt
}
```

(3). t.c file: Same as in UP_RTOS for task management, except for the following modifications. Instead creating task on different CPUs, the initial task of each CPU creates a task to execute INIT_task() on the same CPU. Each task creates another task to execute taskCode(), also on the same CPU. All the tasks have the same priority 1. Each task executes an infinite loop, in which it delays for some time to simulate task computation and then blocks itself on a semaphore associated with the CPU. The local timer of each CPU periodically calls V to unblock the tasks, allowing them to continue. Whenever a ready task is entered into the local scheduling queue of a CPU, it may preempt the current running task on the same CPU.

```
/*********** t.c file of MP_RTOS C10.5 ************/
int run_task(){ // same as before
int kdelay(int d){ // delay }
int taskCode()
{
  int cpuid = get_cpuid();
  while(1){
```

```
        printf("TASK%d ON CPU%d\n", running->pid, cpuid);
        kdelay(running->pid*10000);
        reset_sem(&sem[cpuid]);        // reset sem.value to 0
        P(&sem[cpuid]);
    }
}
int INIT_task()
{
    int cpuid = get_cpuid();
    kfork((int)taskCode, 1, cpuid);
    while(1){
        printf("task%d on CPU%d\n", running->pid, cpuid);
        kdelay(running->pid*10000);
        reset_sem(&sem[cpuid]);        // reset sem.value to 0
        P(&sem[cpuid]);
    }
}

int APstart()
{
     int cpuid = get_cpuid();
     slock(&aplock);
     printf("CPU%d in APstart: ", cpuid);
     config_int(29, cpuid); // need this per CPU
     ptimer_init(); ptimer_start();
     ncpu++;
     running = &iproc[cpuid];
     printf("%d running on CPU%d\n", running->pid, cpuid);
     sunlock(&aplock);
     kfork((int)INIT_task, 1, cpuid);
     run_task();
}
int main()
{
     // initialize device drivers, GIC and kernel: same as before
     kprintf("CPU0 continue\n");
     kfork((int)f1, 1, 0);
     run_task();
}
```

10.8.2.1 Demonstration of MP_RTOS

Figure 10.6 shows the outputs of running the MP_RTOS system.

10.8.3 Nested Interrupts in SMP_RTOS

In a UP RTOS kernel, only one CPU handles all the interrupts. To ensure fast responses to interrupts, it is vital to allow nested interrupts. In a SMP RTOS kernel, interrupts are usually routed to different CPUs to balance the interrupt processing load. Despite this, it is still necessary to allow nested interrupts. For instance, each CPU may use a local timer, which generates timer interrupts with high priority. Without nested interrupts, a low priority device driver may block timer interrupts, causing the timer to lose ticks. In this section, we shall show how to implement nested interrupts in the SMP_RTOS kernel by the sample program C10.6. The principle is the same as in UP_RTOS, except for the following minor difference.

All ARM MPcore processors use the GIC for interrupts control. In order to support nested interrupts, the interrupt handler must read the GIC's interrupt ID register to acknowledge the current interrupt before enabling interrupts. The GIC's interrupt ID register

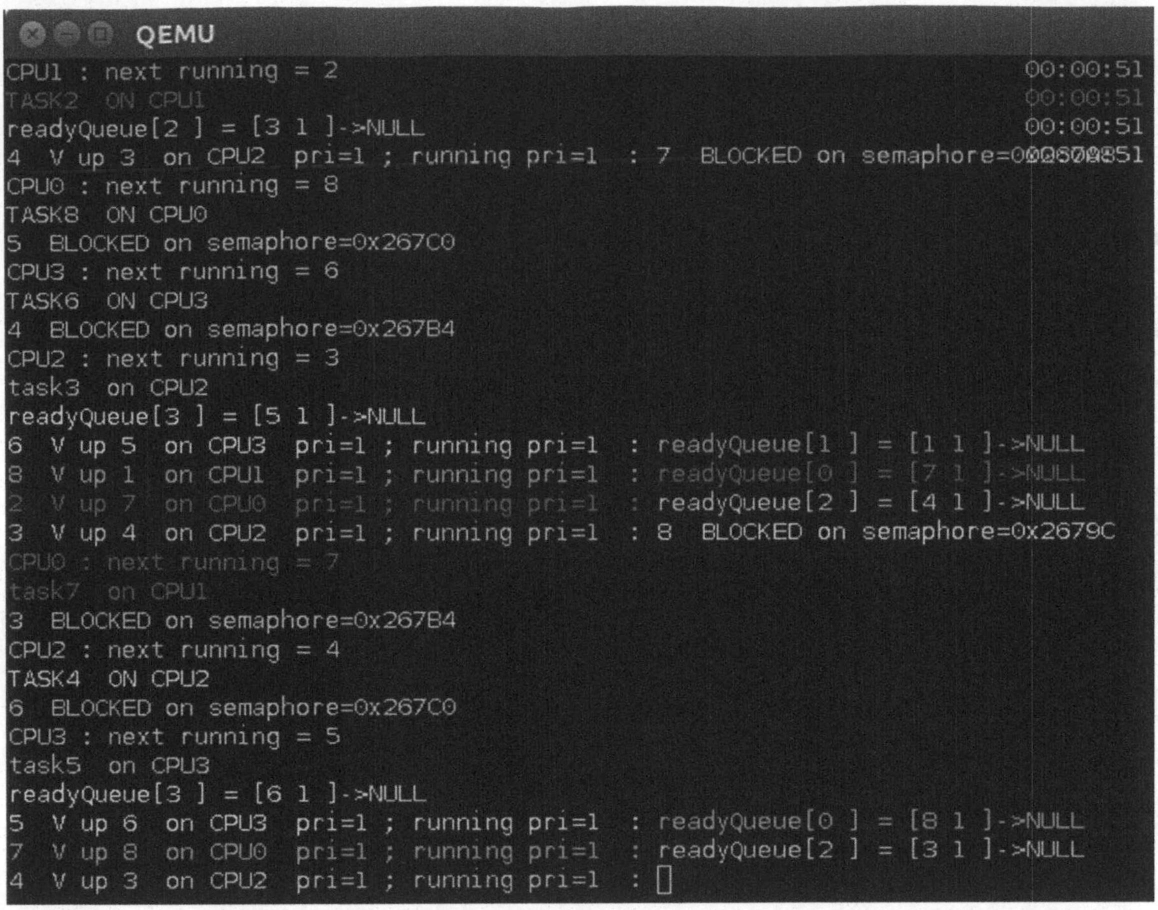

```
QEMU
CPU1 : next running = 2                                                  00:00:51
TASK2  ON CPU1                                                           00:00:51
readyQueue[2 ] = [3 1 ]->NULL                                            00:00:51
4  V up 3  on CPU2  pri=1 ; running pri=1  : 7  BLOCKED on semaphore=0[illegible]
CPU0 : next running = 8
TASK8  ON CPU0
5  BLOCKED on semaphore=0x267C0
CPU3 : next running = 6
TASK6  ON CPU3
4  BLOCKED on semaphore=0x267B4
CPU2 : next running = 3
task3  on CPU2
readyQueue[3 ] = [5 1 ]->NULL
6  V up 5  on CPU3  pri=1 ; running pri=1  : readyQueue[1 ] = [1 1 ]->NULL
8  V up 1  on CPU1  pri=1 ; running pri=1  : readyQueue[0 ] = [7 1 ]->NULL
2  V up 7  on CPU0  pri=1 ; running pri=1  : readyQueue[2 ] = [4 1 ]->NULL
3  V up 4  on CPU2  pri=1 ; running pri=1  : 8  BLOCKED on semaphore=0x2679C
CPU0 : next running = 7
task7  on CPU1
3  BLOCKED on semaphore=0x267B4
CPU2 : next running = 4
TASK4  ON CPU2
6  BLOCKED on semaphore=0x267C0
CPU3 : next running = 5
task5  on CPU3
readyQueue[3 ] = [6 1 ]->NULL
5  V up 6  on CPU3  pri=1 ; running pri=1  : readyQueue[0 ] = [8 1 ]->NULL
7  V up 8  on CPU0  pri=1 ; running pri=1  : readyQueue[2 ] = [3 1 ]->NULL
4  V up 3  on CPU2  pri=1 ; running pri=1  :
```

Fig. 10.6 Demonstration of MP_RTOS system

can only be read once. The interrupt handler must use the interrupt ID to invoke the irq_chandler(int intID) in C. For special SGI interrupts (0-15), they should be handled separately since they have high interrupt priorities and do not need any nesting.

```
   .set GICreg, 0x1F00010C  // GIC intID register
   .set GICeoi, 0x1F000110  // GIC EOI   register
irq_handler:
   sub   lr, lr, #4
   stmfd sp!, {r0-r12, lr}
   mrs   r0, spsr
   stmfd sp!, {r0}           // push SPSR
// read GICreg to get intID and ACK current interrupt
   ldr   r1, =GICreg
   ldr   r0, [r1]
// handle SGI intIDs direcly
   cmp   r0, #15             // SGI interrupts
   bgt   nesting
   ldr   r1, =GICeoi
   str   r0, [r1]            // issue EOI
   b     return
nesting:
// to SVC mode
   msr   cpsr, #0x93         // SVC mode with IRQ off
   stmfd sp!, {r0-r3, lr}
   msr   cpsr, #0x13         // SVC mode with IRQ interrupts on
   bl    irq_chandler        // call irq_chandler(intID)
```

```
    msr   cpsr, #0x93        // SVC mode with IRQ off
    ldmfd sp!, {r0-r3, lr}   //
    msr   cpsr, #0x92        // IRQ mode with interrupts off
return:
    ldmfd sp!, {r0}
    msr   spsr, r0
    ldmfd sp!, {r0-r12, pc}^
int irq_chandler(int intID)  // called with interrupt ID
{
   int cpuid = get_cpuid();
   int cpsr = get_cpsr() & 0x9F; // CPSR with |IRQ|mode| mask

   intnest[cpuid]++;          // inc intsest by 1
   if (intID != 29)
      printf("IRQ%d on CPU%d in mode=%x\n", intID, cpuid, cpsr);
   if (intID == 29){          // local timer of CPU
       ptimer_handler();
   }
   if (intID == 44){          // UART0 interrpt
       uart_handler(0);
   }
   if (intID == 49){          // SDC interrupt
       sdc_handler();
   }
   if (intID == 52){          // KBD interrupt
       kbd_handler();
   }
   *(int *)(GIC_BASE + 0x110) = intID; // issue EOF
   intnest[cpuid]--;          // dec intnest by 1
}
```

Figure 10.7 shows the outputs of running the program C10.6, which demonstrates SMP_RTOS with nested interrupts. As the figure shows, all interrupts are handled in SVC mode.

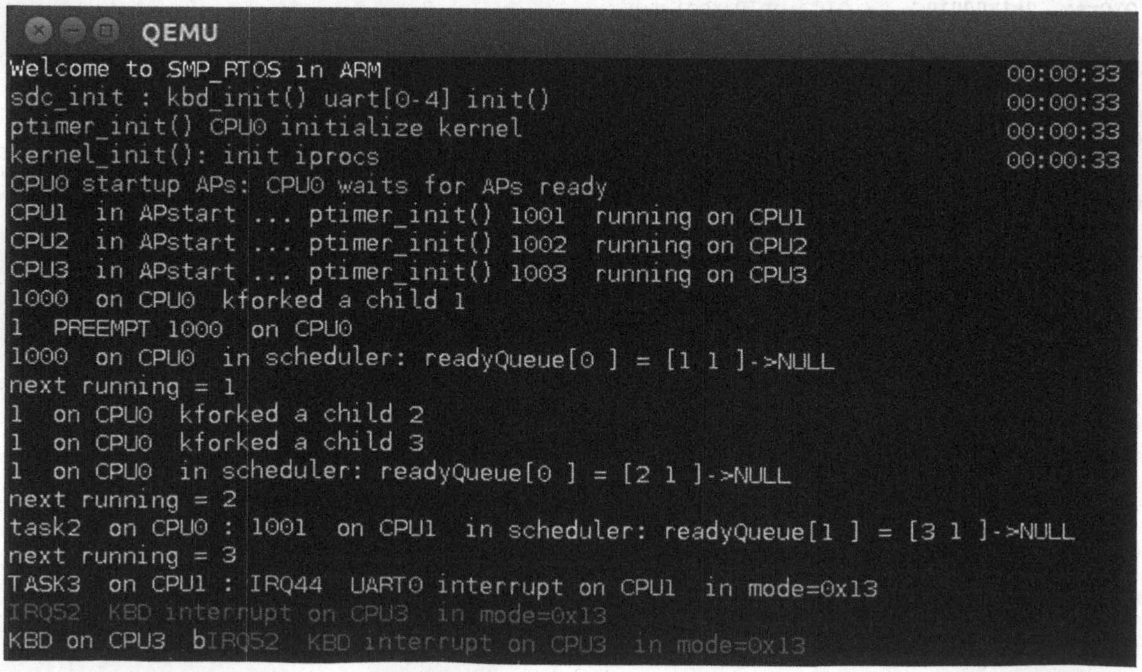

```
QEMU
Welcome to SMP_RTOS in ARM                                              00:00:33
sdc_init : kbd_init() uart[0-4] init()                                  00:00:33
ptimer_init() CPU0 initialize kernel                                    00:00:33
kernel_init(): init iprocs                                              00:00:33
CPU0 startup APs: CPU0 waits for APs ready
CPU1  in APstart ... ptimer_init() 1001  running on CPU1
CPU2  in APstart ... ptimer_init() 1002  running on CPU2
CPU3  in APstart ... ptimer_init() 1003  running on CPU3
1000  on CPU0  kforked a child 1
1  PREEMPT 1000  on CPU0
1000  on CPU0  in scheduler: readyQueue[0 ] = [1 1 ]->NULL
next running = 1
1  on CPU0  kforked a child 2
1  on CPU0  kforked a child 3
1  on CPU0  in scheduler: readyQueue[0 ] = [2 1 ]->NULL
next running = 2
task2  on CPU0 : 1001  on CPU1  in scheduler: readyQueue[1 ] = [3 1 ]->NULL
next running = 3
TASK3  on CPU1 : IRQ44  UART0 interrupt on CPU1  in mode=0x13
IRQ52  KBD interrupt on CPU3  in mode=0x13
KBD on CPU3  bIRQ52  KBD interrupt on CPU3  in mode=0x13
```

Fig. 10.7 Nested interrupts in SMP_RTOS

10.8.4 Preemptive Task Scheduling in SMP_RTOS

In order to meet task deadlines, every ROTS must support preemptive task scheduling to ensure high priority tasks can be executed as soon as possible. In this section, we shall show how to implement preemptive task scheduling in the SMP_RTOS kernel and demonstrate task preemption by the sample program C10.7. The task preemption problem can be classified into two main cases.

Case 1: When a running task makes another task of higher priority ready to run, e.g. when creating a new task with a higher priority or unblocking a task with a higher priority.

Case 2: When an interrupt handler makes a task runnable, e.g. when unblocking a task with a higher priority or when the current running task has exhausted its time quantum in round-robin scheduling by timeslice.

In a UP kernel, implementing task preemption is straightforward. For the first case, whenever a task makes another task ready to run, it simply checks whether the new task has a higher priority. If so, it invokes task switching to preempt itself immediately. For the second case, the situation is only slightly more complex. With nested interrupts, the interrupt handler must check whether all nested interrupts processing have ended. If so, it can invoke task switching to preempt the current running task. Otherwise, it defers task switching until the end of nested interrupts processing.

In SMP, the situation is very different. In a SMP kernel, an execution entity, i.e. a task or an interrupt handler, executing on one CPU may make another task ready to run on a different CPU. If so, it must inform the other CPU to reschedule tasks for possible task preemption. In this case, it must issue a SGI, e.g. with interrupt ID=2, to the other CPU. Responding to a SGI_2 interrupt, the target CPU can execute a SGI_2 interrupt handler to reschedule tasks. We illustrate SGI based task switch by the sample program C10.7. For the sake of brevity, we only show the relevant code segments.

10.8.5 Task Switch by SGI

```
/************ irq_handler() in ts.s file *************/
irq_handler:                    // IRQ interrupts entry point
   sub   lr, lr, #4
   stmfd sp!, {r0-r12, lr}    // save all Umode regs in kstack
   bl     irq_chandler        // call irq_handler() in C in svc.c file
// return value=1 means swflag is set => swtich task
   cmp   r0, #0
   bne   doswitch
   ldmfd sp!, {r0-r12, pc}^ // return
doswitch:
   msr   cpsr, #0x93
   stmfd sp!, {r0-r3, lr}
   bl    ttswitch              // switch task in SVC mode
   ldmfd sp!, {r0-r3, lr}
   msr   cpsr, #0x92           // return by context in IRQ stack
   ldmfd sp!, {r0-r12, pc}^
/************** t.c file *****************************/
int send_sgi(int ID, int targetCPU, int filter)
{
  int target = (1 << targetCPU); // set CPU ID bit
  if (targetCPU > 3) // > 3 means to all CPUs
     target = 0xF;    // turn on all 4 CPU ID bits
  int *sgi_reg = (int *)(GIC_BASE + 0x1F00);
  *sgi_reg = (filter << 24) | (target<<16) | ID;
}
int SGI2_handler()
{
  int cpuid = get_cpuid();
  printf("%d on CPU%d got SGI_2: set swflag\n", running->pid, cpuid);
```

```
    swflag[cpuid] = 1;
}
int irq_chandler()
{
    int intID = *(int *)(GIC_BASE + 0x10C);
    int cpuid = get_cpuid();
    if (intID == 2){
        SGI2_handler();
    }
    // other intID cases: same as before
    *(int *)(GIC_BASE + 0x110) = (cpuid<<10)|intID; // issue EOI
    if (swflag[cpuid]){ // return 1 for task switch in irq_handler
        swflag[cpuid] = 0;
        return 1;
    }
    return 0;
}
int kdelay(int d){ // delay to simulate task computation }
int f2()
{
    int cpuid = get_cpu_id();
    while(1){
      printf("TASK%d RUNNING ON CPU%d: ", running->pid, cpuid);
      kdelay(running->pid*10000);  // simulate task compute time
    }
}
int f1()
{
    int cpuid = get_cpu_id();
    while(1){
      printf("task%d running on CPU%d: ", running->pid, cpuid);
      kdelay(running->pid*100000); // simulate task compute time
    }
}
int main()
{
    // start up APs: same as before
    printf("CPU0 continue\n");
    kfork((int)f1, 1, 1); // create tasks on CPU1
    kfork((int)f2, 1, 1);
    kfork((int)f1, 1, 2); // create tasks on CPU2
    kfork((int)f2, 1, 2);
    while(1){
      printf("input a line: ");
      kgetline(line); printf("\n");
      send_sgi(2, 1, 0);  // SGI_2 to CPU1
      send_sgi(2, 2, 0);  // SGI_2 to CPU2
    }
}
```

In the sample program C10.7, after starting up the APs, the main program executing on CPU0 creates two sets of tasks on both CPU1 and CPU2, all of the same priority 1. All the tasks execute an infinite loop. Without external inputs, each CPU would continue to execute the same task forever. After creating tasks on different CPUs, the main program prompts for an input line from the keyboard. Then it sends a SGI with intID=2 to the other CPUs, causing them to execute the SGI2_handler. In the SGI2_handler, each CPU turns on its switch task flag and returns a 1. In the irq_handler code, it checks

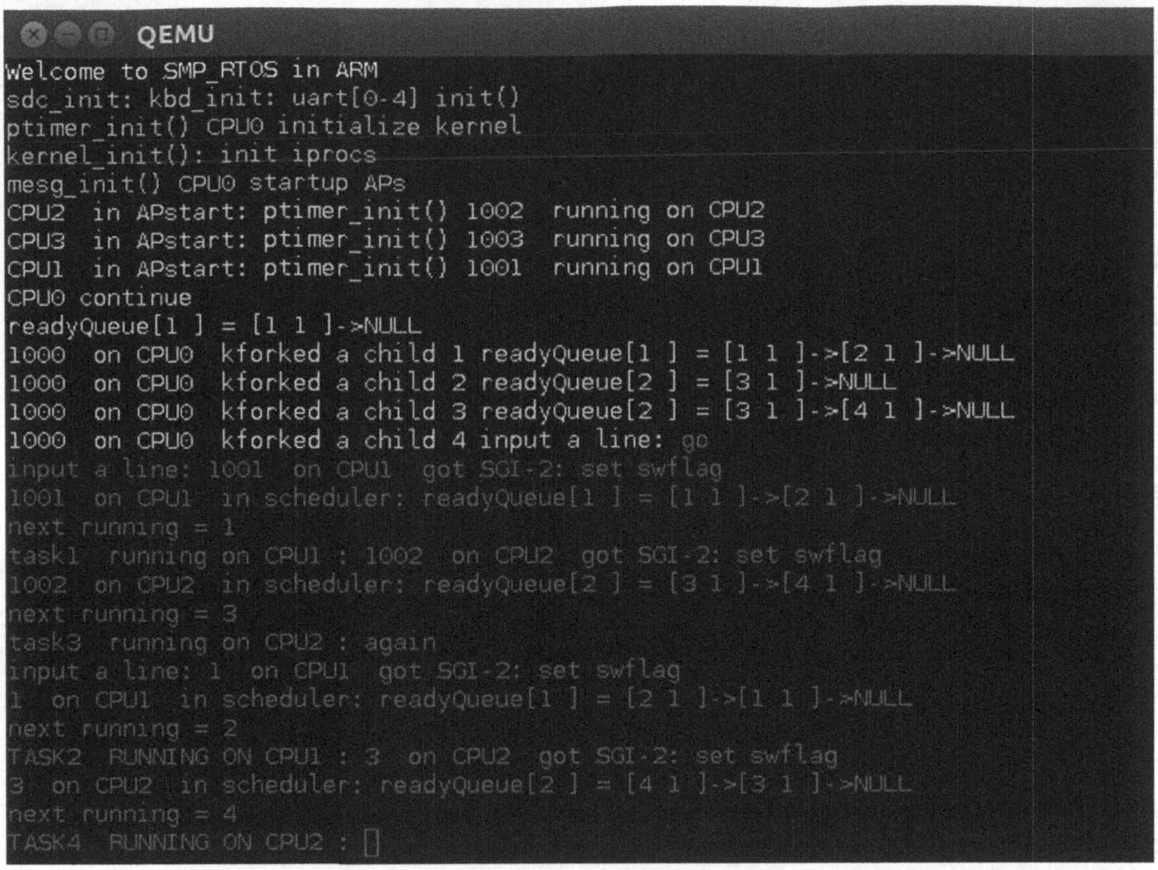

```
QEMU
Welcome to SMP_RTOS in ARM
sdc_init: kbd_init: uart[0-4] init()
ptimer_init() CPU0 initialize kernel
kernel_init(): init iprocs
mesg_init() CPU0 startup APs
CPU2  in APstart: ptimer_init() 1002  running on CPU2
CPU3  in APstart: ptimer_init() 1003  running on CPU3
CPU1  in APstart: ptimer_init() 1001  running on CPU1
CPU0 continue
readyQueue[1 ] = [1 1 ]->NULL
1000  on CPU0  kforked a child 1 readyQueue[1 ] = [1 1 ]->[2 1 ]->NULL
1000  on CPU0  kforked a child 2 readyQueue[2 ] = [3 1 ]->NULL
1000  on CPU0  kforked a child 3 readyQueue[2 ] = [3 1 ]->[4 1 ]->NULL
1000  on CPU0  kforked a child 4 input a line: go
input a line: 1001  on CPU1  got SGI-2: set swflag
1001  on CPU1  in scheduler: readyQueue[1 ] = [1 1 ]->[2 1 ]->NULL
next running = 1
task1  running on CPU1 : 1002  on CPU2  got SGI-2: set swflag
1002  on CPU2  in scheduler: readyQueue[2 ] = [3 1 ]->[4 1 ]->NULL
next running = 3
task3  running on CPU2 : again
input a line: 1  on CPU1  got SGI-2: set swflag
1  on CPU1  in scheduler: readyQueue[1 ] = [2 1 ]->[1 1 ]->NULL
next running = 2
TASK2  RUNNING ON CPU1 : 3  on CPU2  got SGI-2: set swflag
3  on CPU2  in scheduler: readyQueue[2 ] = [4 1 ]->[3 1 ]->NULL
next running = 4
TASK4  RUNNING ON CPU2 : 
```

Fig. 10.8 Task switch by SGI in SMP_RTOS

the return value. If the return value is nonzero, indicating a need for task switch, it changes to SVC mode and calls ttswitch() to switches task. When the switched out task resumes, it returns to the original interrupted point by the saved context in IRQ stack. Figure 10.8 shows the outputs of task switching by SGI. As the figure shows, each input line sends SGI-2 to both CPU1 and CPU2, causing them to switch tasks.

10.8.6 Timesliced Task Scheduling in SMP_RTOS

In a RTOS, tasks may have the same priority. Instead of waiting for such tasks to give up CPU voluntarily, it may be desirable or even necessary to schedule them by round-robin with timeslice. In this scheme, when a task is scheduled to run, it is given a timeslice as the maximum amount of time the task is allowed to run. In the timer interrupt handler, it decrements the running task's timeslice periodically. When the running task's timeslice reaches zero, it is preempted to run another task. We demonstrate task scheduling by timeslice in the SMP_RTOS kennel by an example. The example program, C10.8, is the same as C10.7 except for the following modifications.

```
/********** kernel.c file of C10.8 ***********/
int scheduler()
{
  int pid; PROC *old=running;
  int cpuid = get_cpuid();
  if (running->pid < 1000){ // regular tasks only
    if (running->status==READY)
      enqueue(&readyQueue[cpuid], running);
```

```
  }
  running = dequeue(&readyQueue[cpuid]);
  if (running == 0)          // if CPU's readyQueue is empty
     running = &iproc[cpuid]; // run the idle task
  running->timeslice = 4;     // 4 seconds timeslice
  swflag[cpuid] = 0;
  sunlock(&readylock[cpuid]);
}
/********* ptimer.c file of C10.8 ***************/
void ptimer_handler()         // local timer handler
{
   int cpuid = get_cpuid();
   // update tick, display wall clock: same as before
   ptimer_clearInterrupt();   // clear timer interrupt
   if (tp->tick == 0){        // per second
     if (running->pid < 1000){// only regular tasks
        running->timeslice--;
        printf("%dtime=%d ", running->pid, running->timeslice);
        if (running->timeslice <= 0){
           printf("%d on CPU%d time up\n", running->pid, cpuid);
           swflag[cpuid] = 1; // set swflag o switch task at end of IRQ
        }
     }
   }
}
/************* t.c file of C10.8 ****************/
int f3()
{
  int cpuid = get_cpuid();
  while(1){
    printf("TASK%d ON CPU%d\n", running->pid, cpuid);
    kdelay(running->pid * 100000); // simulate computation
  }
}
int f2()
{
  int cpuid = get_cpuid();
  while(1){
    printf("task%d on cpu%d\n", running->pid, cpuid);
    kdelay(running->pid * 100000); // simulate computation
  }
}
int f1()
{
  int pid, status;
  kfork((int)f2, 1, 1);     // create task on CPU1
  kfork((int)f3, 1, 1);     // create task on CPU1
  while(1){
     pid = kwait(&status); // P1 waits for ZOMBIE child
  }
}
int run_task(){ // same as before }
int main()
{
   // start APs" same as before
```

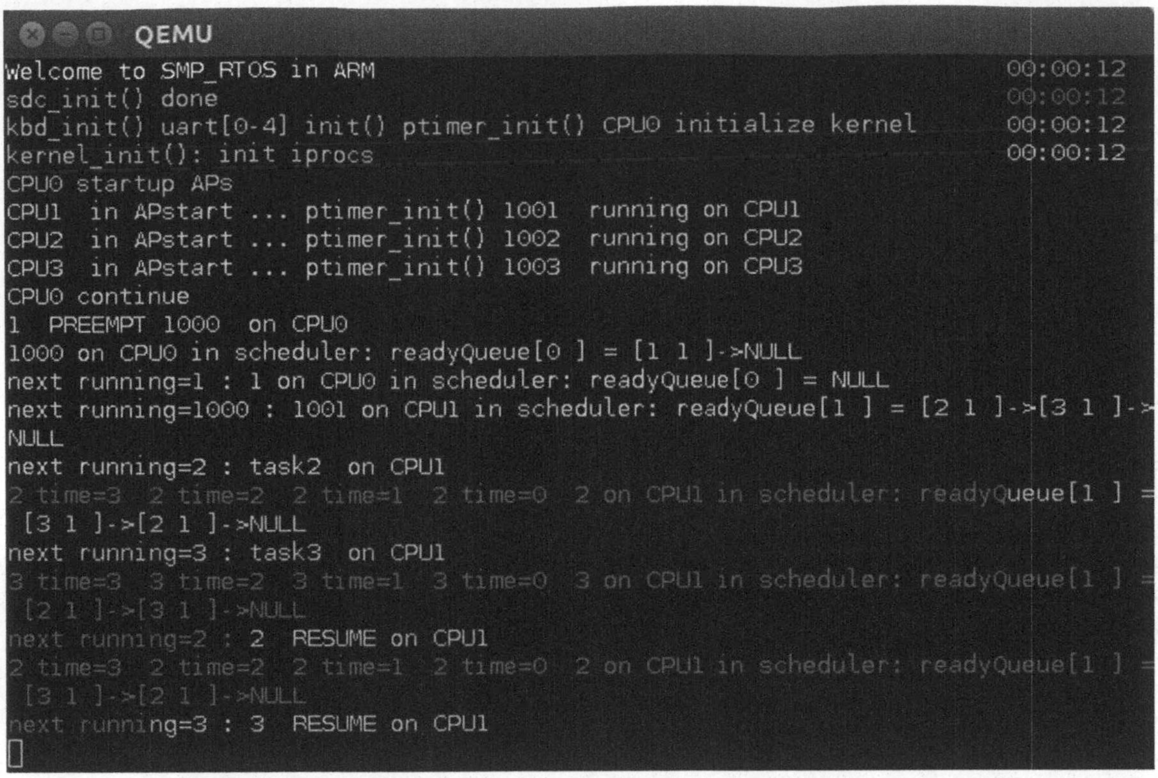

```
QEMU
Welcome to SMP_RTOS in ARM                                              00:00:12
sdc_init() done                                                         00:00:12
kbd_init() uart[0-4] init() ptimer_init() CPU0 initialize kernel        00:00:12
kernel_init(): init iprocs                                              00:00:12
CPU0 startup APs
CPU1  in APstart ... ptimer_init() 1001  running on CPU1
CPU2  in APstart ... ptimer_init() 1002  running on CPU2
CPU3  in APstart ... ptimer_init() 1003  running on CPU3
CPU0 continue
1  PREEMPT 1000  on CPU0
1000 on CPU0 in scheduler: readyQueue[0 ] = [1 1 ]->NULL
next running=1 : 1 on CPU0 in scheduler: readyQueue[0 ] = NULL
next running=1000 : 1001 on CPU1 in scheduler: readyQueue[1 ] = [2 1 ]->[3 1 ]->
NULL
next running=2 : task2  on CPU1
2 time=3  2 time=2  2 time=1  2 time=0  2 on CPU1 in scheduler: readyQueue[1 ] =
 [3 1 ]->[2 1 ]->NULL
next running=3 : task3  on CPU1
3 time=3  3 time=2  3 time=1  3 time=0  3 on CPU1 in scheduler: readyQueue[1 ] =
 [2 1 ]->[3 1 ]->NULL
next running=2 : 2  RESUME on CPU1
2 time=3  2 time=2  2 time=1  2 time=0  2 on CPU1 in scheduler: readyQueue[1 ] =
 [3 1 ]->[2 1 ]->NULL
next running=3 : 3  RESUME on CPU1
```

Fig. 10.9 Timesliced task scheduling in SMP_RTOS

```
    printf("CPU0 continue\n");
    kfork((int)f1, 1, 0);   // create task 1 on CPU0
    run_task();
}
```

In the sample program C10.8, the initial task of CPU0 creates task1 to execute f1() on CPU0. Task1 creates task2 and task3 on CPU1 with the same priority. Then task1 waits for any child task to terminate. Since both task2 and task3 have the same priority, without timeslicing, CPU1 would continue to run task2 forever. With timeslicing, task2 and task3 would take turn to run, each runs for a timeslice=4 seconds. Figure 10.9 shows the outputs of running the C10.8 program, which demonstrates timesliced task scheduling.

10.8.7 Preemptive Task Scheduling in SMP_RTOS

Based on the above discussions, we now show the implementation of preemptive task scheduling in the SMP_ROTS kernel.

10.8.7.1 Reschedule Function for Task Preemption

```
int reschedule(PROC *p)
{
   int cpuid = get_cpuid();  // executing CPU
   int targetCPU = p->cpuid; // target CPU
   if (cpuid == targetCPU){   // if same CPU
      if (readyQueue[cpuid]->priority > running->priority){
         if (intnest[cpuid]==0){ // PREEMPTION immediately
            ttswitch();
```

```
        }
        else{                        // defer until end of nested IRQs
           swflag[cpuid] = 1;
        }
     }
   }
   else{                      // different CPU
      send_sgi(intID=2, CPUID=targetCPU, filter=0);
}
}
```

10.8.7.2 Task Creation with Preemption

```
int kfork(int func, int priority, int cpu)
{
   int cpuid = get_cpuid();  // executing CPU
   // create a new task p, enter into readyQueue[cpuid] as before
   reschedule(p);              // invoke resechedule
   return p->pid;
}
```

10.8.7.3 Semaphore with Preemption

```
int V(struct semaphore *s)
{
  PROC *p = 0;
  int SR = int_off();
  slock(&s->lock);
  s->value++;
  if (s->value <= 0){
     p = dequeue(&s->queue);
     p->status = READY;
     slock(&readylock[p->cpuid]);
      enqueue(&readyQueue[p->cpuid], p);
     sunlock(&readylock[p->cpuid]);
  }
  sunlock(&s->lock);
  int_on(SR);
  if (p)              // if has unblocked a waiter
     reschedule();
}
```

10.8.7.4 Mutex with Preemption

```
int mutex_unlock(MUTEX *s)
{
  PROC *p = 0;
  int SR = int_off();
  slock(&s->lock)      // acquire spinlock
  // ASSUME: mutex has waiters: unblock one as new owner
  p = dequeue(&s->queue);
```

```
  p->status = READY;
  s->owner = p;
  slock(&readylock[p->cpuid]);
    enqueue(&readyQueue[p->cpuid], p);
  sunlock(&readylock[p->cpuid]);
  sunlock(&s->lock);  // release spinlock
  int_on(SR);
  if (p)              // if has unblocked a waiter
     reschedule();
}
```

10.8.7.5 Nested IRQ interrupts with Preemption

```
int SGI2_handler()
{
   int cpuid = get_cpuid();
   PROC *p = readyQueue[cpuid];
   if (p && p->priority > running->priority){
        swflag[cpuid] = 1;  // set task switch flag
   }
}
int irq_chandler(int intID) // called with interrupt ID
{
  int cpuid = get_cpuid();
  intnest[cpuid]++;           // inc intsest count by 1
  if (intID == 2){            // SGI_2 interrupt
     SGI2_handler();
  }
  // other IRQ interrupt handlers: same as before
  *(int *)(GIC_BASE + 0x110) = (cpuID<<10)|intID; // issue EOI
  intnest[cpuid]--;           // dec intnest count by 1
}
int checkIRQnesting()// return 1 if end of IRQ nesting && swflag is set
{
   int cpuid = get_cpuid();
   if (intnest[cpuid] == 0 && swflag[cpuid])
      return 1;
   return 0;
}
/*** irq_handler with nested interrupts AND SGI_2 ****/
    .set GICreg, 0x1F00010C  // GIC intID register
    .set GICeoi, 0x1F000110  // GIC EOI   register
irq_handler:
   sub   lr, lr, #4
   stmfd sp!, {r0-r12, lr}
   mrs   r0, spsr
   stmfd sp!, {r0}  // push SPSR
// read GICreg to get intID
   ldr   r1, =GICreg
   ldr   r0, [r1]
// handle SGI intID=0 direcly
   cmp   r0, #0              // SGI 0 used during startup
   bgt   nesting
   ldr   r1, =GICeoi
```

```
    str   r0, [r1]              // issue EOI
    b     return
nesting:                        // to SVC mode
    msr   cpsr, #0x93           // SVC mode with IRQ off
    stmfd sp!, {r0-r3, lr}
    msr   cpsr, #0x13           // SVC mode with interrupts on
    bl    irq_chandler          // irq_chandler return 1: task switch
    msr   cpsr, #0x93           // SVC mode with IRQ off
    bl    checkIRQnesting       // return 1 if OK to switch task
    cmp   r0, #0
    bgt   doswitch
    ldmfd sp!, {r0-r3, lr}
    b     return
doswitch:
    stmfd sp!, {r0-r3, lr}
    bl    ttswitch              // switch task is SVC mode
    ldmfd sp!, {r0-r3, lr}
return:
    msr   cpsr, #0x92           // IRQ mode with interrupts off
    ldmfd sp!, {r0}
    msr   spsr, r0
    ldmfd sp!, {r0-r12, pc}^
/************ end of irq_handler *******************/
```

10.8.8 Priority Inheritance

In the SMP_RTOS kernel, spinlocks, mutexes and semaphores are used for process synchronization. Spinlocks are non-blocking, i.e. they do not cause context switch, so there is no need for priority inheritance. Mutexes and semaphores are blocking, which may cause task switch, so they must support priority inheritance. Implementation of priority inheritance in SMP is similar to that of UP_RTOS, except for the following differences. When a task is about to become blocked on a mutex or semaphore, it may have to boost the priority of the mutex or semaphore owner, which may be running on a different CPU. If the mutex or semaphore owner and the current task are on the same CPU, the situation is the same as in UP_RTOS. Otherwise, the current task must send a SGI to the CPU of the mutex or semaphore owner, causing it to adjust task priority and reschedule tasks on the target CPU. The scheme can be implemented as follows. When a task is about to become blocked on a mutex or semaphore:

(1). Write [pid, priority] to a dedicated shared memory, where pid is the PID of the mutex/semaphore owner and priority is the new priority.
(2). Issue SGI_3 to the CPU of the mutex/semaphore owner task
(3). SGI_3_handler on the target CPU:
    Get [pid, priority] from shared memory;
    Raise target task's priority and reschedule task

The following shows the implementation of priority inheritance for mutexes in SMP_ROTS. The implementation of priority inheritance for semaphores is similar.

```
volatile int PID[NCPU];  // shared memory for SGI_3 handlers
int mutex_lock(MUTEX *s)
{
  PROC *p;
  int cpuid = get_cpuid();
  int SR = int_off();
```

```
  slock(&s->lock);    // spinlock
  if (s->state==UNLOCKED{  // mutex is in unlocked state
     s->state = LOCKED;
     s->owner = running;
     sunlock(&s->lock);
     int_on(SR);
     return 1;
  }
  /************** mutex already locked ******************/
  if (s->owner == running){ // already locked by this task
      printf("mutex already locked by you!\n");
      sunlock(&s->lock);
      int_on(SR);
      return 1;
  }
  printf("TASK%d BLOCK ON MUTEX: ", running->pid);
  running->status = BLOCK;
  enqueue(&s->queue, running);
  // boost owner priority to running's priority
  if (running->priority > s->owner->priority){
     if (s->owner->cpuid == cpuid){ // same CPU
        s->owner->priority = running->priority; // boost owner priority
     }
     else{ // not the same CPU
        PID[s->owner-cpuid] = s->owner->pid; // owner's pid
        send_sgi(3, s->owner->cpuid, 0); // send SGI to recehdule task
     }
  }
  sunlock(&s->lock);
  tswitch();  // switch task
  int_on(SR);
  return 1;
}
int mutex_unlock(MUTEX *s)
{
  PROC *p;
  int cpuid = get_cpuid();
  int SR = int_off();
  slock(&s->lock);          // spinlock
  if (s->state==UNLOCKED || s->owner != running){
     printf("%d mutex_unlock error\n", running->pid);
     sunlock(&s->lock);
     int_on(SR);
     return 0;
  }
  // caller is owner and mutex was locked
  if (s->queue == 0){       // mutex has no waiter
     s->state = UNLOCKED;  // clear locked state
     s->owner = 0;         // clear owner
     running->priority = running->rpriority;
  }
  else{  // mutex has waiters: unblock one as new owner
     p = dequeue(&s->queue);
     p->status = READY;
     s->owner = p;
```

```
        slock(&readylock[p->cpuid]);
          enqueue(&readyQueue[p->cpuid], p);
        sunlock(&readylock[p->cpuid]);
        running->priority = running->realPriority; // restore priority
        if (p->cpuid == cpuid){  // same CPU
           if (p->priority > running->priority){
              if (intnest==0)     // not inside IRQ handler
                 ttswitch();      // preemption now
              else{
                 swflag[cpuid] = 1; // defer preemption until IRQ end
        }
        else // caller and p on different CPUs => SGI to p's CPU
           send_sgi(2, p->cpuid, 0); // send SGI to reschedule task
    }
    sunlock(&s->lock);
    int_on(SR);
    return 1;
}
```

10.8.9 Demonstration of the SMP_RTOS System

The SMP_RTOS system integrates all the functionalities discussed above to provide the system with the following capabilities.

(1). A SMP kernel for task management. Tasks are distributed to different CPUs to improve concurrency in SMP.
(2). Preemptive task scheduling by priority and timeslice
(3). Support nested interrupts
(4). Priority inheritance in mutex and semaphore operations
(5). Task communication by shared memory and messages
(6). Inter-processor synchronization by SGI
(7). A logging task for recording task activities to a SDC

We demonstrate the SMP_RTOS system by the sample program C10.9 For the sake of brevity, we only show the t.c file of the system.

```
/******** t.c file of SMP_RTOS demo. system C10.9 ********/
#include "type.h"          // system types and constants
#include "string.c"
struct semaphore s0, s1;   // for task synchroonization
int irq_stack[4][1024], abt_stack[4][1024], und_stack[4][1024];
#include "queue.c"         // enqueue/dequeue functions
#include "pv.c"            // semaphores with priority inheritance
#include "mutex.c"         // mutex with priority inheritance
#include "uart.c"          // UARTs driver
#include "kbd.c"           // KBD driver
#include "ptimer.c"        // local timer driver
#include "vid.c"           // LCD display driver
#include "exceptions.c"    // exceptions handlers
#include "kernel.c"        // kernel init and task schedulers
#include "wait.c"          // kexit() and kwait() functions
#include "fork.c"          // kfork() function
#include "sdc.c"           // SDC driver
```

```
#include "message.c"        // message passing

int copy_vectors(){ // same as before }
int config_gic()   { // same as before }
int config_int(int N, int targetCPU){// same as before }
int irq_chandler(){ // same as before }
int APstart()      { // same as before }

int SGI_handler()   // sgi handler
{
  int cpuid = get_cpuid();
  printf("%d on CPU%d got SGI-2: ", running->pid, cpuid);
  if (readyQueue[cpuid]){ // if readQ nonempty
     printf("set swflag\n");
     swflag[cpuid] = 1;
  }
  else{ // empty readyQ
    printf("NO ACTION\n");
    swflag[cpuid] = 0;
  }
}
int klog(char *line){ send(line, 2) }
char *logmsg = "log information";
int task5()
{
  printf("task%d running: ", running->pid);
  klog("start");
  mutex_lock(mp);
     printf("task%d inside CR\n",  running->pid);
     klog("inside CR");
  mutex_unlock(mp);
  klog("terminate");
  kexit(0);
}
int task4()
{
  int cpuid = get_cpuid();
  while(1){
    printf("%d on CPU%d: ", running->pid, cpuid);
    kfork((int)task5, 5, 2);  // creat task5 task on CPU2
    kdelay(running->pid*10000);
    kexit(0);
  }
}
int task3()
{
  int cpuid = get_cpuid();
  while(1){
    printf("%d on CPU%d: ", running->pid, cpuid);
    klog(logmsg);
    mutex_lock(mp);  // lock mutex
      pid = kfork((int)task4, 4, 2); // create task4 on CPU2
      kdelay(running->pid*10000);
    mutex_unlock(mp);
```

```
    pid = kwait(&status);
  }
}
int task2()  // log task
{
  int r, blk;
  char line[1024];
  printf("log task %d start\n", running->pid);
  blk = 0;   // start write to block 0 of SDC
  while(1){
    printf("LOGtask%d: receiving\n", running->pid);
    r = recv(line);            // receive a log line
    put_block(blk, line);      // write log to SDC (1KB) block
    blk++;
    uprintf("%s", line);       // display log to UART0 on-line
    printf("log: ");
  }
}
int task1()
{
  int status, pid, p3;
  int cpuid = get_cpuid();
  fs_init();
  kfork((int)task2, 2, 1);
  V(&s0); // start up logtask
  kfork((int)task3, 3, 2);
  while(1)
     pid = kwait(&status);
}
int run_task(){ // same as before }

int main()
{
  int cpuid = get_cpuid();
  enable_scu();
  fbuf_init();
  kprintf("Welcome to SMP_RTOS in ARM\n");
  sdc_init();
  kbd_init();
  uart_init();
  ptimer_init();
  ptimer_start();
  msg_init();
  config_gic();
  kprintf("CPU0 initialize kernel\n");
  kernel_init();
  mp = mutex_create();
  s0.lock = s1.lock  = s0.value = s1.value = 0;
  s0.queue = s1.queue = = 0;
  printf("CPU0 startup APs: ");
  int *APaddr = (int *)0x10000030;
  *APaddr = (int)apStart;
  send_sgi(0x0, 0x0F, 0x01);
```

```
00:00:00 : 03 : log information
00:00:00 : 05 : start
00:00:09 : 05 : inside CR
00:00:09 : 05 : terminate
00:00:09 : 03 : log information
00:00:09 : 07 : start
00:00:21 : 07 : inside CR
00:00:21 : 07 : terminate
00:00:21 : 03 : log information
00:00:21 : 09 : start
00:00:35 : 09 : inside CR
00:00:35 : 09 : terminate
00:00:35 : 03 : log information
00:00:35 : 11 : start
```

```
QEMU
Welcome to SMP_RTOS in ARM                                                    00:00:01
sdc_init : kbd_init() uart[0-4] init()                                        00:00:04
mesg_init() CPU0 initialize kernel                                            00:00:04
kernel_init(): init iprocs                                                    00:00:04
CPU0 startup APs: CPU0 waits for APs ready
1000  on CPU0  kforked a child 1  to CPU0 : 1  PREEMPT 1000  on CPU0 : CPU0 : ne
xt running=1
fs_init(): mount root: EXT2 MAGIC=0xEF53  bmap=9  imap=10  iblk=11  OK
1  on CPU0  kforked a child 2  to CPU3 : sendSGI to 3 : 1  on CPU0  kforked a ch
ild 3  to CPU1 : sendSGI to 1 : CPU0 : next running=1000
1001  on CPU1  got SGI-2: set swflag
CPU1 : next running=3
3  on CPU1 : task3  send msg to 2
3  on CPU1  kforked a child 4  to CPU2 : sendSGI to 2 : 1002  on CPU2  got SGI-2
: set swflag
CPU2 : next running=4
4  on CPU2 : 4  on CPU2  kforked a child 5  to CPU2 : 5  PREEMPT 4  on CPU2 : CP
U2 : next running=5
task5  running: task5  send msg to 2
TASK5  BLOCK ON MUTEX: RAISE PROC3  PRIORITY to 5
CPU2 : next running=4
1003  on CPU3  got SGI-2: set swflag
CPU3 : next running=2
open: filename=log: found log ino=130  opened fd = 0x77D0C
log: log: proc2  BLOCKED on 0x1C4B8
CPU3 : next running=1003
```

Fig. 10.10 Demonstration of SMP_RTOS

```
    printf("CPU0 waits for APs ready\n");
    while(ncpu < 4);
    kfork((int)task1, 1, 0);
    run_task();
}
```

Figure 10.10 shows running the SMP_RTOS system. The top part of Fig. 10.10 shows the on-line log information generated by the logging task. The bottom part of Fig. 10.10 shows the startup screen of SMP_RTOS.

In the main() function of the SMP_RTOS system, the initial process on CPU0 creates a task P1, which behaves as the INIT process. P1 creates task2 as the logging task on CPU1, and task3 on CPU2. Then P1 executes a loop to wait for any ZOMBIE child. Task3 first locks up a mutex. Then it creates task4, which creates task5, all on CPU2 but with different priorities. When task5 runs, it tries to acquire the same mutex lock, which is still held by task3. In this case, task5 will be blocked on the mutex but it raises the effective priority of task3 to 5, which demonstrates priority inheritance. In addition, the system also shows task preemption and inter-processor synchronization by SGIs. All the tasks may send log information as messages to the logging task, which invokes the SDC driver directly to write the log information to a (1KB) block of the SDC. It also writes the log information to a UART port to display it on-line. Variation to the logging scheme is listed as a programming project in the Problems section.

10.9 Summary

This chapter covers embedded real-time operating systems (RTOS). It introduces the concepts and requirements of real-time systems. It covers the various kinds of task scheduling algorithms in RTOS, which include RMS, EDF and DMS. It explains the problem of priority inversion due to preemptive task scheduling and shows how to handle priority inversion and task preemption. It includes case studies of several popular real-time OS and presents a set of general guidelines for RTOS design. It shows the design and implementation of a UP_RTOS for uniprocessor (UP) systems. Then it extends the UP_RTOS to a SMP_RTOS, which supports nested interrupts, preemptive task scheduling, priority inheritance and inter-processor synchronization by SGI.

List of Sample Programs

C10.1. UP_RTOS with periodic tasks and round-robin scheduling
C10.2. UP_RTOS with periodic tasks and preemptive scheduling
C10.3. UP_RTOS with dynamic tasks
C10.4. SMP_RTOS kernel for task management
C10.5. MP_RTOS, a localized SMP system.
C10.6. Nested Interrupts in SMP_RTOS
C10.7. Task Preemption by SGI in SMP_RTOS
C10.8. Timesliced task scheduling in SMP_RTOS
C10.9. Demonstration of SMP_RTOS

Problems

1. In the UP_RTOS and SMP_RTOS systems, the logging task saves log information to a SDC directly, bypassing the file system. Modify the programs C10.1 and C10.8 to let the logging task save log information to a file in an (EXT2/3) file system on a SDC. Discuss the advantages and disadvantages of using log files.
2. In the UP_RTOS kernel, priority inheritance in only one-level. Implement priority inherence in a chain of nested requests for mutex or semaphore locks.
3. For the example UP_ROTS system C10.3, perform a response time analysis to determine whether the tasks will meet their deadlines. Verify whether the task can meet their deadlines empirically.
4. In the MP_RTOS system, tasks are bound to separate CPUs for parallel processing. Devise a way to support task migration, which allows tasks to be distributed to different CPUs to balance the processing load.
5. In the SMP_RTOS system, tasks running on different CPUs may interfere with one another since they all execute in the same address space. Use MMU with non-uniform VA to PA mapping to provide each CPU with a separate VA space.
6. In a SMP RTOS, interrupts may be handled directly by ISRs or by pseudo-tasks. Design and implement RTOS systems which employ these interrupts processing schemes and compare their performances.

References

Audsley, N.C: "Deadline Monotonic Scheduling", Department of Computer Science, University of York, 1990
Audsley, N. C, Burns, A., Richardson, M. F., Tindell, K., Wellings, A. J.: "Applying new scheduling theory to static priority pre-emptive scheduling. Software Engineering Journal, 8(5):284–292, 1993.
Buttazzo, G. C.: Counter RMS claims Rate Monotonic vs. EDF: Judgment Day, Real-Time Systems, 29, 5–26, 2005
Dietrich, S., Walker, D., "The evolution of Real-Time Linux", http://www.cse.nd.edu/courses/cse60463/www/amatta2.pdf, 2015
DNX: DNX RTOS, http://www.dnx-rtos.org, 2015
FreeROTS: FreeRTOS, http://www.freertos.org, 2016

Hagen, W. "Real-Time and Performance Improvements in the 2.6 Linux Kernel", Linux journal, 2005
Josheph, M and Pandya, P., "Finding response times in a real-time system"', *Comput. J.*, 1986, 29, (5). pp. 390-395
Jones, M. B. (December 16, 1997). "What really happened on Mars?". Microsoft.com. 1997
Labrosse, J.: Micro/OS-II, R&D Books, 1999
Leung, J.Y T., Merrill, M. L : "A note on preemptive scheduling of periodic, real-time tasks. Information Processing Letters, 11(3):115–118, 1982
Linux: "Intro to Real-Time Linux for Embedded Developers", https://www.linux.com/blog/intro-real-time-linux-embedded-developers
Liu, C. L.; Layland, J. "Scheduling algorithms for multiprogramming in a hard real-time environment", Journal of the ACM 20 (1): 46–61, 1973
Micrium: Micro/OS-III, https://www.micrium.com, 2016
Nutt, G. NuttX, Real-Time Operating system, nuttx.org, 2016
POSIX 1003.1b: https://en.wikipedia.org/wiki/POSIX, 2016
QNX: QNX Neutrino RTOS, http://www.qnx.com/products/neutrino-rtos, 2015
Reeves, G. E.: "What really happened on Mars? - Authoritative Account". Microsoft.com. 1997.
RTLinux: RTLinux, https://en.wikipedia.org/wiki/RTLinux
Sha, L., Rajkumar,R., Lehoczky,J.P.: (September 1990). "Priority Inheritance Protocols: An Approach to Real-Time Synchronization", IEEE Transactions on Computers,Vol 39, pp1175–1185, 1990
VxWorks: VxWorks: http://windriver.com, 2016.
Yodaiken, V.:"The RTLinux Manifesto", Proceedings of the 5th Linux Conference, 1999
Wang, K. C.: "Design and Implementation of the MTX Operating System", Springer Publishing International AG, 2015

Index

K.C. Wang, *Embedded and Real-Time Operating Systems*,
DOI 10.1007/978-3-319-51517-5

Printed in the United States
By Bookmasters